AURORAL PLASMA PHYSICS

Space Science Series of ISSI

Volume 15

The International Space Science Institute is organized as a foundation under Swiss Law. It is funded through recurrent contributions from the European Space Agency, the Swiss Confederation, the Swiss National Science Foundation, and the Canton of Bern. For more information, see the homepage at http://www.issi.unibe.ch/

The titles published in this series are listed at the end of this volume.

AURORAL PLASMA PHYSICS

Edited by

GÖTZ PASCHMANN

International Space Science Institute, Bern, Switzerland and Max-Planck-Institut für Extraterrestrische Physik, Garching, Germany

STEIN HAALAND

International Space Science Institute Bern, Switzerland and Max-Planck-Institut für Extraterrestrische Physik, Garching, Germany

and

RUDOLF TREUMANN

Max-Planck-Institut für Extraterrestrische Physik, Garching, Germany

Reprinted from *Space Science Reviews*, Vol. 103, Nos. 1–4, 2002

KLUWER ACADEMIC PUBLISHERS

DORDRECHT / BOSTON / LONDON

INTERNATIONAL SPACE SCIENCE INSTITUTE

Space Sciences Series of ISSI

A C.I.P. Catalogue record for this book is available from the Library of Congress.

ISBN 1-4020-0963-1

Published by Kluwer Academic Publishers,
P.O. Box 17, 3300 AA Dordrecht, The Netherlands.

Sold and distributed in North, Central and South America
by Kluwer Academic Publishers,
101 Philip Drive, Norwell, MA 02061, U.S.A.

In all other countries, sold and distributed
by Kluwer Academic Publishers,
P.O. Box 322, 3300 AH Dordrecht, The Netherlands.

Printed on acid-free paper

Contents

Foreword

The present 15th volume of the ISSI Space Science Series is devoted to Auroral Plasma Physics. The aurora is arguably the most intriguing phenomenon in space plasma physics. Not only is it the most spectacular manifestation of the Sun-Earth connection chain, but the underlying plasma processes are expected to be ubiquitous in the plasma universe. Recognizing the enormous progress made over the last decade in the understanding of the physics of the auroral acceleration processes, it seemed timely to write a comprehensive and integrated book on the subject. Recent advances concern the clarification of the nature of the acceleration process of the electrons that are responsible for the visible aurora, the recognition of the fundamental role of the large-scale current systems in organizing the auroral morphology, and of the interplay between particles and electromagnetic fields.

The project began in March 1999, as a natural follow-up of the project on Magnetospheric Plasma Sources and Losses that resulted in volume 6 of this series, with a planning meeting by a core-group that coordinated the project. The group consisted of J. E. Borovsky, Los Alamos National Laboratory; C. W. Carlson, University of California, Berkeley; G. Haerendel, Max-Planck-Institut für extraterrestrische Physik, Garching; B. Hultqvist, Swedish Intitute of Space Physics, Kiruna; H. E. J. Koskinen, Finnish Meteorological Institute, Helsinki; W. Lotko, Dartmouth College, Hanover, New Hampshire; K. A. Lynch, University of New Hampshire, Durham and G. Marklund, Royal Institute of Technology, Stockholm. G. Paschmann, ISSI, Bern, was the project leader.

The work was organized through three major workshops, with nearly 30 invited attendants, followed by a series of smaller meetings that concentrated on writing and editing of the manuscript. While the core-group carried much of the work load, as authors, chapters editors and reviewers, a total of of 32 authors finally contributed to the book. The team members must be thanked for their spirit of collaboration and dedication, highlighted by the fact that they all agreed that their names would not appear prominently, but instead would be hidden in a strictly alphabetical list. Thanks are also due to C.-G. Fälthammar, R. Nakamura and J. Vogt for valuable comments on some of the chapters.

This book is intended to be used both as a learning tool for graduate students, and as a reference for auroral researchers. It must be stressed that while scientific understanding of the aurora is advancing rapidly, much remains to be learned. This book represents a snapshot of our current understanding, with emphasis on those aspects that are related to particle acceleration processes. Attempts have been made throughout to indicate those topics for which there are significant gaps in our understanding.

All chapters in this book were written in LaTeX, and all figures, more than 200, were provided as encapsulated PostScript files. BibTeX was used for the references. The index includes the acronyms and abbreviations used in the book.

V. Manno, B. Fasler, G. Indermühle, X. Schneider, D. Taylor, and S. Wenger of the ISSI staff provided the infrastructure for the many meetings of the project team. B. Mory at MPE Garching helped with many of the figures.

The book is dedicated to the memory of Yuri Galperin, whose insight and enthusiasm contributed much to the spirit and success of the project, but who did not live to see its completion.

Bern in July 2002 G. Paschmann, S. Haaland, R. Treumann

Contributing Authors

O. Amm, J. Birn, J. Bonnell, J. E. Borovsky, J. F. Carbary, C. W. Carlson, C. A. Cattell, R. E. Ergun, A. Eriksson, Y. I. Galperin, S. E. Haaland, T. Hallinan, G. Haerendel, B. Hultqvist, T. Karlsson, D. J. Knudsen, H. E. J. Koskinen, W. Lotko, G. Lu, K. A. Lynch, R. L. Lysak, G. Marklund, J. P. McFadden, A. Masson, C.-I. Meng, H. Opgenoorth, G. Paschmann, R. Pottelette, T. I. Pulkkinen, V. A. Sergeev, Y. Song, R. A. Treumann

Chapter 1

Introduction

Bright auroral displays during dark polar nights belong certainly to the most memorable natural phenomena a person can experience. In addition to their emotional impression, the auroral phenomena have excited the curiosity of scientists since the days of Galileo. The rapid expansion of our capacity to observe the auroral processes, both remotely and in situ, has provided an enormous amount of data, which we are trying to organize and explain in the present book. With a better understanding, we can enjoy the beauty of aurorae in a much deeper sense, just like music opens for those who are willing to invest their time to understand the structure and dynamics of the musical system.

In this chapter we introduce the aurora as a visible phenomenon (Section 1.1) and relate it to the properties of the polar ionosphere (Section 1.2). While the visual aurora itself appears in the ionosphere at altitudes around 100 km, it is a phenomenon that is closely connected to much of the volume of the magnetosphere. In Section 1.3 we discuss the magnetospheric context of the aurora. The coupling of the ionosphere to the magnetosphere, and the processes arising from that coupling, are introduced in Section 1.4. This chapter concludes in Section 1.5, dealing with large-scale magnetospheric flows. Section 1.6 briefly introduces the remaining chapters of the book.

1.1 Visible Aurorae

Most great cultures have produced some written notes about heavenly phenomena that can be interpreted as auroral displays. Some Chinese traditions from the mid third millenium BC have been interpreted as references to aurorae (Wang and Siscoe, 1995). Many specific auroral occurrences are described in Greek and Roman literature. In northern latitudes the aurorae have produced a large account of folklore which has survived through mostly oral tradition to our times (for a comprehensive discussion of the old records, see Eather, 1980).

Space Science Reviews **103:** 1–475, 2002.

The term Aurora Borealis most likely was introduced by Galileo in 1619. As Aurora was the Roman name for the Greek goddess of dawn, Eos, the great genius clearly did not yet have any clue of what the aurorae really were.

An important step toward modern understanding was taken by the Swedish scientists Anders Celsius and Olav Hiorter in 1747 when they reported on Hiorter's earlier observations that the magnetic needles reacted to strong auroral displays. Celsius also conducted one of the first large-scale experiments when he compared simultaneous behavior of magnetic needles in London and in Uppsala. All this happened much before Ørsted, Biot, Savart and Ampère established the relationship between the magnetic field and electric current.

Auroral displays take place at altitudes of about 100 km and above, in the same region as the horizontally flowing currents that cause the magnetic deflection on the ground. This was established statistically by the Norwegian scientist Carl Størmer around the beginning of the 20th century. At the same time Kristian Birkeland conducted laboratory studies of how charged particles approach a uniformly magnetized sphere, and Størmer calculated orbits of charged particles by a brute force method much before Alfvén had introduced the guiding center approximation in the 1940's.

The 20th century witnessed enormous progress in all fields of physics. Much of this progress was also necessary to understand aurorae. For example, the explanations for the auroral emission lines required the emergence of atomic physics and quantum mechanics. During the 1930's the understanding that the origin of the processes causing the most impressive auroral displays was in the perturbations on the Sun, led to early theories of magnetic storms by Chapman and Ferraro. In 1951 Biermann argued, based on the structure of cometary tails, that there had to be a continuous solar wind flow. This was conclusively proven to be correct by the Mariner II observations in 1962.

At the dawn of space flight another large effort contributed significantly to the progress of auroral physics. The International Geophysical Year 1957–1958 signaled the beginning of more or less continuous auroral observations with a newly-developed tool, the auroral all-sky camera (ASC). Based on ASC recordings, Feldstein (1963) introduced the concept of the auroral oval and Akasofu (1964) laid the foundation for modern research on auroral substorms. In the same time frame Alfvén proposed the possibility of non-zero magnetic field-aligned electric fields in space plasmas (Alfvén, 1958; Alfvén and Fälthammar, 1963), which have turned out to be essential for the understanding of the acceleration of charged particles in the auroral region.

The space age opened several new windows to auroral studies. In addition to revealing the large scale structure and dynamics of the magnetosphere and making detailed observations of acceleration processes of particles causing the aurorae, it became possible to image the aurorae from above. The first imager was flown on the ISIS 2 spacecraft in 1972 and many others have followed since.

Figure 1.1. Auroral arc on 22-02-1901 above the Utsjoki church at 69° 52′ N; 27° 1′ E. Lithography by Harald Moltke.

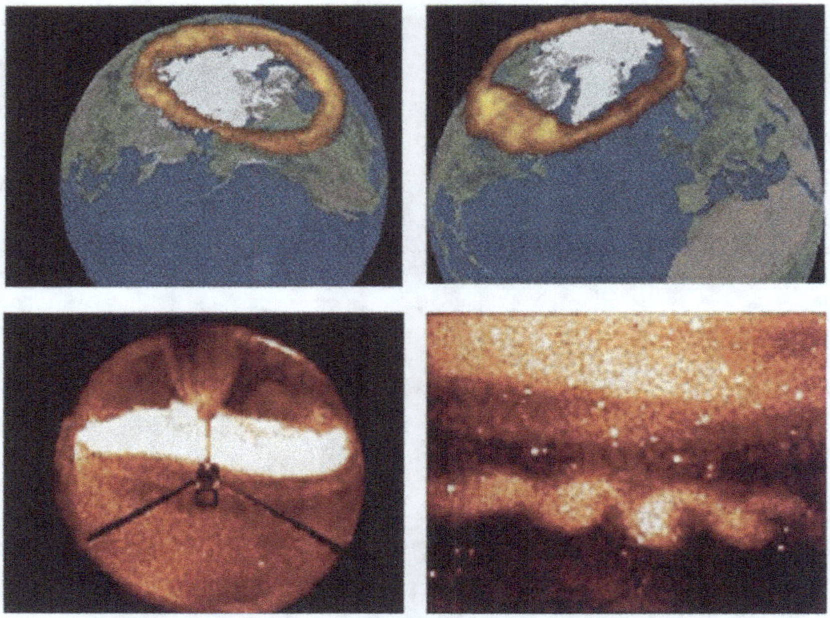

Figure 1.2. Top: auroral oval (left) and the oval with substorm expansion (right), observed by the VIS imager onboard Polar (courtesy L. Frank). Bottom: all-sky camera image of an auroral arc (left), and (right) a narrow-angle (7°) view just below the center of the all-sky image (courtesy T. Trondsen).

The visual aurorae appear in a great variety of forms, such as arcs, spirals, folds, or curls, and on spatial scale sizes from kilometers to the size of the entire auroral oval with a circumference of the order of 10,000 km. Figure 1.2 illustrates the variable scales from the entire auroral oval observable with satellite imagers down to the 100-km scale of all-sky cameras and the 10-km scale of narrow angle cameras.

Sometimes the structures, e.g., arcs, can remain static for tens of minutes, but often there are rapid variations and motions. Understanding the underlying physics of both the quasi-static and the temporally varying auroral phenomena is the topic of this book.

1.2 Ionospheric Framework

Auroral light is produced when electrons and protons impinging on Earth's upper atmosphere excite the atmospheric atoms and molecules. Almost all auroral light consists of emission lines and bands of neutral or ionized N_2, O, O_2, and N. The wavelength of the most copious green emission line was determined by Ångström in 1866 to be 5577 Å, or 557.7 nm. Its origin remained a mystery for almost 60 years until McLennan and Schrum identified it in 1925 as the metastable (forbidden) transition of atomic oxygen (cf. Figure. 1.3) from the ^1S state to ^1D state

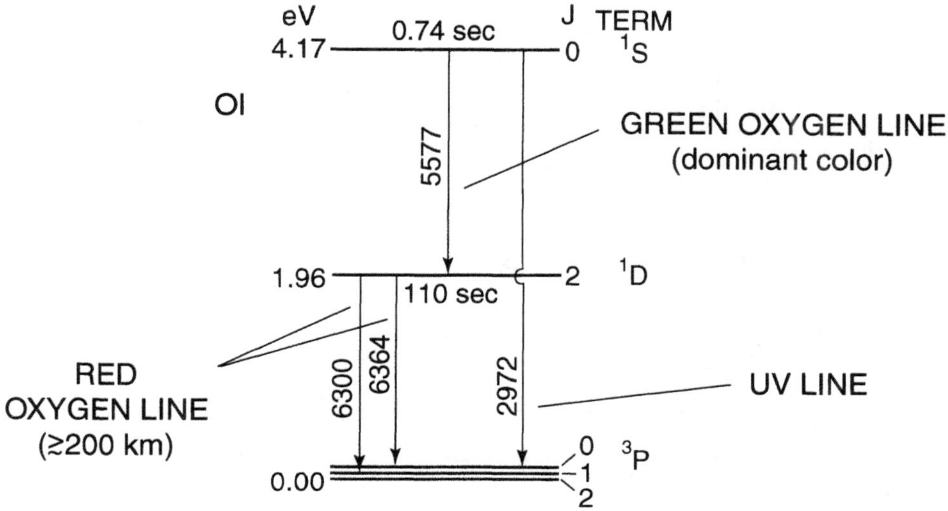

Figure 1.3. Excited states of the oxygen atom that give rise to forbidden transitions prominent in auroral emissions, and their lifetimes (after Roach and Smith, 1967).

(Vallance Jones, 1974). The metastable state has a life-time of 0.74 s. This is much longer than the lifetime of 10^{-7} s of spontaneous emissions like the blue transitions of N_2^+ at 427.8 nm and 391.4 nm, which belong to the most important auroral emission lines. An even longer-lived atomic oxygen transition is the red-line triplet $^1D \rightarrow {}^3P$ at 630.0 - 636.4 nm with a life-time of 110 s. This means that the exited atoms can move a significant distance after their excitation and, consequently, the red auroral light at these wavelengths is more diffuse than the green light at 557.7 nm.

The forbidden oxygen lines with their long lifetimes are prominent in the upper atmosphere because of the low collision frequencies found there. Below about 100 km, however, the green line is quenched by collisional de-excitation of the upper 'S-state', whereas the intermediate state, which gives rise to the two red lines is effectively quenched below 200 km. In combination with the fact that the typical energies of the primary auroral electrons range between 1 and 10 keV, which allows them to penetrate to a height range between 150 and 100 km (Figure 1.4), the green line at 557.7 nm dominates the visual appearance of auroral arcs. For very soft electron spectra (< 1 keV, say) the red emissions above 200 km become prominent. If there are substantial fluxes of electrons well above 10 keV, a lower red border reaching down to 80 km height is observed. Such auroral arcs are invariably very active and exhibit a rayed structure with fast horizontal motions. Comparison of auroral intensities in the crucial lines or molecular bands reveals reliable, though coarse, information on the primary energy spectra. This method has been

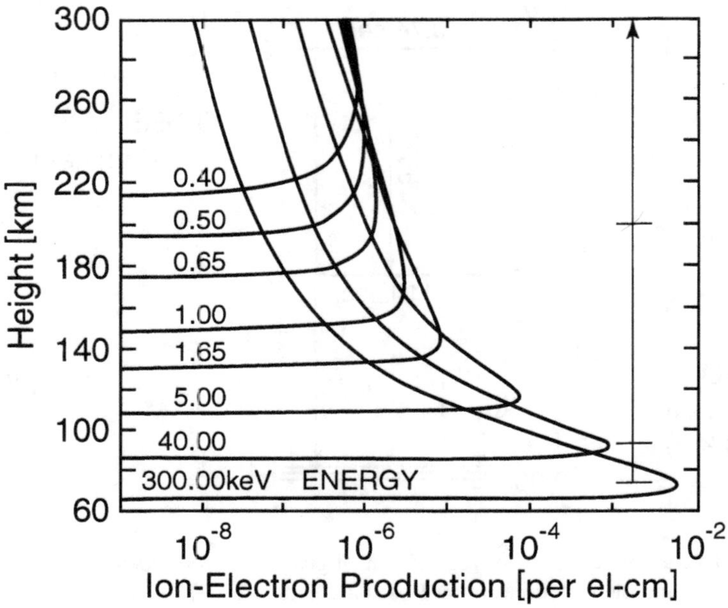

Figure 1.4. Production of ion-electron pairs per unit path length of primary electrons with various initial energies. After Rees (1963).

elaborated by Rees and Luckey (1974) and is frequently applied with the help of meridian scanning photometers and filtered highly sensitive cameras.

A further important type of auroral emissions are the hydrogen emissions H_α (656.3 nm) and H_β (486.nm). They are produced when precipitating protons capture an electron producing an excited hydrogen atom H^*. These emission lines were first measured by Vegard in 1939–41 who also measured the Doppler broadening of these lines. From the Doppler shift it is possible to determine the speed and thus energy of the precipitating protons, which may exceed 20 keV.

The region of the upper atmosphere where the auroral emissions take place is called the ionosphere. The ionosphere is the ionized component of the neutral atmosphere, the various regions of which - troposphere, stratosphere, mesosphere and thermosphere - are shown in Figure 1.5.

In the thermosphere, the solar ultraviolet (UV) light and energetic particles precipitating from the magnetosphere produce ionization increasing with altitude. At the same time the particle density is low enough to make the recombination times of the ionized atoms and molecules sufficiently long to allow a significant fraction of the gas to remain ionized. This produces a conducting layer of the atmosphere known as the ionosphere. The ionosphere begins at ~65 km, has a peak plasma density between 200 and 300 km, and eventually merges with magnetospheric regions at ~1000–2000 km altitudes. Figure 1.6 shows the mid-latitude

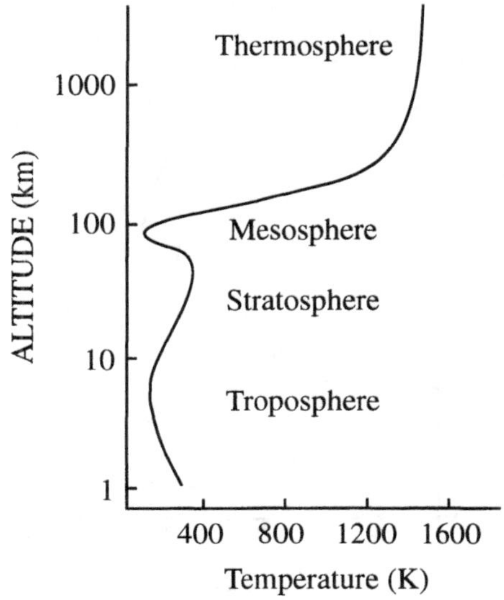

Figure 1.5. Schematic of the temperature profile of the neutral atmosphere. Adapted from Kelley (1989).

ionospheric electron density as a function of altitude for both day and night, and for solar maximum and minimum conditions. The maximum ionisation, called the Chapman layer, occurs where the ionizing radiation (decreasing toward Earth due to absorption) and the number density of particles possible to ionize (increasing toward Earth) both maximize.

Early measurements of the ionosphere using radio-sounding techniques revealed its layered structure (D, E, and F regions, Figure 1.6). Since the plasma density peaks in the F region, information about the plasma density above this height cannot be determined from radio-sounding from the ground because the radio waves are reflected at or below the peak. The higher altitude plasma densities must be measured either by sounding rockets and satellites or using incoherent scatter radars.

The ionosphere shows significant variations with latitude and local time. Since solar UV radiation is the primary source of ionisation, dramatic changes in ionospheric conductivity occur as the atmosphere rotates in and out of sunlight. Near the polar regions, precipitating particles associated with the aurora and energetic particles from solar eruptions can contribute significantly to the ionisation in the ionosphere. The ionosphere also varies with solar cycle because the solar UV input depends on solar activity.

Collisionless plasmas such as those in the solar wind or in the magnetosphere have, in general, little or no conductivity perpendicular to the magnetic field. Ions and electrons drift together across the magnetic field at a velocity where electric

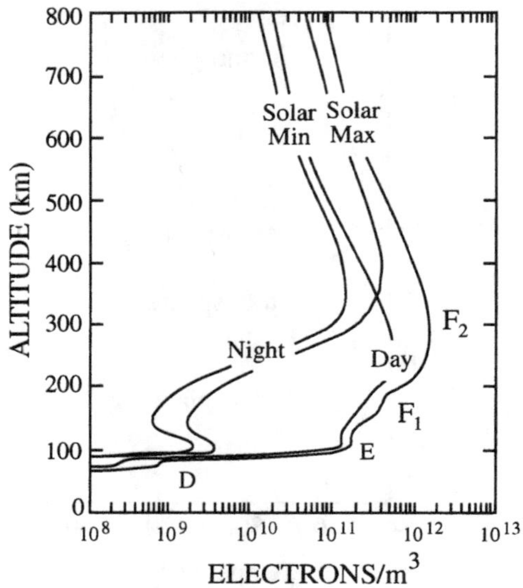

Figure 1.6. Ionospheric density profiles with the major layers (D, E, F1, F2) identified. From Richmond (1987).

and magnetic forces balance, $\mathbf{v}_E = \mathbf{E} \times \mathbf{B}/B^2$. The ionospheric E-region plasma is, however, collisionally tied to the neutrals mainly by ion-neutral collisions. This decouples electron and ion drifts from each other and allows currents to flow perpendicular to the magnetic field. These currents play a key role in the auroral dynamics because they are coupled to the magnetospheric currents through field-aligned current systems discussed extensively in this book.

1.3 Magnetospheric Framework

In contrast to the ionosphere, much of the magnetospheric plasma regions and the surrounding interplanetary space are characterized by very large conductivity. Under such conditions the plasma is strongly coupled to the magnetic field, and the changes of this coupled system can be described and visualized by the 'frozen in' concept, as if magnetic field lines were frozen in the plasma fluid (see Section 3.2 for more details).

The whole near-Earth system is ultimately driven by the expanding solar corona known as the solar wind. Observations at the distance of Earth from the Sun in the ecliptic plane show that the solar wind has a typical density, n, of $\sim$7 cm^{-3}, flow velocity of $\sim$400 km s^{-1}, and electron and proton temperatures, T, of $\sim$100,000 K ($\sim$10 eV). The composition is primarily H^+, with a small ($\sim$4%) He^{++} component and traces of heavier high charge-state ions. The presence of strong magnetic fields in the solar corona coupled with the high conductivity of the solar wind produces a magnetised solar wind with typical field strengths of a few to several tens of

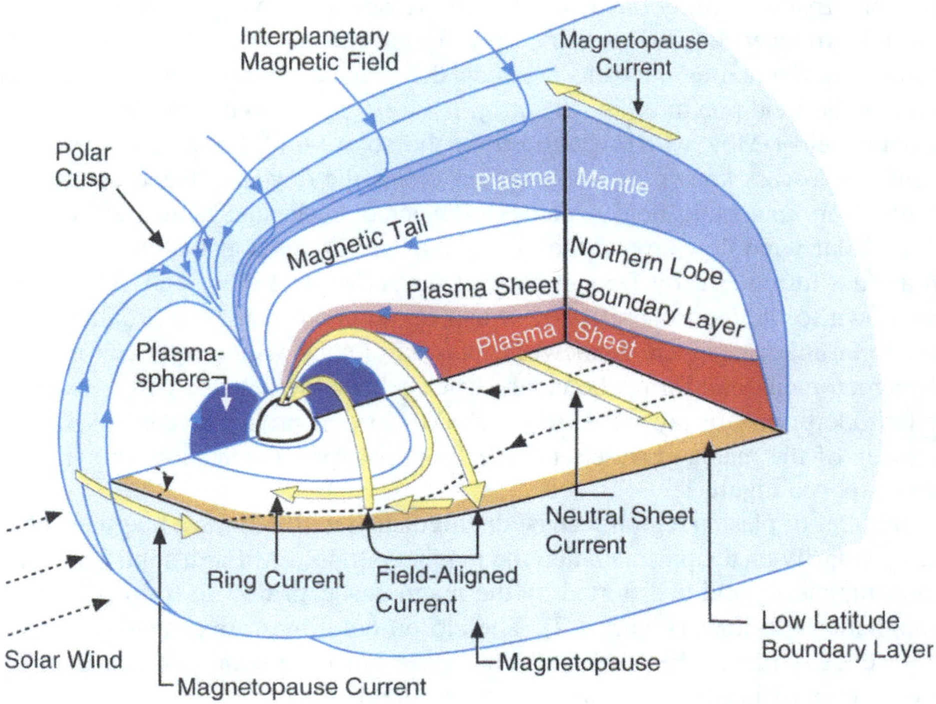

Figure 1.7. The structure of the magnetosphere and the main current systems. Adapted from Kivelson and Russell (1995).

nanoteslas at the distance of Earth's orbit at 1 Astronomical Unit (1 AU equals the distance from the Earth to the Sun). This solar wind plasma interacts with Earth's magnetic field, compressing it on the sunward side and producing an extended tail as it stretches the magnetic field in the direction of the flow. The region where the dynamics are controlled by the magnetic field generated within or near Earth is called the magnetosphere (Figure 1.7), bounded from the solar wind by an outer boundary, the magnetopause.

As the solar wind is supersonic and super-Alfvénic at 1 AU, a bow shock forms upstream of Earth. The solar wind is slowed and compressed before flowing around the magnetopause. In the magnetosheath between the bow shock and the magnetopause, the solar wind plasma and magnetic field are compressed by a factor of ~4 near the subsolar point and the electron and ion temperatures increase to ~50 eV and ~200 eV, respectively. The magnetosheath plasma flows around Earth's magnetic field, compressing the dayside magnetic field and forming an extended magnetic tail.

The magnetopause is defined by a current sheet that separates the solar wind magnetic field from the magnetospheric field. Across this thin boundary, the plasma transitions from a weakly magnetized, compressed solar wind (high β) plasma to a strongly magnetized, low density (low β) plasma on the Earthward side. The plasma β is defined as the ratio between the plasma pressure, $p = nk_BT$, and the magnetic field pressure, i.e., $B^2/2\mu_0$, $\beta = 2\mu_0 p/B^2$. When the interplanetary magnetic field (IMF), which is carried by the solar wind, has a southward component[1], a process known as reconnection opens the dayside magnetopause. The interplanetary magnetic field becomes connected to the magnetic field of Earth and the solar wind flow around the magnetosphere drives a global convective motion as first introduced by Dungey (1961) (as shown in Figure 1.8). This motion maps down to the ionosphere where it can be observed using e.g., scatter radars. In the open magnetosphere framework, plasma of solar wind origin can penetrate the magnetopause and form a boundary layer, whose low latitude part is called the low-latitude boundary layer (LLBL). Dayside auroral processes are linked to the presences of the magnetopause's unique funnel-shaped indentation known as the polar cusp, see Figure 1.7.

A number of plasma regions can be distinguished within Earth's magnetosphere, based primarily on the plasma β and the magnetic topology. Earth's intrinsic quasi-dipolar magnetic field is distorted by the magnetosheath flow to form two classes of magnetic field lines (Figure 1.7): The closed field lines are connected to Earth at both ends, whereas the open field lines have only one connection to Earth and extend outwards into the solar wind magnetic field.

The tail lobes are low-β (<0.01) plasma regions in Earth's magnetotail threaded by magnetic field lines that extend from the polar cap, the region poleward of the auroral oval, to hundreds of R_E (Earth radius, $1\,R_E = 6370$ km) tailward. Most characteristically, the polar cap is usually void of auroral structures and filled with very low-energy electron precipitation of solar wind origin termed polar rain (e.g., Baker et al., 1986). However, occasionally auroral arcs extend poleward of the main auroral oval. During magnetically very quiet periods, auroral precipitation can bifurcate the polar cap forming the so-called theta-aurora.

The tail lobes having opposite magnetic field directions are separated by the plasma sheet. This higher-density (0.1–1 cm^{-3}), hot ($T_i \sim 2-20$ keV, $T_e \sim 0.4-4$ keV) plasma region resides on closed magnetic field lines and contains plasma originating both from the solar wind and from the ionosphere. The plasma sheet also carries an electric current in the dawn to dusk direction consistent with the magnetic field reversal. The center of this current sheet is also referred to as the neutral sheet, because the magnetic field direction reverses and the magnitude becomes very small (<5 nT). A thin layer known as the plasma sheet boundary layer (PSBL) is observed at the interface between the lobes and the plasma sheet. This

[1]When discussing phenomena in the solar wind or magnetosphere, we normally use the geocentric solar ecliptic (GSE) coordinate system. The GSE system has its origin in the center of the Earth, X points towards the sun, Z is the normal to the ecliptic pointing to the north, and the Y axis completes the right handed system.

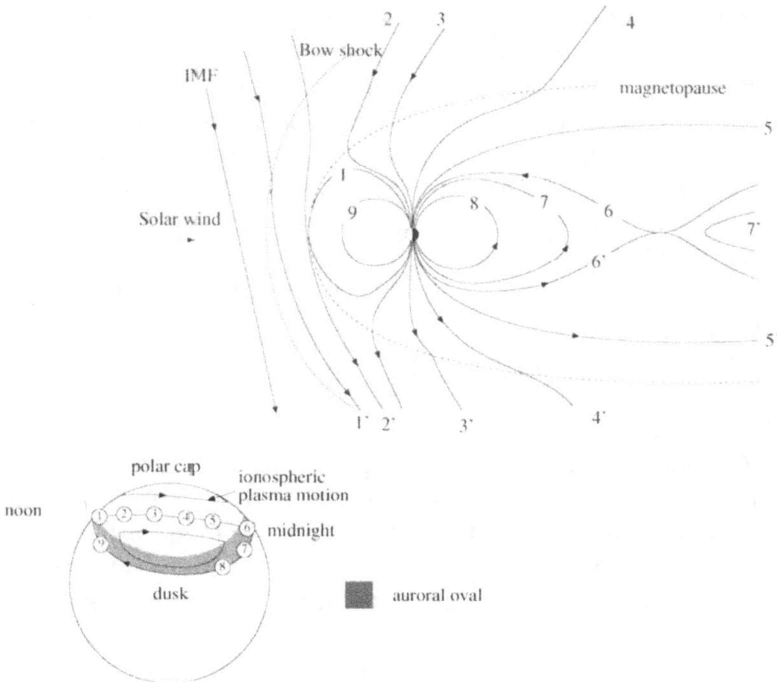

Figure 1.8. Dungey-type convection pattern and its mapping into the auroral ionosphere. From Kivelson and Russell (1995).

region is characterized by field-aligned ion and electron populations. The ionospheric mappings of both the plasma sheet and the plasma sheet boundary layer form the main part of the nightside auroral oval.

The inner boundary of the plasma sheet is often colocated with a sharp transition from hot to cold plasma and an increase in density by 1–2 orders of magnitude. This inner torus-shaped low-latitude region of the magnetosphere is called the plasmasphere. There the plasma flow is dominated by co-rotation with the underlying ionosphere. During magnetically quiet periods, the plasmasphere is filled by cold ($\sim$1 eV) ionospheric plasma on field lines that may extend out to more than 6 R_E at the equator. The plasmasphere maps to subauroral latitudes equatorward of the main auroral oval. Due to their low energies the plasmasphere particles do not produce observable auroral emissions.

The dayside equatorial magnetosphere, extending from the cold plasmasphere to the magnetopause, is often referred to as the dayside plasma sheet. As discussed in Section 1.5, the plasma convects around Earth from the nightside plasma sheet to the dayside plasma sheet and eventually to the reconnection region at the subsolar magnetopause. The dayside plasma sheet consists of closed field lines that map to the low-latitude portion of the dayside auroral oval within a few hours of local noon.

During periods when the solar wind magnetic field has a southward component, there are open field lines within the polar cusps, poleward of the closed field lines and of the dayside plasma sheet. The polar cusps host newly reconnected field lines, where plasma sheet and magnetosheath plasma are accelerated and mixed (Sckopke, 1979). Plasma sheet and magnetosheath plasmas in the cusp eventually convect over the poles of Earth forming a boundary layer known as the plasma mantle. As this plasma convects tailward, it drifts across the tail toward the plasma sheet and eventually fills the tail lobes.

The presence of high-energy particles in space has been known since the first in situ space observations (Van Allen et al., 1958). These particles are trapped in the quasi-dipolar near-Earth magnetic field, and their dynamics can be described by three adiabatic invariants of motion: the magnetic moment associated with periodic gyro motion, the longitudinal invariant associated with periodic bounce motion between magnetic mirror points, and the flux invariant associated with periodic drift motion around Earth. A trapped particle conserving its adiabatic invariants will be confined to a surface specified by the distance, L (in units of Earth radii, R_E), to the equatorial field line as shown by McIlwain (1961).[2]

The trapped energetic particles are often divided into the high-energy (>1 MeV) radiation belt populations and the ring current. The innermost proton belt is formed by the decay of neutrons produced by energetic cosmic ray interactions with the atmosphere. The trapped inner and outer belt electrons have lifetimes of the order of several months. The detailed energization mechanisms for these populations are still debated, although it is clear that they involve short term energization and diffusion of particles onto these field lines. These energetic electrons have peak fluxes at about 2 and 4 R_E at the equator and a power law spectrum that extends beyond several MeV (e.g., Baker et al., 1994). The outer belt electron population often extends beyond the geostationary orbit (6.6 R_E), and hence their ionospheric projection is a belt equatorward of, but partially colocated with, the main auroral oval.

The ring current consists of ions trapped inside 4 - 5 R_E at the equator in the same region as the outer electron belt, and their energy ranges from a few tens of keV up to a few MeV (e.g., Daglis et al., 1999). These ions are primarily injected during geomagnetic storms and cause a significant increase in the local plasma pressure. The westward current carried by this population can decrease the surface magnetic field by as much as 1%; thus the (mid-latitude) surface field can be used to monitor the ring current intensity. For the ring current decay, charge exchange is

[2]The invariant latitude, Λ, often used in this book, is related to the L parameter through $cos^2\Lambda = 1/L$. For a nearly dipolar magnetic field, Λ gives the latitude of the footpoint of a magnetic field line.

an important process, and the ring current decays on time periods ranging from a few hours up to several weeks.

1.4 Magnetosphere-Ionosphere Coupling

In a steady state, the electric fields associated with magnetospheric convection map along the magnetic field to the ionosphere. In the ionosphere these electric fields generate currents that are perpendicular to the magnetic field. Ionospheric currents provide forces on the plasma that balance the drag caused by collisions with the neutrals.

At large scales, the ionospheric currents are coupled to the magnetosphere (Figure 1.7) through upward and downward field-aligned currents, abbreviated FAC. Field-aligned currents are often also called Birkeland currents. The large-scale field-aligned currents are traditionally divided into two classes: the poleward system is called Region 1 and the equatorward system is called Region 2. The main distinction between them is that the Region 2 currents close in the inner magnetosphere, whereas the Region 1 currents connect mostly to the magnetospheric boundaries. Figure 1.9 illustrates how the upward and downward currents are coupled to the horizontal currents in the ionosphere. The Region 1 currents flow near the polar cap boundary, out from the ionosphere in the evening sector and into the ionosphere in the morning sector. On the dayside and near the terminators, these currents are thought to connect to the dayside magnetospheric boundaries (e.g., Janhunen and Koskinen, 1997), but part of the nightside currents are expected to originate from the plasma sheet (Tsyganenko et al., 1993). The total current in this current circuit is 1–2 MA.

The Region 2 currents flow equatorward of the Region 1 currents and in the opposite direction, i.e., into the ionosphere in the evening sector and out of the ionosphere in the morning sector. At high altitudes, these currents merge with the ring current in the inner magnetosphere. The total current in the Region 2 circuit is slightly smaller than in the Region 1 current circuit, typically less than 1 MA. Current closure in the ionosphere produces a drag on the magnetospheric flows. The field-aligned currents are related to also change the magnetic field geometry, increasing the magnetic shear or twist.

The field-aligned currents are primarily carried by electrons, relatively cold ionospheric upgoing electrons in the downward current regions and hot magnetospheric electrons in the upward current regions. For small current densities, the thermal flux of electrons is adequate to carry the current. However, during active periods when magnetospheric flows are fast, the associated current densities become too large to be carried by thermal particles. Electric fields parallel to the magnetic field form to provide current continuity and maintain charge neutrality (except for the very small differences in electron and ion densities that make the electric fields). The region where parallel electric fields form decouples the magnetospheric flows from the ionospheric flows and the magnetic field lines are

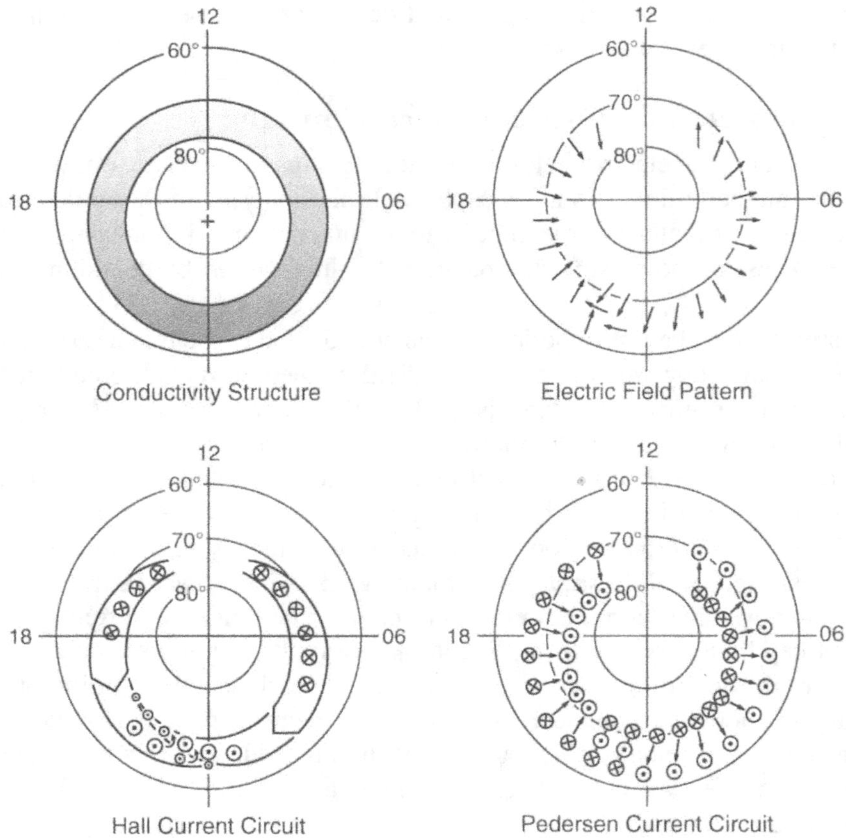

Figure 1.9. Sketch of conductivity, electric field, and Hall and Pedersen currents in the auroral ionosphere. The dots show the upward and the crosses the downward field-aligned currents. From Baumjohann and Treumann (1996).

no longer equipotentials, violating the frozen-in condition. These parallel electric fields are observed in both the upward and downward current regions and are responsible for much of the particle acceleration in the auroral regions. The visible aurorae are primarily due to downward electron acceleration by parallel electric fields in the upward field-aligned current regions. Details of the formation of parallel electric fields are a primary topic of this book.

The elements of ionospheric electrodynamics in the auroral region are sketched in Figure 1.9. The high conductivity region corresponds to the auroral oval, and is the site of the *Auroral Electrojets*. The polar cap encircled by the oval is characterized by antisunward plasma flow which corresponds to the antisunward flow of magnetosheath plasma on the same field lines. The sunward return flow takes place in the region of the auroral oval from where the field lines map to the plasma sheet where the predominant magnetospheric flow is also sunward. Equatorward of the auroral oval is a very weak convection field associated with co-rotation of the plasmasphere.

As the ionospheric ions are collisionally coupled to the neutrals, the current perpendicular to the geomagnetic field has two components: the Hall current which is directed perpendicular to the ionospheric electric field and the Pedersen current which is parallel to the electric field. In Figure 1.9 the coupling of these currents to upward and downward field-aligned currents is indicated by dots and crosses.

The connection between the aurora and the magnetosphere-ionosphere interactions through field-aligned currents was not realised until in situ observations provided a clear picture of the energy exchange mechanism. The first direct observations of the particle fluxes responsible for the aurora were made by a sounding rocket. McIlwain (1960) showed that the precipitating auroral electrons had a nearly monoenergetic peak in the count rate, which suggested acceleration by parallel electric fields. This initial clue of parallel electric fields was discounted by many as unrealistic for a highly conducting plasma although their existence had been predicted by Alfvén (1958). In addition, when the fluxes were plotted as a distribution function, they showed only a weak peak with a broad plateau and a power law at lower energies. These electron distributions were later shown by Evans (1974) to be consistent with acceleration by parallel electric fields if atmospheric secondaries and backscattered primary electrons were included. Upgoing electrons are reflected by the parallel electric field and appear as part of the downgoing component.

As discussed extensively in Chapter 4 of this book, the parallel electric fields are not limited to the upward current regions. The Freja observations revealed that diverging electric field structures of > 1 V m^{-1} exist in the return current region (Marklund et al., 1994, 1997), with a downward electric field component being responsible for the upward acceleration of ionospheric electrons. Later rich observational material of upward accelerated electrons was obtained by the FAST satellite at higher altitudes where their occurrence is more frequent (Carlson et al., 1998a).

The relation between field-aligned potentials and currents was recognized by Knight (1973) who developed a simple analytical expression for the current assuming a Maxwellian distribution of magnetospheric electrons (see Section 3.3.1). In the upward current region and for a relatively wide range of conditions, there is a regime where this relationship can be expressed by a linear formula $j_{\parallel} = K\Phi_{\parallel}$ where $j_{\parallel}$ is the FAC density in the ionosphere and $\Phi_{\parallel}$ is the potential difference between the ionosphere and magnetosphere. Experimentally, the proportionality constant has been found to be in the range 10^{-10}–10^{-8} S m^{-2} (e.g., Olsson et al., 1996a, and references therein).

The spatial distribution of parallel electric fields in both the upward and downward current regions is complex and will be addressed in depth in later chapters. Nevertheless, it is worth a few words at this point to allow the overall structure to be visualised. Parallel potentials in the upward current region have been observed at altitudes as low as ~2000 km and to extend to altitudes >20000 km. The lower-altitude boundary depends on solar illumination but is typically >3000 km in the winter hemisphere and >4000 km in the summer hemisphere.

The distribution of parallel electric fields in the downward current region is less well known experimentally. According to Freja observations (Marklund et al., 1997) parallel potentials appear to extend to lower altitudes in the downward current region than in the upward current regions which is consistent with FAST observations (Carlson et al., 1998a). In the winter hemisphere, divergent (up to 1 $\mathrm{V\,m^{-1}}$) electric field structures and upward accelerated electron beams have been observed below 1500 km but are more frequent at higher altitudes (>2000 km). In the summer hemisphere, they are rarely observed below 4000 km.

While this introductory discussion has been limited to a quasi-static coupling between the ionosphere and magnetosphere, the real physical system is time-dependent and so is the coupling. The time dependent aspects are described in terms of Alfvén waves. The Alfvén wave is a current carrying mode that transmits magnetic stress along the field line. The wave propagates somewhat like a signal in a co-axial cable, where currents flow along the inner and outer conductors and these currents are connected through polarization currents in the dielectric material associated with the transverse electric fields. In the same manner, Alfvén waves carry field-aligned currents that are connected by polarization currents perpendicular to the magnetic field. The polarization currents are consistent with the perpendicular electric fields and plasma acceleration perpendicular to **E** and **B**. The wave propagates at the Alfvén velocity $v_A = B/\sqrt{\mu_0 \rho}$, where ρ is the mass density of the plasma. The Alfvén velocity can vary by several orders of magnitude along a given auroral magnetic field line.

Alfvén waves are the means through which differences in magnetospheric and ionospheric flows find equilibrium. For example, an increase in the reconnection rate at the magnetopause may drive an increase in the magnetospheric flow velocity. The enhanced convection is communicated to the ionosphere via Alfvén waves. Alfvén waves propagate down the magnetic field accelerating the plasma to match the magnetospheric driver. However, when the wave reaches the ionosphere, a portion of the wave will reflect off the sharp impedance change due to high perpendicular ionospheric conductivity. For typical ionospheric and magnetospheric parameters, the reflected wave adds to the incident wave so as to reduce the net electric field and enhance the field-aligned current. The reflected wave travels back to the magnetosphere, increasing the drag, and may reflect off the magnetospheric driver. Several bounces are normally required before the magnetosphere and ionosphere come to equilibrium. Field-aligned currents build up during this period until the magnetospheric generator feels the entire drag of the ionosphere.

1.5 Large Scale Magnetospheric Flows and Dynamics

In order to understand the magnetospheric flow pattern, it is important to recall that plasma motion perpendicular to the magnetic field corresponds to a convection electric field given by $\mathbf{E} = -\mathbf{v} \times \mathbf{B}$. In steady-state ideal magnetohydrodynamics, the magnetic field lines are electric equipotentials. In such a convective motion two plasma elements initially within a magnetic flux tube will at a later time also be within the same magnetic flux tube. This approximation is generally valid in large regions of the magnetosphere, but can be violated at boundaries, in the reconnection regions, and within auroral acceleration regions. A zeroth-order flow pattern forming during southward interplanetary magnetic field was introduced by Dungey (1961) (Figure 1.8). The same year another driving mechanism for the magnetospheric convection, termed viscous interaction, was proposed by Axford and Hines (1961). Even though these models ignore much of the structure of magnetospheric plasma flows, they provide immensely useful frameworks in which the details can be discussed.

Magnetospheric flows are driven by coupling of the solar wind to the magnetosphere across the magnetopause. Reconnection at the dayside magnetopause reduces the dayside magnetic flux on closed field lines and creates new open magnetic flux within the polar cusps. The magnetosheath plasma flow drags the plasma (and the frozen-in magnetic flux) on these flux tubes tailward, which increases the magnetic flux in the tail lobes. Consequently, both the cross-tail current and the magnetopause currents encircling the lobes are enhanced. In the distant magnetotail, where the plasma convects toward the plasma sheet from both lobes, another reconnection region produces new closed flux tubes that convect earthward and around Earth to the dayside plasma sheet. Behind this distant neutral line the reconnected flow joins the downstream solar wind flow. Dayside plasma sheet flux tubes continue to convect toward the dayside magnetopause, where they can again reconnect and start a new cycle through the system. The ultimate driver of the magnetospheric convection is the solar wind kinetic energy flux, which is coupled to the magnetosphere mainly through the reconnection process and to a lesser extent by viscous interaction along the magnetopause.

This steady-state convection picture is continuously interrupted by large-scale reconfiguration processes associated with enhanced energy input during southward interplanetary fields. Most typically, this energy cycle is realised through a sequence of events termed the substorm (Akasofu, 1964). As the interplanetary magnetic field direction varies, there usually are several substorms per day. Originally substorms were divided into two phases: the expansion phase and the recovery phase (Akasofu, 1964). Later McPherron et al. (1973) introduced the concept of growth phase preceding the onset of the expansion phase.

The growth phase starts shortly after the north-south component of interplanetary magnetic field turns southward and enhanced dayside reconnection begins to increase the magnetospheric energy content. In the ionosphere, the growth phase is associated with equatorward motion of the auroral oval. This corresponds to the expansion of the polar cap area consistent with the increasing magnetic flux in the tail lobes. In the evening sector ionosphere equatorward drifting discrete auroral arcs are often observed. In the magnetotail, the growth phase is seen as an increase in the lobe magnetic field consistent with the polar cap area increase. In the inner tail near geostationary orbit the magnetic field normal to the current sheet (B_Z) decreases. These effects are caused by the formation of a thin and intense current sheet that extends from near-geostationary orbit out to 30 or 40 R_E. This thin current sheet typically has a thickness of only about 1000 km (ion Larmor radius scale), and it is embedded within the much thicker plasma sheet. The formation and structure of such thin current sheets are not yet fully understood, but thin current sheets are almost invariably observed during the growth phase evolution.

During the growth phase the solar wind energy is stored in the form of magnetic field energy in the open tail lobes. After about one hour (on average), the energy input leads to the growth of an internal instability which is associated with reconnection in the mid-magnetotail at distances of about 20-30 R_E from Earth. The instability causes a large-scale dipolarization of the tail magnetic field, intense auroral activity, and ejection of a portion of the plasma sheet downstream of Earth. These processes are discussed in more detail in Chapter 8.

Periods of extended (>4 hours), southward (>10 nT) solar wind magnetic fields produce large-scale magnetospheric disturbances, geomagnetic storms, which can last from several hours to several days. The main phase of geomagnetic storms is typically characterized by intense substorm activity and results in a strong enhancement of the ring current in the inner magnetosphere. Because of the high activity level also in the inner magnetosphere, the strongly enhanced auroral activity extends to latitudes much lower than the average position of the auroral oval. The average number of such disturbances is strongly correlated with the solar activity, maximizing near sunspot maximum and the declining phase, and minimizing near solar minimum.

1.6 An Overview of the Book

After this brief introduction the reader is invited to turn to the main body of the book. Chapter 2 discusses auroral forms as they are observed in the ionosphere using remote sensing techniques. It begins with a description of quiet auroral arcs and thereafter deals with diffuse and pulsating aurora, and the proper motion of auroral forms. The Chapter concludes with a summary of the physical characteristics of auroral arcs and an overview of auroral morphology.

In Chapter 3 the most important theoretical building blocks to understand auroral plasma physics are introduced. The chapter discusses the elements of the ionospheric and magnetospheric plasma electrodynamics, magnetic field-aligned electric fields, dynamic magnetosphere-ionosphere coupling, and wave-particle interactions. Chapter 3 is written at such a level that a student equipped with basic plasma physical knowledge should grasp the essential elements needed for further study of auroral plasma physics.

Chapter 4 provides an extensive and up-to-date review of in situ observations of auroral phenomena by sounding rockets and nearly polar-orbiting satellites. The main division of the material in Chapter 4 is between upward, downward and time-varying field-aligned current regions. The wave phenomena are collected in a separate section in order not to confuse the discussion by repeating similar wave phenomena in different environments. As there is much more interesting material available than is possible to include in this volume, a guiding principle has been that not all interesting plasma physics in the auroral region should be categorized as auroral plasma physics. Thus a priority has been given to phenomena that are essential to aurorae as they are observed in the ionosphere.

Statistical aspects of auroral observations are collected to Chapter 5. This chapter also deals with the non-trivial matter of mapping between the source regions in the magnetosphere and the observations in and above the auroral ionosphere.

Chapter 6 discusses the ionospheric electrodynamics of some major auroral forms. These are the eveningside discrete arcs, westward travelling surges, omega bands, auroral streamers and polar cap aurora.

In Chapter 7 the theoretical models are discussed in detail. The topics include static parallel electric fields, boundary layer coupling to the ionosphere, current and voltage generators, resonance models, propagating arc models, transient acceleration, and large-scale features. The Chapter concludes with an attempt to identify the elements of an integrated theory of auroral phenomena.

Chapter 8 deals with the dynamic coupling of auroral phenomena to the magnetosphere. It contains discussions of the magnetosphere as a continuous source of precipitation, magnetospheric energy circulation, and the magnetospheric and ionospheric control of the coupling processes.

Aurorae are not limited to the terrestrial magnetosphere. Chapter 9 discusses the observations of aurorae on the large magnetised planets in the outer solar system and the possibility of similar phenomena on other planets. There are also striking similarities between the plasma physics of the solar atmosphere and the terrestrial magnetosphere, in particular, between explosive events such as solar flares and magnetospheric substorms.

To help with navigation between related parts, the book uses extensive cross-referencing. A subject index that includes acronyms and abbreviations has been appended. For introductions to the concepts and notations of space physics, the reader is referred to the books by Kivelson and Russell (1995) and Baumjohann and Treumann (1996).

Chapter 2

Remote Sensing of Auroral Arcs

In this chapter, we illustrate how remote sensing of auroral luminosity and the high-latitude ionosphere may be used to develop a synoptic interpretation of discrete auroral arcs. In favorable situations, in situ data of arcs may be combined with a full complement of remotely sensed data from visible and ultraviolet imagers on polar-orbiting satellites, ground-based all-sky and meridian-scanning photometer arrays, ground-based magnetometer arrays, coherent backscatter radar measurements of ionospheric convection, and incoherent-scatter radar measurements of the ionospheric density, velocity and temperature distribution.

The length scales of the visual aurora span an enormous range, from the global scale of the auroral oval down to sub-kilometer-scale structures within discrete arcs. The relevant time scales range from hours for stable auroral arcs to fractions of a second for fast moving, coherent structures inside auroral arcs. Discrete arcs are intermediate scale structures, and they may be of the quasi-static variant described in Section 2.1 or intrinsically dynamic. Some aspects of breakup and pulsating aurorae are described in 2.2, together with the pervasive form known as diffuse aurora.

Auroral arcs are the visual manifestation of collisionless energy conversion in the lower magnetosphere, and differences in their spatio-temporal character provide important insights into magnetospheric processes. In determining whether an auroral arc is intrinsically time variable or quasi-steady, it is first necessary to identify its 'proper motion'. Observational procedures for determining proper motion are presented in Section 2.3.

Rapidly evolving and small-scale auroral features are best resolved by optical instruments which provide good sensitivity over a wide field of view. The chapter concludes with a discussion of the morphology of visual auroral features and the relationship between observed length scales of discrete arcs and larger scale auroral precipitation structures. The current state-of-the-art in auroral imaging makes use of all-sky and narrow-field cameras as well as meridian-scanning photometer arrays with spectral filters designed to isolate various atomic excitation lines.

Spectral information is useful in characterizing the species and energy of auroral particle precipitation, and thereby in providing a long-duration and large-area proxy for comparison with the more limited line-of-flight, but also more detailed, in situ particle measurements from satellites and rockets.

2.1 Quiet Auroral Arc

The visible auroral arc is the most readily observed ionospheric signature of coupling between the ionospheric and the magnetospheric plasma. It represents the visible signature of ionization of the upper atmosphere by energetic electrons precipitating from the magnetosphere. In the nightside auroral oval, it usually occurs as a narrow structure tangential to magnetic latitude circles and can be a few hundred meters to several tens of km wide in latitude while extending several thousand km in longitude. The arc often displays a relatively sharp lower boundary at around 100 km altitude, determined by the thermospheric penetration altitude of primary auroral electrons. The altitude extent of auroral arcs and the display of colors usually depend on the range of energies and relative electron fluxes.

Arcs are produced by thin sheets of precipitating electrons, which also carry a considerable upward directed field-aligned current. In simple terms, the relation between auroral particle fluxes and the field-aligned currents stems from the need to maintain current continuity (and charge neutrality) in the presence of a low number density of current carrying electrons and increasing current concentration just above the topside ionosphere resulting from the convergence of the magnetic field lines. The current continuity is achieved by downward acceleration of the electrons in an upward electric potential drop, $\Phi_{||}$. In addition, forms of anomalous resistivity (cf. Section 3.5.1) may add to $\Phi_{||}$ in some regions. In spite of the low densities involved, a high degree of charge neutrality has to be maintained. This yields another contribution to $\Phi_{||}$, in particular at the interface between the magnetosphere and topside ionosphere.

Because of the high degree of incompressibility of the near-Earth magnetic field, the electric field is nearly electrostatic, i.e., describable by continuous equipotential lines, at least within the lower few Earth's radii in altitude. This leads to the topology of U-shaped potential distributions with converging transverse electric fields on either side of the maximum $\Phi_{||}$ (Gurnett, 1972). Figure 2.1 is a sketch of the relation between the upward $j_{||}$ and such electric fields. Furthermore, the picture shows that part of the contour lines continue into the ionosphere owing to the superposed slow ionospheric convection. The much higher transverse electric fields above the topside ionosphere (denser contours) correspond to fast flows of the magnetospheric plasma above arcs in opposing directions, as confirmed by satellite measurements. Figure 2.1 also points out another manifestation of these fast shear flows, namely the counter-motions of rays, folds or curls observable in active arcs. Similar potential contours and flows, but with the direction of the electric field and velocity inverted, can exist in regions of intense downward field-aligned currents.

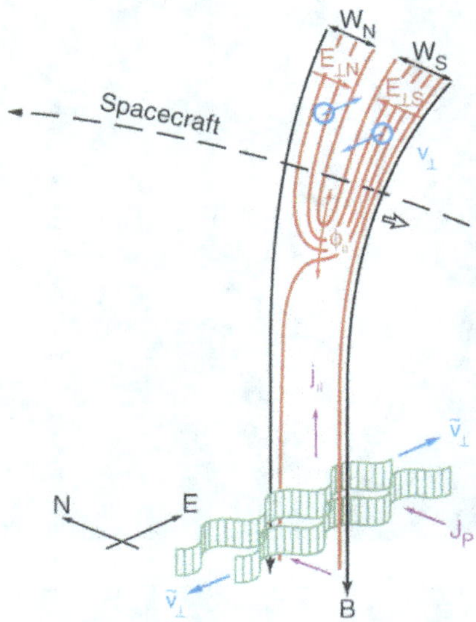

Figure 2.1. Schematic view of parallel currents, $j_{||}$, plasma motions, $v_{\perp}$, electric fields, $E_{\perp}$, and potential contours above an auroral arc, sketched in green. After Haerendel et al. (1996).

They do not create aurora (the electrons are accelerated upward), but rather dark lanes adjacent to or in between arcs, often refered to as black aurora.

It should be noted that the high-altitude electric fields can be strongly structured, not as smooth as Figure 2.1 suggests. This is probably caused by the contribution to $\Phi_{||}$ from the transition from the dilute, hot magnetospheric plasma to the denser, cool plasma of the topside ionosphere that creates a contact potential. Due to ion acceleration and extraction from the latter plasma, its spatial distribution is not smooth. Density holes ('auroral cavity') may be generated by the interaction with high electric fields. In addition, the magnetospheric plasma density is not uniform in the horizontal direction, thus varying levels of $\Phi_{||}$ are needed to maintain a constant field-aligned current.

Prepared by those remarks, we can now look at real data. Figure 2.2 is a rare example of a satellite crossing above multiple parallel arcs and simultaneous optical coverage of these arcs from an aircraft. The arcs can be identified by peaks in energy spectra and flux of the downgoing electron component (so-called 'inverted-V events' - see Section 4.1.1.1). Furthermore, the east-west component of the magnetic perturbation field increases upon transit through a downward current sheet and decreases with upward current (see second panel from bottom). Where the gradient of this field component sharpens, $j_{||}$ intensifies, and arcs should be found. This is seen between 09:19:00 and 09:19:40 UT and between 09:21:00 and 09:21:15 UT. However, not all inverted-V's are clearly attributable to current intensifications, and other parameters, such as a reduced magnetospheric density may

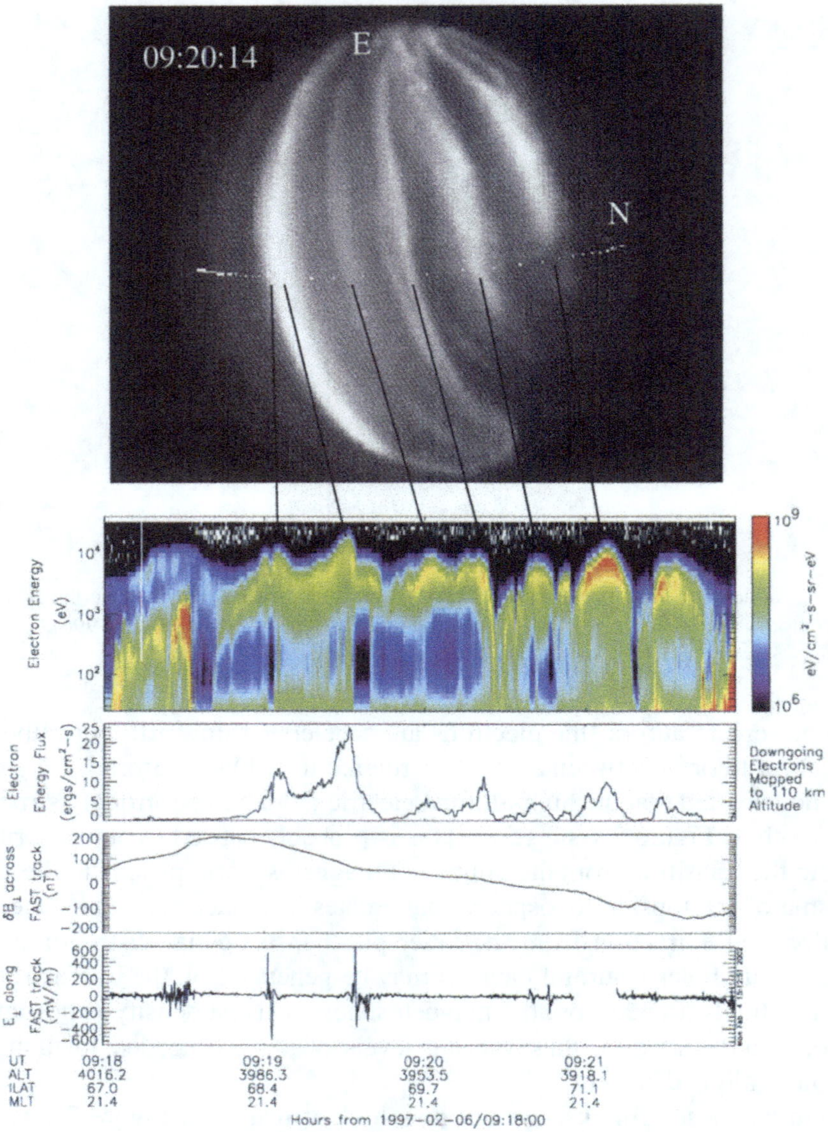

Figure 2.2. The top panel shows multiple arcs seen in the all sky image, taken by a low-light level TV-camera from an aircraft, and the FAST orbit mapped to 110 altitude. The second panel shows the electron energy spectrum measured by FAST. The third panel is the precipitated energy flux on a linear scale mapped down to 110 km. When ion beams were present, the flux was mapped assuming a parallel potential between FAST and 110 km equal to the mean energy of the ion beam. In the fourth panel, the perpendicular, across-FAST-track magnetic field perturbation from the International Geophysical Geomagnetic Reference Field (IGRF) magnetic field model is shown in the field-aligned coordinate system. In the fifth panel, the electric field along the FAST track is displayed. The aurorae, in particular the two arcs to the right (north) did change over the 4 minutes of the pass. After Stenbaek-Nielsen et al. (1998).

be important. Finally, one can see how strongly the electric field (bottom panel) can be structured at times, and that at altitudes of a few 1000 km, the spacecraft is moving in and out of the region containing the accelerating fields (see the inverted V's between 09:20:00 and 09:20:30 UT). Also visible are brief regions of downward current (near 09:18:30 UT and 09:21:30 UT).

Stable evening-sector auroral arcs typically drift equatorward slowly at a speed of several 100 m s^{-1}. They slow down at the equatorward edge of the auroral oval, where a more diffuse and stable auroral band appears at the magnetic projection of the earthward edge of the plasma sheet. In contrast to rapidly expanding, active auroral forms seen during auroral substorms and other sporadic magnetospheric energy releases, the slowly drifting behavior of the auroral arc makes it ideal for studies using ground-based instruments. Studies of other more active auroral forms usually require dense networks of stations both in latitude and longitude or, alternatively, high-resolution radar scans in several directions to resolve the relevant electrodynamics, whereas an auroral arc often provides its own scan by slowly drifting over a magnetometer or an incoherent scatter radar. The optimum antenna pointing direction for arc studies using incoherent scatter radars is magnetic-field aligned, because this position allows the simultaneous measurements of plasma parameters along a narrow flux tube, which is a necessary criterion to derive physically meaningful parameters from the measured line-of-sight profiles of various plasma parameters. The spatial resolution of the 'scan' through the drifting structure of auroral precipitation, associated emissions and three-dimensional current flow is determined by the temporal resolution of the observing instrument and the drift velocity of the arc itself. Most often the electrodynamics of the arc does not change over a time interval comparable to the time it takes for the entire feature to drift across the field line mapping to the ground-based instrument.

As arcs are typically located in the poleward part of the evening sector auroral oval and equatorward part of the morning sector auroral oval, a connection with the large-scale upward-directed Region 1 and Region 2 currents is evident.

Another ground-based diagnostic tool for the study of auroral electrodynamics, the coherent auroral backscatter radar, revealed an interesting and related feature, which has been termed 'radar aurora' (Uspensky et al., 1983). Radar aurora, in the form of a quite narrow but longitudinally extended auroral backscatter, was often found to be adjacent on only one side of visible auroral forms like arcs. And as radar scattering irregularities in the E-region are easily excited within strong electric fields, the radar aurora was readily coupled with zones of enhanced electric fields and strong ionospheric convection. The local-time dependence of the location of this electric field zone with respect to the visible aurora was easily identified: radar aurora typically occurs on the equatorward side of evening sector and on the poleward side of morning sector aurora, with a clear mixed behavior around magnetic midnight and a few hours westward into the late evening sector.

Data from the high resolution and tristatic EISCAT[1] UHF facility (Opgenoorth et al., 1990) and from coherent scatter observations, interpreted in light of past rocket experiments (Timofeev and Galperin, 1991), both indicate that this pattern of relative location of aurora and radar aurora must be associated with the direction of ionospheric current closure in the various local time sectors. In the evening sector, predominantly northward directed convection electric fields lead to northward directed Pedersen current closure of the upward directed field-aligned currents in auroral arcs, and also in the larger scale Region 1 field-aligned currents. Consequently the balancing downward field-aligned currents must be located equatorward of the evening sector aurora. In the absence of strong ionizing precipitation, the closure current is maintained with the help of a strong electric field in the ionosphere, giving rise not only to continuous current flow but also to ion and electron heating and radar aurora. EISCAT observations of depletions in the ionospheric ion density below normal background levels were interpreted as an indication of the extraction of electrons and ions by the flow of excess current.

2.2 Diffuse and Pulsating Aurora

Magnetospheric substorms, in particular those observed in the pre-midnight sector, go through three distinct phases of auroral appearance, a diffuse phase, a rayed phase, and a pulsating phase. The transition from the diffuse to the rayed phase is called auroral 'breakup'. It is followed by a poleward expansion of the aurora. These three phases have been described by Størmer (1955) in the following words: "The succession of events in a common Aurora Borealis display: that it starts as a homogeneous arc, which may persist unchanged for hours except that it may move slowly southward. Then its lower border grows more sharp and suddenly it changes to a rayed arc, followed by curtains, rays, and corona. After this outburst of rayed forms, the aurora fades and changes to diffuse surfaces, pulsating forms, and flames." Figure 2.3 Illustrates the physical appearance of the aurora during these three phases and Figure 2.4 is an example of light intensity pulsations and the respective power spectrum for a particular event, taken from Winckler and Nemzek (1993).

The term 'diffuse' implies that there is no internal structure, that the structure, if any, is unresolvable by the detector, or that the structure is of low contrast and is deemed unimportant. Needless to say, these are subjective criteria. But it is true that, while the aurora is often dominated by a few obvious discrete arcs, there are other regions where the luminosity is fairly low and is spread fairly uniformly over a wide region. These regions occur typically in the equatorward portion of the aurora.

[1] European Incoherent Scatter Radar

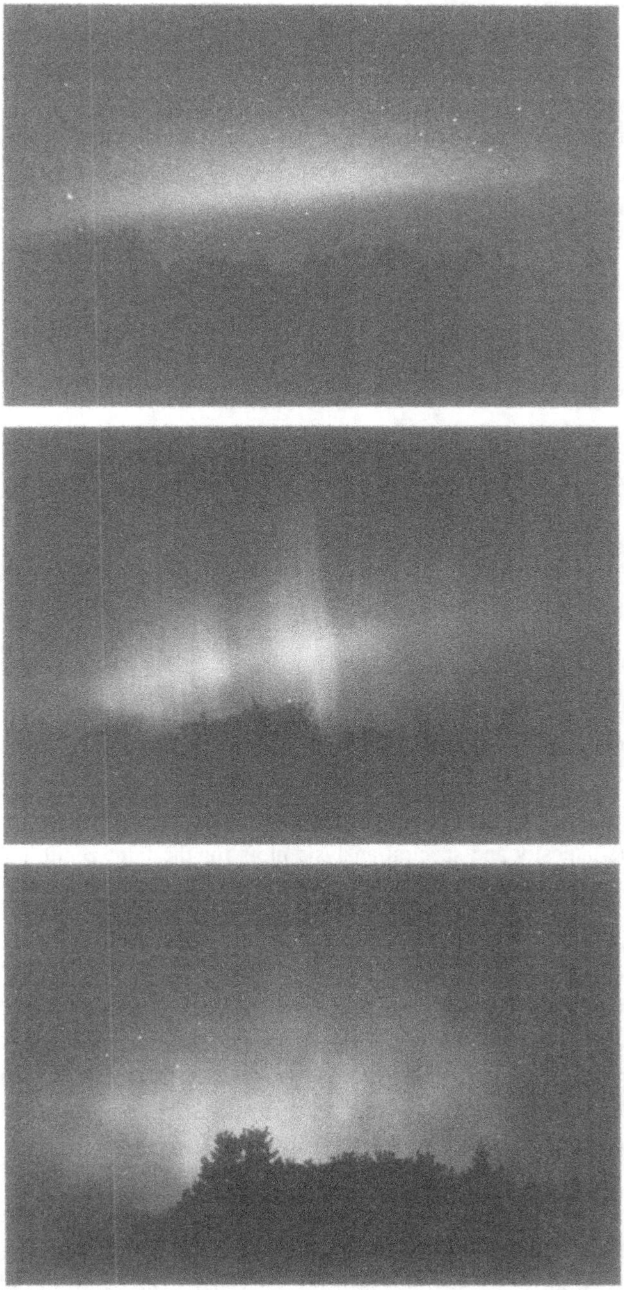

Figure 2.3. Photographic view of three phases of a moderate aurora on 13-14 May, 1991: diffuse aurora (top), rayed aurora (middle), and diffuse surfaces (bottom). From Winckler and Nemzek (1993).

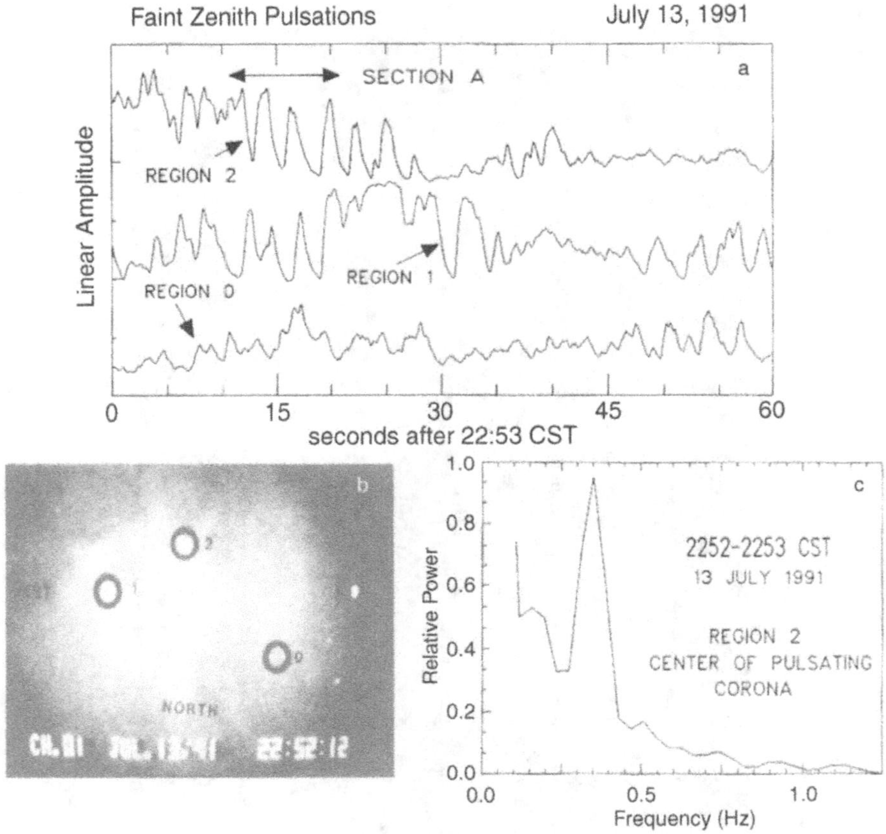

Figure 2.4. Time-intensity and spectral analysis plots for the faint zenith pulsations on 13 July 1991. The regions are defined in the lower left photo. From Winckler and Nemzek (1993).

On the evening side, the diffuse region coincides with a region of proton precipitation. Even if it is structured at high altitude, the proton precipitation spreads widely in the upper atmosphere. This diffusion occurs because the proton goes through a large number of charge-exchange interactions with atmospheric atoms and thus spends a portion of its time as a neutral hydrogen atom that can move freely across field lines, and can also emit light (Lui and Anger, 1973; Akasofu, 1974).

The region of proton precipitation also includes substantial electron precipitation, sometimes peaking in a predominantly electron arc at the poleward boundary of the protons as illustrated in one example of FAST data (Stenbaek-Nielsen et al., 1998). The electron precipitation within the proton aurora can be highly structured, often producing structures far more narrow than are typical of the discrete aurora. These narrow structures are resolvable only when they are located in the local magnetic zenith and then only by high-resolution imagers.

On the morning side, the diffuse aurora is usually well separated from the region of proton precipitation and is dominated by patches of pulsating aurora (Cresswell and Davis, 1966; Royrvik and Davis, 1977). Although occasionally arc-like, these patches are most often highly irregular in shape.

The pulsation phase, which dominates at the equatorward edge of the auroral oval, is particularly intense at sub-auroral latitudes. It typically begins 5-10 min after auroral breakup at late evening hours and develops into post-midnight hours. Episodes of subsidence and re-appearance often alternate with quasi-periods of 10-15 min duration. The pulsation frequencies range from 0.05 to 3 Hz with horizontal scale sizes of 10-100 km.

The origin of pulsating aurora is largely unclear. Particle measurements (Johnstone, 1978) indicate that energetic ($\geq$ 10 keV) electrons, freshly injected into the trapping region during substorms, play a key role. Several theories relate the energy influx to pitch angle diffusion (Davidson, 1990). Upward field aligned motions of the lower border of a pulse have been interpreted as a velocity dispersion effect with the modulating source located near the equator (Bryant, 1983).

The high frequency pulsations discussed above do not have a clear relation with observed geomagnetic pulsations. However, the hypothesis has been advanced (Samson et al., 1991, 1996) that some classes of discrete auroral arcs may be associated with isolated narrow-band field line resonances of longer (Pc 3-5) period (Lotko et al., 1998). In-situ observations of such an event are presented in section 4.4.3.

2.3 Proper Motion of Auroral Arcs

By 'proper motion' of an auroral arc we mean a progression of the luminous band in its normal direction that is not shared by the ionospheric plasma, implying that the auroral acceleration region is not frozen in the plasma. The assessment of proper motions needs two types of measurements, an optical observation of the spatial distribution of luminosity as a function of time and a measurement of plasma drifts or electric fields above the region where the Pedersen currents flow, i.e., where ion motions transverse to the magnetic field are not affected by collisions.

There seem to be two classes of proper motions, fast poleward motions, mainly found during substorm expansions, and much slower ones (of the order of 100 m s^{-1}), which may be directed poleward or equatorward. It may be that the processes underlying the two classes are quite different. Since the measurement of slow plasma drifts in the normal direction is not an easy matter, in particular in the presence of much higher drifts tangential to the auroral arc, slow proper motions are difficult to determine. A very sensitive tool for such investigation was the tracing of the ambient plasma flow by luminous barium plasma clouds. Wescott et al. (1970) and Haerendel and Lüst (1970) presented convincing evidence for such motions,

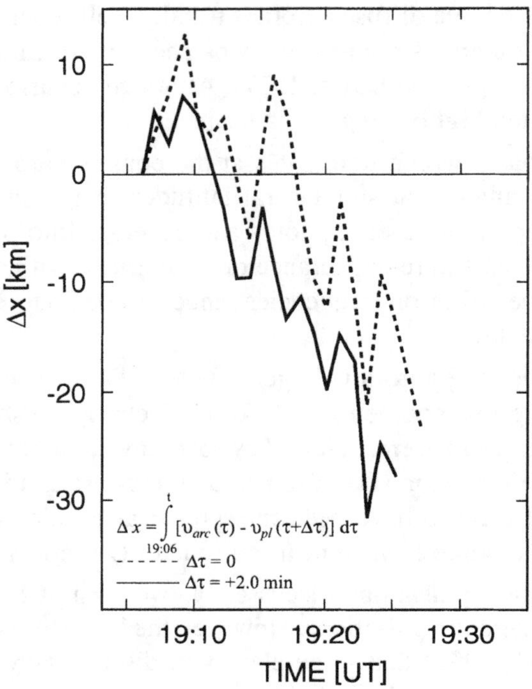

Figure 2.5. Integrated velocity difference between a slowly drifting auroral arc and the surrounding plasma. The dashed line was obtained by taking the two velocities at the same time, the solid line by delaying the plasma velocity by 2 minutes. From Haerendel et al. (1993).

but because of the small number of events and incomplete ionospheric diagnostics no systematic conclusions could be drawn.

A more systematic approach became possible by combined optical and radar measurements with the EISCAT facility in the magnetic zenith of the optical site (Haerendel, 1983; Frey et al., 1996). Figure 2.5 shows the integral of the velocity difference between a slowly drifting isolated arc and the adjacent plasma. The average velocity difference was 35 m s^{-1}, but showed superposed oscillations of a few minutes period with higher velocity amplitudes. A more complete set of measurements above Kiruna was obtained by Frey et al. (1996) by employing sensitive image-intensified CCD cameras and the EISCAT radar with 20 s resolution.

Figure 2.6 shows six cases of plasma motions in relation to various arcs upon their transit through the magnetic zenith of the optical site. This way the errors in determining the arc motions could be minimized. In addition, the region of the dominant Hall current adjacent to the arc is shown as determined from the tangential plasma drift and ground magnetograms. If one sees the arc as one edge of an elongated current circuit with its interior on the side where most of the Hall current is located (see Figure 2.7), one can conclude that more often, but not always, the

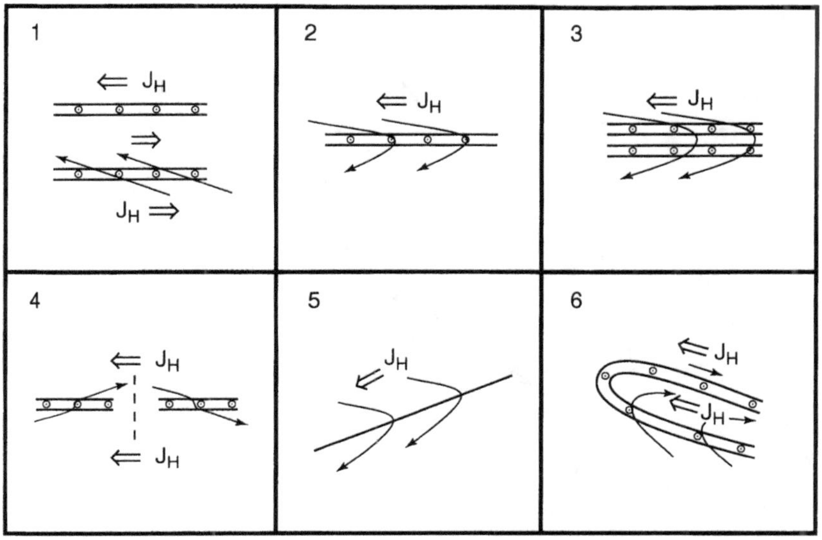

Figure 2.6. Simplifying summaries of the observed flows (solid lines) relative to the auroral arcs, of the arcs' normal velocities in the observer's frame (double arrows) and the location of the dominant Hall current, J_H. The magnetic field direction is out of the plane. From Frey et al. (1996).

arc moves towards the interior of the circuit with typical velocities of the order of 200 m s^{-1}. In Section 7.8 an interpretation in terms of energy tapping from the circuit by the arc and re-supply of energy will be given.

2.4 Physical Characteristics

Arc Widths. The latitudinal width of auroral arcs has been the subject of considerable interest, both observational (Størmer, 1955; Maggs and Davis, 1968; Borovsky et al., 1991; Trondsen and Cogger, 1998; Knudsen et al., 2001) and theoretical (Borovsky, 1993a). Unfortunately, the determination of arc widths is highly dependent both on the observational technique and on the author's choice of a definition for arcs. All-sky cameras lack the resolution to see the finest arcs. Moreover, the thinnest arcs can be measured only when the arc is precisely in the observer's magnetic zenith. Borovsky et al. (1991) have summarized a number of observational difficulties that result in the overestimation of arc thicknesses. Conversely, narrow-field TV measurements in the magnetic zenith tend to de-emphasize any arcs having thicknesses comparable to or greater than the field of view of the camera (typically 10-50 km) as shown in Figure 2.8.

In satellite and rocket data, it is difficult to distinguish between thin arcs and temporal transients. Moreover, as shown by Stenbaek-Nielsen et al. (1998), the almost universal custom of presenting the electron spectrograms on a log scale suppresses the appearance of small features (Figure 2.2). The situation is still more

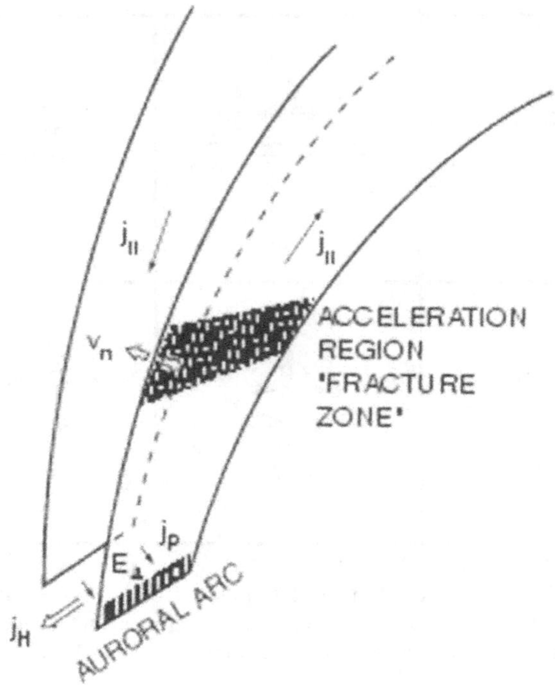

Figure 2.7. Framework for the interpretation of the relative motion of arc and plasma: the arc is seen as the trace of upward field-aligned current, $j_{\parallel}$, which forms a narrow current circuit with a similar downward directed current elongated in the E-W direction and connected via Pedersen currents, J_P. Here v_n is the proper motion, which may be directed to enlarge or reduce the size of the current circuit. After Frey et al. (1996).

confused by the fact that in a typical situation there are a variety of scale sizes present simultaneously. However, whereas the large scales can be readily understood in magnetosphere-ionosphere coupling theories with field-aligned potential drops, the small scales ($\ll 1$ km) are presently not explained by any existing theory (Borovsky, 1993a). In part, the solution may lie in the lack of brightness calibration of low-light-level TV systems and overemphasis of contrast generated by small-amplitude brightness variations. In part there may be manifestations of such small scale potential structures caused by the extraction of cold ions out of the topside ionosphere (Haerendel, 1999). Comparison of the all-sky image with the in situ data demonstrates that optical arc widths are contrast-dependent and exhibit a wide range of scales embedded in wider-scale electron precipitation events.

Arc Height. Although the electrons and protons that produce the aurora may originate in the outer magnetosphere, the visible glow occurs only where sufficient gas exists to produce light. Hence the visible aurora occurs usually below 300 km altitude, although occasionally tall rays are reported to reach substantially higher

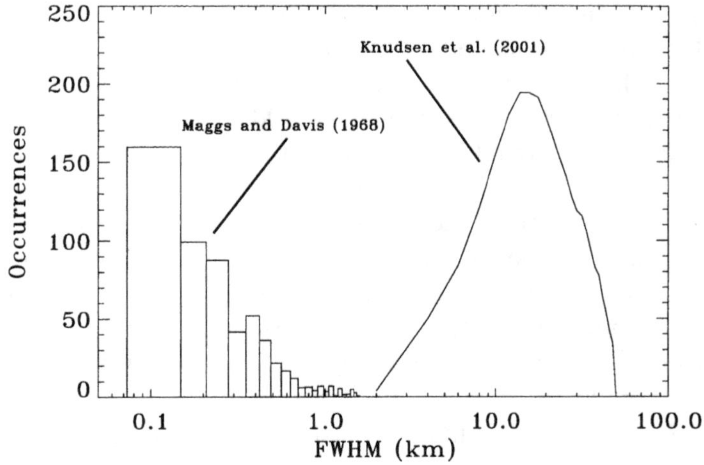

Figure 2.8. Probability of observing an arc of a given thickness, using narrow field (left) and all-sky optics (right). The left-hand edges of both curves are due to intrumental cut-off. The two probabilities are separately normalized. Figure courtesy of Eric Donovan.

altitudes (Størmer, 1946). The altitude of the lower border is determined by the energy of the precipitating electrons and is most commonly about 110 km (Størmer, 1946; Rees, 1963; Banks et al., 1974). Below about 80 km, the collision frequency is so high that precipitating particles lose their energy in a short distance. So the lower border of the auroral curtains is rarely much below 80 km.

Vertical Luminosity Profiles. In principle, given a model of the upper atmosphere and the energy spectrum of the precipitating electrons, it is possible to calculate as a function of altitude the emission rate at various wavelengths, the ionization rate, and the electron density (Rees, 1963; Banks et al., 1974). Owing to the large scale-height of the atmosphere at 100 km, these profiles are smooth and extend over tens of kilometers. Reconstruction of volume emission rates from arc observations at two separate sites has been achieved by Frey et al. (1996). It often appears that the visible auroral luminosity is either confined to or enhanced within thin horizontal layers with thicknesses of the order of 1 km (Stenbaek-Nielsen and Hallinan, 1979; Oguti, 1975; Hallinan et al., 1985). These thin layers of enhanced luminosity also have an unusually high ratio of N_2 1st positive emissions to N_2^+ 1st negative emissions. Hallinan et al. (1997) argued that this implies a source of suprathermal electrons within the emitting volume and suggested that the source is cold ionospheric electrons that are heated locally by waves. The waves in turn would be driven by the precipitating electrons.

Auroral Spectra. The color of the aurora varies with particle energy and with altitude. At the highest altitudes, the atmosphere is dominated by atomic oxygen and the 630-nm red line of atomic oxygen is prominent. Since this is a forbidden transition subject to collisional deactivation of the upper state, there is little emission at 630 nm below 200 km. The altitude of peak emission at 630 nm is about 250 km and the red color extends upward of 300 km (Vallance Jones, 1974). Astronauts at shuttle orbit (around 300 km) describe seeing red above them and green below. The 630 nm emission is caused by electrons of a few eV. Higher energy, primary auroral electrons stimulate 630-nm emissions via low-altitude production of copious secondaries in the few eV range. The 630 nm emission is most prevalent in the polar cusp, where solar wind electrons can penetrate to the atmosphere, and in occasional great red aurorae near sunspot maximum. These are observed at low latitudes and may have given rise to many of the ancient reports that associated aurorae with fire or blood.

As discussed in Section 1.2, the brightest line in the visible spectrum in most aurorae is the 557.7 nm green line from atomic oxygen. It is not significantly quenched at altitudes above 100 km. Much of the 557.7 emission is produced by secondary processes (Vallance Jones, 1974), whose relative importance is not fully understood. Consequently, this emission is rarely modeled despite its importance. Except for extremely energetic precipitation that penetrates below 90 km, the intensity of the green line is roughly proportional to that of the N_2^+ emission at 427.8 nm and is a good indicator of electron energy flux.

The N_2^+ 1st negative bands at 391.4 nm, 427.8 nm, and 470.9 nm are the best understood emissions and are due to simultaneous ionization and excitation by electrons having energies greater than a few hundred eV. The average energy loss of the primary auroral electrons per electron-ion pair produced is 35 eV. Closely associated with these are the red Meinel bands from N_2^+. These are rarely measured since they mostly overlap the 1st positive bands from neutral N_2. The blue 1st negative bands, the red Meinel bands, and the red 1st positive bands combine to produce a magenta color. This is normally overwhelmed by the green 555.7 nm line. However, the magenta can become dominant when the green is collisionally quenched (below 90 km) or if the auroral forms are moving rapidly so that the green is lagging behind and is spatially smeared out (Hallinan et al., 1998).

Of special importance is the 486.1 nm H_β line from Hydrogen. Through charge exchange, a precipitating proton captures an electron and becomes neutral hydrogen in an excited state, giving rise to an H_β photon. Subsequently the hydrogen atom becomes repeatedly ionized and neutralized through additional collisions, producing more H_β emissions. This emission is particularly strong within the diffuse aurora in the evening sector and mixes with the discrete aurora in the midnight and morning sectors. Since the protons collide with atmospheric constituents, the proton precipitation also produces the standard atmospheric emissions. In addition, the proton precipitation is usually accompanied by electron precipitation.

The above discussion covers the main emissions in the visible and near IR, the basis of ground-based observations. In space-borne applications, such as satellite imaging, the ultraviolet region of the spectrum becomes important. The chief emissions in this region are the 130.4 nm and 135.6 nm lines from atomic oxygen and the Lyman-Birge-Hopfield (LBH) bands from N_2.

2.5 Auroral Morphology

Auroral plasmas are often defined as much by their boundaries in space and time as by their internal properties or electrodynamic behavior. These boundaries are readily apparent in the visible forms produced by the interaction of the auroral plasmas with the predominantly neutral atmosphere. We speak of the auroral oval, theta and polar aurora, auroral arcs, poleward expansions and poleward moving auroral forms, fan arcs, moving rays, spirals, curls, drifting patches, flickering, pulsating patches, omega bands, streamers and so on. More generally, the aurora is often divided, primarily on the basis of its appearance in imagery, into discrete aurora and diffuse aurora. This distinction, while somewhat dependent on the resolution and field of view of the sensor, remains in wide use even though it may not correlate precisely with other physical characteristics.

Auroral luminosity also exhibits characteristic patterns that correlate with magnetic local time[2] and latitude, auroral activity state, and interplanetary conditions of the observations. The statistical patterns of auroral distributions are considered in Chapter 5. The auroral morphology described in this section provides a useful visual context for interpreting the plasma dynamic processes discussed in subsequent chapters and their interrelationships. Auroral morphology is actually a subject in itself. Interested readers are referred to Davis (1992) for further details.

2.5.1 Distortions in Auroral Arcs

The basic building block of discrete aurorae is the auroral arc or curtain described in previous sections of this chapter. While arcs can extend a thousand kilometers or more along an L shell, they also exhibit considerable fine structure and can distort into other more complex shapes. Structures described as arcs can have latitudinal widths ranging from less than 100 m to greater than 100 km. Where simultaneous optical and in situ data are available, the visible arcs correspond well to the electron energy flux into the atmosphere and to the inverted V structures (Haerendel, 1994; Stenbaek-Nielsen et al., 1998; Hallinan and Stenbaek-Nielsen, 2001). Distortions such as loops or spirals or simple brightenings along the arc typically drift westward pre-midnight and eastward post-midnight. Smaller scale structures such as rays race along the arc in both directions. The arcs can also drift poleward or equatorward, or the luminosity can progress by the formation of new arcs along

[2]Magnetic local time (MLT) is defined in essentially the same way as 'ordinary' local time, but with the Earth's rotation axis replaced by the dipole axis.

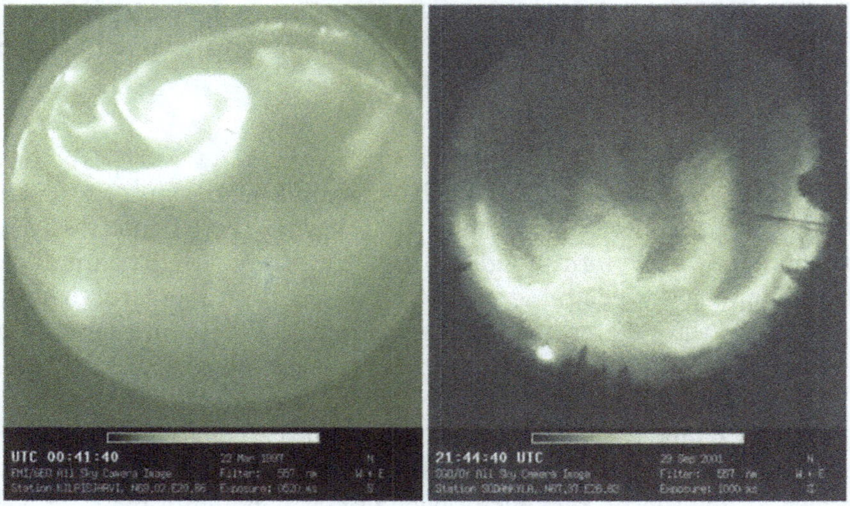

Figure 2.9. Left: an auroral spiral observed by an all sky camera located in Kilpsijärvi, Finland. Right: all sky image of Omega bands above Sodankylä, Finland. Figures courtesy of the Finnish Meteorological Institute.

with the decay of the existing arcs. At any given time, there may be several visible arcs distributed across the auroral oval.

Spirals. During active conditions, an arc or a set of arcs can develop a large spiral form, as shown Figure 2.9, or even a series of spirals strung out along the arc (Haerendel, 1994; Stenbaek-Nielsen et al., 1998; Hallinan et al., 2001). The spirals range in size from a few hundred km to over 1500 km. The spirals occur particularly during auroral breakup where they are described as a 'Westward Traveling Surge' (see Section 6.2) in the evening sector. The are also frequently observed in the afternoon sector of the dayside aurora (Elphinstone et al., 1993). Although often associated with the Westward Traveling Surge, the spirals can also form and remain in place (Rostoker et al., 1987b).

Development of a series of spirals along an arc resembles a Kelvin-Helmholtz vortex street, but the apparent wrapping of the spirals has a clockwise sense of rotation (looking up in the northern hemisphere) opposite to that expected for shear flow vortices (cf. the paragraph on curls below). This behavior has been interpreted in terms of a magnetic shear instability (versus velocity shear instability) resulting from the intense upward field-aligned current in the arc (Hallinan, 1976). In this case, the vortices develop as a function of distance along the magnetic field rather than as a function of time. The observed rotational sense of the spirals in either hemisphere is consistent with an upward current. The formation of a velocity shear driven vortex street versus a magnetic shear driven spiral street is evidently

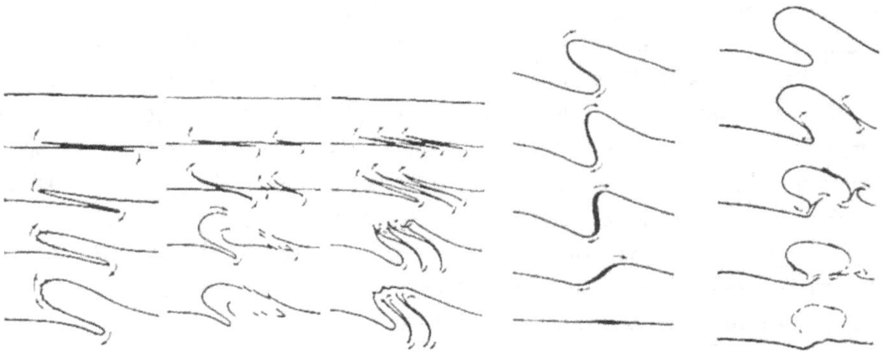

Figure 2.10. Various arc splitting (a-c) and rotational unfolding (d-e) patterns. From Oguti (1975).

regulated by the ionospheric conductivity, with low/high values supporting velocity/magnetic shear instability (Lysak and Song, 1996).

Deformations with a spiral form also arise from 'fractional splitting of a degenerated arc fold followed by counterclockwise rotational unfolding' (Oguti, 1974). The origins of fractional splitting are not known. When spirals form via this mode, the luminous structure shown in Figure 2.9 evolves dynamically with a spiral-like intensification developing as the arc fold unwinds in a counterclockwise sense (looking up in Northern Hemisphere). Figure 2.10 shows sketches of various ways of arc spliting and rotational unfoldings as seen when viewing the arcs along the magnetic field (Oguti, 1975).

Loops. The arc can also be bent into a large-scale loop as shown in the lower left of Figure 2.11 opening either to the East or the West. At times these loops may be related to spirals. But at other times, they appear as distinct and stable shapes that persist for tens of minutes and drift hundreds of km east or west (Davis, 1962; Snyder et al., 1974). Sometimes there is a set of nested loops.

Omega Bands. In the morning sector, the arcs often appear distorted into the shape of the Greek letter Ω or even a series of Ω's strung out along the arc (Akasofu, 1964), see Figure 2.9. Omega bands are described in more detail in Section 6.3. The omegas are typically transient features, lasting only a few minutes. They can form seemingly out of nothing or can occur as a brightening along the boundary of a region of diffuse aurora. The formation of an omega usually involves a considerable brightening of the aurora and is often accompanied by rapid internal ray motions along the omega. The omegas invariably drift toward the East. Omega bands clearly have all the properties of discrete aurorae. However, there are also at times long poleward fingers or torch like forms in the diffuse aurora that are sometimes referred to in the literature as omega bands (Akasofu, 1974).

Folds. On a smaller scale (20 km), the arcs distort into a series of folds depicted in the upper left of Figure 2.11 (Hallinan and Davis, 1970; Wagner et al., 1983). The axis of the folds is skewed so that (as seen looking up in the Northern Hemisphere) it goes from southeast to northwest. i.e., the axis is rotated clockwise from the north-south line. The appearance of the folds is reminiscent of the buckling seen in rocks that have been subjected to compressional shear. The folds can be stationary or can drift along the arc. Seen from the side, the varying optical path length through the arc gives the appearance of rays. When the auroral curtain has fairly stable rays, such as those that can be photographed with a time exposure, it is most likely that the rays are actually folds.

Curls. Like folds, curls are small-scale distortions in an auroral arc (upper right of Figure 2.11). The arc forms a series of counterclockwise vortices by a continuous fluid-like motion (looking up in the Northern Hemisphere). The vortices are identical in appearance to a Kelvin- Helmholtz vortex street in a layer of fluid shear. Hallinan and Davis (1970) proposed that they result from the Kelvin-Helmholtz instability acting on a magnetized field-aligned sheet of excess negative space charge. Equivalently, the flux tubes terminating in the arc are characterized by a convergent electric field. The vortices are most often in motion, and probably account for most examples of active ray motions in auroral arcs. Velocities of 10-20 km s^{-1} are common (Hallinan and Davis, 1970), with speeds up to 90 km s^{-1} (Trondsen and Cogger, 1998). When the aurora is overhead, an arc is often seen to consist of several narrower arcs and it is these small arc-elements that form the curls. When curls are present in two or more arc elements, they are observed invariably to move in a counterclockwise sense relative to a point between the elements (Hallinan and Davis, 1970; Vogt et al., 1999). The motions are thought to be $\mathbf{E} \times \mathbf{B}$ motions, and thus give an indication of the perpendicular electric field above the acceleration region (Vogt and Glassmeier, 2000; Hallinan and Stenbaek-Nielsen, 2001; Hallinan et al., 2001).

Flickering Aurora. Within discrete arcs, there are often irregularly shaped patches, a few kilometers in diameter, that appear, flicker on and off for a few cycles, and disappear (Beach et al., 1968; Berkey et al., 1980). The frequency is usually in the 5-15 Hz range and is thought to correspond to an electromagnetic O^+ ion cyclotron wave at about 1 R_E altitude (Temerin et al., 1986; McFadden et al., 1987; Lund et al., 1995; Arnoldy et al., 1999). The flickering elements are seen in the general region of the arc, but seem quite independent of the other structure within the arc such as the narrow arc elements and the Kelvin-Helmholtz vortices (Kunitake and Oguti, 1984). Seen from the side, they appear as flickering rays.

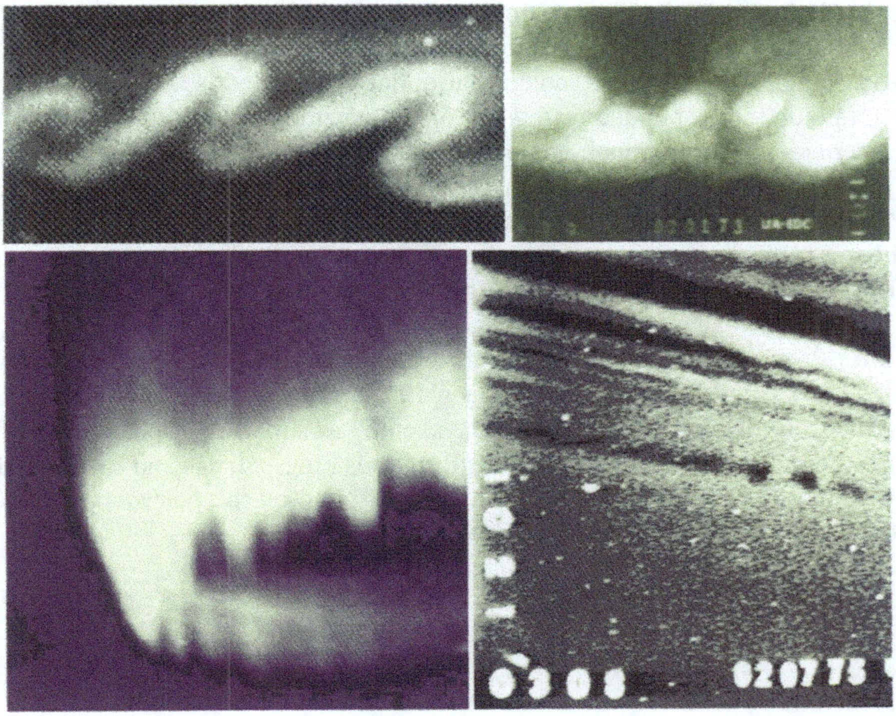

Figure 2.11. Various small-scale auroral forms. Upper left: auroral folds over the northern hemisphere (after Wagner et al., 1983). Upper right: auroral curls (after Hallinan, 1981). Lower left: an auroral loop (after Hallinan et al., 1985). Lower right: a black auroral arc with clockwise curls (after Kimball and Hallinan, 1998b)

2.5.2 Black Aurora

Sometimes it is easiest to describe the appearance of the aurora by calling attention to small voids in an otherwise glowing region (Royrvik and Davis, 1977; Davis, 1978; Trondsen and Cogger, 1997; Kimball and Hallinan, 1998a,b). There seem to be at least two distinct classes of black aurora. The first class is black arcs. These are narrow east-west black filaments in an otherwise fairly uniform glow as shown in the lower right panel of Figure 2.11. By definition, they are seen in the diffuse aurora. They can occur either in the evening or the morning sector.

Occasionally the black filaments develop Kelvin-Helmholtz vortex streets (black curls). The curls have a clockwise rotation sense (opposite to that of the 'white' curls). The rotational sense implies that the black curls correspond to field-aligned sheets of excess positive space charge. Marklund et al. (1994) has noted in the Freja electron data the existence of narrow regions of dropouts in the precipitated electron energy flux. These regions also correspond to electric field reversals indicating a divergent field (positive space charge) and to upward accelerated electron beams (Marklund et al., 1997). While it seems probable that these in situ features correspond to the optical black aurora, no optical data were available. More recently, Peticolas et al. (2001) noted that optical black arcs seen conjugate to the FAST satellite corresponded to regions of weak pitch-angle diffusion surrounded by larger areas of strong pitch-angle diffusion. On occasion, clockwise black curls and counterclockwise white curls are observed simultaneously (Kimball and Hallinan, 1998b). On one occasion, they observed a close pairing between a black vortex street and a white vortex street as seen in Karmann vortex streets.

'Black patches' are also voids in the luminosity, but have a distinctly different morphology. They are elliptical in shape (1 km by 6 km) and drift eastward 1.5 km s^{-1} on average. On occasion, they have luminous cores making them appear as black rings (Trondsen and Cogger, 1997). They occur predominantly in the early recovery phase of a substorm in the evening and midnight sectors. They form along a boundary between westward drifting pulsating patches and eastward drifting nonpulsating forms. The black patches are found on the poleward edge of the pulsations (Kimball and Hallinan, 1998b).

Chapter 3

Theoretical Building Blocks

The previous chapter provided a comprehensive overview of the visible auroral forms, emphasizing our main interest in understanding auroral arcs. The visible aurora is merely the tip of an iceberg leading into the plasma physical phenomena that are taking place in the near-Earth environment. Out along the magnetic field lines that connect to the auroral, it is a rich and exotic world. The visible aurorae are produced by particle precipitation from the magnetosphere into the atmosphere, but the structure is produced by particle acceleration through complex electric fields. At least two major regions are essential in producing auroral forms: a generator region located far out in the magnetosphere, where magnetospheric energy is tapped to power the aurora, and an acceleration region located some 1000s of km above the atmosphere, where electrical energy is converted into particle kinetic energy.

The present chapter gives the main physical concepts needed for such an understanding. The text assumes in many cases that the reader has some basic knowledge in plasma physics. Section 3.1 offers an overview of the electrical properties of the ionosphere and gives a definition of its conductivities. Section 3.2 presents an MHD description of hot plasmas and looks into how the magnetosphere can drive auroral currents. Section 3.3 deals with the static electric fields in the acceleration region high above the ionosphere. Section 3.4 treats the problem of dynamic coupling of the magnetosphere to the ionosphere. Section 3.5, finally, deals with interactions between particles and waves in some processes important for the aurora.

3.1 The Conducting Ionosphere

As discussed in Section 1.2, the ionosphere is the boundary region between the magnetosphere and the atmosphere where collisions between the ionized plasma particles and the neutral atmospheric particles are important. This region has a sharp lower edge around ~90 km and extends upward to ~300 km altitude. The upper boundary is not sharp, but rather constitutes a diffuse continuous transition from the collisional ionosphere to the collisionless plasma of the magnetosphere.

The density of neutral particles in this region is much larger than the plasma density, and so the plasma does not affect the dynamics of the neutral atmosphere. This holds at least on time scales relevant for the auroral processes. On the other hand, the dynamics of the electrons and ions in the ionosphere are strongly affected by collisions with neutral particles. Such collisions cause a breakdown in the $\mathbf{E} \times \mathbf{B}$ motion which the particles perform in the absence of collisions. This disruption of $\mathbf{E} \times \mathbf{B}$ drifts allows electrical currents to flow perpendicular to the magnetic field $\mathbf{B}$ in response to an electric field $\mathbf{E}$. Perpendicular currents in the ionosphere play an important role in closing the field-aligned currents that are imposed from the magnetosphere. A comprehensive discussion of the ionosphere, both in the auroral zone and at lower latitudes can be found in other texts (Banks and Kockarts, 1973; Rees, 1989; Kelley, 1989).

3.1.1 Ionospheric Conductivity

In the collisionless plasma above the ionosphere, particles (1) move freely along the magnetic field and (2) $\mathbf{E} \times \mathbf{B}$ drift across the magnetic field. The plasma in this region is very cold (a few eV in energy), and there are essentially no net currents perpendicular to the magnetic field. In the ionosphere, a region containing electrons and ions and also the neutral atoms and molecules of the upper atmosphere, momentum-exchanging collisions of electrons and ions with neutrals (1) interfere with the free motion of the electrons and ions along $\mathbf{B}$ and (2) disrupt the $\mathbf{E} \times \mathbf{B}$ drifts of the electrons and ions across $\mathbf{B}$. The first effect adds a resistivity to currents flowing along $\mathbf{B}$ in the ionosphere (i.e., it lowers the infinite conductivity along $\mathbf{B}$ to some finite value) while the second effect allows currents to flow across $\mathbf{B}$.

A sketch of how this latter process works appears in Figure 3.1. In the top panel of this figure the $\mathbf{E} \times \mathbf{B}$-drifts of an electron and an ion in the absence of collisions are shown. Since, in the time-averaged sense, the electron and ion move with the same velocity, there is no current flowing in response to the electric field. In the bottom panel the ion is allowed to undergo collisions with a background neutral gas: each collision (marked with a cross) is taken to completely absorb the momentum of the ion, and the ion must restart its $\mathbf{E} \times \mathbf{B}$ motion. In this case the ion and electron no longer have the same time-averaged velocity and so no longer move together. Owing to the collisions, current now flows in response to the electric field. The current component that is in the direction of $\mathbf{E}_\perp$ is known as the Pedersen current (shown in the figure), the current component that flows in the $\mathbf{E} \times \mathbf{B}$ direction is known as the Hall current. While collisions reduce the field-aligned electrical conductivity compared to the ideally conducting magnetosphere where no neutrals exist, they allow for transverse current flow by scattering the particles in the perpendicular direction. This contributes to perpendicular conductivity.

The electric field $\mathbf{E}$ and the electrical current density $\mathbf{j}$ in the ionosphere are coupled through Ohms's law $\mathbf{j} = \sigma \cdot \mathbf{E}$, where the electrical conductivity σ is a tensor. Here the electric field is defined in the frame moving with the neutral gas, which

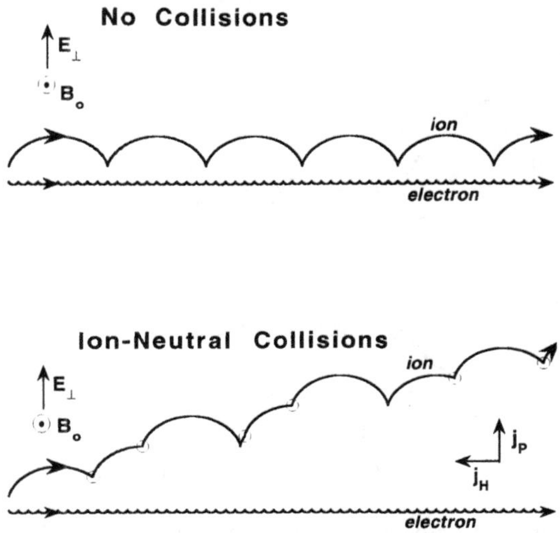

Figure 3.1. A sketch of the disruption of $\mathbf{E} \times \mathbf{B}$ drifts by collisions and the resulting perpendicular current. In the top panel an ion and an electron undergo $\mathbf{E} \times \mathbf{B}$ drift without collisions. In the bottom panel the electron $\mathbf{E} \times \mathbf{B}$ drifts without collisions, but the ion undergoes momentum-exchange collisions with the neutral gas at the places in the orbit marked with circles. In the presence of collisions, the electron and ion no longer drift together. (Note that the ion and electron gyroradii are not to scale). Figure provided by J. Borovsky.

in the absence of winds is the frame corotating with the Earth. With scalar conductivities (see below), this Ohm's law splits into three components: the field-aligned (parallel) current $\mathbf{j}_\parallel = \sigma_\parallel \mathbf{E}_\parallel$, the Pedersen current $\mathbf{j}_P = \sigma_P \mathbf{E}_\perp$ flowing parallel to the transverse electric field $\mathbf{E}_\perp$, and the Hall current $\mathbf{j}_H = \sigma_H \hat{\mathbf{b}} \times \mathbf{E}$ which flows transverse to both the magnetic and electric fields.[1] These three current components are mediated by the three conductivities, $\sigma_\parallel, \sigma_P, \sigma_H$. If there is a neutral wind $\mathbf{v}_n$ in the ionosphere, it will drive Pedersen and Hall currents even in the absence of an electric field in the frame of the observer. Accounting for this, Ohm's law is written out as (Boström, 1973)

$$\mathbf{j} = \sigma_\parallel E_\parallel \hat{\mathbf{b}} + \sigma_P(\mathbf{E}_\perp + \mathbf{v}_n \times \mathbf{B}) - \sigma_H(\mathbf{E}_\perp + \mathbf{v}_n \times \mathbf{B}) \times \hat{\mathbf{b}} \tag{3.1}$$

where $\mathbf{v}_n$ is the velocity of the neutral gas and $\mathbf{E}$ is the electric field in the corotating frame. The parallel, Pedersen and Hall conductivities are defined in the rest frame of the neutral particles and are given by the following expressions (e.g., Boström,

[1] Throughout the book, unit vectors are denoted by a hat, e.g., $\mathbf{B}/B \equiv \hat{\mathbf{b}}$

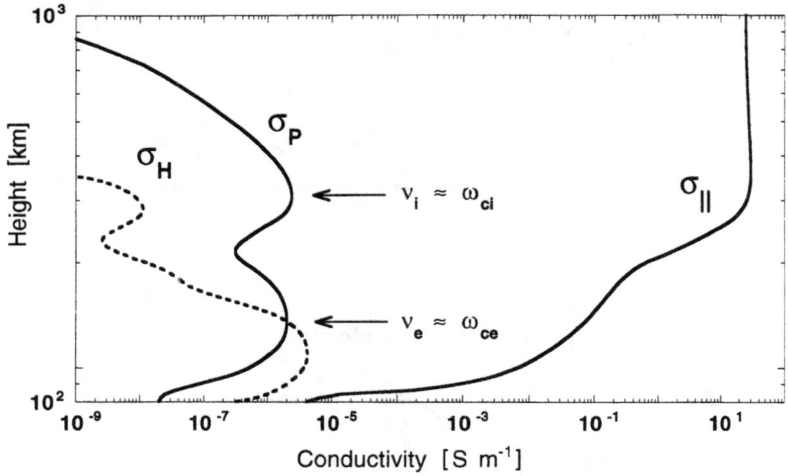

Figure 3.2. A sketch of the conductivities as a function of altitude in the ionosphere (after Boström, 1973). Note that these conductivity profiles vary considerably with local time, latitude, solar activity, and auroral activity. Note also that both axes of the graph are logarithmic.

1973; Kelley, 1989)

$$\sigma_{\parallel} = \sum_s \frac{n_s q_s^2}{m_s \nu_s}, \quad \sigma_P = \sum_s \frac{n_s q_s^2}{m_s} \frac{\nu_s}{\nu_s^2 + \omega_{cs}^2}, \quad \sigma_H = -\sum_s \frac{n_s q_s^2}{m_s} \frac{\omega_{cs}}{\nu_s^2 + \omega_{cs}^2} \qquad (3.2)$$

where n_s, q_s, m_s and ν_s indicate density, charge, mass, and particle-neutral collision frequency, respectively, of particles of species s, and $\omega_{cs} = q_s B / m_s$ is the gyrofrequency of species s. The electron gyrofrequency ω_{ce} is negative in this notation.

A sketch of the altitudinal dependences of the perpendicular conductivities in the ionosphere is shown in Figure 3.2. Above roughly 1000 kilometers, the ion-neutral collision frequencies ν_i and ν_e become nearly zero, and so the Pedersen and Hall conductivities vanish. For σ_H this is not directly obvious from Eq. (3.2); it follows from cancellation of electron and ion contributions.

We note, however, that it is not meaningful to extend these expressions far into the magnetosphere, where the effects of neutrals become negligible and the conductivity should be based on (however, rare) ion-electron collisions. As discussed in Section 3.2, Ohm's law in the magnetosphere becomes significantly modified. The appropriate frame to define conductivities in this region is the plasma rest frame. In this frame parallel and perpendicular current persist even in the absence of significant electric fields, so that both, field-aligned and perpendicular conductivities become very large.

As can be seen in Figure 3.2, the Pedersen conductivity in the ionosphere has two maxima, one at a higher altitude where $\nu_i \approx \omega_{ci}$ and one at a lower altitude where

$\nu_e \approx \omega_{ce}$. In the upper layer the Pedersen current is mainly carried across **B** by ions while the electrons are magnetized and in the lower layer the Pedersen current is mainly carried across **B** by electrons while the ions are collisionally immobilized. The Hall current has local maxima at slightly different altitudes.

The region in which ionospheric currents flow is limited to a layer of a few hundred kilometers in thickness, much smaller than the dimensions of the entire magnetosphere. Hence from a magnetospheric physics point of view, the ionosphere constitutes a thin conducting layer and it makes sense to neglect vertical currents and vertical electric fields inside this layer and refer to 'height-integrated conductivities' $\Sigma_{P,H} = \int dz\, \sigma_{P,H}$, where z is the coordinate in the vertical direction (Fejer, 1953; Swift, 1972). (Note that these are more properly called 'ionospheric conductances'.) These enter a height-integrated ionospheric Ohm's law

$$\mathbf{J} = \Sigma_P(\mathbf{E} + \mathbf{v}_n \times \mathbf{B})_h - \Sigma_H[(\mathbf{E} + \mathbf{v}_n \times \mathbf{B}) \times \hat{\mathbf{b}}]_h \tag{3.3}$$

where $\mathbf{J}$ is the height-integrated (horizontal) current density $\mathbf{J} = \int \mathbf{j}\, dz$. It has units Am^{-1}. The subscript 'h' refers to the horizontal component of a vector. These height-integrated expressions rely on the assumption that the electric and magnetic fields and the neutral wind velocity $\mathbf{v}_n$ do not vary significantly over the ionospheric height. Note that, because of $\sigma_\parallel \to \infty$ in the magnetosphere, the notion of an integrated parallel conductivity is meaningful only in the ionosphere. We have further used the fact that, in the auroral zone, the deviation of the magnetic field direction from the vertical axis is small, only $\sim 10°$, and neglected a small contribution from the horizontal component of the parallel electric field in Eq. (3.3).

The height-integrated model is useful to describe the closure of parallel currents by ionospheric currents. The current-continuity equation, in the absence of significant temporal variations of the charge density, $\nabla \cdot \mathbf{j} = 0$, can be broken up into its horizontal and vertical parts

$$\nabla_h \cdot \mathbf{j}_h = -\frac{\partial j_z}{\partial z} \tag{3.4}$$

Integrating Eq. (3.4) over the ionospheric height from bottom to top gives

$$\nabla_h \cdot \mathbf{J}_h = -(j_z^{top} - j_z^{bot}) \simeq -j_\parallel^{top} \cos\chi \tag{3.5}$$

where in the last step we assume that no current flows out the bottom of the ionosphere into the non-conducting atmosphere, and $j_z = j_\parallel \cos\chi$ is used, where χ is the angle between the magnetic field and the vertical axis. Note that χ is close to 180° in the northern auroral zone, so that $\cos\chi$ is negative and close to -1. Using expression (3.3) and taking $\cos\chi = -1$ and $\mathbf{v}_n = 0$, the field-aligned current entering (or exiting) the top of the ionosphere can be written as

$$j_\parallel = \nabla_h \cdot (\Sigma_P \mathbf{E}_h - \Sigma_H \mathbf{E}_h \times \hat{\mathbf{b}}) \tag{3.6}$$

(In the Earth's rest frame $\mathbf{E}_h$ should be replaced by $\mathbf{E}_h + \mathbf{v}_n \times \mathbf{B}$.) This equation can be rewritten as

$$j_\parallel = \Sigma_P \nabla_h \cdot \mathbf{E}_h + \mathbf{E}_h \cdot \nabla_h \Sigma_P + (\mathbf{E}_h \times \nabla_h \Sigma_H) \cdot \hat{\mathbf{b}} \tag{3.7}$$

This equation illustrates the interesting fact that if the ionospheric conductivities are uniform, the second and third term in this expression vanish, and therefore only the Pedersen current contributes to closing the field-aligned currents.

3.1.2 Conductivity from Electron Precipitation

The plasma density in the ionosphere evolves according to the conservation of particle number by the continuity equation

$$\left(\frac{\partial}{\partial t} + \mathbf{v} \cdot \nabla\right) n = S - R\left(n^2 - n_0^2\right) \tag{3.8}$$

where n_0 represents the equilibrium density due to solar radiation, R is the recombination coefficient, which is approximately 2×10^{-7} cm^3 s^{-1} (Brown, 1966), and S is a source term due to electron precipitation. One electron-ion pair is produced for every 35 eV of precipitating electron energy (Rees, 1963). This implies that the source term can be written as $S = \epsilon/(35H\,\text{eV})$, where $\epsilon = nWv_{e\parallel}$ is the precipitating energy flux of electrons of energy W and parallel velocity $v_{e\parallel}$, and H is the effective height of the ionosphere. The steady-state solution of Eq. (3.8) implies that the density is given by

$$n = \sqrt{n_0^2 + S/R} \tag{3.9}$$

For strong electron precipitation the n_0^2 term can be ignored and the equilibrium density becomes proportional to the square root of the energy flux. Because the collision frequencies in Equations (3.2) depend on the neutral density but not significantly on the plasma density (see, e.g., Appendix A of Kelley, 1989), the perpendicular conductivities are simply proportional to the plasma density, and variations in the plasma density will change the conductivities. Consequently, the height-integrated conductivities also become proportional to the square root of the energy flux. Observations suggest that $\Sigma_P \approx 5.2\,\text{S}\,\epsilon^{1/2}$ (Harel et al., 1981), where the energy flux ϵ is given in erg cm^{-2} s^{-1}, and $\Sigma_H = 0.55W^{0.55}\Sigma_P$, where the energy W of precipitating electrons is given in keV . The Hall conductivity increases with electron energy: higher-energy electrons deposit their energy at lower altitudes where the Hall conductivity is more important. A more elaborate useful dependence on electron energy W and precipitation flux ϵ extracted from measurements yields (Vickrey et al., 1981; Spiro et al., 1982; Reiff, 1984)

$$\Sigma_P = \frac{20W}{4+W^2}\sqrt{\epsilon}, \qquad \Sigma_H = W^{\frac{5}{8}}\Sigma_P \tag{3.10}$$

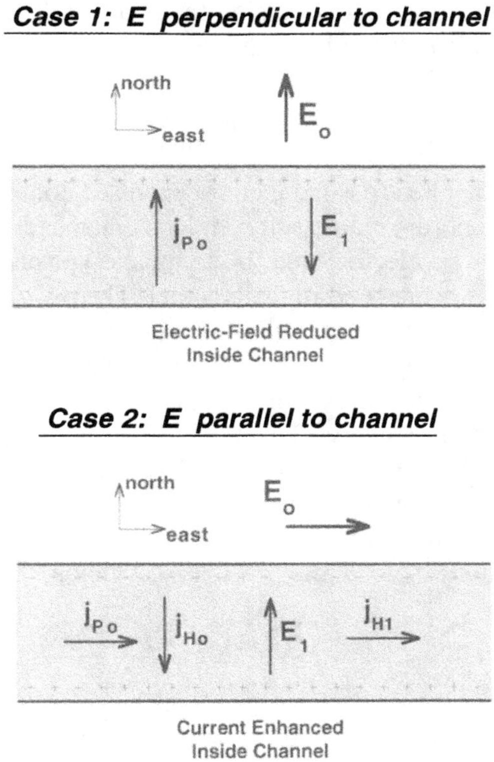

Figure 3.3. Two examples of channels of enhanced conductivity in the ionosphere, caused by east-west aligned regions of electron precipitation. In both cases the reader is looking down onto the northern high-latitude ionosphere from above, with the ambient magnetic field pointing into the page. In the top panel the ambient electric field in the ionosphere is perpendicular to the channel, and in the bottom panel the ambient electric field in the ionosphere is parallel to the channel. Figure provided by J. Borovsky.

High auroral conductivities have a strong effect on the ionospheric current system. When large-scale ionospheric electric fields $\mathbf{E}_0$ in the ionosphere interact with patches of high ionospheric conductivity, subtle polarization effects occur. Assume that a channel of high conductivity is caused by an east-west-aligned region of electron precipitation. Figure 3.3 shows two cases of such an interaction, for the electric field being normal to the conductivity channel and for the field being aligned with the channel. For a normal-directed electric field (top panel), the primary electric field $\mathbf{E}_0$ drives Pedersen currents $\mathbf{j}_{P,0}$ across the channel. Because the Pedersen current is stronger within the channel, layers of polarization charge build up at the channel edges. These charge layers produce a secondary polarization electric field $\mathbf{E}_1$ that partially opposes the applied field $\mathbf{E}_0$. The total electric field $\mathbf{E} = \mathbf{E}_0 + \mathbf{E}_1$ in the channel is thus smaller than the primary field $\mathbf{E}_0$. The Pedersen current in the high-conductivity channel driven by $\mathbf{E}$ is equal to the Pedersen current due to

$\mathbf{E}_0$ in the weakly conducting region outside the channel. The north-south electric field inside the channel is reduced in magnitude according to

$$E_{in} = E_{out} \frac{\Sigma_{P,out}}{\Sigma_{P,in}} \tag{3.11}$$

Eq. (3.11) can also be applied to a channel of decreased conductivity with $\Sigma_{P,in} < \Sigma_{P,out}$. In this case the north-south electric field is enhanced within the channel.

For an aligned east-west electric field (bottom panel) primary Pedersen and Hall currents $\mathbf{j}_{P,0}$ and $\mathbf{j}_{H,0}$ are generated in the channel. The primary Hall current flows normal to the channel and is stronger within the channel than outside. This enhancement of the Hall current causes a buildup of charge layers on the edges of the channel producing a secondary polarization electric field $\mathbf{E}_1$ that is directed across the channel. The secondary electric field drives a secondary Hall current $\mathbf{j}_{H,1}$ along the channel that adds to the primary Pedersen current $\mathbf{j}_{P,0}$. Current continuity in y direction inside and outside the channel yields $E_0\Sigma_{H,out} = E_0\Sigma_{H,in}$ - $E_1\Sigma_{P,in}$. Solving for E_1 and plugging into the expression for the total east-west current within the channel $j_{ch} = E_0\Sigma_{P,in} + E_1\Sigma_{H,in}$ yields

$$j_{ch} = \Sigma_{P,in} \left(1 + \frac{\Sigma^2_{H,in}}{\Sigma^2_{P,in}} - \frac{\Sigma_{H,in}\Sigma_{H,out}}{\Sigma^2_{P,in}} \right) E_0 \tag{3.12}$$

This expression shows that if the height-integrated Hall conductivity within the channel $\Sigma_{H,in}$ exceeds the height-integrated Pedersen conductivity $\Sigma_{P,in}$, then the total current within the channel, commonly referred to as the Cowling channel, will be greatly enhanced. The factor in front of the electric field E_0 is the 'Cowling conductivity', which is a Hall-amplified Pedersen conductivity in the current channel. In the auroral electrojet region it can be up to an order of magnitude larger than the ordinary Pedersen conductivity. Under such conditions the current is concentrated inside the channel.

3.2 Magnetospheric Current Sources

The ionospheric auroral currents flow at the bottom boundary of the auroral magnetosphere as part of a large-scale three-dimensional current system that intensifies during auroral disturbances. The sources of this current system are located in the magnetosphere, in the plasma sheet and its boundary layers. In this section we will identify some of these sources. Before turning to this task we here present basic elements of the description of the magnetospheric plasma.

3.2.1 Generalized Ohm's Law

The collisionless magnetospheric plasma consists of electrons and ions that are coupled via their self-consistently generated fields. Macroscopically these two species can be described as two separate charged fluids, but it is also appropriate to describe such a system by their common momentum balance, including the forces exerted by the currents on the plasma, and by a generalized Ohm's law that governs differences between ion and electron flows, that is electric currents, and their relation to the electric field and the dynamical properties of the plasma.

In this case the momentum balance equation of the plasma fluid becomes

$$\rho \frac{d\mathbf{v}}{dt} = \rho \left(\frac{\partial \mathbf{v}}{\partial t} + \mathbf{v} \cdot \nabla \mathbf{v} \right) = \mathbf{j} \times \mathbf{B} - \nabla p \tag{3.13}$$

which includes pressure gradient forces and the Lorentz force (e.g., Krall and Trivelpiece, 1973; Baumjohann and Treumann, 1996). It results from the sum of the ion and electron equations of motion, where $\rho = mn$ is the mass density, $m = m_i + m_e$ the mass, $n = (m_i n_i + m_e n_e)/m$ the density, $\mathbf{v} = (m_i n_i \mathbf{v}_i + m_e n_e \mathbf{v}_e)/mn$ the plasma velocity, $\mathbf{j} = en(\mathbf{v}_i - \mathbf{v}_e)$ the current density (for a proton-electron plasma), and $p = p_e + p_i$ the plasma pressure, assumed to be isotropic. Here we have already used the fact that the differences between ion and electron charge densities are typically very small ('quasi-neutrality assumption'), so that the net charge density and associated electric forces $e(n_i - n_e)\mathbf{E}$ are negligible.

The momentum equation (3.13) is complemented by a relation between the current and electric field which is known as Ohm's law. Due to the scarcity of collisions between electrons and ions in the tenuous, fully ionized magnetospheric plasma, Ohm's law deviates from the simple Ohm's law $\mathbf{j} = \sigma \mathbf{E}$ known from electrodynamics. It follows essentially from subtracting the ion and electron momentum transport equations:

$$\mathbf{E} + \mathbf{v} \times \mathbf{B} = \eta \mathbf{j} + \frac{1}{ne} (\mathbf{j} \times \mathbf{B} - \nabla \cdot \mathsf{P}_e) + \frac{m_e}{ne^2} \left[\frac{\partial \mathbf{j}}{\partial t} + \nabla \cdot (\mathbf{j}\mathbf{v} + \mathbf{v}\mathbf{j}) \right] \tag{3.14}$$

where P_e is the full anisotropic electron pressure tensor in the plasma rest frame. This equation is termed the generalized Ohm's law, because it consists of Ohm's law formulated for a moving plasma, with additional terms associated with large-scale electric fields that do not vanish in the plasma rest frame. For most of the magnetosphere most of the time, and specifically in the auroral acceleration region, the electrons satisfy the gyro-drift approximation[2]. In this approximation the pressure tensor is 'gyrotropic', that is, isotropic in the plane perpendicular to the magnetic field, as described by

$$\mathsf{P}_e = p_\perp \mathsf{I} + (p_\parallel - p_\perp) \hat{\mathbf{b}}\hat{\mathbf{b}} \tag{3.15}$$

where I is the unit tensor, and $p_\perp$ and $p_\parallel$ are the perpendicular and parallel components, respectively, of the pressure tensor with respect to the magnetic field direction. Deviations from gyrotropy, however, are important occasionally. They may

[2]In the gyro-drift approximation the particle motion is averaged over its gyro orbit and thus replaced by the motion of its gyro center.

lead to the appearance of off-diagonal terms in the pressure tensors in a field-aligned coordinate system. Such terms are crucial, for instance, for the electric field enabling collisionless magnetic reconnection (e.g. Hesse and Winske, 1998) as observed at the magnetopause and in the magnetospheric tail.

The above Ohm's law essentially represents the equation of motion of the electron fluid seen from the plasma rest frame. The approximations made in deriving this equation from the equations of motion of the electron and ion fluids are quasi-neutrality and the neglect of the ratio m_e/m_i of the electron mass versus ion mass.

We should note at this point that Ohm's law when taking only the first term on the right of Eq. (3.14) into account is similar to the Ohm's law that applies in the ionosphere. In the ionosphere the resistivity is due to the interaction between neutrals and charged particles. Naturally such a resistivity is of tensorial character. In the magnetosphere there are no neutrals, and the resistivity is solely due to the Coulomb interaction between electrons and ions. For typical plasma sheet conditions the mean free path for Coulomb collisions is huge, exceeding by far the dimensions of the magnetosphere. Therefore this term can justifiably be neglected. Nevertheless it is useful to retain it as a simple representation of collective effects from wave particle interactions that may be approximated as anomalous resistivity (Section 3.5). Since these mechanisms are not well understood yet, only scalar resistivity is included in Eq. (3.14), for simplicity.

In order to estimate the magnitude of the various terms in Eq. (3.14) it is useful to normalize all quantities, using, for instance, a characteristic (average) magnetic field strength B_c and a characteristic speed, which in a magnetized plasma is taken to be the Alfvén speed $v_A = B_c/\sqrt{\mu_0 \rho}$. These two quantities define a characteristic transverse-electric-field magnitude $E_c = v_A B_c$. One also choses a characteristic macroscopic reference length scale L_c, which is the scale-size for the currents. Together with the Alfvén speed, this scale defines the time reference scale $\tau_A = L_c/v_A$.

In these units one finds that the first term on the right-hand side of Eq. (3.14) is of the order $\tau_A/\tau_D = 1/S$, where τ_D is a diffusion time associated with the resistivity η. The quantity $S = \mu_0 L_c v_A/\eta$ is referred to as the Lundquist number. Thus this term can be neglected for large $S \gg 1$ which holds in the collisionless plasma of the auroral magnetosphere due to the lack of significant collisions between the particles.

The second or Hall-current term is the order of λ_i/L_c, where $\lambda_i = c/\omega_{pi}$ is the ion inertial length (or ion skin depth) and ω_{pi} is the ion plasma frequency. It can be ignored whenever one is interested in spatial scales $L_c \gg \lambda_i$ that are much longer than the ion inertial scale.

The third, pressure gradient, term scales similarly as the second one when the pressure is normalized to the magnetic pressure $p_c = B_c^2/2\mu_0$ based on the reference field B_c, except for the additional factor T_e/T_i. (In the magnetosphere T_e/T_i is of the order of 0.1–0.2.) As long as the electron pressure does not vary in the direction transverse to the magnetic field over distances comparable to the ion inertial length this term does not contribute to the dynamics. It becomes, however, important in generating parallel electric fields along the magnetic field whenever field-aligned pressure gradients exist.

Finally, the last term on the right-hand side of Eq. (3.14) scales as λ_e/L_c, where $\lambda_e = c/\omega_{pe} = \sqrt{m_e/m_i}\lambda_i$ is the electron inertial length or electron skin depth. It becomes important only on very short spatial scales the order of the electron inertial length, e.g., in dispersive Alfvén wave activity.

The evaluation of these dimensionless factors for typical magnetospheric parameters shows that the terms on the right-hand side of Eq. (3.14) are much smaller than those on the left-hand side, when the characteristic length and time scales are given by the typical macroscopic structures $\geq 1\,\mathrm{R_E}$ and the typical magnetohydrodynamic (MHD) wave periods of tens of seconds to minutes. This assumption forms the basis for the 'ideal MHD' approximation, which states that the electric field in the plasma rest frame is negligibly small. In this case one may write

$$\mathbf{E} + \mathbf{v} \times \mathbf{B} \approx 0 \tag{3.16}$$

The consequence of this approximation is the property that moving plasma elements that are connected by a magnetic field line at one time remain connected by the field line also at a later time. The plasma motion and the associated magnetic field changes can be visualized as magnetic field lines 'frozen' into the plasma. This 'frozen-in flux' assumption has the important consequence that it prohibits a change in magnetic topology and plasma transport across magnetic boundaries (separatrices), that is, magnetic reconnection. It also forbids any electric fields parallel to the magnetic field. The system of equations (3.13) and (3.16), together with Maxwell's equations and equations describing mass and entropy conservation, comprise ideal MHD.

The breakdown of ideal MHD and the presence of parallel electric fields requires that one or more of the terms on the right-hand side of Eq. (3.14) become relevant in the auroral magnetosphere. In the magnetospheric tail and magnetopause reconnection regions, and also in the auroral magnetosphere, the term in Eq. (3.14) most likely to cause deviation from the ideal MHD approximation (3.16) is the second term on the right-hand side, i.e. the Hall term, as Hall effects become important when the characteristic gradient scale length L_c approaches the ion inertial length $\lambda_i = c/\omega_{pi}$. With $n \sim 0.5\,\mathrm{cm^{-3}}$, the ion inertial length is $\lambda_i \sim 300\,\mathrm{km}$. If only the Hall term is included on the right hand side of (3.14), it can be combined with the $\mathbf{v} \times \mathbf{B}$ term, yielding

$$\mathbf{E} + \mathbf{v}_e \times \mathbf{B} = 0 \tag{3.17}$$

where $\mathbf{v}_e$ is the electron bulk flow velocity. In the approximation of Eq. (3.17), the magnetic field is frozen in the electron fluid, which still neglects parallel electric fields and prohibits a change in magnetic topology and magnetic reconnection.

The most relevant terms contributing to the parallel electric field in the generalized Ohm's law are the electron pressure term and the electron inertial term. Basic scaling estimates, as well as particle simulations, demonstrate that for collisionless magnetic reconnection the dominant term is the electron pressure gradient term, or more precisely, gradients of non-gyrotropic contributions to the electron pressure tensor. Both the electron pressure term and the electron inertia term are considered for auroral acceleration, as discussed in later sections. The MHD (single-fluid) description is inappropriate for the auroral acceleration region, since kinetic terms are necessary to describe parallel electric fields. Approximating the kinetic effects by an anomalous resistive term may be useful for some simple models but bypasses the physics of this region. The same holds for the reconnection sites.

3.2.2 Field-Aligned Current Generation

Electric currents play a key role in various forms of coupling mechanisms that affect plasmas in space and laboratories; they provide momentum coupling by means of Lorentz forces $\mathbf{j} \times \mathbf{B}$, energy coupling through Joule dissipation $\mathbf{j} \cdot \mathbf{E}$, and electric coupling through induction. Field-aligned currents are the major means of coupling the magnetosphere to the ionosphere, and setting the scene for the formation of aurorae. The large-scale currents of the coupled ionosphere-magnetosphere system are generated in the magnetosphere.

We must caution the reader that the question of how the currents are *generated* or *driven* does not necessarily address the cause and effect. The problem of current generation in plasmas is generally a self-consistent problem. That means, the charged particle motions in distorted electric and magnetic fields are associated with electric currents that generate magnetic fields or distort preexisting fields (such as the Earth's dipole field), which in turn govern the particle motions and currents. Current generation may also involve spatial as well as temporal aspects.

From a macroscopic point of view, particularly when ideal MHD is a valid description of the plasma, it is often useful to visualize the distortions of the magnetic field by plasma motions using the frozen-in field description (see Section 3.2.1). The currents can then be calculated from the curl of the magnetic field. This view is illustrated in the following subsection. On the other hand, the self-consistency implies that the particle motions in these fields add up to the current inferred from $\nabla \times \mathbf{B}$. When the particle motions are adiabatic, that is, satisfy the gyro-drift approximation, this addition not only requires all appropriate particle drifts, e.g., curvature drifts, gradient drifts, or polarization drifts, but also inclusion of the magnetization current (Northrop, 1963).

3.2.2.1 Currents and Plasma Motions

Currents can be generated, intensified or reduced, by plasma motions in the magnetosphere. In a fixed frame of reference the electric field is dominated by the $-\mathbf{v} \times \mathbf{B}$ term in Ohm's law. This plasma motion transports and distorts the frozen-in magnetic field and builds up currents (e.g. Song, 1998) as described by combining Ampère's and Faraday's laws

$$\nabla \times \mathbf{B} = \mu_0 \mathbf{j} \tag{3.18}$$

$$\frac{\partial \mathbf{B}}{\partial t} = -\nabla \times \mathbf{E} \tag{3.19}$$

leading to

$$\frac{\partial \mathbf{j}}{\partial t} = -\frac{1}{\mu_o} \nabla \times [\mathbf{B}\, \nabla \cdot \mathbf{v} + \mathbf{v} \cdot \nabla \mathbf{B} - \mathbf{B} \cdot \nabla \mathbf{v}] \tag{3.20}$$

We note that in Ampère's law (3.18) the displacement current is neglected. This assumption basically excludes high-frequency plasma and electromagnetic waves and is consistent with the assumption of quasi-neutrality.

The easiest way to demonstrate the build-up of field-aligned currents from plasma motion is to look at small perturbations of an initially uniform magnetic field. In that case the first order terms of Eq. (3.20) yield

$$\frac{\partial \mathbf{j}_\parallel}{\partial t} = \frac{1}{\mu_o} \mathbf{B} \cdot \nabla \Omega_\parallel \tag{3.21}$$

where $\Omega_\parallel$ is the field-aligned component of the flow vorticity $\Omega = \nabla \times \mathbf{v}$. Eq. (3.21) demonstrates that a clockwise twisting of a magnetic flux tube (as viewed in the direction of the magnetic field) that decreases in the same direction leads to the build-up of a field-aligned current opposite to the field direction.

Figure 3.4 is an illustration of field-aligned current generation by motion in the magnetosphere. In the top panel it shows a channel of earthward flow in the magnetosphere that leads to dipolarization of the magnetic field near midnight. The finite extent of the flow channel in z makes the central field lines more dipolar. Since the magnetic field farther away from midnight remains stretched, the flow shear in y thus leads to a magnetic shear $\partial B_z/\partial y$, corresponding to a current in the x direction. The bottom panel shows how both a localized azimuthal and earthward flow can generate magnetic shear $\partial(B_\phi/B_x)/\partial z$ or increase a pre-existing magnetic shear. In each of these cases it is important that the flow channel is limited in z, so that the field lines farther out are not (or less) affected.

Each of the mechanisms illustrated in Figure 3.4 corresponds to field-aligned currents that flow toward the Earth on the dawn side and away from the Earth on the dusk side, characteristic of Region 1 currents (see Section 1.4). We should further note that the magnetic shear associated with earthward plasma flow in the center of the tail, illustrated in the top panel of Figure 3.4, could also be related to tailward flow that stretches field lines near the flanks of the tail. This illustrates how anti-sunward flow in the low-latitude boundary layer can contribute as a source of Region 1 type field-aligned currents (see also Section 7.2).

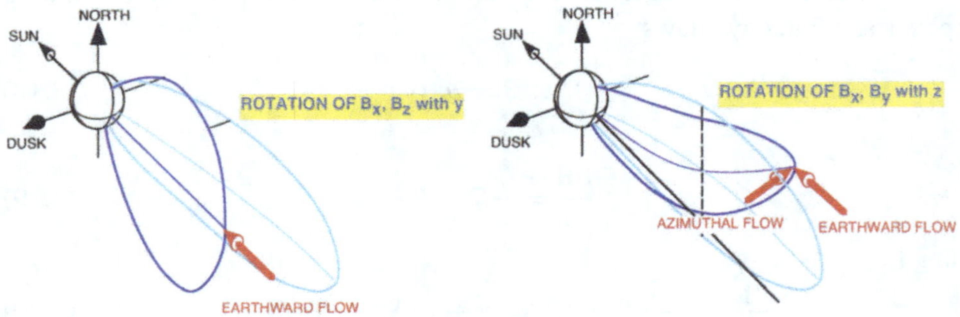

Figure 3.4. Typical flows in the magnetotail that build up or enhance field-aligned currents. In the left panel earthward flow leads to a more dipolar shape of the field lines near midnight, while field lines away from midnight remain more stretched. As a consequence the direction of the vector comprised of the components B_x and B_z rotates along the y direction at the indicated location. This rotation is a manifestation of field-aligned currents. In the right panel a rotation of the B_x, B_y field vector is accomplished by bending a field line in comparison to another one at slightly higher latitude. This bending may result from azimuthal flow or from an intensification of a preexisting bending caused by earthward flow. Figure provided by J. Birn.

3.2.2.2 Currents and Force Balance

In steady, or nearly steady situations, when $\partial \mathbf{B}/\partial t \approx 0$, Faraday's law (3.19) does no longer provide information about the currents. In those cases (but also in more general cases) it is often useful to investigate the current balance and the conversion (or diversion) of perpendicular into parallel currents. This 'generation' of field-aligned currents is investigated starting from the momentum balance Eq. (3.13). By taking the cross product of this equation with the magnetic field **B**, one obtains an expression for the perpendicular current

$$\mathbf{j}_\perp = -\frac{1}{B^2}\left(\rho\frac{\mathrm{d}\mathbf{v}}{\mathrm{d}t} \times \mathbf{B} + \nabla p \times \mathbf{B}\right) \tag{3.22}$$

Current conservation $\nabla \cdot \mathbf{j} = 0$ is then used to derive field-aligned currents as a function of the divergence of the perpendicular current, $\nabla \cdot \mathbf{j}_\parallel = -\nabla \cdot \mathbf{j}_\perp$, or

$$\nabla \cdot \mathbf{j}_\parallel = \mathbf{B} \cdot \nabla \frac{j_\parallel}{B} = -\nabla \cdot \left[\frac{\mathbf{B}}{B^2} \times \left(\rho\frac{\mathrm{d}\mathbf{v}}{\mathrm{d}t} + \nabla p\right)\right] \tag{3.23}$$

The current associated with the inertial term (first term on the right-hand side of Eq. (3.22)) is often termed the 'polarization current', because it is formally equivalent to the polarization drift in a time varying electric field.

When the inertial term is small, integration of Eq. (3.23) along the magnetic field from the ionosphere out to the equatorial plane in the magnetosphere leads to an expicit expression for the parallel current density (e.g. Vasyliunas, 1970; Schindler and Birn, 1978)

$$\left.\frac{j_{\parallel}}{B}\right|_{eq}^{ion} = -\frac{\mathbf{B}_{eq}}{B_{eq}^2} \cdot \nabla p_{eq} \times \nabla V = -\mathbf{j}_{\perp} \cdot \nabla V \tag{3.24}$$

This equation is often referred to as the 'Vasyliunas equation'. Here V is the differential flux tube volume (i.e., the volume of a magnetic flux tube of unit magnetic flux). This volume is given by

$$V = \int_{eq}^{ion} \frac{\mathrm{d}s}{B} \tag{3.25}$$

where the integral is extended along a magnetic field line from the equatorial plane to the ionosphere. If, for simplicity, we assume that $j_{\parallel}$ vanishes in the equatorial plane, Eq. (3.24) yields the parallel current density at the ionosphere (or some other chosen reference location). Of course, as mentioned above, this approach does not imply any mechanism of how the currents are generated. It simply addresses diversion from the perpendicular to parallel currents and vice versa. We should further emphasize that, although Eq. (3.24) contains fields evaluated at the equatorial plane, this does not imply that the current diversion takes place at this plane or in its vicinity.

The field-aligned current generation described by Vasyliunas' equation (3.24) requires that contours of constant pressure p and constant flux tube volume V in the equatorial plane are not aligned with each other. Fig. 3.5 illustrates deformations of these contours that correspond to the generation of Region 1 type field-aligned currents. Such deformation can be associated with a reduction of the plasma pressure or of the magnetic flux tube volume around midnight (or both). The former is more appropriate in low-β regions, where the magnetic field deformations are small. In contrast, in high-β regions, both changes of p and V are comparable, or the change of the flux tube volume from changes in the magnetic field may even be the dominant one; this was found, for instance, in simulations of reconnection in the magnetotail (see Section 7.7.3).

Alternatively, when $\mathbf{B}$, ρ, and p are nearly constant, the inertia term dominates in Eq. (3.23). This assumption is the basis for the viscous boundary layer model developed by Sonnerup (1980) and Lotko et al. (1987), treated in more detail in Section 7.2. In this case, the current divergence equation (3.23) reduces to

$$B\frac{\partial}{\partial s}\left(\frac{j_{\parallel}}{B}\right) = \frac{\rho \mathbf{B}}{B^2} \cdot \nabla \times \left(\frac{d\mathbf{v}}{dt}\right) = \frac{\rho}{B}\frac{d\Omega_{\parallel}}{dt} \tag{3.26}$$

where $\Omega_{\parallel}$ is the parallel component of the vorticity and s is again the coordinate along the magnetic field direction, given by the unit vector $\hat{\mathbf{b}}$. Eq. (3.26) shows that a change in the vorticity along a streamline will give rise to a field-aligned current. Note that in the main two-cell convection pattern, the vorticity is parallel to the

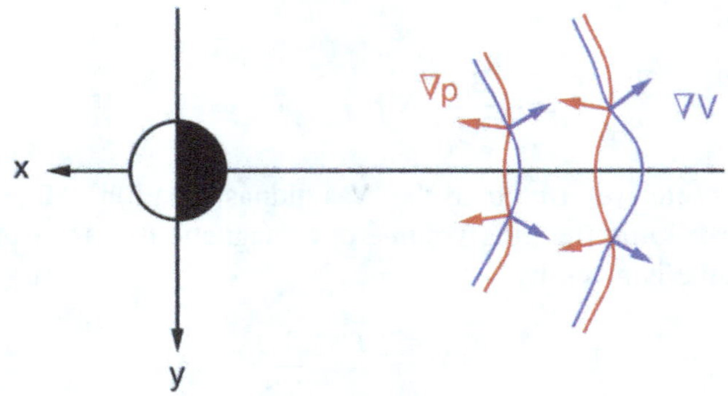

Figure 3.5. Section of contours of constant plasma pressure p (red) and flux tube volume V (blue) in the equatorial plane, giving rise to field-aligned currents of Region 1 type. The arrows indicate the directions of the gradients of p and V. Figure provided by J. Birn.

magnetic field on the dusk side, and anti-parallel at dawn. Thus, a slowing down of this convection is consistent with field-aligned currents in the northern hemisphere that are anti-parallel to the magnetic field at dusk and parallel at dawn. Going from the equator to the southern hemisphere, the integration with respect to s is in the negative sense, and so the current directions are parallel at dusk and anti-parallel at dawn. These results can be summarized by noting that the currents flow out of the ionosphere at dusk and into the ionosphere at dawn in both the northern and southern hemispheres. These are thus consistent with the Region 1 field-aligned current systems.

The result that a *slow down* of the main two-cell convection pattern is associated with Region 1 type field-aligned currents appears to be in contrast to the effects of flow discussed in the previous subsection, where it was concluded that the *buildup* of the field-aligned currents of Region 1 sense results from a flow pattern that is qualitatively consistent with parts of the two-cell convection pattern, i.e., earthward in the tail region around midnight then diverted dawnward and duskward around the Earth toward the Sun. How can we solve this apparent contradiction? The answer lies in the fact that these views are derived from different equations, which are satisfied self-consistently, and consider different aspects of the interaction between flows and currents. Faraday's law relates temporal changes of the currents to spatial gradients of the flow, whereas the current diversion aspect considers the instantaneous force balance, which involves the effect of the Lorentz force $\mathbf{j} \times \mathbf{B}$ on acceleration or deceleration of the flow.

The distortion of the magnetic field from the standard two-cell circulation in the magnetosphere initially leads to the build-up of a current system that not only includes field-aligned currents of Region 1 type, but also perpendicular currents. These currents provide Lorentz forces that would decelerate the original flow (or even reverse the flow pattern, if it is not maintained by external forces), unless

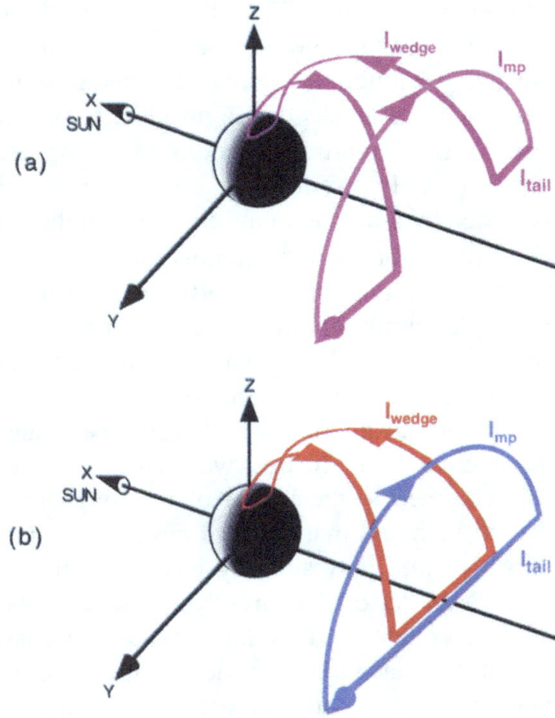

Figure 3.6. The substorm current wedge circuit represented in two different ways. The cross-tail current (I_{tail}) and the tail-ionosphere current system (I_{wedge}) are closed by a magnetopause current (I_{mp}) in the simplest fashion. Figure provided by J. Birn.

they are compensated by pressure gradients. We emphasize here that it is not the slow-down of convection that causes the currents to flow but rather the $\mathbf{j} \times \mathbf{B}$ forces causing the slow down of the flow. This picture can be applied to the substorm current wedge discussed below, as well as to the more persistent Region 1 type currents associated with tailward flow in the low-latitude boundary layer, discussed in the previous section.

3.2.2.3 Current Circuits

Current systems in the magnetosphere and the currents in the coupled ionosphere-magnetosphere system may be represented by current circuits. Circuit models are useful for visualization, but they are questionable for predictions, because in a distributed plasma currents are not confined to circuits. A change of the current in one place may lead to a local perturbation involving currents with local closure, rather than a change of the current in the other parts of the original circuit.

Consider the circuits in Fig. 3.6. This figure is a schematic illustration of the inferred currents associated with substorm onset and expansion, called the 'substorm current wedge' (e.g. McPherron et al., 1973). In its simplest form, the substorm current wedge consists of a cross-tail current that is diverted to the ionosphere on the dawnside, flows westward through the ionosphere, and returns to the tail and its original duskward direction on the duskside (Figure 3.6, top). It can also be represented as a superposition of two circuits, with a closed ionosphere-magnetosphere circuit that opposes the cross-tail current in the tail (Figure 3.6, bottom). In each case the tail current is closed by a magnetopause current in the simplest fashion.

As an instantaneous representation both models are equivalent, however, each one is suggestive of a different relationship between ionospheric and magnetospheric currents. In the single-circuit picture of the top panel of Fig. 3.6, an increase of the ionospheric current would be related to an increase of the duskward tail currents connected to the outside. In contrast, in the two-circuit picture in the bottom panel, an increase of the ionospheric current would be related to an increased dusk-to-dawn current, which opposes the duskward current in the tail. In this case, the current changes are confined to the inside of the magnetosphere-ionosphere system, although it should be noted that the two circuits are not independent but coupled inductively. Each picture may actually have its merit. The single-circuit picture can be viewed as a model of the directly-driven component of the ionospheric currents, strongly correlated with solar-wind-driven changes. In contrast, the two-circuit picture can be viewed as a model of the unloading component of the ionospheric currents, associated with magnetotail dynamic processes that are not directly driven by the solar wind.

Current-source regions are often identified by tracing magnetic-field lines. This method, however, may fail. In the vicinity of the Earth, where the magnetic field energy by far exceeds the plasma thermal energy (low plasma β), the currents are strongly field-aligned, so that current lines and magnetic field lines are largely identical. However, the field lines from the auroral region typically connect to source regions in the outer magnetosphere and magnetotail where the plasma β is high. Here the currents perpendicular to the magnetic field are appreciable or even dominant. Hence, currents no longer simply flow along magnetic field lines and magnetic-field-line tracing differs from current tracing (see Section 5.2 for the description of mapping).

3.2.2.4 Short Scale Effects

So far, we have concentrated on the large-scale mechanisms for the generation or diversion of currents, considering plasma flow and plasma pressure, dominated by the ions. These currents are related to the basic auroral current systems, such as Region 1 or 2 or to the substorm current wedge (McPherron et al., 1973) but not to individual arcs. There might, however, also be a relation between tail phenomena and auroral arcs on smaller scales, corresponding to, say, 10 km latitudinal width in the ionosphere, which maps roughly to a few hundred km in the north-south direction in the near tail. This scale is comparable to an ion inertial length or an ion

Larmor radius. At such small scales, the MHD description is no longer adequate to describe all the relevant physical processes, and the full plasma kinetic theory should be used. In particular, plasma waves and instabilities and wave-particle interactions may play a dominant role in the plasma dynamics at these scales.

The mechanisms discussed in the previous sections, however, can also be applied to smaller scales, when the ion terms are suitably replaced by electron terms. For instance, we already discussed in Section 3.2.1, that the electron flow replaces the plasma flow in Ohm's law (3.16), when only Hall effects are taken into account. Similarly, the pressure gradient term in Eq. (3.13) can be generalized to include anisotropy and ion gyro-viscosity; it already includes the electron pressure contribution.

Several persistent substructures in the tail exist that have a suitable location for an association with arcs. These are the *inner edge of the electron plasma sheet*, the *thin growth phase current sheet* in the near tail, the *near-earth reconnection site*, and the *plasma sheet boundary layer*. While the association of these more persistent structures with auroral arcs has not been well established, some sporadic magnetospheric features, specifically, short-duration flow bursts in the tail plasma sheet, indeed show a close relationship with small-scale auroral features, such as north-south aligned arcs (Henderson et al., 1998), poleward boundary intensifications (Lyons et al., 1999b), and auroral streamers (Nakamura et al., 2001a, see also Section 6.4). These structures will be discussed in the later chapters of this book.

3.2.3 Energy Release, Transport and Conversion

In the auroral zone, particle acceleration in parallel electric fields above the ionosphere and current dissipation by Joule heating $\mathbf{j} \cdot \mathbf{E}$ in the ionosphere require an external energy supply to maintain the current system, although not necessarily in a steady fashion. In the coupled ionosphere-magnetosphere system, the energy disspation in the ionosphere and in the acceleration region requires a generator in the magnetosphere.

It is instructive to consider the individual equations that govern separate forms of energy. Taking the scalar product of Eq. (3.19) with $\mathbf{B}$ and using Eq. (3.18), one finds that the magnetic energy transport is governed by

$$\frac{\partial}{\partial t}\frac{B^2}{2\mu_0} = -\nabla \cdot \left(\frac{\mathbf{E} \times \mathbf{B}}{\mu_0}\right) - \mathbf{j} \cdot \mathbf{E} \tag{3.27}$$

This equation is commonly called 'Poynting's theorem'. Here electromagnetic energy flow is contained in the Poynting flux vector $\mathbf{E} \times \mathbf{B}/\mu_0$. On the left, the electric field energy density $\frac{\epsilon_0}{2} E^2$ does not appear because of our neglect of the displacement current in Ampere's law, Eq. (3.18). In many cases, particularly for high-frequency waves, the displacement current term cannot be neglected. In such cases, the electric field energy $\frac{1}{2}\epsilon_0 E^2$ should also be included on the left.

Equation (3.27) can be further rewritten using Ohm's law (3.14)

$$\begin{aligned} \frac{\partial}{\partial t}\frac{B^2}{2\mu_0} &= -\nabla \cdot \left(\frac{\mathbf{E}\times\mathbf{B}}{\mu_0}\right) + \mathbf{j}\cdot(\mathbf{v}\times\mathbf{B}) - \eta\, j^2 \\ &= -\nabla \cdot \left(\frac{\mathbf{E}\times\mathbf{B}}{\mu_0}\right) - \mathbf{v}\cdot(\mathbf{j}\times\mathbf{B}) - \eta\, j^2 \end{aligned} \tag{3.28}$$

For simplicity, only the Ohmic term from Eq. (3.14) is included. More generally, this term would be replaced by Joule dissipation $\mathbf{j}\cdot\mathbf{E}'$, where $\mathbf{E}'$ is the electric field in the plasma rest frame given by the right-hand side of Eq. (3.14).

When heat flux is neglected, the thermal energy transport is governed by

$$\frac{\partial u}{\partial t} = -\nabla \cdot [(u+p)\mathbf{v}] + \mathbf{v}\cdot\nabla p + \eta\, j^2 \tag{3.29}$$

Here, $u = \frac{3}{2}\,p$ is the thermal energy density of the plasma, and isotropic pressure is assumed. The first two terms on the right describe adiabatic, i.e., entropy conserving, compression or expansion. Consistent with (3.28), again only the Ohmic dissipation is included. This term provides a direct transfer from magnetic to thermal energy.

The transport equation for the kinetic energy of the plasma bulk flow follows from taking the dot product of the momentum equation (3.13) with $\mathbf{v}$ and using the continuity equation (mass conservation) given by

$$\frac{\partial \rho}{\partial t} = -\nabla \cdot \rho\mathbf{v} \tag{3.30}$$

This yields

$$\frac{\partial}{\partial t}\frac{\rho}{2}v^2 = -\nabla \cdot \left(\frac{\rho}{2}v^2\mathbf{v}\right) + \mathbf{v}\cdot(\mathbf{j}\times\mathbf{B} - \nabla p) \tag{3.31}$$

Equations (3.28), (3.29), and (3.31) can be added together to yield the conservation law of total energy

$$\frac{\partial}{\partial t}\left(\frac{B^2}{2\mu_0} + u + \frac{\rho}{2}v^2\right) = -\nabla \cdot \left[\frac{\mathbf{E}\times\mathbf{B}}{\mu_0} + (u+p)\mathbf{v} + \frac{\rho}{2}v^2\mathbf{v}\right] \tag{3.32}$$

The first terms on the right-hand sides of Eqs. (3.28), (3.29) and (3.31) represent the conservation of each of the different forms of energy, while the remaining terms represent the conversion of one form of energy into another. In Eq. (3.28), $\mathbf{j}\cdot\mathbf{E}$ governs the conversion of magnetic energy. Positive $\mathbf{j}\cdot\mathbf{E}$ corresponds to a load. Negative $\mathbf{j}\cdot\mathbf{E}$ corresponds to a generator. Since Ohmic dissipation is positive, a generator requires plasma motion against the Lorentz force. As demonstrated by Eq. (3.31), the associated energy can be supplied either by kinetic bulk flow energy or by mechanical work $-\mathbf{v}\cdot\nabla p$ that derives from thermal energy or compression via Eq. (3.29). It should be noted that the Joule dissipation or generator term is not altered if the Hall-electric field is included from Eq. (3.14).

One part of the $\mathbf{j} \cdot \mathbf{E}$ term, Ohmic (or Joule) heating, goes directly into thermal energy in (3.29), while the remaining part represents the acceleration (or deceleration) by Lorentz forces, which affects kinetic bulk flow energy. The combination of (3.29) and (3.31) shows that, in approximate force balance, this kinetic energy also is transferred into thermal energy via mechanical work. Equations (3.28), (3.29) and (3.31) also illustrate that the conversion of internal (i.e., thermal) energy to magnetic energy requires as an intermediate step the conversion to kinetic bulk flow energy.

Consider again the current circuit of Fig. 3.6 (top). This figure suggests how the ionospheric load is connected with a generator in the anti-sunward flow driven by the solar wind at the high-latitude magnetopause or in the low-latitude boundary layer (outer parts of the cross-tail current regions in Fig. 3.6). At the high-latitude magnetopause the magnetic field for an open magnetosphere is southward, so that the electric field is duskward, opposite to the magnetopause current. Similarly, the northward magnetic field in the closed field line regions of the low-latitude boundary layer corresponds to a dawnward electric field, which again is opposite to the cross-tail current. In either case the ultimate energy source driving this current system is the kinetic energy of the solar wind. We note that in the energy discussion it is not appropriate to split currents into several circuits as done in the bottom part of Fig. 3.6, because of the nonlinearity of the current terms in the energy equations.

3.2.4 Magnetospheric Driving of the Ionosphere

In the previous sections we have presented basic elements of the physics that govern magnetospheric currents relevant for the coupling to the ionosphere. In this section we use these elements to provide a simplified picture of how the magnetosphere couples with the ionosphere and how the two regions become decoupled. First of all, using the results of Sections 3.1.1 and 3.2.3, we emphasize that, in the Earth's rest frame and neglecting neutral winds, the ionosphere is normally a load and not a generator in the coupled system, and $\mathbf{j} \cdot \mathbf{E}$ is positive.

The simplest model illustrating the basic physics of how the magnetosphere drives flow and currents in the ionosphere is a magnetic flux tube connecting the two regions, as illustrated in the top part of Figure 3.7. In an idealized fashion, the coupling mechanism may also be discussed using a straight flux tube as in the bottom part of Figure 3.7.

In a first step let us consider the ionosphere as massive and ideally conducting, so that magnetic field lines are tied to the ionosphere. The magnetospheric plasma also is assumed to be an ideal conductor so that the picture of frozen-in magnetic flux applies, governed by Eq. (3.16). (We note, however, that the discussion below also applies for smaller scales, when Hall effects become important but the magnetic field lines can still be considered as frozen in the electron fluid, governed by

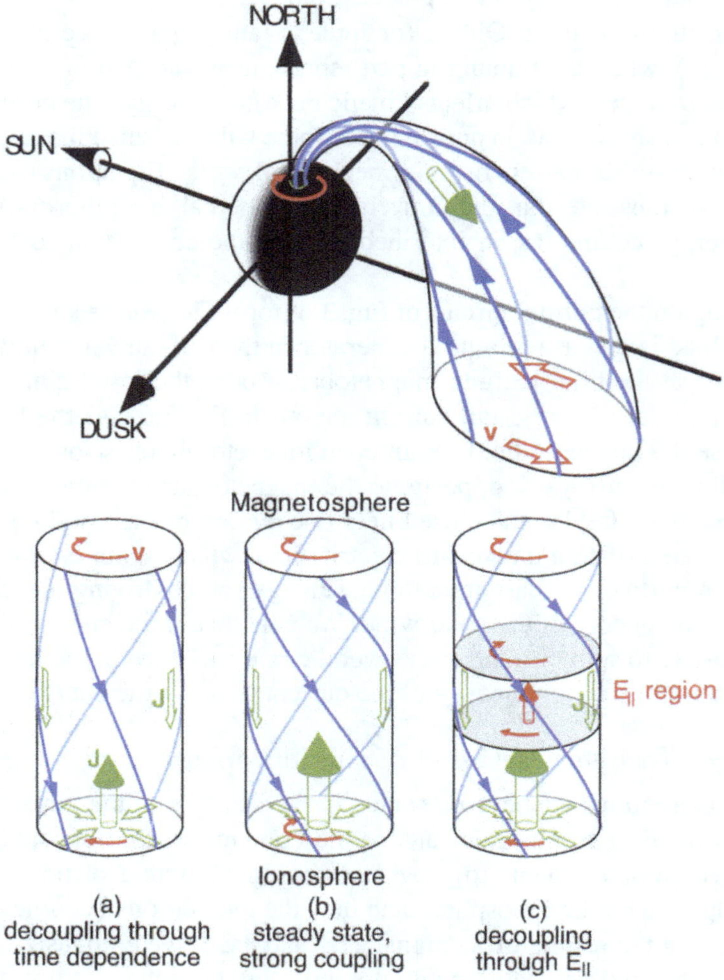

Figure 3.7. Magnetic flux tube connecting the magnetosphere with the ionosphere. Twist or shear motion (red arrows) in the magnetosphere distorts the magnetic field lines (blue) and builds up field-aligned current (green arrow), coupling the two regions. These currents close through ionospheric surface currents and currents on the outside of the flux tube, as illustrated by the open green arrows (bottom). The bottom part further illustrates three phases of the coupling; (a) a build-up phase in which the magnetospheric vorticity exceeds the corresponding motion on top of the ionosphere; in this case the electric fields are partially or fully decoupled, due to time dependent (inductive) effects; (b) a steady state, in which the top and bottom parts of the flux tube rotate at the same angular speed and the electric fields and flows in the two regions are strongly coupled; and (c) a state in which this coupling is reduced through parallel electric fields (open red arrow) between the two regions, so that the vortical motion and the electric field are reduced at the bottom. Figure provided by J. Birn.

Eq. (3.17)). A vortical or shear motion in the magnetosphere then leads to a twist or magnetic shear of the flux tube, representing a field-aligned electric current (upward green arrow in Figure 3.7a). A clockwise rotation (as viewed in the direction of the magnetic field) leads to upward field-aligned current inside the flux tube. If we consider the shear or twist as localized, the field-aligned current inside the flux tube is compensated by an oppositely directed surface current. These currents are connected through ionospheric currents, which we may also consider as surface currents (open green arrows in Figure 3.7a). Since there is no dissipation in this system, the currents can keep flowing after some finite initial twisting, provided the forces in the magnetosphere maintain the torque on the flux tube.

In the next step let us consider finite resistivity in the ionosphere. In that case field lines can slip along the ionospheric surface and the magnetic flux tube will become untwisted on a time scale determined by this resistivity, again assuming that there is no relaxation in the magnetosphere. Under these conditions, a continued driving in the magnetosphere will lead to a steady state in which the slippage at the ionosphere exactly corresponds to the driving vortical motion in the magnetosphere. For a fixed driving motion the magnitude of the field-aligned currents in this coupled system depends on the resistivity: the higher the ionospheric conductivity, the stronger the twist and the higher the field-aligned currents get.

So far we have considered the ionospheric resistivity as fixed. However, the conductivity depends, among other influences, on electron precipitation. Outward flowing field-aligned currents correspond to precipitating electrons, which increase the ionospheric conductivity and therefore strengthen the field line tying at the footpoints of these current regions. This leads to an intensification of the twist and the field-aligned currents. When the ionospheric resistance decreases, an effective field-aligned resistance becomes relatively more important, such that parallel electric fields and deviations from ideal MHD (or ideal electron MHD) increase. This effect is described in Section 3.3 and illustrated by Fig. 3.10. As a consequence of this breakdown of ideal MHD in the parallel electric field region the motion of the ionosphere at the magnetic footpoints becomes fully or partially decoupled from the magnetospheric motion. This may be visualized as a slippage of magnetic field lines in the parallel electric field region; it is illustrated in Figure 3.7c.

The simplest completion of the current circuits in Fig. 3.7 requires currents at the magnetospheric end of the flux tubes that flow radially outward, opposite to the $-\mathbf{v} \times \mathbf{B}$ electric field associated with the vortical flow. The magnetospheric end of the flux tube hence represents a generator, while the ionospheric end and the region of parallel electric field represent a load. In the view described above, we have considered relatively slow variations, neglecting propagation and reflection of perturbations at the Alfvén velocity, or rather considering a time when reflections, interference, and wave damping have led to a new quasi-equilibrium. For a time-dependent case, this propagation, reflection and interference must be taken into account (see Section 3.4).

The top part of Fig. 3.7 can be applied to large-scale and small-scale coupling. On the largest scale it illustrates the association between the dusk part of a two-cell convection pattern and the upward field-aligned current of Region 1. We note that magnetospheric convection is not confined to the nightside, but extends into the dayside as well. As discussed previously, a suitable steady driver for this pattern could be the anti-sunward flow in the low-latitude boundary layer (see Section 7.2). Similarly, tailward flow in the open field line region of the plasma mantle would drive the same sense of field-aligned current in or near the plasma sheet boundary layer. The general association of these large-scale convection patterns with the upward and downward field-aligned currents into the ionosphere has been well established. However, the direct correspondence between these magnetospheric regions and the aurora remains unclear. This is in part due to the absence of reliable magnetic field models that connect these two regions, and in part to the decoupling effects from time dependence and parallel electric fields illustrated in Figure 3.7. For instance, it seems possible that a more localized part of the magnetospheric two-cell convection pattern with steep gradients could be a suitable driver for the quiet evening arc, which will be discussed in Section 6.1, but this has not been established.

The top part of Figure 3.7 can also be used to illustrate the build-up of field-aligned currents by earthward flows ('bursty bulk flows') inside the magnetotail. This association is indeed established; it will be discussed in more detail in Sections 6.4 and 8.3.

As mentioned earlier, the association between flows and field-aligned currents is not restricted to large scales. On a scale comparable to the ion inertia scale, field lines can still be considered frozen into the electron fluid. Electron shear motions hence can cause magnetic shear, even when the ions do not participate in this motion. In this limit, the magnetospheric counterpart of the shear flow or vorticity associated with an auroral arc (e.g., Hallinan, 1981) (see also Section 4.1.3) would be a thin current layer (or current layers) carried by electron $\mathbf{E} \times \mathbf{B}$ drift, rather than a plasma bulk flow (see also Chapter 6 and Section 7.7.5).

3.3 Parallel Electric Fields

Some of the most important problems of auroral plasma physics concern the acceleration of electrons to tens of keV kinetic energies, energies that are much higher than their initial thermal energies. Observations have provided compelling evidence that this acceleration is achieved by field-aligned electric fields. A theoretical challenge is to describe how parallel electric fields can arise in a collisionless plasma in a self-consistent fashion.

3.3.1 Current-Voltage Relation in the Upward $j_\parallel$ Regions

The aurora is caused primarily by the precipitation of magnetospheric electrons into the atmosphere. These precipitating electrons excite the airglow and carry the

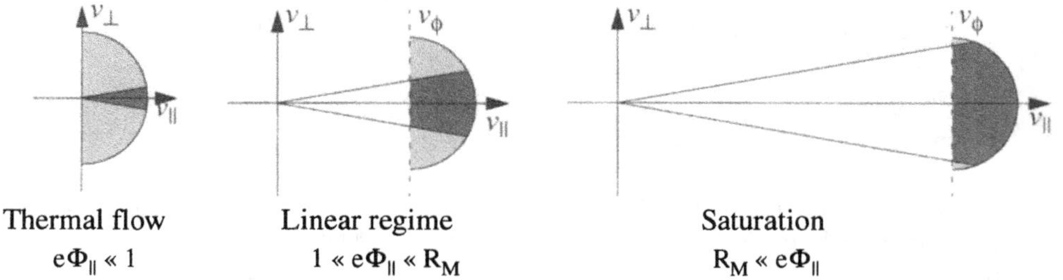

Figure 3.8. The three regimes of the Knight relation. The dark shaded area contains particles with initial velocities $v_\parallel^2 > (R_M - 1)v_\perp^2$ that can travel from the plasma sheet to the ionosphere. The leftmost diagram shows the thermal flow is restricted to a small region of phase space which has access to the ionosphere. Increasing $v_\parallel$ will allow more of the distribution to flow to the ionosphere until the flow saturates as depicted by the rightmost diagram.

field-aligned currents. Magnetospheric electrons tend to be trapped in the magnetic field of the magnetosphere; to reach the atmosphere an electron must have a pitch angle, defined as $\theta = \arctan(v_\perp/v_\parallel)$, that is within the 'loss cone' $\theta < \theta_{lc}$ (see Figure 3.8). As seen from the equatorial magnetosphere where the magnetic field strength is B_{eq} the angular size of the loss cone is

$$\theta_{lc} = \arcsin\left(\left[\frac{B_{eq}}{B_{ion}}\right]^{1/2}\right) \tag{3.33}$$

where B_{ion} is the magnetic field strength in the upper atmosphere (ionosphere). In the auroral zone $B_{ion} \sim 50{,}000$ nT. For an isotropic distribution of magnetospheric electrons at the equator, the fraction $\tilde{f}$ that are in the loss cone is (for small angles)

$$\tilde{f} = \frac{\pi\,\theta_{lc}^2}{2\pi} \sim \frac{B_{eq}}{B_{ion}} \tag{3.34}$$

For example, if the magnetospheric magnetic field is 20 nT (and with $B_{ion} = 50{,}000$ nT), the loss cone has a half angle $\theta_{lc} = 1.1°$ and the fraction of magnetospheric electrons that are in the loss cone is $\tilde{f} \sim 2 \times 10^{-4}$.

Owing to the mirroring of magnetospheric particles in the strong magnetic field above the ionosphere, there is a diode effect to the ability of the magnetospheric plasma to carry field aligned currents to the ionosphere. For upgoing currents, carried by downward electrons, the mirror force impedes the electron motion, while for downward current, the mirror force aids the motion of upgoing electrons. For upward field-aligned currents, the maximum allowed field-aligned current density at the equatorial magnetosphere can be expressed as $j_{eq\parallel} \approx \tilde{f}\,j_{mth}$, where $j_{mth} = en(k_B T_e/2\pi m_e)^{1/2}$ is the maximum current density that can be carried in the magnetospheric plasma in the absence of loss-cone effects. This maximum current density is known as the Bohm current (Chen, 1984). A field-aligned current with no divergence is constant, so the current density $j_{\|}$ scales with B. The Bohm limiting current can be expressed as $j_{ion\|}/B_{ion} = j_{eq\|}/B_{eq}$, which yields

$$j_{ion\|} \approx j_{mth} \tilde{f} \frac{B_{ion}}{B_{eq}} \tag{3.35}$$

For $\tilde{f} \approx B_{eq}/B_{ion}$, this current is $j_{ion\|} \approx j_{mth}$. For the distant plasma sheet (appropriate for high-latitude aurora) n_e is in the range 0.05 - 0.3 cm^{-3} and T_e is in the range 0.1 - 1 keV, which yields j_{mth} in the range 0.01 - 0.3 μA m^{-2}. For the near-Earth plasma sheet (appropriate for low-latitude aurora) n_e is in the range 0.3 - 2 cm^{-3} and T_e is in the range 0.5 - 2 keV, which yields j_{mth} in the range 0.2 - 3 μA m^{-2}. These values are comparable to the average Region 1 and 2 current densities; however, the current density can be much larger in strong auroral arcs. Field-aligned electric fields in the auroral magnetosphere act to increase these upward-current-density limits by electrostatically enlarging the atmospheric loss cone, i.e., by using an electric field to put more of the magnetospheric electrons into orbits that will hit the atmosphere. An upward-directed parallel electric field increases the parallel velocity of a downgoing electron, decreasing its pitch angle as shown in Figure 3.8. Such an electric field, if located high above the atmosphere, can take electrons that were not in the local loss cone and put them into the local loss cone.

Along these lines, a theoretical relationship between the field-aligned potential and the parallel current density was recognized by Knight (1973) and Antonova and Tverskoy (1975b). The increase in field-aligned current density from a parallel potential applied in the magnetosphere can be calculated by including the parallel potential $\Delta\Phi_{\|}$ in the distribution function, $f(v_{\|}, v_{\perp})$, and integrating over the region $v_{\|}^2 > (R_M - 1)v_{\perp}^2$ in velocity space

$$j_{\| ion} = 2\pi e R_M \int_{v_0}^{\infty} \mathrm{d}v_{\|} \int_0^{\alpha v_{\|}} v_{\perp} v_{\|} \mathrm{d}v_{\perp} f(v_{\|}, v_{\perp}) \exp\left(\frac{e\Delta\Phi_{\|}}{k_B T_e}\right) \tag{3.36}$$

where $R_M = B_{ion}/B_{top}$ is the 'mirror ratio', and B_{top} is the magnetic field strength at the topside of the potential drop, $\alpha = (R_M - 1)^{-1}$, and $v_0 = (2e\Delta\phi_{\|}/m_e)^{1/2}$.

The above integration results in a current-voltage relation, the so-called 'Knight relation'

$$j_{\| ion} = j_{mth} \left\{ R_M - (R_M - 1) \exp\left[-\frac{e\Delta\Phi_{\|}}{k_B T_e (R_M - 1)} \right] \right\} \tag{3.37}$$

The above expression assumes an isotropic Maxwellian distribution and does not include electron outflow from the ionosphere or the contribution of ions to the current.

Figure 3.9 plots the above current-voltage relation for different values of R_M for a magnetospheric source of electrons with T_e = 1 keV and n_e = 0.3 cm^{-3}, which are mid-range numbers for the electron plasma sheet. For this population, the Bohm

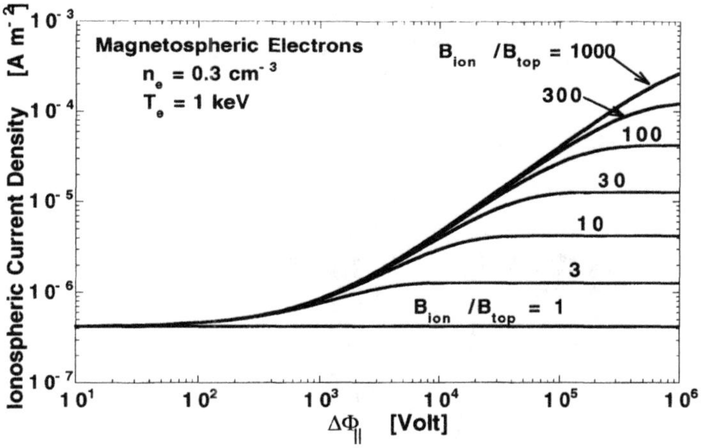

Figure 3.9. The field-aligned current density $j_\parallel$ out of the top of the ionosphere is plotted as a function of the potential difference between the magnetosphere and ionosphere for various values of the mirror ratio $R_M = B_{ion}/B_{top}$, where B_{ion} is the field strength in the upper atmosphere and B_{top} is the field strength at the magnetospheric edge of the potential drop. The magnetospheric plasma is taken to have $T_e = 1$ keV and $n = 0.3$ cm^{-3}. Note the linear region of the Knight relation which is almost independent of the mirror ratio. Figure provided by J. Borovsky.

current is $j_{mth} = 0.42\ \mu\mathrm{A\,m}^{-2}$. If $e\Delta\Phi_\parallel/k_B T_e \ll 1$, the Knight relation predicts a minimum current flow from the magnetosphere. If $e\Delta\Phi_\parallel/k_B T_e \gg R_M$, the current saturates at $j_{mth} R_M$.

The Knight relation is nearly linear (as for Ohmic currents) over a broad range $1 \ll e\Delta\Phi_\parallel/k_B T_e \ll R_M$ of parallel potential drops, so the above expression can be simplified to

$$j_{\parallel\mathrm{ion}} = K\Delta\Phi_\parallel \qquad \text{where} \qquad K = \frac{e^2 n}{\sqrt{2\pi m_e k_B T_e}} \tag{3.38}$$

is the Knight conductance. This simple linear form of the Knight relation is extremely useful.

3.3.2 Current-Voltage Relation in the Downward $j_\parallel$ Regions

Similar to the upward-current region, there is evidence for the presence of parallel electric fields in the downward-current regions of the aurora as discussed in Section 4.2. The association between current and voltage is less well established in the downward-current region than it is in the upward-current region. Rather than as a need to electrostatically enlarge the loss cone, the need for the parallel electric field is argued to arise when current is drawn through the low-density plasma above the ionosphere (Jasperse, 1998; Temerin and Carlson, 1998b; Rönnmark, 1999; Lynch et al., 2002). In order to keep the electron current density high where the ion density

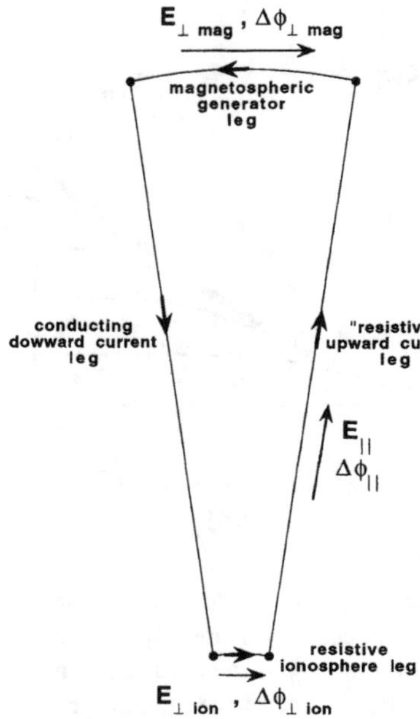

Figure 3.10. A sketch of a current loop driven by a perpendicular potential drop $\Delta\Phi_{\perp mag}$ in the magnetosphere (*top*). The direction of current flow is indicated by the bold arrows. The potential drop in the magnetosphere results in a perpendicular potential drop $\Delta\Phi_{\perp ion}$ in the resistive ionosphere (*bottom*) and a parallel potential drop $\Delta\Phi_{\parallel}$ in the upward-current leg of the current loop where magnetospheric electrons mirror while carrying current. After Borovsky (1993a).

is low, high electron velocities are required and therefore a potential drop forms to accelerate the electrons. This electric field produces the upgoing electron beams seen in these downward-current zones. The current-voltage relationship in this downward-current region will be a complicated function of the densities and temperatures of all the different populations along the field line: ionopheric ions and electrons and magnetospheric ions and electrons. Since the parallel electric field that forms will affect the particle populations, there are strong feedbacks which must be part of any current-voltage model. The $j_{\parallel}E_{\parallel}$ energy dissipation will go into electron-beam production, downward ion acceleration, wave generation, and ion and electron heating. The ion heating causes the ion scale height to change, which changes the required electric field for a given current density.

3.3.3 The Nature of $E_{\parallel}$

Owing to the current limitation, there are theoretical arguments supporting the need for a parallel electric field in regions of upward field-aligned current (see Section 3.3.1). The observational evidence is compelling that electrostatic potential structures accelerate electrons along the magnetic field to make downgoing auroral-arc electron beams (see Section 4.1) and to make upward electron beams (Section

4.2). The functional form that the field-aligned electrostatic potential drop takes is a long-standing question in auroral research (Shawhan et al., 1978; Swift, 1981; Borovsky, 1993b). It is not known whether the potential drop varies slowly over a long distance, whether it is concentrated sharply over a short distance, or whether it consists of multiple small drops. In this Section a number of theoretical models for static $E_\parallel$ structures are discussed. Models which rely on fluctuation plasma waves or moving 'weak double layers' to produce a broadly distributed net potential drop are deferred to Section 3.5.1, which deals with 'anomalous resistivity'.

Theoretical electrostatic structures in collisionless plasmas are obtained by solving the Vlasov-Poisson system of equations (Montgomery and Joyce, 1969)

$$\frac{\partial f_i}{\partial t} + \mathbf{v} \cdot \nabla f_i = e\, m_i\, (\nabla \Phi - \mathbf{v} \times \mathbf{B}) \cdot \nabla_v f_i \tag{3.39}$$

$$\frac{\partial f_e}{\partial t} + \mathbf{v} \cdot \nabla f_e = -\, e\, m_e\, (\nabla \Phi - \mathbf{v} \times \mathbf{B}) \cdot \nabla_v f_e \tag{3.40}$$

$$\nabla^2 \Phi = -e\, (n_i - n_e) \tag{3.41}$$

where $f_i(\mathbf{v})$ and $f_e(\mathbf{v})$ are the ion and electron phase-space densities, Φ is the electrostatic potential (where $\mathbf{E} = -\nabla\Phi$), $\mathbf{v}$ is the velocity of an element of phase space, and $\nabla_v = \left(\partial/\partial v_x,\ \partial/\partial v_y,\ \partial/\partial v_z\right)$. The ion and electron number densities are given by $n_i = \int f_i \mathrm{d}\mathbf{v}$ and $n_e = \int f_e \mathrm{d}\mathbf{v}$. For time-stationary solutions, $\partial f_i/\partial t = 0$ and $\partial f_e/\partial t = 0$ are taken. In general, the system of Eqs. (3.39) - (3.41) are computationally solved subject to the relevant boundary conditions and sometimes utilizing various approximations. Depending on the approximations used, different types of solutions are obtained.

Quasi-neutral Solutions. When the assumption of quasi-neutrality is used to restrict $n_i \approx n_e$ everywhere in space, then weak-$E_\parallel$ fields smoothly distributed over 1000's-of-km distances are obtained (e.g., Chiu and Schulz, 1978; Chiu and Cornwall, 1980; Jasperse and Grossbard, 2000). These quasi-neutral electric-field structures are sensitive to the temperature anisotropies of the magnetospheric particle populations and are supported by the interaction of the various charge species with the magnetic mirror force in the converging field lines above the auroral ionosphere.

Electrostatic Shock. Swift (1975, 1979) attempted to solve the time-stationary Vlasov equations for oblique (neither exactly parallel to **B** nor exactly perpendicular to **B**) electric-field structures under the restriction that the spatial gradient of Φ would be weak compared with ion gyroradii. It was hoped that electrostatic potential jumps with thicknesses that scaled with ion gyroradii would be obtained. Unfortunately, no solutions that satisfied physical boundary conditions could be found. A recent numerical model by R. Ergun, described in Section 7.1, does, however, reproduce the evolution of such electrostatic shocks quite well.

Double Layers. When the Poisson-Vlasov equations are solved without approximation, strong electric fields concentrated over distances of 10's of km are found (Block, 1972; Borovsky, 1993a). These strong-electric-field solutions are known variously as double layers, strong double layers, classical double layers, or oblique double layers and are the concentrated potential jumps seen in collisionless plasmas in the laboratory (e.g., Langmuir, 1929). These potential drops, which have thicknesses that scale with the Debye length of the magnetospheric electrons, are supported by a balance of particle acceleration across the structure and particle inflow into the structure. In the auroral magnetosphere, possibly so-called weak double layers, (cf. Section 4.3) are of importance. They are microscopic structures related to solitary structures, ion and electron phase space holes (cf. Section 3.5.4).

3.3.4 Static Magnetosphere-Ionosphere Coupling

Owing to the field-aligned potential drops needed to drive field-aligned currents, perpendicular voltages imposed by the magnetosphere may or may not map to the ionosphere. If they do not map well, then the $\mathbf{E} \times \mathbf{B}$ flows of the magnetosphere are disconnected from the ionosphere and the magnetospheric flows will slip relative to the ionospheric flows; if they map well, then the magnetosphere and ionosphere move together, as discussed in Section 3.2.4. It will turn out that large-scale potentials map well and the small-scale potentials disconnect (Lyons, 1980; Weimer et al., 1985).

Assume that a perpendicular static electric field of limited spatial size is imposed in the magnetosphere. This electric field will drive a current that closes in the ionosphere, as sketched in Fig. 3.10. That electric field is associated with a potential drop in the magnetosphere $\Delta\Phi_{\perp \mathrm{mag}}$. The imposed potential $\Delta\Phi_{\perp \mathrm{mag}}$ leads to a perpendicular potential drop $\Delta\Phi_{\perp \mathrm{ion}}$ in the ionosphere plus a field-aligned potential drop $\Delta\Phi_{\parallel}$ in the upward leg of the current system, as shown in the figure. If the magnetosphere-ionosphere coupling is perfect, then $\Delta\Phi_{\perp \mathrm{ion}} = \Delta\Phi_{\perp \mathrm{mag}}$ and $\Delta\Phi_{\parallel} = 0$. If the coupling is poor, then $\Delta\Phi_{\perp \mathrm{ion}} \ll \Delta\Phi_{\perp \mathrm{mag}}$. 'Static coupling' implies $\partial\rho/\partial t + \nabla \cdot \mathbf{j} \approx \nabla \cdot \mathbf{j} = 0$, which can be written

$$\frac{\partial j_{\parallel}}{\partial z} = -\nabla_{\perp} \cdot \mathbf{j}_{\perp} \tag{3.42}$$

where z is the coordinate parallel to $\mathbf{B}$. This is a repeat of Eq. (3.4), but written in magnetic field-aligned coordinates. The current in the ionosphere closes in a relatively thin layer in altitude. Integrating the above expression over the thin layer in z, one gets

$$j_{\parallel} = \pm \nabla_{\perp} \cdot \mathbf{J}_{\perp} \tag{3.43}$$

where $\mathbf{J}_\perp$ is the (upward) height-integrated ionospheric current, given in Eq. (3.3). Here, the + (-) sign refers to the northern (southern) hemisphere.

For this static case one can derive a simple relation between the electric and magnetic field variations, which will be used in Section 4.4.1. Taking for $\mathbf{J}_\perp$ only the Pedersen current, Eq. (3.43) yields $j_\parallel = \pm\nabla_\perp(\Sigma_P \mathbf{E}_\perp)$, where the plus (minus) sign refers to the northern (southern) hemisphere, respectively.

From Ampère's law one has $j_\parallel = \frac{1}{\mu_0}(\nabla \times \mathbf{b})_\parallel$ where $\mathbf{b}$ is the magnetic field variation associated with $j_\parallel$. Assuming a situation where the fields vary predominantly in the x-direction, the two expressions for $j_\parallel$ can be combined to yield

$$\frac{1}{\mu_0}\frac{\partial b_y}{\partial x} = \pm\frac{\partial}{\partial x}(\Sigma_P E_x) \tag{3.44}$$

from which immediately follows that the ratio between the electric and magnetic fields is

$$\frac{E_x}{b_y} = \pm\frac{1}{\mu_0 \Sigma_P} \tag{3.45}$$

Returning to Eq. (3.43) and applying to its left hand side the Knight relation (3.38) in the form of $j_\parallel = K\Delta\Phi_\parallel$, and to its right-hand side the Ohm's law (3.3) for the case of uniform conductivities, $J_\perp = \Sigma_P E_\perp$, yields

$$K\Delta\Phi_\parallel = \Sigma_P \nabla_\perp E_{\perp\mathrm{ion}} \tag{3.46}$$

where the height-integrated Pedersen conductivity Σ_P is taken to be uniform horizontally. If the width of the potential structure in the ionosphere is w_{ion}, then $E_{\perp\mathrm{ion}} \sim \Delta\Phi_{\perp\mathrm{ion}}/w_{\mathrm{ion}}$ and $\nabla_\perp E_{\perp\mathrm{ion}} \sim \Delta\Phi_{\perp\mathrm{ion}}/w_{\mathrm{ion}}^2$. This gives

$$\Delta\Phi_\parallel = \frac{\Sigma_P}{K}\frac{\Delta\Phi_{\perp\mathrm{ion}}}{w_{\perp\mathrm{ion}}^2} \tag{3.47}$$

Applying Kirchhoff's law to the current loop in Figure 3.10 gives $\Delta\Phi_{\perp\mathrm{mag}} = \Delta\Phi_{\perp\mathrm{ion}} + \Delta\Phi_\parallel$. Using this to algebraically eliminate $\Delta\Phi_\parallel$ leads to the expression

$$\Delta\Phi_{\perp\mathrm{ion}} = \left(1 + \frac{\Sigma_P}{K}\frac{1}{w_{\perp\mathrm{ion}}^2}\right)^{-1} \Delta\Phi_{\perp\mathrm{mag}} \tag{3.48}$$

As can be seen from this expression, if $w_{\perp\mathrm{ion}} \gg (\Sigma_P/K)^{1/2}$, then $\Delta\Phi_{\perp\mathrm{ion}} \approx \Delta\Phi_{\perp\mathrm{mag}}$ and the magnetosphere and ionosphere are well coupled. If $w_{\perp\mathrm{ion}} \ll (\Sigma_P/K)^{1/2}$, then $\Delta\Phi_{\perp\mathrm{ion}} \ll \Delta\Phi_{\perp\mathrm{mag}}$ and the magnetospheric electric field does not couple to the ionosphere, rather a parallel potential drop $\Delta\Phi_\parallel \approx \Delta\Phi_{\perp\mathrm{mag}}$ dominates in size over $\Delta\Phi_{\perp\mathrm{ion}}$. The above equation contains an electrostatic scale length $\lambda_{ES} = \sqrt{\Sigma_P/K}$ in the ionosphere, which is sometimes termed the magnetosphere-ionosphere coupling scale length. For typical values of $\Sigma_P = 10\,\mathrm{S}$ and $K = 10^{-9}\,\mathrm{S\,m^{-2}}$, this scale length is 100 km. Solutions to this type of problem for isolated potential structures and the associated current system yield somewhat more complicated results than the concept that magnetospheric electric fields do not penetrate to the ionosphere on scale sizes smaller than λ_{ES} (Borovsky and Bonnell, 2001).

The electrostatic scale length λ_{ES} is large and therefore cannot explain the scale size of individual auroral arcs. However, as suggested by (Lyons, 1981), it is probably relevant for the scale size of large-scale inverted-V events, which are discussed in Section 4.1.1.1.

3.4 Dynamic Magnetosphere-Ionosphere Coupling

The considerations discussed previously have all been based on a steady-state picture of magnetosphere-ionosphere coupling. While steady auroral arcs are often observed, it is more typical for the aurora to be dynamic: auroral features move and change their brightness. Since the changes in the aurora are associated with changes in the auroral currents, the dynamics of current variations is important. Within the context of MHD, when currents change, the information about the change propagates along the magnetic field at the Alfvén velocity, i.e. the electric fields that adjust the currents propagate as Alfvén waves, analogous to wave propagation in transmission lines (Goertz and Boswell, 1979). As it turns out, when the perpendicular wavelength of these Alfvén waves becomes small, an ideal-MHD description breaks down and ideal MHD must be replaced by a two-fluid model, for instance including the generalized Ohm's law (3.14), or by using kinetic theory. In this section, we will first consider the fluid description of shear Alfvén waves, both from an ideal MHD point of view and from a two-fluid point of view. We will then discuss the effects of inhomogeneity on the propagation of these waves. For a recent review of Alfvén wave coupling, see Vogt (2002).

3.4.1 Ideal MHD Alfvén Waves

Of the three MHD wave modes (fast, slow, and intermediate), only the shear Alfvén (intermediate) wave carries a field-aligned current. This wave is magnetically incompressible, meaning that the wave alters the direction but not the magnitude of the magnetic field. Thus, the simplest model for this wave involves the ideal, incompressible MHD equations. If we assume that the background magnetic field $\mathbf{B}_0$ is directed along the z direction, these equations can be written in terms of the velocity perturbation $\mathbf{v}$ and the magnetic perturbation, which we will denote[3] by a lower case vector, $\mathbf{b}$. Then the relevant linearized equations become:

$$\rho \frac{\partial \mathbf{v}}{\partial t} = \mathbf{j} \times \mathbf{B}_0 = \frac{1}{\mu_0} \mathbf{B}_0 \cdot \nabla \mathbf{b} \tag{3.49}$$

[3]Not to be confused with the unit vector along $\mathbf{B}$, which we denote by $\hat{\mathbf{b}}$.

$$\frac{\partial \mathbf{b}}{\partial t} = \nabla \times (\mathbf{v} \times \mathbf{B}_0) = \mathbf{B}_0 \cdot \nabla \mathbf{v} \tag{3.50}$$

Note that these equations have been simplified by assuming that the background magnetic field is uniform, and by making the incompressibility assumptions that $\mathbf{B}_0 \cdot \mathbf{b} = 0$ and $\nabla \cdot \mathbf{v} = 0$. Fourier transforming these equations in the usual manner leads to the dispersion relation $\omega = k_{||} v_A$ where the Alfvén speed is defined by $v_A = B_0/\sqrt{\mu_0 \rho}$. Note that the velocity and magnetic field perturbations are related as $\mathbf{v} = \mp \mathbf{b}/\sqrt{\mu_0 \rho}$ or $\mathbf{v}/v_A = \mp \mathbf{b}/B_0$, where in this and the following expressions, the top sign applies for propagation parallel to $\mathbf{B}_0$ and the bottom sign for propagation anti-parallel to $\mathbf{B}_0$.

To look at some interesting properties of the Alfvén wave, note first that the perpendicular electric field in ideal MHD is given by

$$\mathbf{E}_\perp = -\mathbf{v} \times \mathbf{B}_0 = \pm \frac{\mathbf{b} \times \mathbf{B}_0}{\sqrt{\mu_0 \rho}} = \pm v_A \mathbf{b} \times \hat{\mathbf{b}} \tag{3.51}$$

Hence, $\mathbf{b}$ and $\mathbf{E}_\perp$ are in phase (or anti-phase), and the ratio $E_\perp/b$ is just v_A, and the electric field is perpendicular to both the magnetic field perturbation $\mathbf{b}$ and to $\mathbf{B}_0$. The perpendicular current in the wave can be calculated from Ampere's law

$$\mathbf{j}_\perp = \frac{1}{\mu_0} (\nabla \times \mathbf{b})_\perp = \frac{1}{\mu_0} \frac{\partial}{\partial z} \left(\hat{\mathbf{b}} \times \mathbf{b} \right) \tag{3.52}$$

where z is along $\mathbf{B}_0$. Since $\mathbf{B}_0$ is assumed to be uniform, using Eq. (3.51) in Eq. (3.52) we can write a relation between the perpendicular current and the perpendicular electric field

$$\mathbf{j}_\perp = \mp \frac{1}{\mu_0 v_A} \frac{\partial \mathbf{E}_\perp}{\partial z} \equiv \mp \Sigma_A \frac{\partial \mathbf{E}_\perp}{\partial z} \tag{3.53}$$

where the last step defines an effective conductance known as the Alfvén conductance,

$$\Sigma_A = \frac{1}{\mu_0 v_A} \tag{3.54}$$

This quantity can be numerically estimated by noting that Σ_A is 0.8 S for an Alfvén speed of 1000 km s^{-1}. The perpendicular current $j_\perp$ in an Alfvén wave (Eq. 3.54) is a polarization-drift current.

Using Σ_A, the ratio between b and $E_\perp$, can be written as

$$\frac{E_\perp}{b} = \pm v_A = \pm \frac{1}{\mu_0 \Sigma_A} \tag{3.55}$$

Since Σ_A is much less than 0.1 S, while the Pedersen conductance, $\Sigma_P \approx 10$ S, the magnitude of the $E_\perp/b$ ratio is generally much larger for Alfvén waves than it is in the static case (see Eq. 3.45).

If we then consider the parallel component of the current, Ampere's law implies that

$$j_{\parallel} = \frac{1}{\mu_0}(\nabla \times \mathbf{b})_{\parallel} = \pm\Sigma_A \nabla_{\perp} \cdot \mathbf{E}_{\perp} \tag{3.56}$$

where Eq. (3.53) is used in the last step. (Note that $E_{\parallel} = 0$ in ideal MHD.) Unlike any of the wave properties listed before, the field-aligned current depends on the perpendicular scale-length of the wave. In the limit of infinite perpendicular wavelength, $\nabla_{\perp} \to 0$, there is no field-aligned current; however, in any real system, there must be a finite size to the wave in the perpendicular direction, and so the field-aligned current is non-zero.

If the assumption of incompressibility is violated, the wave magnetic amplitude is no longer constant, $db/dt \neq 0$, and the wave is accompanied by a density variation, $\partial n/\partial t \neq 0$. Such waves are called compressional Alfvén waves.

3.4.2 Reflection of Alfvén Waves

As happens at any boundary between plasmas with different values of the Alfvén velocity, the wave is partly reflected and partly transmitted when incident on the ionosphere.

A plane shear Alfvén wave of frequency ω and wave number k_z propagating downward toward the ionosphere can be represented by

$$E_x(z,t) = E_0 \exp(i\omega t + ik_z z) \tag{3.57}$$

$$b_y(z,t) = \frac{E_0}{v_A}\exp(i\omega t + ik_z z) \tag{3.58}$$

When the waves are reflected from the ionosphere the incident and reflected waves are superimposed to yield

$$E_x(z,t) = E_0\left[\exp(i\omega t + ik_z z) + R\exp(i\omega t - ik_z z)\right] \tag{3.59}$$

$$b_y(z,t) = \frac{E_0}{v_A}\left[\exp(i\omega t + ik_z z) - R\exp(i\omega t - ik_z z)\right] \tag{3.60}$$

where R is the ionospheric reflection coefficient, which will be derived below.

For non-zero R, the ratio of the Fourier transforms $\tilde{E}_x(\omega)/\tilde{b}_y(\omega)$ now depends on frequency and position. In general, it is complex for non-zero R, implying a phase shift between E_z and b_y.

The reflection coefficient can be derived by considering the parallel current variation. Assume that the ionospheric conductivity is uniform. The field-aligned current is related to the perpendicular electric field in the ionosphere by Eq. (3.7), which can be written in the case where the Pedersen conductivity is dominant as

$$j_{\parallel} = \Sigma_P \nabla_{\perp} \cdot \mathbf{E}_{\perp} \tag{3.61}$$

Now, consider the northern hemisphere situation where the background magnetic field is into the ionosphere. The incident (downgoing) Alfvén wave will have the plus sign in Eq. (3.56), while the reflected (upgoing) wave will have the minus sign. Writing both sides of Eq. (3.61) in terms of a sum of the two waves, we have

$$j_{\|}^{inc} + j_{\|}^{ref} = \Sigma_A \nabla_\perp \cdot \left(\mathbf{E}_\perp^{inc} - \mathbf{E}_\perp^{ref}\right) = \Sigma_P \nabla_\perp \cdot \left(\mathbf{E}_\perp^{inc} + \mathbf{E}_\perp^{ref}\right) \tag{3.62}$$

which can be written as

$$\nabla_\perp \cdot \left[(\Sigma_A - \Sigma_P)\,\mathbf{E}_\perp^{inc} - (\Sigma_A + \Sigma_P)\,\mathbf{E}_\perp^{ref}\right] = 0 \tag{3.63}$$

Assuming a localized wave with fields that go to zero at some boundary, we can integrate Eq. (3.63) and write it in terms of a reflection coefficient for the electric field

$$R \equiv \frac{E_x^{ref}}{E_x^{inc}} = \frac{\Sigma_A - \Sigma_P}{\Sigma_A + \Sigma_P} \tag{3.64}$$

This reflection coefficient was first derived by Scholer (1970) and was used by Mallinckrodt and Carlson (1978) and Goertz and Boswell (1979) in their modeling of Alfvén wave interactions with the ionosphere.

Eq. (3.64) is rich in consequences. First of all, it can be seen that the reflection coefficient goes to zero if the Pedersen and Alfvén conductivities are equal. This has a direct analogy with the termination of a transmission line by a resistor: if the transmission-line impedance is matched by the load, there is no reflection. In general, however, these conductivities are not matched. The most typical case is that the Pedersen conductivity, which is generally 1-10 S in the ionosphere, is larger than the Alfvén conductivity, which is usually less than 0.5 S (Lysak and Hudson, 1987; Lysak, 1990), i.e. $\Sigma_P > \Sigma_A$. It is easy to see that in this case, the reflection coefficient is negative, implying that the electric field in the reflected wave is in the opposite direction as the electric field in the incident wave. Thus if these two waves are superimposed, the net effect is to reduce the electric field strength. Indeed, in the extreme limit of infinite ionospheric conductivity ($\Sigma_P \gg \Sigma_A$), the reflection coefficient is -1, and the reflected electric field cancels the incident field. In this case the ionosphere can be said to short out the perpendicular electric field.

On the other hand, the reflected wave magnetic field has a relative minus sign with respect to the incident field, and so if the reflected electric field is in the opposite direction as the incident field the reflected magnetic field reinforces the incident wave magnetic field. Thus the high conductivity ionosphere tends to enhance the perpendicular magnetic field and thus the field-aligned current. Of course the opposite conclusions can be drawn in the case when $\Sigma_P < \Sigma_A$, such as could happen at night or in the winter in the absence of any particle precipitation, when the conductivity can be as low as 0.1 S. This type of low-conductivity ionosphere tends to restrict the flow of current and enhance the electric field.

When there is a mismatch between the conductivities, an Alfvén wave cannot promptly establish a steady-state current system. If one were to arbitrarily launch an Alfvén wave from the magnetosphere and allow it to propagate to the ionosphere, owing to the impedance mismatch the wave would begin bouncing back and forth. This is analogous to the ringing of a transmission line. This ringing is seen in the magnetosphere-ionosphere circuit as the Pi2 oscillations at the onset of a magnetospheric substorm. Suppose a fixed electric field were imposed at the magnetospheric end of a flux tube in a generator region (Goertz and Boswell, 1979; Lysak and Dum, 1983). This electric field would propagate to the ionosphere as an Alfvén wave, and reflection would occur according to the reflection coefficient Eq. (3.64), reducing the electric field (assuming $\Sigma_P > \Sigma_A$) and increasing the field-aligned current. The reflected wave propagates back to the generator, in which the perpendicular electric field is fixed, and another Alfvén wave is launched back toward the ionosphere. Each time the wave interacts with the ionosphere, the net result is that the ionospheric electric field is increased by a relative amount equal to $1-|R|$. Thus if R is very close to -1, a great number of bounces are needed to impose an electric field on the ionosphere, while if R is nearly zero, fewer bounces are required. A similar scenario is found if a fixed magnetic perturbation is imposed at the generator, although in this case the electric field and field-aligned current overshoot their initial values and approach a steady state in an oscillatory manner (Lysak and Dum, 1983). A detailed study of the differences between these voltage and current generators is found in Lysak (1985) and will be discussed further in Section 7.3.

Equation (3.64) represents what is known as the 'thin slab model' of the ionosphere. It is based on the assumptions that 1) the ionospheric thickness and gradient scale lengths are much smaller than the parallel Alfvén wavelength $\lambda_{\parallel}$, 2) that perpendicular electric fields map unattenuated through the ionosphere, and 3) that the magnetic field lines above the ionosphere are perfectly conducting in the parallel direction. The first assumption breaks down for wave periods less than $\sim$10 s (Lysak, 1991; Knudsen et al., 1992). The second breaks down for horizontal wavelengths $\lambda_{\perp}$ <2 km (Lessard and Knudsen, 2001). The following discussion addresses the case for finite field line conductivity above the ionosphere, as described by the Knight relation.

Replacing the infinite parallel conductivity of ideal MHD with finite parallel conductivity modifies the Alfvén-wave reflection coefficient. In essence, this replacement modifies the load in a transmission-line problem and so modifies the reflection coefficient of the transmission-line signal. In fact, it makes the load more resistive. Vogt and Haerendel (1998) have derived a modified reflection coefficient by including the Knight field-aligned current-voltage relation Eq. (3.38) in both the upgoing and downgoing field-aligned currents. This modification can be included by introducing the approximate current-voltage relation first given by Knight (1973), which states that the field aligned current density can be related to the potential drop by the relation $j_{\parallel} = K\Delta\Phi_{\parallel}$. This relation contains the magnetosphere-ionosphere coupling length given by $\lambda_{ES} = \sqrt{\Sigma_P/K}$, (see Section

3.3.4) which is the order of 100 km for typical parameters. Under these conditions, the Pedersen conductivity in Eq. (3.64) should be replaced by an effective Pedersen conductivity

$$\Sigma_{P,\text{eff}} = \frac{\Sigma_P}{1 + k_\perp^2 \lambda_{ES}^2} \tag{3.65}$$

Eq. (3.65) indicates that at smaller scale lengths, the effective Pedersen conductivity is reduced. Using Eq. (3.65) to replace Σ_P in Eq. (3.64), multiplying the numerator and denominator by $(1 + k_\perp^2 \lambda_{ES}^2)$, Eq. (3.64) becomes

$$R = \frac{\Sigma_A - \Sigma_P + k_\perp^2 \frac{\Sigma_P \Sigma_A}{K}}{\Sigma_A + \Sigma_P + k_\perp^2 \frac{\Sigma_P \Sigma_A}{K}} \tag{3.66}$$

In Figure 3.11, for the case Σ_A = 0.5 S, Σ_P = 5 S, and $k = 6 \times 10^{-10}$ S m^{-2} (T_e = 500 eV and n_e = 0.5 cm^{-3}), the reflection coefficient (Eq. (3.66)) is plotted as a function of the gradient scale size $L_\perp = k_\perp^{-1}$ of the Alfvén wave in the ionosphere. As can be seen in the figure, for large structures the reflection coefficient is negative and matches the value given by Eq. (3.64), which in this case is R = -0.82. For very small spatial scales the reflection coefficient is $R \rightarrow 1$. For $\Sigma_A \ll \Sigma_P$, the $R = 0$ crossover occurs at $L_\perp \approx \lambda_A$, where

$$\lambda_A = \left(\frac{\Sigma_A}{K}\right)^{1/2} \tag{3.67}$$

is defined to be a 'transient length scale'. The scale length λ_A is typically in the 10 km range. For waves that have smaller perpendicular wavelengths, the Alfvén waves are reflected such that the electric field is enhanced. Should a case $\lambda_A \gg \lambda_P$ arise, the reflection coefficient would be $R \approx 1$ for all values of $L_\perp$; in that case the ionosphere is already a high-resistivity load and adding further resistivity owing to finite field-aligned conductance makes little difference.

It should be noted that the Alfvén speed can vary considerably over the auroral field line. Near the ionosphere, the Alfvén speed is only a few hundred kilometers per second, but in the auroral acceleration region at 4000-6000 km altitude, this speed can increase to over 100000 km s^{-1}. It then decreases more slowly to the 1000 km s^{-1} range. Thus, the Alfvén conductivity is generally in the 0.01-1.0 S range. Which value of this conductivity is appropriate to take depends on the frequency of the wave. For low-frequency waves with periods of minutes, the wavelength of the wave is longer than the distance to the peak of the Alfvén speed, and the Alfvén conductivity in the outer magnetosphere, which is generally somewhat less than 1 S, is appropriate. For waves with periods of about 1 s, this inhomogeneity of the Alfvén speed must be taken into account in detail, and the simple reflection coefficient given by Eq. (3.64) does not hold and the detailed structure of the Alfvén speed profile determines the reflection (Lysak, 1991). Further discussion of this issue will be presented in Section 7.4.

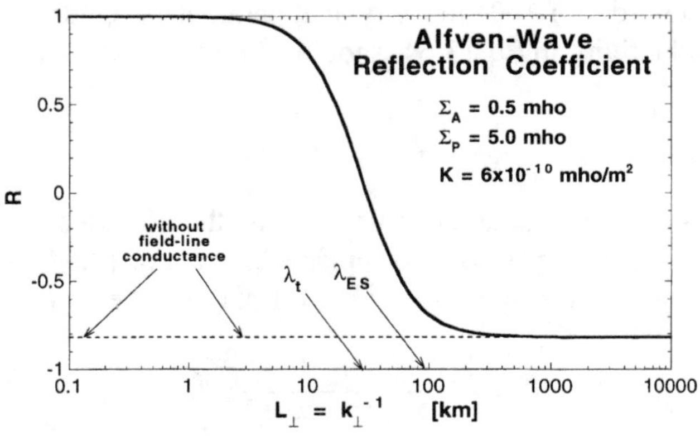

Figure 3.11. The Alfvén-wave reflection coefficient for long period waves incident on a thin slab ionosphere through a region of finite mirror resistivity, plotted as a function of the perpendicular scale size $L_\perp$ (in the ionosphere) of the Alfvén wave. Figure provided by J. Borovsky.

For a weak inhomogeneity, one can use a WKB-like approximation[4], which implies that the wave dispersion follows the local dispersion relation. This approximation holds as long as the wavelength $\lambda \ll L$ is small compared to the scale size of the inhomogeneity. Under this approximation, the wave is not reflected. The Poynting flux integrated over the flux tube cross section is constant. Since the flux tube cross section is inversely proportional to the background magnetic field, the quantity $E_x b_y \propto B_0$, where E_x and b_y are the components of the wave field, and $\mathbf{B}_0$ is the background field. These wave components are related to each other by Eq. (3.51), and we can write $E_x \propto B_0/\rho^{1/4}$ and $b_y \propto \rho^{1/4}$, with ρ the background mass density.

This procedure is not valid when the scale length of the inhomogeneity is smaller than the wavelength of the Alfvén wave. Then Alfvén waves are reflected by the gradients. Consider the extreme case of a discontinuity in the Alfvén speed. If the wave is incident from the side with Alfvén speed v_{A1}, and is transmitted to the side with speed v_{A2}, then an argument analogous to that for the ionospheric reflection yields

$$R = \frac{E_{ref}}{E_{inc}} = \frac{\Sigma_{A1} - \Sigma_{A2}}{\Sigma_{A1} + \Sigma_{A2}} \tag{3.68}$$

Thus, propagation into increasing Alfvén speed corresponding to decreasing Alfvén conductivity leads to an enhancement of the wave electric field and a decrease in the magnetic field, and vice versa. This model has been used to calculate the propagation of Alfvén waves in inhomogeneous media (Mallinckrodt and Carlson, 1978).

[4]The WKB approximation states that a wave in an inhomogeneous plasma can be treated as a locally homogeneous wave with spatially-dependent amplitude and phase, where the amplitude and phase variation is weak on a scale length comparable to the wavelength of the wave (e.g., Stix, 1992).

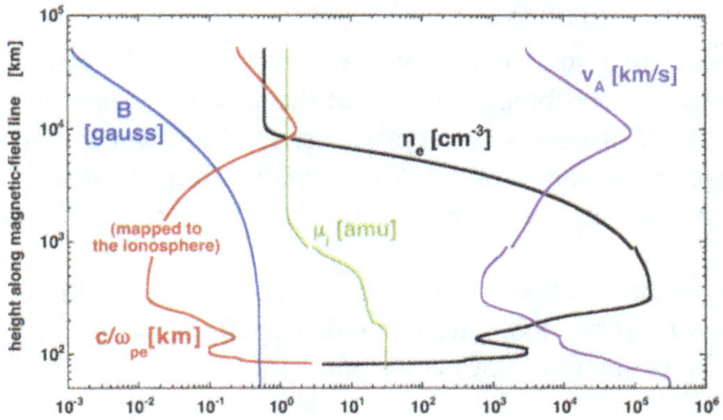

Figure 3.12. Typical profiles for the Alfvén velocity v_A, the electron density n_e, the magnetic-field strength B, and the mean mass of an ion μ_i versus height along an auroral flux tube. At low altitudes the parameters are obtained from the International Reference Ionosphere (Rawer et al., 1978) for Alaska Winter midnight. In the collisional bottomside of the ionosphere and in the atmosphere, the Alfvén velocity goes to the speed of light. The equatorial end of the flux tube is at a height of 50,500 km (which is a length of 8.0 R_E along the field line); there the magnetospheric plasma is taken to be 10% oxygen ions and 90% protons. This figure also shows the variations of the electron inertial length (or skin-depth), c/ω_{pe}. Figure provided by J. Borovsky.

In the topside ionosphere, there is a strong inhomogeneity due to the sharp exponential decrease in the plasma density with increasing altitude above a few hundred kilometers. This decrease takes place with a typical scale height of less than 1000 km. The magnetic field varies weakly on such spatial scales, and the Alfvén speed increases exponentially with a comparable scale height above the ionosphere. Above about 6000 km altitude, the density decrease becomes gradual. The decrease in the magnetic field leads to a slow decrease in the Alfvén speed. A typical set of profiles of the Alfvén speed and related parameters are shown in Figure 3.12.

The Alfvén speed profile gives rise to a resonant cavity along the field line, known as the ionospheric Alfvén resonator (Trakhtengertz and Feldstein, 1984, 1991; Lysak, 1991, 1993). The characteristic frequency of the cavity is $\omega_n = \xi_n v_{AI}/2h$, with v_{AI} the Alfvén speed in the ionosphere and h the density scale height. The ξ_n's form a set of dimensionless numbers, which for an idealized profile of the Alfvén speed can be related to the zeroes of Bessel functions (Lysak, 1991). For typical parameters, these frequencies are in the 1 Hz range. This structure of the Alfvén speed in the topside ionosphere therefore indicates that there is a characteristic wave frequency in the 1 Hz range for Alfvén waves. Such a frequency can often be seen in observations of electric and magnetic fields above the auroral zone as discussed in Section 4.3.

3.4.3 Two-Fluid Shear-Alfvén Waves

When the perpendicular scale length of the Alfvén waves becomes too small, the ideal MHD approximation breaks down and the waves become dispersive. (For a general discussion of dispersive Alfvén waves, see Stasiewicz et al. (2000)). This leads to a change in the nature of the Alfvén waves, in particular the introduction of a parallel wave electric field. One commonly refers to these waves as kinetic Alfvén waves.

Under these conditions, the momentum equation, Eq. (3.49) remains the same, but Faraday's law must be modified to include the effects of parallel electric fields. Thus, the induction equation (3.50) is written

$$\frac{\partial \mathbf{b}}{\partial t} = -\left(\frac{\partial}{\partial z}\left[\hat{\mathbf{b}} \times \mathbf{E}_\perp\right] + \nabla_\perp \times \mathbf{E}_{||}\right) \tag{3.69}$$

The parallel electric field is determined by the generalized Ohm's law

$$E_z = \frac{m_e}{ne^2}\frac{\partial j_z}{\partial t} - \frac{1}{ne}\frac{\partial p_e}{\partial z} \tag{3.70}$$

where we have assumed an isotropic electron pressure. Here, an equation for the electron pressure is needed. Noting that the electron motion is predominantly along the magnetic field (with the exception of the $\mathbf{E} \times \mathbf{B}$ drift, which does not have a divergence in this simplified geometry), and assuming isothermal electrons, the linearized equation of electron continuity $\partial n_e/\partial t + n_0 \nabla \cdot \mathbf{v}_e = 0$ can be multiplied by T_e to yield

$$\frac{\partial p_e}{\partial t} = -n_0 T_e \frac{\partial \mathrm{v}_z}{\partial z} = \frac{T_e}{e}\frac{\partial j_z}{\partial z} \tag{3.71}$$

Taking the time derivative of Eq. (3.70) yields

$$\frac{\partial E_z}{\partial t} = \frac{m_e}{ne^2}\frac{\partial^2 j_z}{\partial t^2} - \frac{T_e}{ne^2}\frac{\partial^2 j_z}{\partial z^2} \tag{3.72}$$

To find the two-fluid Alfvén wave equation, we take the time derivative of Eq. (3.69). The resulting time derivative of the perpendicular field is evaluated by using Eqs. (3.51) and (3.49), and the second term is evaluated by taking the curl of Eq. (3.72). Assuming that the plasma parameters are uniform in space, the wave equation becomes

$$\frac{\partial^2 \mathbf{b}}{\partial t^2} = v_A^2 \frac{\partial^2 \mathbf{b}}{\partial z^2} + \frac{c^2}{\omega_{pe}^2}\frac{\partial^2}{\partial t^2}\nabla_\perp^2 \mathbf{b} - c^2 \lambda_D^2 \frac{\partial^2}{\partial z^2}\nabla_\perp^2 \mathbf{b} \tag{3.73}$$

in terms of the magnetic perturbation. If we Fourier analyze in the usual manner, the dispersion relation can be written as

$$\omega^2 = k_{||}^2 v_A^2 \frac{1 + k_\perp^2 \rho_s^2}{1 + k_\perp^2 \lambda_e^2} \tag{3.74}$$

where we introduce the electron inertial length (electron skin depth) $\lambda_e = c/\omega_{pe}$ and the ion acoustic gyroradius, $\rho_s = c\lambda_D/v_A = c_s/\omega_{ci}$ where $c_s = \omega_{pi}\lambda_D$ is the sound speed (recall that $c/v_A = \omega_{pi}/\omega_{ci}$), and $\lambda_D = v_{eth}/\omega_{pe}$, is the electron Debye-length. (Note that there could be two different polarizations of this wave that have identical dispersion relations since Eq. (3.73) does not distinguish between perturbations in the x and y directions.)

The two-fluid effects change the relationship between the various quantities in Alfvén waves. Using Faraday's law with $\mathbf{E}_\perp = -\mathbf{v} \times \mathbf{B}_0$, the relation between the velocity and the magnetic field perturbation can be written as

$$\mathbf{v} = -\frac{B_0}{\mu_0 \rho}\frac{k_\parallel}{\omega}\mathbf{b} = \mp\sqrt{\frac{1+k_\perp^2\lambda_e^2}{1+k_\perp^2\rho_s^2}}\frac{\mathbf{b}}{\sqrt{\mu_0\rho}} \tag{3.75}$$

where the top sign indicates propagation parallel to the background magnetic field direction, and the bottom sign indicates anti-parallel propagation. Similarly, the perpendicular electric field becomes

$$\mathbf{E}_\perp = \pm v_A\sqrt{\frac{1+k_\perp^2\lambda_e^2}{1+k_\perp^2\rho_s^2}}\,\mathbf{b} \times \hat{\mathbf{b}} \tag{3.76}$$

Equations (3.72) and (3.52) give the parallel electric field in terms of the magnetic perturbation. With Eq. (3.76), together with the dispersion relation (3.74) written in terms of the parallel derivative rather than the parallel wave number (using $\partial/\partial z = ik_\parallel$), this can then be used to give the parallel electric field in terms of the perpendicular field:

$$E_\parallel = \left(\rho_s^2 - \lambda_e^2\frac{1+k_\perp^2\rho_s^2}{1+k_\perp^2\lambda_e^2}\right)\frac{\partial}{\partial z}\nabla_\perp \cdot \mathbf{E}_\perp \tag{3.77}$$

This expression holds for both directions of propagation. The first term in Eq. (3.77) gives the electron inertial effect and the second term gives the effect of finite electron pressure. These effects counteract each other in that the direction of the parallel electric field is opposite for these two cases. In the inertial limit, the parallel electric field acts to accelerate the electrons in the direction required to carry the field-aligned current while it acts to retard the electrons in the warm plasma limit. It should be asked which of these two effects is most important. From the dispersion relation (3.74), it can be seen that the inertial effect is more important when λ_e is large, while the pressure effect is important when ρ_s is large. The ratio between these two terms is

$$\frac{\rho_s}{\lambda_e} = \frac{c\lambda_D/v_A}{c/\omega_{pe}} = \frac{\sqrt{k_B T_e/m_e}}{v_A} = \sqrt{\beta_e\frac{m_i}{m_e}} \tag{3.78}$$

where we have defined the electron plasma beta $\beta_e = \mu_0 n k_B T_e / B_0^2$. Thus, in cold and strongly magnetized plasmas, the electron inertial length is dominant, whereas in warm or weakly magnetized plasmas, the electron pressure is the most important effect. Lysak and Carlson (1981) showed that on auroral field lines, the region below about 4-5 R_E is inertia dominated, while above that altitude the pressure dominates.

Physically, Eq. (3.78) shows that the pressure effect is dominant when the electron thermal speed is greater than the Alfvén speed. Conversely, in the inertia-dominated regime the wave speed is much greater than the electron-thermal speed. From the dispersion relation (3.74), it can be seen that the pressure effect increases the wave phase speed while the inertia effect decreases it; thus, roughly speaking, it could be said that the electrons accelerate the wave in the pressure regime and that electron inertia effects slow down the Alfvén wave in the cold regime. It may be noted that in the inertial regime, the electrons act as a cold fluid to the Alfvén wave and can be accelerated in bulk. It can be said that in this regime, the Alfvén wave carries a parallel electric field in order to accelerate the electrons up to the point where they can carry the current required by the Alfvén wave.

Of course, the fluid model described above does not consider all of the possible kinetic effects, such as ion gyroradius effects and Landau damping. Lysak and Lotko (1996) gave a derivation of the kinetic theory including these effects. The electron response in the kinetic theory is given by the usual plasma dispersion function $Z(z)$ (Fried and Conte, 1961; Ichimaru, 1986). In the hot electron limit, the $Z(z)$ function can be expanded in the small argument limit with the result that

$$\left(\frac{\omega}{k_{\|} v_A}\right)^2 \approx 1 + k_\perp^2 \left(\rho_s^2 + \rho_i^2\right) \tag{3.79}$$

This expression is in agreement with the warm plasma limit of Eq. (3.74), with the addition of an additional term due to the ion gyroradius effect that was not included in the two-fluid picture. Landau damping on the electrons can be important in this limit, but it was found by Lysak and Lotko (1996) that the damping rate $\nu/\omega < 0.1$ whenever the wave number satisfies both $k_\perp \rho_s < 1$ and $k_\perp c/\omega_{pe} < 1$.

For cold electrons, we can expand the Z function in the large argument limit, yielding

$$\left(\frac{\omega}{k_{\|} v_A}\right)^2 = \frac{1 + k_\perp^2 \rho_i^2}{1 + k_\perp^2 c^2/\omega_{pe}^2} \tag{3.80}$$

which reduces to the cold plasma limit of equation (3.74) when the ion gyroradius is ignored. Landau damping is unimportant in this limit since the Landau resonance is on the tail of the electron distribution. The kinetic model also gives small corrections to the field ratios as given by (3.76) and (3.77). Plots of these ratios are given by Lysak (1998).

Figure 3.12 shows the profile of the electron density which is inversely proportional to the electron inertial length mapped to the ionosphere. This parameter maximizes in the 3000-6000 km altitude range. By the arguments in the preceding section, this is the region where the electron inertial length is the most important, and therefore, where the parallel electric field would be most important. It should be noted that this also corresponds to the point where the drift velocity, v_D, of the current-carrying electrons must be the largest, since the current density $j_\parallel = nev_D \propto B_0$, and therefore $v_D \propto B/n$. It seems likely that no matter what the detailed mechanism is for the parallel electric field, these mechanisms will operate more efficiently in this range, which corresponds to the auroral acceleration region. For further details concerning Alfvén waves on auroral field lines, see Chapter 7.

3.5 Wave-Particle Interactions

Wave-particle interactions play an important role in plasma dynamics. Plasma-wave instabilities are discussed in many elementary texts (e.g. Krall and Trivelpiece, 1973; Gary, 1993; Treumann and Baumjohann, 1997). In the kinetic theory of a magnetized plasma, these instabilities are commonly associated with wave-particle resonances that can occur when the particle satisfies either the Landau resonance condition

$$\omega - k_\parallel v_\parallel = 0 \tag{3.81}$$

or the cyclotron resonance condition

$$\omega - k_\parallel v_\parallel - \frac{l\omega_{ce}}{\gamma_{rel}} = 0 \tag{3.82}$$

In this last expression, ω_{ce} represents the non-relativistic cyclotron frequency, γ_{rel} is the relativistic Lorentz factor, and l is an integer. In many applications, this relativistic factor can be set to unity; however, in the case of radio emission from the auroral plasma this relativistic effect is important.

In general, resonances lead to a very strong interaction between field fluctuations and particles having resonant energies, which in the non-relativistic case is given by $\frac{1}{2}mv_{\parallel,res}^2 = \frac{1}{2}m(\omega - l\omega_{ce})^2/k_\parallel^2$. In the course of this interaction energy and momentum can either be transferred from the particles to the field fluctuations or vice versa. In the first case one speaks of the excitation of fluctuations or 'plasma instability', in the second case of 'wave damping', 'plasma heating' or 'resonant acceleration' depending on which point of view one takes. From the wave point of view, in the second case, the wave loses energy and thus is damped, from the plasma point of view the plasma is heated when wave energy is transferred to the bulk of the plasma while particles are resonantly accelerated when only the small group of resonant particles absorbs the wave energy.

There are important resonant instabilities driven by currents, anisotropic particle distributions, beams, and shear motions. These are ion-cyclotron waves of electrostatic and electromagnetic nature, ion-acoustic waves, electromagnetic whistlers (electron-cyclotron waves), drift-waves, and lower-hybrid waves.

Wave-particle interactions result in the irreversible exchange of energy and momentum between the waves and the resonant particles. In the case of Landau resonance, the interaction affects the parallel momentum of the particles. Thus, this term can be associated with an anomalous resistivity in the plasma, which has been suggested as a mechanism to support parallel electric fields in the auroral plasma. The interaction at the cyclotron resonance leads to an exchange of energy associated with the perpendicular motion of the particles. This interaction can lead to the transverse heating of ions or electrons in the auroral plasma, as well as to the excitation of radio emissions such as auroral kilometric radiation (see Section 4.3.1.1).

3.5.1 Anomalous Resistivity

Among the terms on the right-hand side of the generalized Ohm's law (Eq. 3.14), the resistive term is sometimes taken as the one which breaks the frozen-in condition and allows for the existence of parallel electric fields. Outside the ionosphere classical collisional times are extraordinarily long. Collisions must therefore necessarily be *anomalous* and mediated by the presence of wave fields. The idea is that rather than particles in the plasma scattering off each other via two-body interactions, they scatter off each other collectively by scattering in the collective fields of plasma waves. If the scattering by plasma waves acts to transfer momentum between the electron population and the ion population, then a pseudo resistivity can arise. In such a case, when averaging over times long compared with wave periods, this interaction can be interpreted as friction which reduces the difference in the bulk motion between electrons and ions, which reduces current $ne|\mathbf{v}_i - \mathbf{v}_e| = |\mathbf{j}| = \eta_{\mathrm{a}}^{-1}|\mathbf{E}|$. Anomalous resistivity is defined by a Drude formula

$$\eta_a = \frac{\nu_a}{\epsilon_0 \omega_{pe}^2} \tag{3.83}$$

where ν_a is the anomalous ion-electron collision frequency. It depends on the wave-particle interaction and is a function of the self-excited wave energy spectrum W_k.

To obtain an idea of the relation between the collision frequency and the wave spectrum one may realize that the classical Spitzer-Braginskii collision frequency can be written (e.g., Treumann and Baumjohann, 1997) as

$$\nu_c = \omega_{pe} W_{\mathrm{th}} \ll \omega_{pe}, \qquad W_{\mathrm{th}} \simeq (n\lambda_D^3)^{-1} \tag{3.84}$$

It turns out to be directly proportional to the thermal fluctuation level $W_{\mathrm{th}} \ll 1$ in non-magnetized plasma waves. This level is in general very low and thus inefficient in scattering the electrons in collisionless plasmas. Whenever the fluctuation level is increased in the presence of an instability, ν_a can exceed ν_c by many orders of magnitude. A very general expression for the anomalous collision frequency is thus

$$\nu_a = \omega_{pe} W_{\mathrm{turb}} \tag{3.85}$$

$$W_{\mathrm{turb}} = \frac{\epsilon_0}{2}\frac{\langle \epsilon(\omega_k, \mathbf{k})|\mathbf{E_k}|^2\rangle}{nk_B(T_e + T_i)} \gg W_{\mathrm{th}} \tag{3.86}$$

where $\mathbf{E}_k$ is the turbulent plasma wave electric field, $\epsilon(\omega_k, \mathbf{k})$ is the nonlinear plasma response function, and the angular brackets indicate averaging over the whole spectrum of fluctuations. Usually the response function is calculated from linear theory.

In weak-turbulence theory the unstable modes evolve according to the nonlinear wave equation for the spectral energy density, W_k of the waves,

$$\frac{\partial W_k}{\partial t} = 2\gamma_k(\omega_k, t)W_k + A_k W_k^2 + \ldots \tag{3.87}$$

where the number of relevant terms in the expansion on the right depends on the dispersion of the wave. The coefficient A_k (and all following coefficients) are complicated functions describing the mutual interaction of wave modes of different wave numbers $k, k', \ldots$ and/or the interaction between waves and particles. Assuming W_k and the nonlinear growth rate are known, in equilibrium the anomalous collisional change $m_e n\nu_a \mathbf{v}_D = -en\mathbf{E}$ in plasma momentum should balance the change in wave momentum

$$\Delta\mathbf{p}_{\mathrm{turb}} = \int \frac{\mathrm{d}^3k}{8\pi^3}\frac{\mathbf{k}}{\omega_k}2\gamma(\mathbf{k})W_k \tag{3.88}$$

It then follows immediately that

$$\nu_a = \frac{1}{nm_e v_D^2}\int \frac{\mathrm{d}^3k}{8\pi^3}\frac{\mathbf{k}\cdot\mathbf{v}_D}{\omega_k}2\gamma(\mathbf{k})W_k \tag{3.89}$$

which is the weak-turbulence expression for the anomalous collision frequency. Note that this expression implies knowledge of the turbulent wave spectrum, the growth rate, and also the reaction of the wave spectrum on the current drift velocity which itself is a function of the collisional interaction. For practicality, this formula is useful only in the 'quasi-linear approximation', where the growth rate is taken in linear approximation. In that case the above formula is rewritten:

$$\nu_a \approx \frac{\epsilon_0|\mathbf{E}|^2}{2nm_e v_D}\left[\frac{\partial\omega\epsilon(\omega)}{\partial\omega}\right]_{\max}\left[\frac{k(\omega)\gamma(\omega)}{\omega}\right]_{\max} \tag{3.90}$$

where the index 'max' indicates that all quantities have to be taken at the wave number $k_{\max}(\omega)$ that corresponds to the largest growth rate of the wave.

In the low-β plasma of the auroral zone, the important wave modes for anomalous resistivity are ion-acoustic (IAW) and electrostatic ion-cyclotron (EIC) waves and, when there are sufficiently steep density gradients, lower-hybrid drift waves. Auroral-zone field-aligned current densities are strong enough to excite the ion-acoustic and ion-cyclotron modes. Of these, the ion-cyclotron mode has a lower threshold (Kindel and Kennel, 1971) than the ion-acoustic mode, which needs the electron-ion drift velocity to exceed the ion sound speed, $v_D > c_{ia}$. Hence, the ion-cyclotron wave will be excited before the ion-acoustic wave. Scattering of electrons in the ion-cyclotron wave could cause anomalous resistivities. Theoretical calculations (Dum and Dupree, 1970) indicate that the wave amplitudes saturate and that they provide a very low anomalous collision frequency, only

$$\nu_a^{\mathrm{ic}} \simeq 0.56\,\omega_{ci}(k_{\parallel}v_e/\omega_{ci}) \tag{3.91}$$

Ion-acoustic waves have higher threshold but grow faster than ion-cyclotron waves and may reach higher saturation levels. Thus their contribution to anomalous resistivity might still be substantial. The canonical formula for anomalous collisions caused by ion-acoustic waves has been derived by Sagdeev (1966)

$$\nu_a^{\mathrm{ia}} \simeq (\pi/k\lambda_D)(v_e/v_D)\langle W_{\mathrm{ia}}\rangle \tag{3.92}$$

After estimating the saturation level $\langle W_{\mathrm{ia}}\rangle$ of the ion-acoustic waves, the maximum anomalous collision frequency is found to be $\nu_{a,max} \sim 0.1\omega_{pe}$ which is a substantial fraction of the plasma frequency and yields very high anomalous resistivities. This requires electron-ion drift speeds $v_D \sim 4c_{ia}$. Note that computer simulations show that ion-acoustic waves evolve, for low drift velocities, into ion phase-space holes, which are very important in the dynamics of the auroral plasma and which are discussed at the end of this section.

3.5.2 Transverse Ion Heating

A common phenomenon on auroral field lines is the presence of ions heated transversely to the magnetic field (Chang et al., 1986). These ions frequently appear in a form known as ion conics, wherein the ion distribution takes the form of a cone in velocity space. This conic distribution is consistent with a perpendicular heating of ions followed by the action of the magnetic mirror force accelerating them away from the Earth.

The perpendicular heating of the ions can be treated in terms of the weak turbulence theory described in the previous section, where the ion heating results from the exchange of energy between the ions and the waves. The ions can interact with low-frequency plasma waves via the cyclotron resonance given by Eq. (3.82) or by wave-particle interaction or by stochastic kicks.

Applying the weak turbulence theory to this cyclotron interaction, we can calculate the ion heating rate in analogy with Eq. (3.89):

$$\frac{\partial W_{\perp}}{\partial t} = \frac{1}{n}\int \frac{d^3k}{(2\pi)^3}\sum_l \frac{l\omega_{ci}}{\omega} 2\gamma_l(\mathbf{k})W_k \tag{3.93}$$

In this expression, the quantity γ_l represents the cyclotron damping rate associated with the $l\omega_{ci}$ cyclotron harmonic. It should be noted that the expression for this quantity is weighted by a Bessel function of order l, which emphasizes the lower l values, particularly the fundamental.

A difficulty in applying Eq. (3.93) is that it depends on the details of the wave spectrum and the distribution function of the ions. A simplified expression that can be directly compared with observations was derived by Chang et al. (1986) and Ball and André (1991) in the limit of long perpendicular wavelengths. Using the usual linear cyclotron damping expression in Eq. (3.93), and expanding the Bessel functions for small $k_\perp$ gives a result

$$\frac{dW_\perp}{dt} = \frac{q^2}{2m} S_L(\omega_{ci}) \tag{3.94}$$

where $S_L(\omega_{ci})$ is the spectral energy density at the ion-cyclotron frequency, expressed in $V^2m^{-2}Hz^{-1}$. Since this quantity is easily measured by satellite wave receivers, this expression is useful in estimating the ability of waves to heat ions. This expression represents an upper limit to the ion heating from waves at the cyclotron frequency, since the finite wavelength corrections reduce the heating rate below this value. On the other hand, this expression does not include heating at the harmonics of the gyrofrequency, since these contributions vanish in the long wavelength limit.

It is worth noting that although this result was found from quasi-linear theory assuming an incoherent spectrum of waves, a similar formulation can be made for coherent electric field pulses.

The ion-conic distribution results from the combination of ion heating and the magnetic mirror force. The evolution of an ion in this situation can be described by the coupled equations of motion (e.g., Lysak et al., 1980)

$$\frac{dW_\perp}{dt} = \left(\frac{dW_\perp}{dt}\right)_{WPI} + \frac{v_\parallel W_\perp}{B}\frac{\partial B}{\partial z} \tag{3.95}$$

$$m\frac{dv_\parallel}{dt} = qE_\parallel - \frac{W_\perp}{B}\frac{\partial B}{\partial z} \tag{3.96}$$

where the effect of a parallel electric field has been included. Note that the last terms in each of these equations represent the transfer of perpendicular to parallel velocity by the magnetic mirror force. A number of interesting consequences arise from these equations. First of all, note that if the ion heating region is localized in altitude and if the resulting perpendicular temperature is much greater than the parallel temperature and if there are no parallel electric fields, the ions will all have a pitch angle of $\sin\alpha \approx \sqrt{B/B_0}$, where B is the magnetic field at the observation point and B_0 is the field in the heating region. This allows one to infer the altitude of heating along the auroral field line. Secondly, if an upward parallel electric field is present, the ions will all be accelerated parallel to the field. If the parallel energy gained by the ions is less than the perpendicular energization, an 'elevated ion conic' will be formed. If the parallel energy is much greater than the perpendicular energy, the ions will form a beam distribution.

When the ion heating occurs in a region with downward parallel electric fields, the mirror force and the parallel electric field will be in opposite directions and the heated ions can be trapped in the region below the parallel electric field. In this case, the heated ions will be held in the ion heating region until they reach the point where the mirror force (which is proportional to the perpendicular ion energy) can overcome the parallel electric field, creating a 'pressure cooker' effect (Gorney et al., 1985). Examples of these various types of ion heating will be presented in Section 4.2.2.

Ion acceleration or heating can also be achieved non-resonantly by the interaction of ions with extremely low frequency broadband electric fluctuations. In the auroral magnetosphere such fluctuations have a power spectral density maximum below 1 Hz at altitudes of 1-2 R_E in correlation with the appearance of ion conics of energy reaching several 10 keV (Hultqvist, 1988; Hultqvist et al., 1988; Block and Fälthammar, 1990). They also have very high amplitudes. It has been suggested that such fluctuations pump the magnetic moment of ions nonresonantly (Lundin et al., 1989). By solving the ion equations of motion in such large amplitude fields for stochastic fluctuations, it has been shown that ion energies up to 100 keV can be reached nonresonantly (Hultqvist, 1996).

3.5.3 Pitch Angle Diffusion

Another important effect in the auroral plasma is the pitch-angle scattering of particles by plasma waves. Pitch angle scattering can be caused by the resonant interaction of particles with waves. In the absence of scattering, charged particles conserve their first adiabatic invariant $\mu = mv_\perp^2/2B = mv^2 \sin^2 \alpha/2B$, where α is the pitch angle. When they come into resonance with waves, the adiabatic invariant may be violated. As a result, particles may undergo diffusion in velocity space, i.e. diffusion in pitch-angle space and energy space. In many cases diffusion in energy is negligible while diffusion in pitch angle can be significant. In particular, when the plasma waves doing the scattering are electromagnetic rather than electrostatic, pitch angle diffusion tends to be dominant.

Pitch angle scattering is a stochastic process which changes the particle distribution function $f(v, \alpha, t)$. If the particle orbits undergo many small-angle scatterings, it can be described by a Fokker-Planck equation (Schulz and Lanzerotti, 1974)

$$\frac{\partial f}{\partial t} = \frac{1}{\sin \alpha} \frac{\partial}{\partial \alpha} \left(D_{\alpha\alpha} \sin \alpha \frac{\partial f}{\partial \alpha} \right) \tag{3.97}$$

The pitch angle diffusion coefficient $D_{\alpha\alpha}$ has dimension radian^2/time. One can write

$$D_{\alpha\alpha} = \frac{\langle (\Delta\alpha)^2 \rangle}{\tau_\alpha} \tag{3.98}$$

where τ_α is the characteristic pitch-angle-scattering time. $D_{\alpha\alpha}$ depends on the wave spectrum in resonance with the particles. For circularly polarized electromagnetic waves of amplitude b in resonance, one can take $D_{\alpha\alpha} \approx \langle b^2 \rangle / 4B_0\tau_b$ for a rough estimate, where τ_b is the particle bounce time in the magnetospheric magnetic field. More detailed and more precise considerations are given in Schulz and Lanzerotti (1974) and Lyons and Williams (1984).

Pitch-angle diffusion scatters magnetospheric particles into the atmospheric loss cone, which can produce auroral emissions. Theories of diffuse electron aurora (Villalon and Burke, 1995; Chen and Schultz, 2001), pulsating aurora (Johnstone, 1983; Demekhov and Trakhtengerts, 1994), and proton aurora (Ashour Abdalla and Thorne, 1978; Kozyra et al., 1997) have been built on the concept of pitch-angle scattering by plasma waves in the magnetosphere. However, as discussed in Blanc et al. (1999), the origin of the diffuse precipitation is an unsolved problem. Pitch angle scattering as a source of auroral precipitation is further discussed in Section 8.1.

3.5.4 Phase-Space Holes

The nonlinear wave-particle interactions between particles and waves can give rise to 'phase-space holes', which are large-amplitude solitary potential structures with trapped-particle populations. First indications of the possibility of such strange entities were obtained by Bernstein et al. (1957), who found that potential structures of arbitrarily large amplitude could be constructed in a plasma if trapped-particle populations could be specified. These nonlinear solutions have become known as BGK (Bernstein-Greene-Kruskal) equilibria or BGK modes (Montgomery, 1971; Krall and Trivelpiece, 1973). One such BGK mode of interest in the auroral zone is solitary potential structures whose trapped-particle populations appear as a ring or hole in phase space (Dupree, 1982; Schamel, 1982a,b; Turikov, 1984; Muschietti et al., 1999a,b).

The formation of a phase-space hole occurs in the nonlinear phase of a plasma instability. For instance, an electron-ion drift can give rise to a two-stream instability and as the growing plasma waves obtain sufficient amplitude they can trap substantial amounts of plasma particles. This can result in quasi-stable, long-lived localized potential structures which trap low-energy particles and so generate holes in the particle phase-space distribution. Candidates of such instabilities under auroral conditions are associated with electron-ion drift, electron beams, or electron-acoustic populations (e.g., Gray et al., 1991; Dubouloz et al., 1993; Pottelette et al., 1999; Miyake et al., 1998).

Simulations of ion-acoustic phase-space holes excited by a field-aligned current performed by Berman et al. (1985) found that the generation of ion holes (which are negatively charged solitary structures) lowers the instability threshold below the ordinary ion-acoustic threshold. Thus ion holes evolve in the weak-current state and should be the normal state of a current-carrying plasma. Gray et al. (1991)

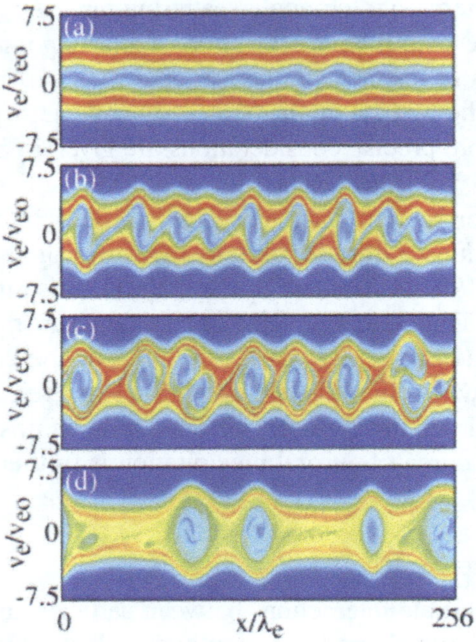

Figure 3.13. A time sequence of snapshots of the electron phase-space distribution ($4\omega_e{}^{-1}$ between frames) from a 1D-Vlasov simulation. From Goldman et al. (1999).

performed similar simulations finding that for stronger currents $v_D > c_{ia}$ the instability evolves into a large number of ion holes.

A series of phase-space plots of the electrons in a computer simulation showing the formation of electron holes in an electron-electron two-stream interaction is shown in Figure 3.13. As can be seen in the series of snapshots, as the structures form, there is a spatial localization and ordering of the electrons until the distribution splits into a population of electrons trapped in the solitary structures, and a population of electrons that freely passes the structures. Each solitary structure is positively charged, with a depletion of electrons in their centers, and trapped electrons oscillating within the positively charged structure.

A description of a time-stationary one-dimensional electron phase-space hole can be obtained analytically from the Bernstein-Green-Kruskal approach (Muschietti et al., 1999b). Defining $w = v^2/2 - \phi$ to be the normalized total energy of an electron (which is a constant of motion), with kinetic energy normalized to the electron thermal energy $k_B T_e$, and potential normalized to $k_B T_e/e$, letting f_t, f_p be the trapped and passing electron distributions, and assuming f_p to be known,

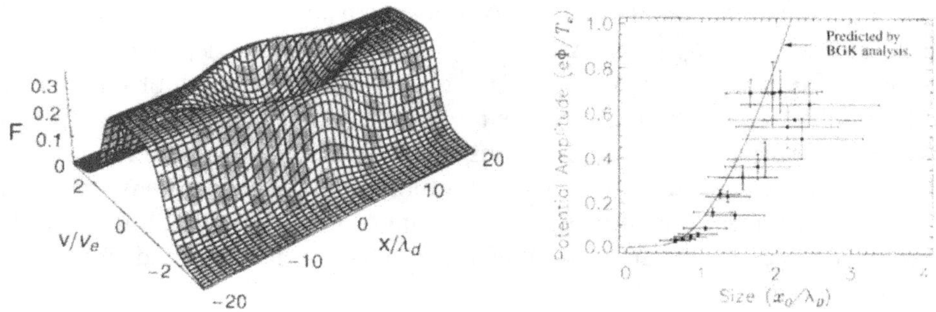

Figure 3.14. Left: distribution function of a model BGK electron hole in phase space (x, v). Since the assumption was that the structure is symmetrical around v and x the distribution is centered at the origin. The flat top becomes core depleted by the imposed balance condition. In the region of large potential the passing electrons are accelerated, as indicated by the bulge in the distribution. Right: width-velocity relation in the model BGK electron holes. The width increases with potential in contrast to the ordinary KdV solitons where it should decrease with increasing potential. The dots and bars show the agreement with measurement in the auroral zone. From Muschietti et al. (1999b,a).

Poisson's equation is written in non-dimensional form

$$\frac{1}{2}\phi'' - \int_0^\infty \mathrm{d}w \frac{f_p(w)}{\sqrt{w+\phi}} + 1 = \int_{-\phi}^0 \mathrm{d}w \frac{f_t(w)}{\sqrt{w+\phi}} \tag{3.99}$$

where the derivative (prime) is with respect to the spacial coordinate, x, and x is normalized to the Debye length λ_D. Using a Laplace transformation technique to invert this equation, one finds for the trapped distribution

$$f_t(w) = \frac{1}{\pi}\int_0^{-w} \mathrm{d}\phi \frac{[\mathrm{d}G(\phi)/\mathrm{d}\phi]}{\sqrt{-w-\phi}} \tag{3.100}$$

where $G(\phi)$ is the right-hand side of Eq. (3.99). Muschietti et al. (1999b) used the following model distribution for the passing component:

$$f_p(w) = 6\sqrt{2}/\pi(8+w^3) \tag{3.101}$$

This is a modified Lorentzian distribution, flat topped and at rest in the frame of the potential structure. The trapped distribution can then be found analytically by performing the above differentiation and integration. Figure 3.14 shows the shape of the electron distribution function in (x, v)-space. As can be seen, the potential in the hole is maintained by the deficit of trapped electrons.

The width of the potential structure is a function of the amplitude of the potential. From the condition that the distribution stays positive at minimum, i.e., at

$w = -\Phi_{min}$, one determines the width Δ. For small (large) potentials $\Phi_{min} \leq 1$ the width scales as $\Delta \propto \Phi_{min}^{1/4}$ and $\Delta \propto \Phi_{min}^{1/2}$, respectively. This dependence is illustrated in the right part of Figure 3.14. One observes that the width of the structure increases in both cases with an increasing potential and increasing velocity. This behavior is confirmed by the observations included in this figure and is in contrast to ordinary Korteweg – de Vries (KdV) solitons (ion-acoustic solitons) where the width should decrease the stronger the potential of the soliton (Nicholson, 1983). One may conclude that in this case BGK electron holes are strongly different from solitons.

Electron holes are not necessarily stable against perturbations. They will have a limited life time. First investigations of their stability can be found in the above cited particle simulation studies papers. Muschietti et al. (2000) recently found that their BGK electron holes in an external magnetic field will disintegrate into smaller structures whenever the electron bounce-to-gyro frequency ratio exceeds one, where the bounce frequency is the frequency of the trapped particle population inside the potential of the electron hole.

Chapter 4

In Situ Measurements in the Auroral Plasma

In this chapter, we turn our attention to observations of the plasma physical processes occurring within the auroral acceleration regions. This includes acceleration processes: electric field formation, particle acceleration; dissipation processes: wave-particle interactions, wave growth, radiation; and energy transfer processes: magnetospheric/ionospheric coupling, Alfvén waves. In Chapter 2, we summarized the visible manifestations of these processes; here, using the theoretical tools provided in Chapter 3, we study the underlying plasma physics.

There is a multitude of ways in which one could arrange such a study: length and frequency scale, acceleration-dissipation-transfer, magnetic local time, to name just a few; and we have chosen an organization which reflects the organization of the processes themselves. Figure 4.1 outlines the electric current structures which delineate the auroral zones. Following this figure, we can organize our discussion by processes within upward current regions, downward current regions, and time-varying current regions. Although much of the auroral zone can be fit within one (or more) of these three categories, there are many features which cannot, so we place these in a separate Specific Regions category. Within these broad groupings, we can classify all the in situ-observed processes under discussion in this chapter. By in situ, we mean that the spacecraft is in or near the region where the process takes place, as opposed to the remote sensing observations of Chapter 2.

Figure 4.2 provides an example of in situ and auroral image data illustrating the regions and processes listed in Figure 4.1. The UV auroral image, from the high-altitude Polar spacecraft, shows the auroral oval at the time of the in situ data shown. The overlaid trajectory shows that the FAST spacecraft is moving poleward across the nightside auroral oval, towards the polar cap. In the in situ data, we can see the accompanying signatures of each of the regions listed in Figure 4.1: upward current regions designated with blue markers in the top panel; downward current regions in green, and Alfvénic regions in red. The polar cap is to the right.

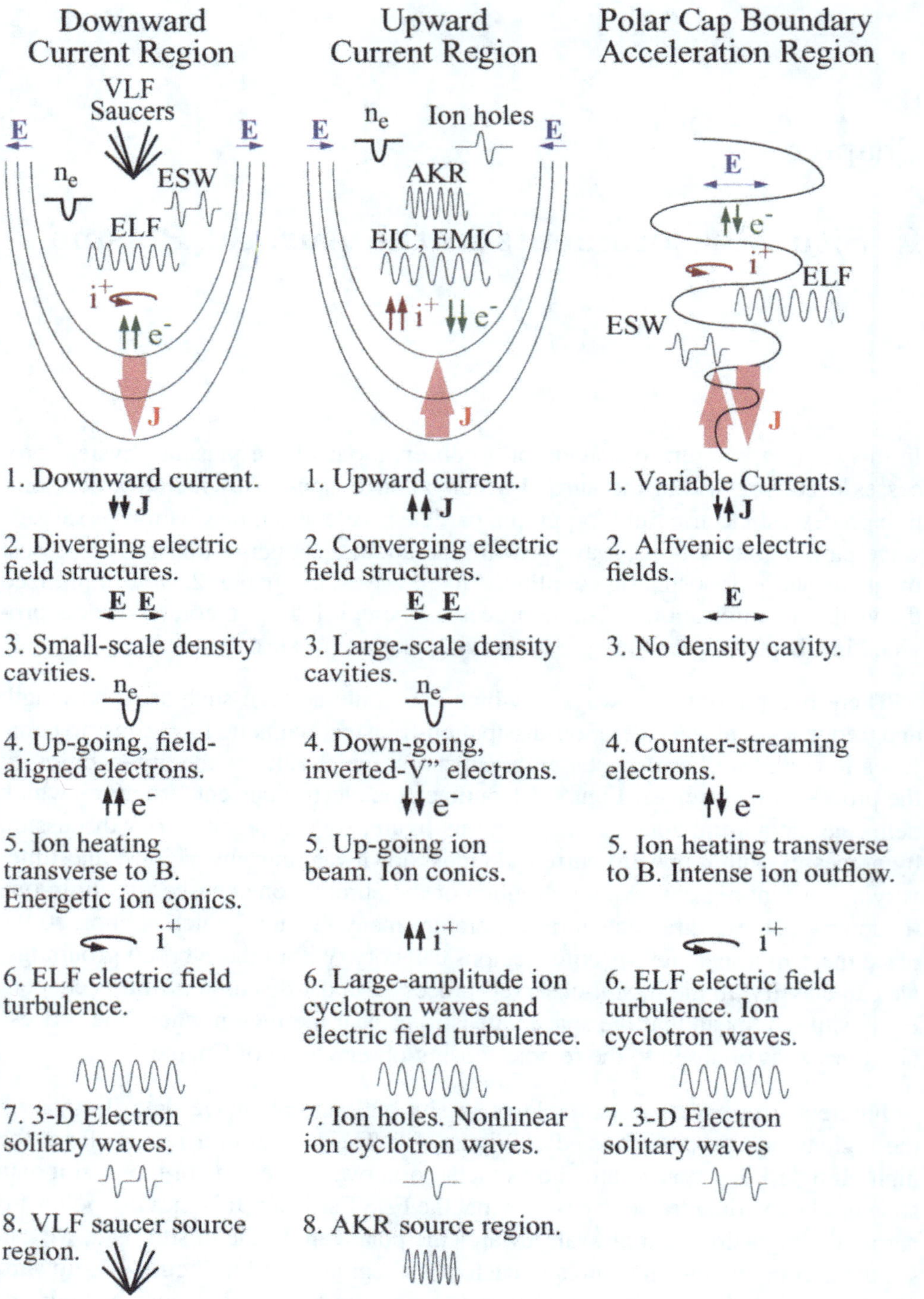

Figure 4.1. An illustration of the current systems in the ionosphere and magnetosphere, providing the motivation for ordering the observational chapter by current sense. The numbered items refer to key characteristics of these regions. Adapted from Carlson et al. (1998b).

Lengths	Phenomena	Processes
100-1000 km	Region 1, 2 current systems	large-scale coupling
10-100 km	inverted-V arc systems	downward electron acceleration
10-30 km	reverse arc systems	upward electron acceleration
0.1-1 km	narrow arcs, filaments, dark aurora	ionospheric-magnetospheric coupling
10 m	ion-gyroradius scales	diffusive processes
1 m	Debye-length scales	non-neutrality, HF-waves
Frequencies		
0-10 Hz	DC	convection, parallel acceleration
0.001-10 Hz	MHD, Alfvén waves	ionospheric-magnetospheric coupling
1-1000 Hz	ELF	ion heating
1 kHz - 1 MHz	VLF	whistler hiss
1-10 MHz	MHz	Langmuir & UH waves
above 300 kHz	AKR	radiation

Table 4.1. Length and frequency scales of various auroral electrodynamic processes and phenomena. DC refers to direct coupling of the E-field sensors to amplifier inputs; that is, they respond down to zero frequency. In contrast, wave measurements are high-pass filtered.

In the upward current region[1], we see precipitating electrons which are narrow in energy and broad in pitch angle; in the downward current region, the (upgoing) electrons are broad in energy[2] and narrow in pitch angle. In the Alfvénic region, the electrons are variable and counterstreaming. Each region has characteristic ion outflow, with differing energy and pitch angle structure and outflow magnitude. The bottom two panels illustrate the huge variety of wave activity associated with the auroral zone. While the example shown here is taken from the nightside, many of these processes occur, albeit in differing combinations and strengths, throughout the auroral oval.

Table 4.1 summarizes, in order of scale size and frequency, many of the processes to be discussed. Table 4.2 lists the various satellites which have provided auroral data over the last half century.

In the following sections, will work our way through the quasistatic processes found in upward current (Section 4.1) and downward current regions (Section 4.2),

[1]The sense of the field-aligned currents is inferred from the cross-track magnetic field perturbation, dB, in the following way: assuming an extended current sheet in the east-west (or y) direction, and a crossing in the northward (x) direction, Ampère's law reduces to $j_z \propto \frac{\partial B_y}{\partial x}$. Thus a positive (negative) slope of dB indicates a current parallel (antiparallel) to $\mathbf{B}$.

[2]In this and many figures to follow, spectrograms are color-coded by differential directional energy flux density, eV(cm^{2}s sr eV)$^{-1}$, loosely referred to as energy flux.

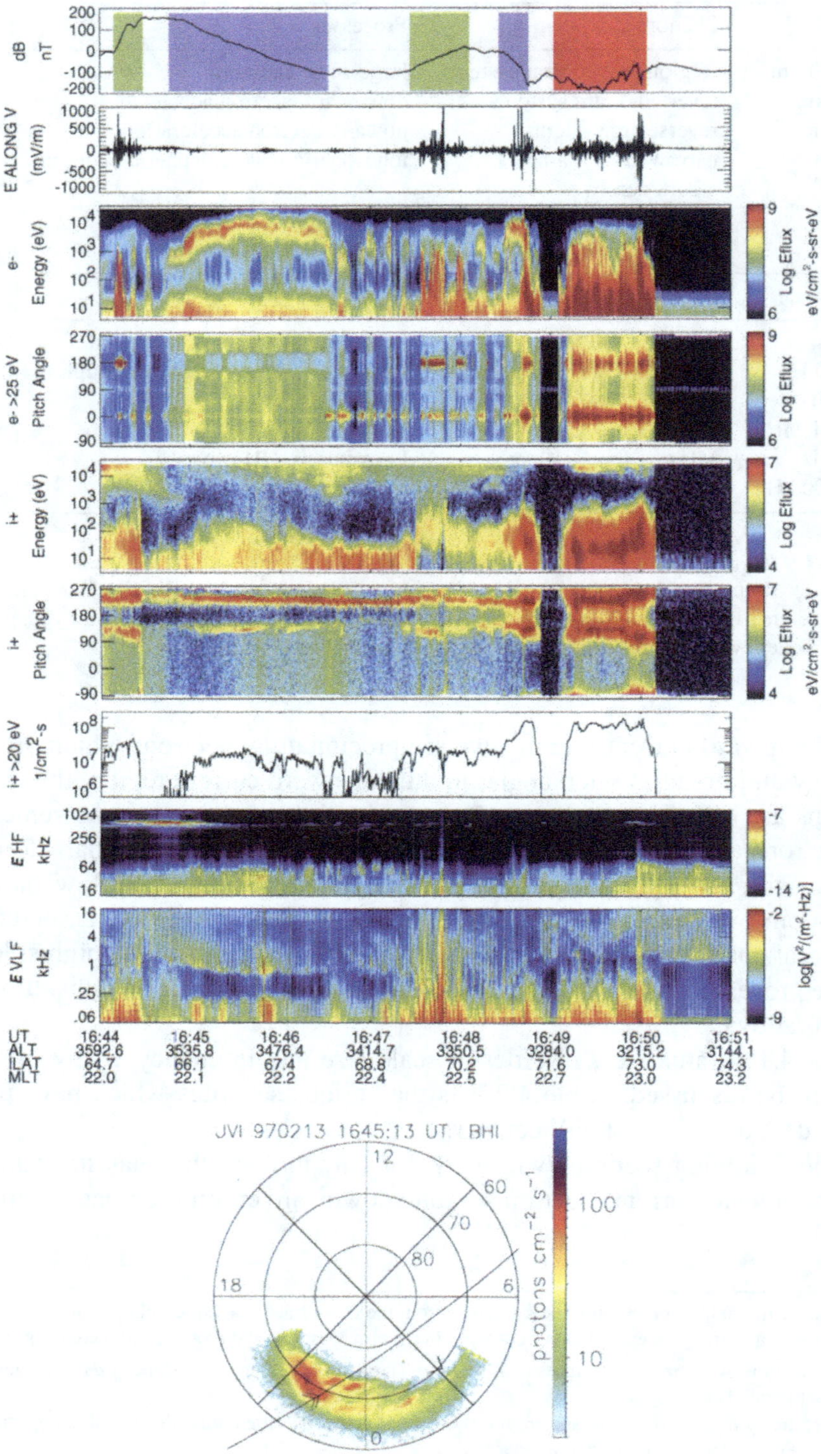
dB nT
E ALONG V (mV/m)
e- Energy (eV)
Log Eflux eV/cm²-s-sr-eV
e- >25 eV Pitch Angle
Log Eflux
i+ Energy (eV)
Log Eflux
i+ Pitch Angle
Log Eflux eV/cm²-s-sr-eV
i+ >20 eV 1/cm²-s
E HF kHz
E VLF kHz
log[V²/(m²-Hz)]
UT 16:44 16:45 16:46 16:47 16:48 16:49 16:50 16:51
ALT 3592.6 3535.8 3476.4 3414.7 3350.5 3284.0 3215.2 3144.1
ILAT 64.7 66.1 67.4 68.8 70.2 71.6 73.0 74.3
MLT 22.0 22.1 22.2 22.4 22.5 22.7 23.0 23.2
JVI 970213 1645:13 UT BHI
photons cm⁻² s⁻¹

Figure 4.2. Opposite page: An auroral pass as seen by the FAST and Polar spacecraft. The Polar UVI image (bottom) shows the auroral oval at the time of the FAST in situ measurements; the FAST trajectory footpoint is overlaid on the image, showing the northward passage across the nightside oval. The top panel shows the magnetic field perturbation relative to the IGRF magnetic field model, with the inferred field-aligned currents indicated in green (downward) and blue (upward), and the Alfvénic currents in red. The DC electric field fluctuations in the second panel show the electrostatic shock structures associated with the auroral acceleration region. The next four panels show ion and electron spectrograms versus energy and pitch-angle. The third panel from the bottom shows integrated ion outflow. The bottom two panels show wave activity from near-DC to MHz frequencies. Figure provided by C. W. Carlson.

Spacecraft	Launched	Orbit [km , km, °]	Primary auroral science
Alouette 1	1962	996, 1031, 81	topside ionospheric study
DMSP	1965	830, 830, 99	particles, fields and imaging
Injun 5	1968	665, 2525, 81	precipitation, VLF
ESRO-1	1968	300, 1550, 90	auroral physics
ISIS-2	1971	1358, 1428, 88	first imager
S3-3	1976	246, 7856, 98	auroral electric fields and particles
DE-1	1981	568, 23289, 90	coordinated with DE-2, global aurora
DE-2	1981	309, 1012, 90	low altitude aurora
Arcad 3	1981	380, 1920, 83	low altitude auroral processes
Viking	1986	822, 14000, 99	auroral acceleration, global aurora, waves
Akebono	1989	300, 8000, 75	high altitude auroral physics
Freja	1992	596, 1764, 63	low altitude, high resolution auroral physics
Fast	1996	348, 4159, 83	auroral physics at very high resolution
Polar	1996	5000, 50551, 86	global aurora
Interball-Auroral Probe	1996	774, 19202, 63	coordinated with Interball Tail Probe
Astrid-2	1998	978, 1013, 83	low altitude auroral physics microsatellite
Cluster	2000	17200, 120500, 90	four-spacecraft constellation

Table 4.2. Satellites that have provided important auroral in situ measurements, with orbits (perigee and apogee altitude; inclination) and prime scientific goals. Note that DMSP refers to a series of spacecraft.

and then tackle the wave-particle interactions within those regions (Section 4.3). We then treat Alfvénic phenomena (Section 4.4), prevalent near the polar cap boundary but present in all regions of the aurora. In Section 4.5, finally, we will deal with observations pertaining to specific auroral regions, including those connected to distinct magnetospheric boundaries.

4.1 Upward Current Region Processes

The upward current regions are characterized by the precipitation of the current-carrying energetic electrons that produce the bright, discrete visible aurora. In the example overview pass shown in Figure 4.2, upward current regions (marked in blue) occur from 16:44:30 UT to 16:46:30 UT and near 16:49 UT. Compare these latitudinal bands of precipitating accelerated electrons with the matching auroral emissions shown in the accompanying UVI image. The precipitating electron structures are referred to as 'arcs', with some distinctions between the in situ observer's use of the word, and the ground-based observer's use of the word.

In the analysis of ground-based data, 'auroral arc' has come to denote a localized brightening in visible light, normally with a relatively large east-west extent and much narrower north-south width. Discrete arcs may or may not be embedded in regions of more diffuse auroral light. In analysis of in situ data, 'arcs' refer to regions of electron precipitation with signatures of acceleration by parallel electric fields. In contrast to such arcs, 'diffuse aurora' are regions with no acceleration features where the electrons merely precipitate. Disagreements between these definitions generally arise when a broad region of electron precipitation contains localized peaks in the electron energy flux indicating electron acceleration. From the ground the localized energy peaks appear as discrete auroral arcs in a diffuse background, whereas in situ measurements show a single acceleration structure of much broader north-south extent than the visible arcs. In the discussion below, we will generally refer to an auroral arc as the large-scale acceleration structure and not distinguish between various peaks in precipitation that would be interpreted as separate arcs from the ground.

4.1.1 Electron Acceleration

The defining, directly detectable, in situ feature encountered in an auroral arc is an enhanced flux of electrons with several-keV energy. Since under non-arc conditions such electrons are not available locally in the auroral zone, they signify an acceleration process at work, at altitudes between about 2000 km and 10,000 km as it turns out. The current carried predominantly by these downward accelerated electrons defines the 'upward current region' under discussion.

4.1.1.1 The Typical Auroral Arc

Consider now the properties of an auroral arc as it manifests itself in the spacecraft observations in the upward current region. Figure 4.3 is an example of part of an auroral satellite pass, focusing on an upward-current arc event.

The topmost panel shows the DC magnetic field signature indicative of the spacecraft's northward passage through a sheet-like region of upward-directed electric current. The presence of the current sheet – the longitudinally extended, latitudinally restricted region of electron precipitation which forms the auroral arc –

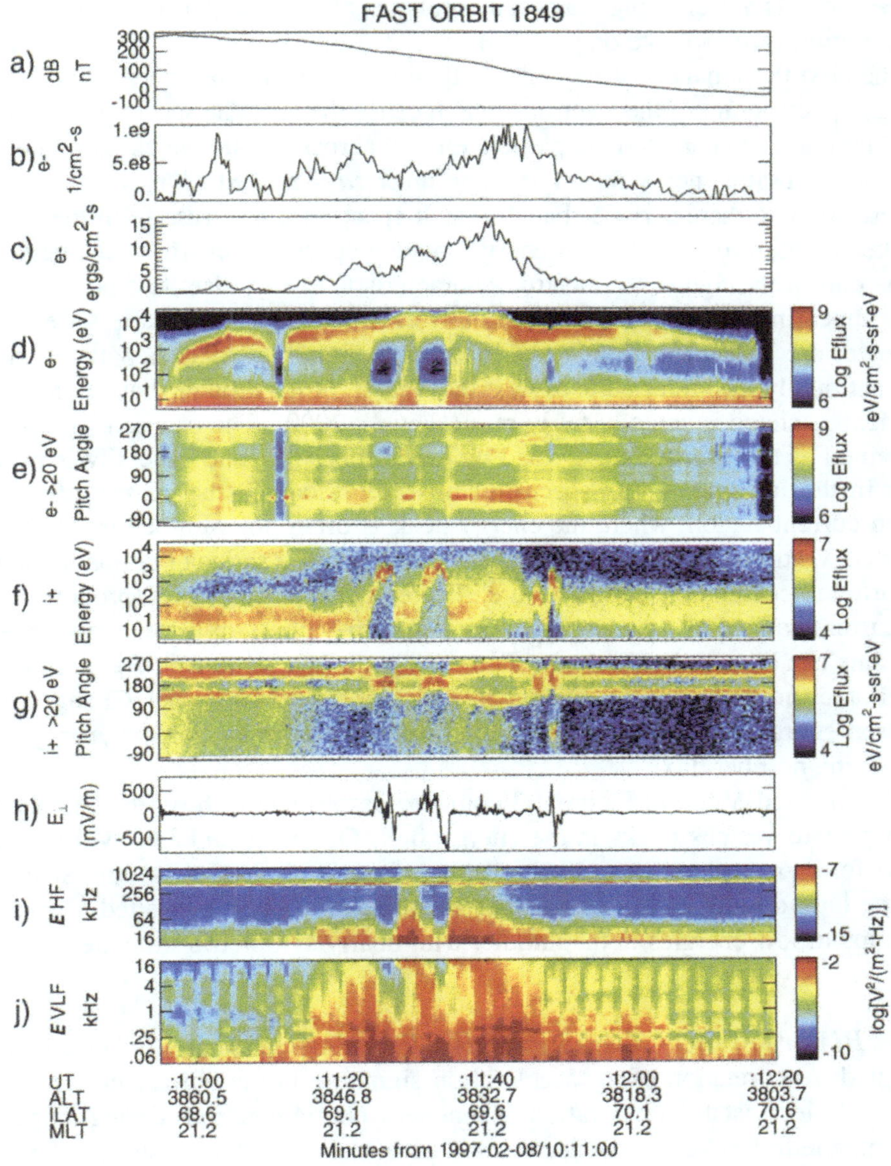

Figure 4.3. An 80-second excerpt from a spacecraft auroral pass concentrating on an upward current inverted-V region. From top to bottom, the panels show: line plots of a) the east-west magnetic deflection, b) the electron number flux, and c) the electron energy flux; energy flux spectrograms versus energy and pitch angle, first for electrons d) and e), then for ions f) and g); a line plot of the perpendicular electric field h), and finally the high and low frequency waves i) and j). From McFadden et al. (1999a).

is seen here as a decreasing magnitude of the eastward component of the magnetic perturbation (see footnote on page 95).

The next three panels of Figure 4.3 show the number flux, energy flux[3], and an energy spectrogram of the electrons which cause the arc. The spectrogram shows a clear increase in characteristic particle energy across the arc which correlates best with precipitating energy flux. The name *inverted-V* arc, which is commonly used to describe the characteristic shape of such spectrograms, arose from the roughly Λ-like signature of the electron energy in such spectrograms: from the beginning to the end of the figure, the electron characteristic energy rises to a peak and then falls, appearing as an upside-down, usually called 'inverted V', shape. Often, much narrower arc structures, from $1 - 10\,\mathrm{km}$, have similar particle signatures and are conventionally also referred to as inverted-V arcs. Note, though, that a purely symmetrical Λ shape is actually fairly rare (Newell, 2000). The *characteristic energy* is defined as the energy flux divided by the number flux and generally is near the peak in the inverted-V flux spectrum; this definition will be more useful in the return current region, where the energy peak is often not well-defined. Often the inverted-V structures have multiple peaks, indicating a series of dramatic changes in particle acceleration within an arc. The electron number flux (proportional to the current carried by the electrons), is often nearly constant across the inverted-V arc. This can be seen by the steady decrease in the magnetic field B_{E-W} in the top panel, and the relatively flat number flux plot in the second panel. Changes in the electron energy flux are usually due to changes in the characteristic energy rather than in the number flux.

The inverted-V arc in Figure 4.3 could be associated with several visible arcs, belonging to various peaks in the energy flux. Often a visible arc will contain a broad fold. Associated in situ observations from a northward-moving spacecraft cutting latitudinally across such a fold show a broad double-inverted-V structure that dips to low energies in the middle, with nearly symmetric structures on either side.

Short History of Field-aligned Potential Drop Interpretation. The commonly accepted interpretation of the acceleration signature in the upward current region is that of electrostatic acceleration of magnetospheric electrons through a parallel (i.e., magnetic-field-aligned) electrostatic potential drop. The existence of parallel electric fields ($E_{||}$) in the auroral acceleration region was first suggested by the monoenergetic appearance of electron spectra measured on a sounding rocket (McIlwain, 1960). The case was strengthened by the interpretation of sounding rocket observations of low-energy electrons as secondaries and backscatter produced by auroral primary electrons and reflected by the parallel potential above the sounding rocket (Evans, 1974). This initial particle data evidence for $E_{||}$ was completed

[3]In this and many figures to follow, energy flux is given in erg cm^{-2}s^{-1}, which is the same as mW m^{-2}.

by the first observation of upgoing ion beams on the S3-3 satellite (Shelley et al., 1976) and upward acceleration of an artificial barium ion cloud (Haerendel et al., 1976).

The existence of an $E_{||}$ had also been inferred indirectly from optical observations of discrete arcs. From the apparent vorticity in such arcs, Hallinan and Davis (1970) had concluded that at high altitudes, where these motions are driven, there must be convergent perpendicular electric fields ($E_{\perp}$) with peak values that (when referenced to the 100 km level) were in excess of 1000 mV m^{-1}. As such large fields are not observed in the lower ionosphere, the perpendicular fields apparently did not map down to low altitudes, meaning that there had to be an $E_{||}$ in between.

The first direct measurements of very large (500–1000 mV m^{-1}) perpendicular electric fields in the auroral acceleration region were made on the S3-3 satellite (Mozer et al., 1977). Again, parallel electric fields were required to explain why these large perpendicular fields did not map to the lower ionosphere. The S3-3 satellite also made direct measurements of parallel electric fields of hundreds of mV m^{-1} (Mozer and Torbert, 1980). The DE 1 and 2 satellites demonstrated that high-altitude $E_{\perp}$ does not map down: Weimer et al. (1985) compared the perpendicular fields at two altitudes, 4500 and 900 km, measured nearly simultaneously by the DE 1 and 2 satellites, and found that the fields were larger and more structured at the higher altitude.

Clearly, despite the collisionless character of the upper auroral ionosphere, there is overwhelming observational evidence that parallel electric fields exist in auroral arc systems. We now consider in detail the observational features of the inverted-V arc.

Electron Spectra. Typical electron energy spectra observed in inverted-V arcs have been termed *monoenergetic* because of the relatively narrow energy peak in differential energy flux, which is proportional to count rate in many auroral particle detectors. Figure 4.4 (top panels) shows three examples of inverted-V arc electron energy-pitch angle plots. The examples are taken from the time period shown in Figure 4.3, with the first example from below the potential drop region, the second from within the potential drop region, and the third showing counterstreaming field-aligned wave-accelerated electrons as well. The examples show prominent monoenergetic flux peaks at 2, 8, and 15 keV (middle panels.) These peaks are an artifact of the plotting units, however, since the distribution functions of most auroral populations are (at least marginally) stable, and decrease monotonically (or plateau) with energy. Count rate, or differential energy flux, is proportional to $W_e^2 f(v)$, where $W_e = m_e v^2/2$ is the (nonrelativistic) electron energy and $f(v)$ is the distribution function. When plotted in units of phase space density[4] (lower

[4] As we specify densities in cm^{-3} and velocities in km s^{-1}, phase space densities in this and other plots are given in unusual units, s^{3}cm^{-3}km^{-3}. The standard unit would be s^{3}cm^{-6} or s^{3}m^{-6}.

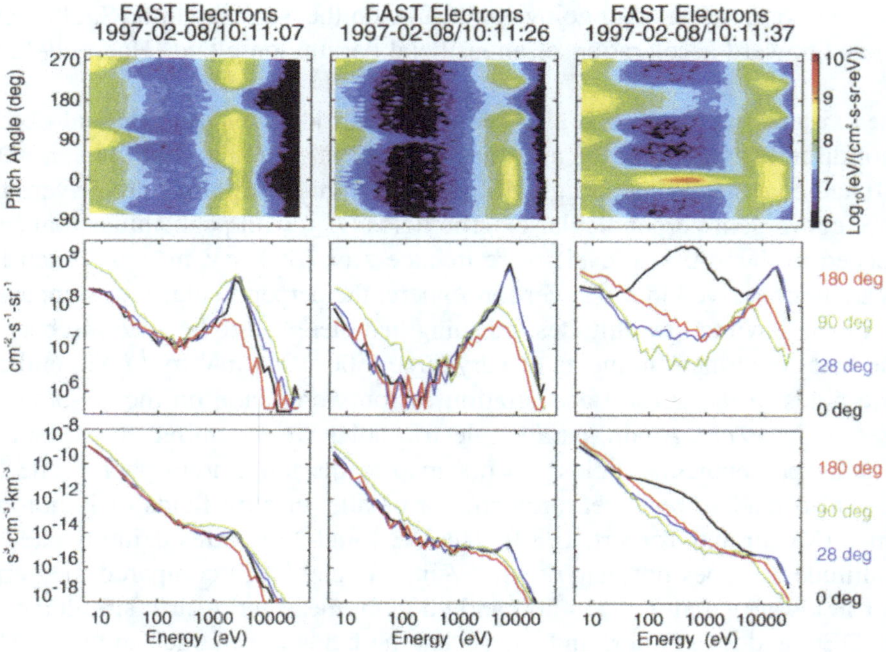

Figure 4.4. Three example electron spectra from the time period shown in Figure 4.3, each plotted as (top) energy, pitch-angle contour plot of differential energy flux, (middle) lineplots at several pitch angles of differential energy flux, and (bottom) lineplots of the distribution function for several pitch angles. Figure provided by J. P. McFadden.

panels of Figure 4.4), the peaks apparent in the flux spectra are typically reduced to a plateau or at most a small enhancement in $f(v)$; positive slopes in $f(v)$ can indicate an electron distribution that is unstable to wave growth.

For observations below the potential drop, this plateauing of the unstable slope in the accelerated distribution function is attributed to wave-particle interactions and the presence of lower energy populations to be discussed below. In comparison, for observations *within* the potential drop (middle column example), notice the relative suppression of these lower energy counts. The ionospheric plasma, backscattered electrons, and atmospheric secondaries that will be discussed below are excluded from this region. The low energy flux which remains (below 100 eV in the middle example), consists of solar UV photoelectrons emitted from the spacecraft surface, plus a small contribution from secondary electrons generated when the primary auroral electron beam population hits the spacecraft.

The part of the electron energy spectrum above the count rate peak provides valuable information about the temperature of the source population. This part of the spectrum can be modeled by a Maxwellian distribution that has passed through a fixed field-aligned voltage drop (Kaufmann et ál., 1978). Since the field-aligned part of the Maxwellian is thus only multiplied by a constant, $\exp(-e\Phi/k_B T_s)$, the source temperature T_s can be extracted from the slope of the field-aligned distribution, i.e.,

$$k_B T_s = -\left[\frac{\partial}{\partial W_e} \ln f(v_{\|}, v_{\perp}{=}0)\right]^{-1}$$

Occasionally, two or more separate source populations can be identified above the spectral peak, with the colder population often forming a field-aligned beam at the peak (because the change in parallel velocity is large compared to the thermal velocity.) The temperature and density of the source populations vary with auroral activity, latitude and local time. Accelerating voltages are observed to scale with the source temperature (Burch et al., 1976), as expected from the Knight relation to which we turn in Section 4.1.1.2 and the largest accelerating potentials are observed on the nightside where the source populations are hotter.

For observations made below the acceleration region, at energies below the peak count rate energy, a prominent low energy power law component extends down to the lowest energies measured. An example of this low energy population can be seen in the first example of Figure 4.4. At the lowest energies, it runs into the spacecraft-produced photoelectron population dominant at 90°. Note that when the observation is made at an altitude within the potential drop region (the auroral cavity), this low energy population is suppressed, as in the second example of Figure 4.4; we will return to this important distinction later. This power law feature at first appears to pose a problem for simple acceleration of plasma through a voltage drop, since all of the accelerated plasma should have energies greater than the fall through the potential drop (Reasoner and Burke, 1973). However, magnetospheric primary electrons accelerated through a parallel electric field towards the lower ionosphere produce an upgoing population of atmospheric secondary and backscattered primary electrons (Evans, 1974). These electrons then reflect back down from the parallel electric field above, and become part of the downgoing population, in turn producing additional upgoing electrons. Known electron secondary electron production rates and scattering rates can be used to calculate the resulting equilibrium spectra below the potential drop that results from a fixed input population. The resulting electron spectrum, which consists of an accelerated Maxwellian distribution together with a power law secondary distribution, is an excellent model of the spectra observed below inverted-V potential drops. Differences between measured and modeled spectra can be explained by a variety of waves that erode away unstable features on the electron distribution (see the VLF hiss discussion of Section 4.3.1), or that accelerate the secondary population as in the third example of Figure 4.4. This quantitative understanding of the formation of observed electron spectra was critical in establishing the validity of the potential-drop model of auroral arcs.

Electron Pitch Angle Distributions. The fifth panel of the example upward current pass (Figure 4.3) shows the pitch angle distribution[5] and of the electron populations, and Figure 4.4 shows detailed energy-pitch angle distributions for three representative times. Note that in the northern hemisphere pitch angles of 0° correspond to downward motion. The inverted-V distributions are generally relatively isotropic in pitch angle, with the exception of the presence of an upgoing loss cone. Hot magnetospheric electrons form a wide precipitating beam beneath the potential drop. This beam is stabilized in energy by generation of VLF emissions, and by the secondary and backscatter populations described above. It also is modified adiabatically in pitch angle by mirror force effects which cause the electrons to change their pitch angle in order to conserve their magnetic moment as they propagate into the increasing magnetic field at low altitudes. Particles precipitating along the field lines are lost to the atmosphere by collisions with neutral atoms before the mirror force is able to reflect them. This missing population in the upgoing hemisphere of phase space (surrounding 180° pitch angle in the northern hemisphere, and 0° in the southern) is the result of the atmospheric absorption of the downgoing electrons that are within the loss cone (see Section 3.3.1).

The width of the electron loss cone is changed measurably by the existence of the parallel potential drop $\Delta\Phi$. This modification provides a probe of the potential drop magnitude *below* the observation point, whereas the potential drop magnitude above the observation point can be inferred from the energy of the precipitating beam. The derivation of the potential below the spacecraft from the width of the electron loss cone is done with an empirical relationship, which compares the expected loss cone (with no potential drop at a given altitude) to the measured actual loss cone width. The potential drop is then given by (e.g., McFadden et al., 1999a)

$$\Delta\Phi_{LC} = \frac{W_e}{e}\left[\cos^2\theta_{LC} + \left(1 - \frac{B_{\rm mir}}{B_{\rm sc}}\right)\sin^2\theta_{LC}\right] \tag{4.1}$$

where $\Delta\Phi_{LC}$ is the potential drop, W_e is the electron energy, $B_{\rm mir}$ is the magnetic field at the mirror altitude (100 km), $B_{\rm sc}$ is the magnetic field at the spacecraft, and θ_{LC} is the loss cone angle at W_e. The value of the potential drop $\Delta\Phi_{LC}$ inferred from the loss cone width can be compared to the energy of ion beams (discussed below) rising upward through the potential drop.

Electron Bursts. A variant of the classical inverted-V population consists of *suprathermal bursts* of energetic electrons (often referred to as STEB - see Section 7.6). These are temporally highly variable electron precipitation events (Johnstone and Winningham, 1982). These bursts represent a class of narrow precipi-

[5] Pitch angles are ordinarily defined to lie between 0° and 180°. But if measurements are made in a plane containing the magnetic field direction, it is convenient to describe the particles' velocity direction as an angle ranging from 0° to 360°. This convention is used in many of the plots in this book.

tation structures that contain an enhanced field-aligned flux, but do not exhibit a pronounced inverted-V 'monoenergetic' energy peak. While some of the precipitation events originally identified as suprathermal bursts may have been narrow inverted-V structures that were unresolved by the early satellites, there are electron bursts that are indeed different from the inverted-V precipitation discussed in the present subsection. They can be found also in the downward current region and can represent the energy-broadened signature of field-aligned electron beams that are produced by downward parallel electric fields (see Section 4.2), or populations produced by Alfvén wave acceleration mechanisms (see Section 4.4).

4.1.1.2 Knight Relation and $E_{||}$

The visible auroral arcs lie almost entirely on magnetic field lines that carry *upward* currents. Indeed, as we can see by comparing the auroral image in Figure 4.2 to the overview in Chapter 1, the visible auroral oval is the manifestation of the noon-to-midnight Region 1 current system, and the midnight-to-noon Region 2 current system. These field-aligned currents are part of the global current circuit, as we saw in Chapter 1. They are driven by magnetospheric plasma convection electric fields which map (partially) to the lower ionosphere, where electron-neutral collisions causing finite Hall and Pedersen conductivities allow for current closure perpendicular to **B**, as we saw in Chapter 3. The field-aligned currents in the resulting ionosphere-magnetosphere current circuit are responsible for the momentum transfer between the magnetospheric and ionospheric flows.

The observed relationship between the accelerating voltage and the temperature of the source population provides important insight into the physics underlying the formation of an auroral arc. This current-voltage relationship, sometimes termed the 'Knight relation', was discussed in Section 3.3.1. We restate the underlying physics that the formation of field aligned electric potential drops is required since the current density which can be carried by the downgoing electrons in the upward current region is restricted, both by a dearth of magnetospheric plasma and by the converging magnetic field geometry. When large field-aligned currents are demanded by the mapping to the conductive ionosphere of magnetospheric perpendicular electric fields and/or perpendicular plasma pressure gradients, parallel electric fields form on the upward current field lines. These parallel electric fields both enhance the allowable field-aligned current (by enlarging the loss cone), and decouple a portion of the magnetospheric potential drop from the lower ionosphere, thereby reducing the ionospheric response. The parallel electric field normally required to maintain an imposed current from a given magnetospheric particle distribution is approximately proportional to the imposed current requirement as demonstrated by several studies (Antonova and Tverskoy, 1975a; Bosqued et al., 1986). Since the parallel field forms to meet a given imposed current requirement, and since the available current is a function of the parameters of the available magnetospheric source of particles, a relation between the source temperature and the magnitude of the field-aligned potential drop should be expected. It is important to note that the Knight relationship gives no information about the altitude profile of the potential drop; it only relates the total drop to the source population temperature. We will see in Section 7.1 that the altitude profile (how $\Delta\Phi$ is distributed along the field line) is a function of the details of the various particle populations found on a magnetic flux tube.

4.1.2 Ion Beams and the Density Cavity

The precipitating accelerated electron signature tells us about the potential drop on the field line above the observation point. Conversely, upgoing beams of accelerated ions can tell us about the potential drop on the field line below the observation point. As with the analysis of auroral electron beams and magnetospheric source populations, analysis of these ion beams provides information about the upper ionospheric source population.

4.1.2.1 Properties of Accelerated Ion Beams

In Figure 4.3, two ion beams are seen as functions of energy (panel 6) and pitch angle (panel 7), between UT 10:11:20 and UT 10:11:40. The energy spectra show that the ion beams are nearly monoenergetic, while the pitch angle spectra show that they propagate upward, highly field-aligned at near 180° pitch angle. Where the ion beams are seen, a fraction of the inverted-V potential drop (equivalent to the ion beam energy) must lie along the field line below the observation point.

Energy Spectra. Figure 4.5 shows typical ion data from within the upward current region. The data are shown in the same format as in Figure 4.4. The central example here shows an ion beam taken from the time interval of Figure 4.3, at 10:11:27 UT. The ion differential energy flux is sharply peaked in energy and pitch angle, as are the monoenergetic electron peaks within and below the potential drop region. As in the case of the electrons, the sharp peak maybe partially due to plotting the differential energy flux which is proportional to $\sim W_i^2 f(v)$.

When the intensities are plotted in terms of phase space density (bottom center panel of Figure 4.5), it can be seen that these ions still exhibit a beam like distribution. However, this is likely a result of the beam composition where the beam in Figure 4.5. is almost entirely protons. When O^+ is present, the H^+ are often plateaued in distribution function down to energies as much as a factor of 16 below the peak (McFadden et al., 1999a). This plateau is formed by the accelerated protons being slowed to the velocity of the equally accelerated oxygen ions. Figure 4.6 shows, for a different example, an observation of ion beams separated by species. The ions at and below the spectral peak are primarily H^+, and those above the spectral peak are predominantly O^+. The proton velocity of the low energy cut-off is roughly equal to the velocity of the O^+ beam. Observed differences in ion beam energy with ion mass are not obviously consistent with the potential drop

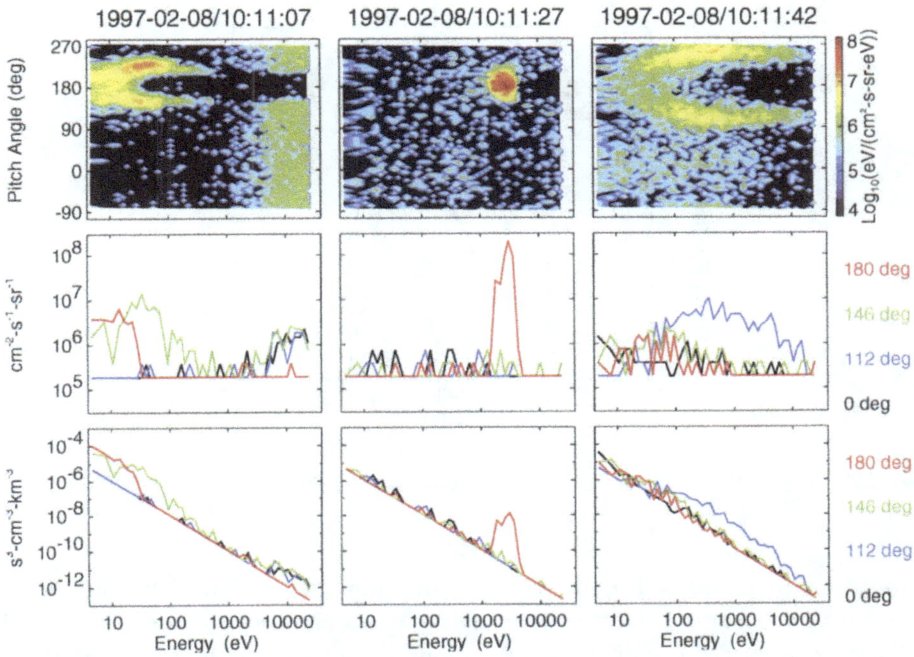

Figure 4.5. FAST observations showing spectra from an ion conic, an ion beam, and another ion conic, plotted as (top) energy, pitch-angle contour plot of differential energy flux, (middle) lineplots at several pitch angles of differential energy flux, and (bottom) lineplots of distribution functions for several pitch angles. The examples are taken from the time period shown in Figure 4.3.

model, which should produce identical energies for the different beam species. The mass-dependent energy differences are attributed to the action of an instability, most probably the two-stream instability, which leads to momentum exchange between H^+ and the heavier ions, producing a plateau on the H^+ distribution and an energetic tail in the O^+ and He^+ distributions.

Pitch Angle Distributions. The top central panel of Figure 4.5 illustrates the distribution function, versus energy and pitch angle, of an ion beam within the acceleration region. The ion beam is confined to a narrow pitch angle range along B that may vary in pitch angle at larger energies. When O^+ is present, the lower energy H^+ portion of the beam is generally narrower in pitch angle than the O^+ beam. This may indicate that ionospheric O^+ is typically hotter than the H^+ entering the bottom of the potential drop. However, if the bottom of the acceleration region contains a strong oblique double layer, finite ion gyroradius effects may provide some perpendicular energization of O^+.

Magnetospheric ions often can be colocated with ion beams, forming a broad plateau at $\sim$10 keV and possessing an atmospheric loss cone. In this example, this feature is more prevalent in the first (ion conic) example.

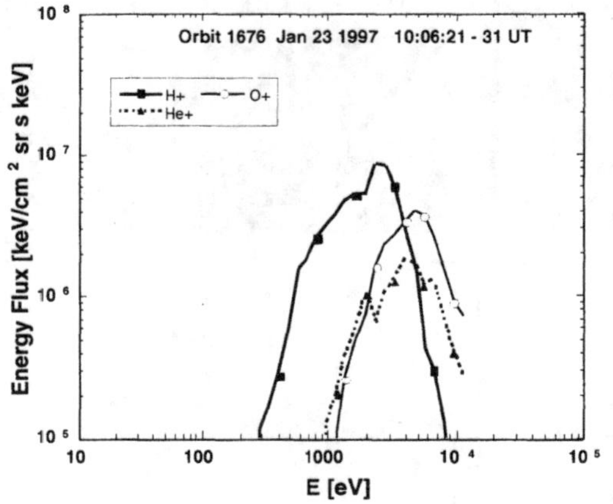

Figure 4.6. Observations of ion beams separated by species. From Möbius et al. (1998).

Temporal Variations. Rapid (< 0.1 s) modulations of ion beam flux and energy are often observed within the inverted-V acceleration region. These turbulent structures are seen with strong variations in the electric field, including electrostatic shocks, intense electrostatic ion cyclotron (EIC) waves, and ion solitary waves (ISW), to be discussed later in this chapter. During times of the most intense wave turbulence the ion beams are observed to broaden in pitch angle from a nominal $\sim 10°$ to as much as $\sim 45°$ This broadening is caused by wave-particle interaction effects, but the nature of the turbulence and the associated energy and momentum exchanges are still a subject of study.

Ion Conics as the Source for Ion Beams. Figure 4.5 also shows two examples of the ion distributions found below, rather than in, the $\Delta\Phi$ region. These distributions, with $T_\perp$ larger than $T_\parallel$, are termed *ion conics* because of their distinctive contour shape in phase space; they were discussed in Section 3.5.2. They are ions which have been heated transversely (to **B**), typically by electromagnetic ion-cyclotron (EMIC) waves. The mirror force gives them upward motion, producing a significant injection of ions into the regions of parallel electric fields where they become beams. We will consider conics in greater detail in the downward current region, where they play a more significant role.

4.1.2.2 The Auroral Density Cavity

The ion beams are a clear observational marker that the spacecraft has moved within a region of upward parallel electric field. Within the $\Delta\Phi$ region, the strong acceleration of the cold ionospheric plasma, changing the relatively low energy ionospheric ion population into a streaming hot beam, causes a drastic change in the plasma environment along the field line. There is a huge density drop compared to the lower altitude part of the flux tube, since the ionospheric ions are accelerated to high velocities and the cold ionospheric electrons are prevented (by the upward-pointing electric field) from moving into the region. Note the second example electron spectrum of Figure 4.4. The portion of the flux tube containing the upward parallel field, the ion beams, and magnetospheric electrons and ions, is known as the *auroral density cavity*.

Cavities contain primarily only hot plasma (Persoon et al., 1988) — upgoing ion beams and precipitating magnetospheric electrons. The technical problems of measuring low energy electrons and ions makes direct observation of the thermal electron and ion distribution functions very difficult, so the comparison of measured hot-electron densities and total plasma densities is often based on wave cutoffs (Ergun et al., 1998a; Strangeway et al., 1998). Within the density cavities, these comparisons show that thermal electrons and ions constitute only very minor components of the cavity plasma. Quasi-neutrality is a strong requirement, so measuring accurately only one species is adequate.

Figure 4.7 shows an example of measured hot electron and ion densities across a cavity, where the measured ion to hot electron density ratio is ~1 (cf. the bottom panel). These data are consistent with the absence of any cold plasma, but observational limitations can only put an upper limit of less than 20% on the possible cold plasma density. Strong exclusion of cold plasma is additional support for the conclusion that the acceleration region contains strong field-aligned electric potential drops which expel the low energy plasma from the cavity. These observations can be used in parameterizing wave dispersion relations for wave particle interactions within the auroral density cavity.

4.1.3 Electric Field Signature and Structure

From the evidence presented above it is clear that magnetic-field-aligned electric fields ($\mathbf{E}_{\parallel}$) are the primary means by which auroral particles are accelerated, typically up to keV energies, during their short transit time through the potential drop. It is interesting to consider how such parallel fields can be created and supported in a collisionless plasma. The requirement for, and theoretical explanation of, these parallel electric fields in the convergent magnetic field geometry of the auroral zone have been discussed in Chapter 3. The two basic elements of the theory are (1) a balance between the magnetic mirror force and the electrostatic field force, and (2) an externally demanded field-aligned current requirement. Simple treatments of this kind do not determine the altitude distribution of the field-aligned potential drop, only its magnitude.

4.1.3.1 Electrostatic Shocks

The first observations of quasistatic electric field structures associated with auroral particle acceleration were made by the S3-3 satellite at altitudes between 3000-8000 km (Mozer et al., 1977). The term *electrostatic shock* was borrowed from

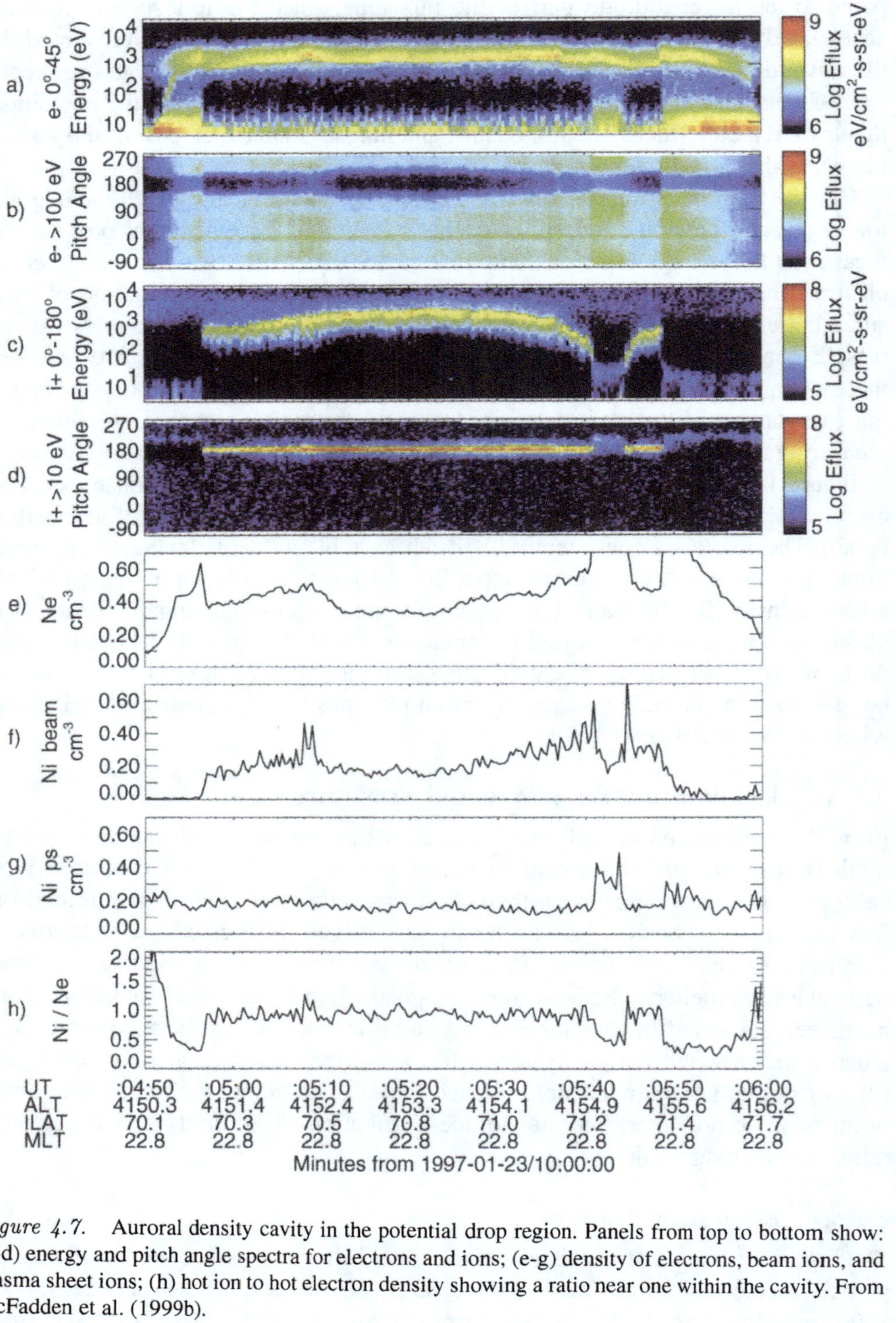

Figure 4.7. Auroral density cavity in the potential drop region. Panels from top to bottom show: (a-d) energy and pitch angle spectra for electrons and ions; (e-g) density of electrons, beam ions, and plasma sheet ions; (h) hot ion to hot electron density showing a ratio near one within the cavity. From McFadden et al. (1999b).

theoretical work to describe the observed electric field structures, which have large amplitude ($> 100\,\mathrm{mV\,m^{-1}}$) are primarily perpendicular to **B**, and are several kilometers thick. The term 'electrostatic' is appropriate since they can have negligible magnetic signatures. However, the term 'shock' is a historical misnomer; in a collisionless magnetized plasma, perpendicular electric fields are not shorted out, and these perpendicular structures are not truly shocks.

Importantly, since correspondingly large electric fields are not observed in the lower ionosphere, these observed high-altitude potential surfaces must close at intermediate altitudes, thereby producing electric fields parallel to **B**. This gives the same type of electrostatic potential configuration which bounds regions of different plasma density and temperature and connects shear flows in the magnetosphere. Perpendicular electrostatic shocks and their associated parallel fields are found in both the upward and downward auroral current regions. Figure 4.1 is based upon just such a scenario. The electric fields bordering a region of upward current are *convergent*, in that they point inward toward the low-density cavity containing the upgoing ion beams and precipitating electrons. Clear examples of these convergent shocks are seen in Figure 4.3, in panel (h), each time the spacecraft enters and exits the ion beam region. *Divergent* electric fields are found in the downward current region, primarily on the borders of the up-going electron beams (see Section 4.2). Note that in either case the potential structures are enforced from above; on open field lines, therefore, they do not necessarily close electrostatically above the potential drop region. Figure 4.8 illustrates a higher altitude version of these observations, showing pairs of perpendicular electrostatic shocks seen on the Polar spacecraft. Note that each pair of shocks is separated, surrounding a region of low field. The auroral density cavity is bordered by the electrostatic shock region, but the potential drop is not distributed uniformly across the cavity. Instead, it forms a concentrated layer at the outer edges. Note also that the shock at 04:14:07 is oblique with a significant parallel component.

Shear Flows. Another point to note here is that this configuration of $\mathbf{E}_\perp$ sets up the shear flow patterns so characteristic of auroral motions. For instance, in the northern hemisphere where **B** is downward, a northward moving spacecraft crossing an inverted-V structure (convergent electric fields) will see first a northward $\mathbf{E}_\perp$ and then a southward one. The $\mathbf{E} \times \mathbf{B}$ motion consistent with these fields is first westward and then eastward. While the electric and magnetic field structures may be quasistatic in the Earth's reference frame, the plasma itself has strong flows.

4.1.3.2 Direct Measurements of $E_\parallel$

Direct identification of the parallel electric fields within the acceleration region has proven difficult. The relatively large perpendicular electric field components in the acceleration region make unambiguous identification of small parallel electric fields nearly impossible. Small differences in antenna response, relatively poor res-

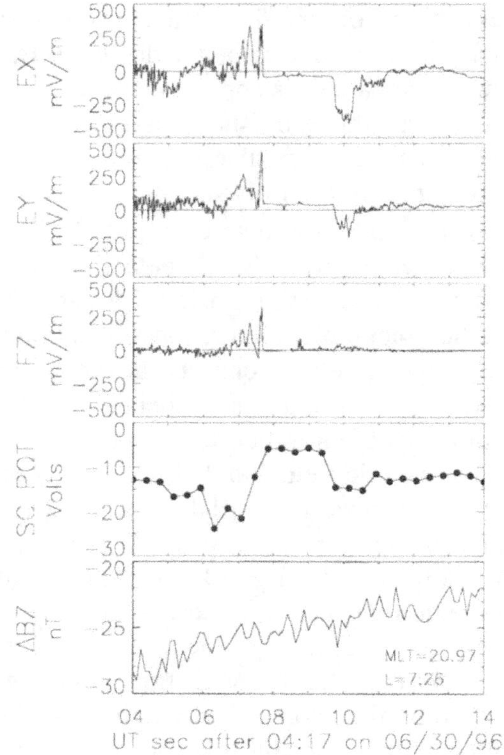

Figure 4.8. Electrostatic shocks as measured by the Polar spacecraft. The electric field components are in a magnetic field aligned coordinate system, with z along **B**. From Mozer and Kletzing (1998).

olution of spacecraft spin axis electric field component measurements, and spacecraft shadowing of antennae all complicate the measurement (e.g., Mozer and Kletzing, 1998).

Observations by the Polar spacecraft provide the most definitive examples yet of direct parallel electric field measurements. The example in Figure 4.8 includes both perpendicular electrostatic shocks and electric field structures with significant parallel electric fields. These observations suggest that parallel electric fields are not simply weak and extended in altitude, but take the form of strong oblique *double layers* near the bottom of the acceleration region. These double layers (cf., Chapter 3) have been described with a variety of names, but all the descriptions are variations of *Debye sheaths*, forming from a balance between electron thermal pressure and the parallel electric field. These double layers typically contain 10 to 50 % percent of the potential drop within the lower few Debye lengths of the parallel electric field region, with the rest of the potential drop located at higher altitudes. This potential distribition can be inferred from Figure 4.7, where ion beams are seen to abruptly turn on to a substantial fraction of the electron beam energy. The remainder of the parallel potential drop may be spread smoothly throughout the density cavity or confined to smaller structures (Mozer and Hull, 2001), but there is no direct evidence, either for this structuring or for a smooth altitude profile.

4.1.3.3 Nonuniform Altitude Profile of the Potential Drop

While direct measurement of the parallel electric field is difficult, much information about the parallel potential profile – and therefore, the parallel field – can be derived from particle data. The precipitating electron data show the magnitude of the potential drop above the observation point. The ion beam data and the electron loss cone data show its magnitude below the observation point. Using all these data in combination, we begin to visualize the altitude structure of the field-aligned potential drop. This is not a simple linear function of altitude. It rather forms sheath structures as discussed above.

If the concept illustrated in Figure 4.1 is indeed correct, we can compare the perpendicular potential drop that the spacecraft sees as it moves across the top of the potential structure, with the ion beam and electron loss cone measurements of the parallel potential drop below the spacecraft. If the electrostatic shocks are static on the transit time of a spacecraft, and if the lower ionospheric convection fields are small, we can measure the potential along the spacecraft trajectory by integrating the electric field along the path of the spacecraft. This perpendicular potential jump should be in agreement with the parallel potential jump inferred from the particle data.

Panel e of Figure 4.9 shows just such a comparison, with reasonably good agreement of the potential estimates. The fifth panel compares the two particle signature measurements of the potential drop below the spacecraft: the ion beam energy, and the electron loss cone broadening. The bottom panel compares the ion beam energy to the potential along the spacecraft trajectory as measured by the integrated $\mathbf{E}_{\perp}$. Comparisons of this kind offer some of the strongest evidence for the potential drop model of auroral acceleration and demonstrate that these structures are generally static on ion (and spacecraft) transit times. Errors in the determination of the loss cone angle and assumptions about the lower ionospheric convection electric fields can undermine this comparison. However, an important feature to note is the concentration of the perpendicular potential drop (the perpendicular electric field) to small regions at the edges of the density cavity. This is a transverse-to-**B** manifestation of the nonuniformity with altitude of the field-aligned potential drop.

The concentration of the potential drop at the lower boundary is evident in panels a and b of Figure 4.9, which show electron energy and angle distributions as the FAST spacecraft skims along the bottom of an inverted-V acceleration region. As the spacecraft moves into the acceleration region, corresponding decreases in electron energy and increases in the atmospheric loss cone (black line, panel b) can be seen, together with (panels c and d) ion beams forming near the bottom of

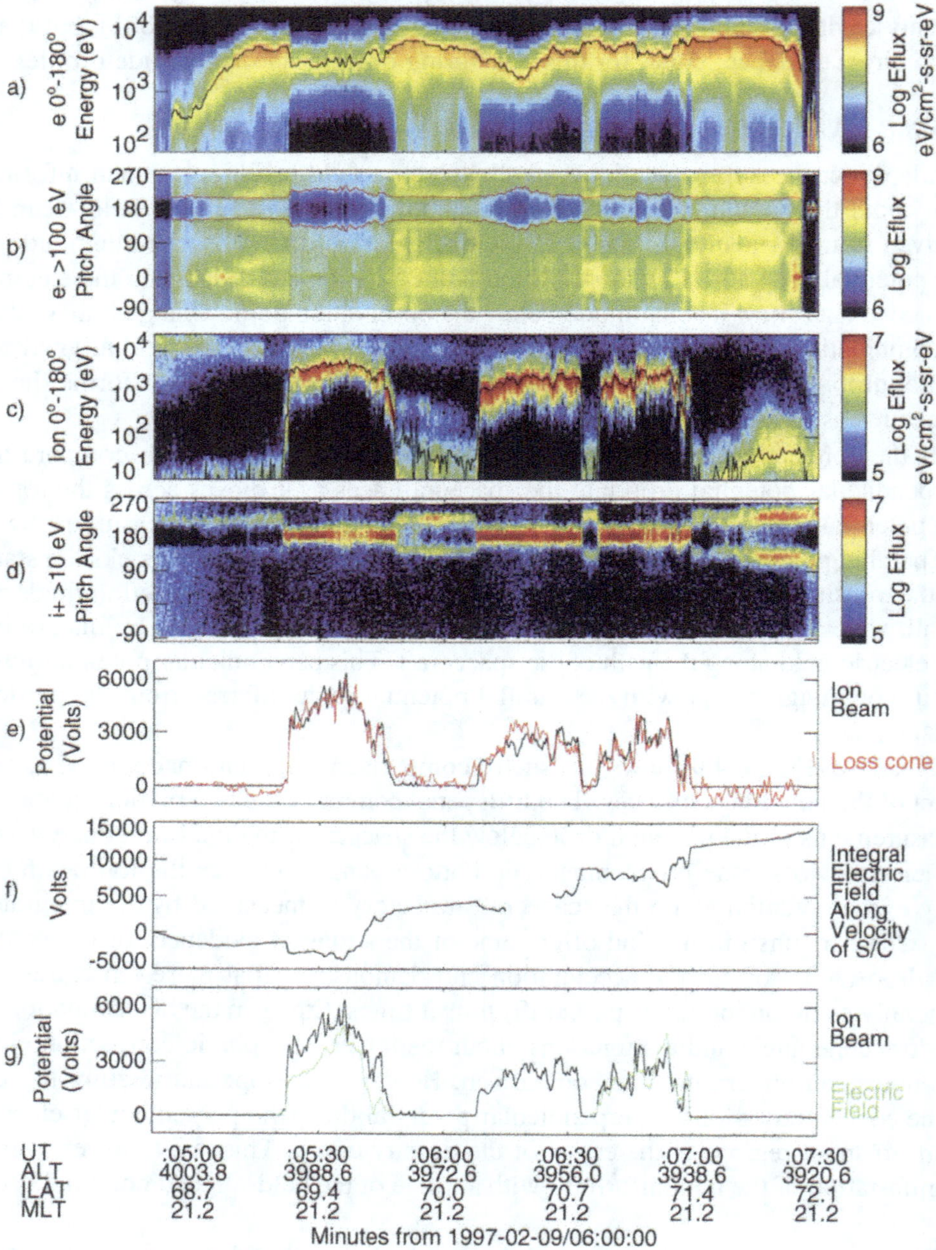

Figure 4.9. From top to bottom: a) electron energy spectrogram with characteristic energy overtrace; b) electron pitch angle spectrogram with loss cone angle overtrace; c) ion energy spectrogram and characteristic energy; d) ion pitch angle spectrogram; e) inferred potential from ion beam and electron loss cone; f) integrated $E_\perp$; g) comparison of the integration of $E_\perp$ with upgoing ion beam energies. From McFadden et al. (1999a).

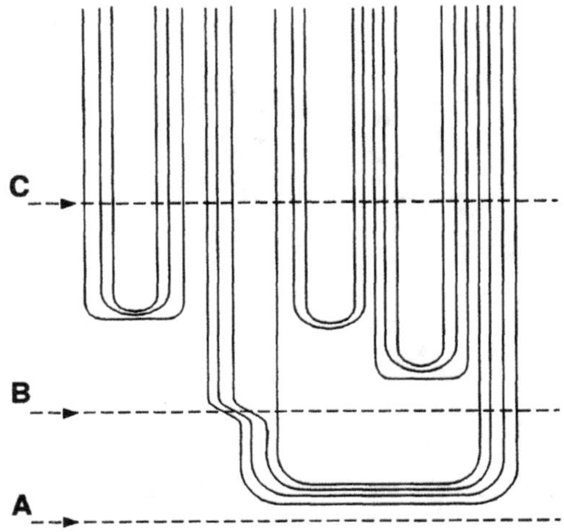

Figure 4.10. Illustration of sheath structure of inverted-V potential drop altitude profile. The paths marked A, B, and C illustrate the passage of a spacecraft through the region at different altitudes. From Mozer and Hull (2001).

the acceleration region. The sudden switching on and off of these beams shows that the lower border of the acceleration region is both *thin* and *corrugated*, allowing the spacecraft to pass above and below the thin acceleration region. When the spacecraft is below the acceleration region, the ion beams disappear and heated ion conic distributions are observed. The conics flow up from the ionosphere, forming beams as they pass through the sheath potential at the lower border of the acceleration region. The corrugation and sudden changes at this lower edge show again, as do the perpendicular electrostatic shocks at higher altitudes, that the potential drop is concentrated at the outer edge of the structure.

4.1.3.4 Quasineutrality, Secondaries and Sheaths

Figure 4.10 illustrates a refinement of the simple schematic shown in Figure 4.1. The acceleration region consists of a low-density cavity bordered by strong quasi-static electric fields along its sides (the electrostatic shocks), and along the lower edge (the sheath). The electric fields, both parallel and perpendicular, point into the cavity. Typically a significant fraction (~10-50%) of the parallel potential drop resides in the sheath at the lower border, with the rest of the parallel potential drop spread out over a 1-2 R_E altitude range above this border (the illustration exaggerates the situation.)

Within this framework we can understand a number of the observational features discussed so far. The strong sheath at the lower edge maintains charge neutrality by

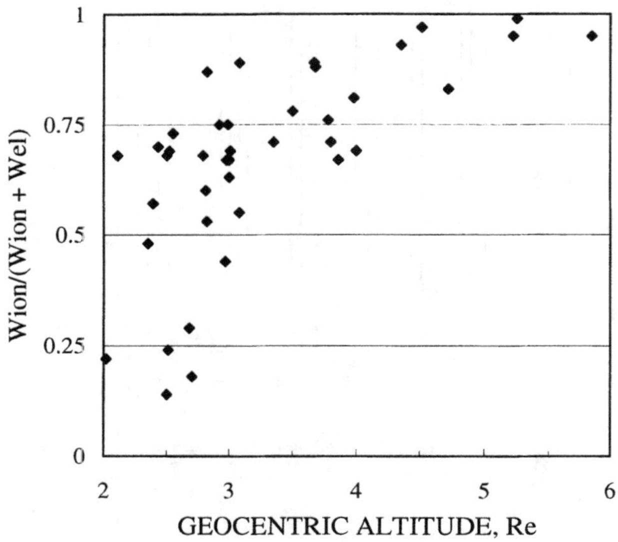

Figure 4.11. Percentage of potential drop below the observing spaceraft, in terms of the ratio between upgoing ion energy and the sum of upgoing ion and downgoing electron energies. From Mozer and Hull (2001).

keeping low energy ionospheric electrons out of the acceleration region, illustrating what we have seen in the electron distributions (Figure 4.4) and in the neutral balance between the hot populations (Figure 4.7.) The path marked 'B' through the irregular lower edge illustrates the rapid entry and exit of the FAST spacecraft in and out of the acceleration region (Figure 4.9.) The concentration of $\mathbf{E}_\perp$ near the edges of the cavity illustrates the strong electrostatic shocks shown in Figure 4.8 at 04:17:07-08 and 04:17:10 UT.

We end our discussion here with an observational conclusion: Figure 4.11 illustrates the statistical summary of the altitude profile of the potential drop, as measured by many orbits of the Polar spacecraft, covering the altitude range from 1000 km to 4000 km. These data are consistent with three mechanisms for generating $\mathbf{E}_{||}$ in the presence of an imposed $\mathbf{j}_{||}$ (Mozer and Hull, 2001): (1) a high altitude field required to accelerate magnetospheric electrons to meet the imposed current requirement, (2) a low-altitude sheath field required to maintain charge neutrality by accelerating ions upward and preventing the entry of ionospheric backscatter electrons, and (3) a mid-altitude field required to maintain overall local charge neutrality.

Many theoretical predictions have been made of strong localized parallel electric fields in the form of double layers (Block, 1972) or oblique electrostatic shocks (Swift, 1975) in the context of auroral electron acceleration. These structures can be modeled as time-stationary solutions to the Vlasov-Poisson equations, supported by field-aligned electric currents. Recent numerical simulations suggest that, under proper auroral conditions, possibly a second double layer may form near the top of the acceleration region (Ergun et al., 2000). These theoretical and computer simulation efforts will be discussed in detail in Section 7.1, showing that the parallel electric fields required for maintaining imposed current densities, naturally collapse into narrow regions of strong double layers parallel to the magnetic field. These Debye-scale double layers form at the boundaries between plasmas of differing properties, providing a transition that maintains both current continuity and quasi-neutrality.

4.1.4 Limitations of the Quasistatic Arc Model

The quasistatic picture presented so far in this chapter provides a framework for understanding many of the observational features of a static auroral arc. Of course, most discrete arcs are anything but static, but may appear so on electron transit time scales (~ 1 s). The basic physics describing the electron acceleration that creats visible aurora is well explained by this quasistatic picture, with a few notable exceptions. However, much of the appeal of auroral physics lies in the complications and limitations of this simple picture.

The first exception is acceleration by kinetic Alfvén waves, discussed in detail in Section 4.4 and Section 3.4. This most dramatic form of variable, transient electron acceleration is observed near the poleward boundary of the nightside auroral oval. This precipitation produces the brightest auroral forms during substorms, and is driven by Alfvénic Poynting flux input from the magnetosphere. The electron precipitation generated by this wave acceleration mechanism is quite different from that produced in inverted-V arcs. The electrodynamics of this process are an important part of the auroral system. Another exception is wave acceleration of low energy electrons below (and within) the potential drop region. We shall discuss this 'flickering' aurora in Section 4.3. Wave activity at ion cyclotron frequencies can accelerate the ionospheric secondary, backscatter, and thermal electrons to energies sufficient to generate visible aurora.

While the U-shaped potential model has been very successful in explaining the low-altitude characteristics of particle observations in the upward current regions above quasi-static auroral arcs, the details of where these potential contours close, either in the conjugate ionosphere or elsewhere in the magnetosphere, remains open. Electrostatic shocks, which are the high altitude signature of U-shaped potential structures, are observed even at > 9 $R_E$on auroral field lines by the Polar spacecraft. However there has not been a definitive studies showing that these potential structures extend to in the opposite ionosphere although conjugate visible auroral are observed. In fact, the very nature of this time varying system implies that potential models are inadequate in describing it. Some results suggest that there is sometimes a mismatch between the inferred potentials determined by integrating the electric fields along spacecraft trajectories at high and low altitudes (Janhunen et al., 1999), or a mismatch between inferred potentials and those deduced from particle distributions (Janhunen et al., 2001), suggesting that O-shaped potential contours can form at altitudes below $\sim 9\ R_E$(Janhunen and Olsson, 2000). However, these conclusions are controversial. The solution may require consideration of time variations present at both high and low altitudes. A more detailed understanding of how the magnetosphere-ionosphere coupling changes in time, including the effect of both low frequency MHD waves and wave-particle interactions on particle distributions, is required for a true understanding of the physics of auroral arcs. We will cover these wave interactions, and their relationship to the more quasi-static picture after studying the other half of the quasi-static aurora, that of the downward current region.

4.2 Downward Current Region Processes

In the example auroral pass shown in Figure 4.2 at the beginning of this chapter, downward current regions (marked in green) occur near 16:44 UT and from 16:47:30 to 16:48 UT. These regions are defined by the sense of the electric current, and are clearly delineated at these altitudes by intense *upgoing* field-aligned accelerated cold ionospheric electrons. Since the upward current region corresponds to the region of bright auroras, this region is sometimes referred to as the primary current while the downward current region is considered the return current. In the example below, we will compare and contrast a downward current region and its accompanying fields and particle observations, to what we have already learned from the study of the upward current region. There are many clear parallels, but also some interesting differences. The upgoing energetic electrons are not rare but appear with a probability similar to that of the upward current region ion beams. There is good agreement between the measured potential of divergent electrostatic shocks and the characteristic energy of the corresponding electrons, just as there was between the convergent shocks and ion beams of the upward current regions. The upgoing electrons' downward currents show excellent quantitative agreement with the measured magnetic field deflections, indicating again that the observable energetic particles carry most of the auroral current, and that these field-aligned currents are well-modeled by sheets with significant longitudinal extent. We will see below, however, that the downward current region has some unique and interesting plasma physics processes, and its equilibrium state is not as well understood as that of the inverted-V region. The wave-particle interactions of the return current region are rather more complex than the more readily separable problems of the inverted-V region. As in the previous section, we begin our discussion with a quasistatic picture of electrostatic fields accelerating electrons. We will see quickly though that in these return current regions, the wave-particle interactions play a far more active role in controlling the equilibrium state.

4.2.1 Upgoing Electrons

4.2.1.1 A Typical Return Current Region Signature

Figure 4.12 shows high time resolution observations from an active northern auroral crossing by the FAST satellite near magnetic midnight. The top panel displays the (mostly) westward component of the perturbation magnetic field (the difference from an IGRF model field.) The negative slope of this trace indicates an upward field-aligned current throughout the first half of the pass, which then reverses, showing that a more intense and structured downward current exists throughout the northern half of the pass. This pattern matches the statistical Region 1 and 2 current systems expected near midnight (cf., Figure 1.9).

Electron data are displayed in the next 5 panels. Panels (b) and (d) are energy-time spectrograms for downgoing and upgoing electrons and panel (c) is the corresponding pitch-angle spectrogram with 0° corresponding to downgoing fluxes. The total upgoing and downgoing particle and energy fluxes are shown in panels (e) and (f).

The upward current region (19:06:00–19:07:10 UT) is filled with a broad region of inverted V electron precipitation, with an energy peak near 1 keV. This example represents a moderately intense auroral arc, with downgoing particle fluxes of 2-3x10^8cm^{-2}s^{-1} and a peak downgoing energy flux of 1 erg cm^{-2}s^{-1}, except for the narrow region of 10 erg cm^{-2}s^{-1} at its northern edge. In comparison, the upgoing electrons in the return current region (located between 19:07:20 UT and the polar cap boundary at 19:09:00 UT) have peak upgoing particle fluxes ranging between 10^9 and 10^{10}cm^{-2}s^{-1}, about 10 times greater than typical inverted V downgoing fluxes, and also have a substantial upward energy flux, peaking at 5 erg cm^{-2}s^{-1}. The upgoing electron energy spectrum is very broad, extending from the lowest energy step at 4 eV up to an abrupt high energy cutoff at a few keV, and has large variations. A weaker and less energetic population of downgoing field-aligned fluxes is also present; both this and the accompanying upgoing field-aligned electron population have similarly narrow pitch angle distributions, with perpendicular temperatures less than 1 eV. The measured density of upgoing electrons varies between 3 and 15 electrons cm^{-3}. No evidence for significant additional cold plasma is found, as we will see below in comparisons with magnetic field deflections.

4.2.1.2 Electron Acceleration and Thermalization

Figure 4.13 illustrates the details of the electron distribution seen during the event shown in Figure 4.12. At this time (19:07:43.9 UT), field-aligned electrons are seen moving both upward and downward. The left panel of Figure 4.13 plots the differential energy flux as a function of energy and pitch angle at this time, showing the extreme field-alignment, and the difference between the upgoing and downgoing populations. This energy-pitch-angle image gives a qualitative impression of

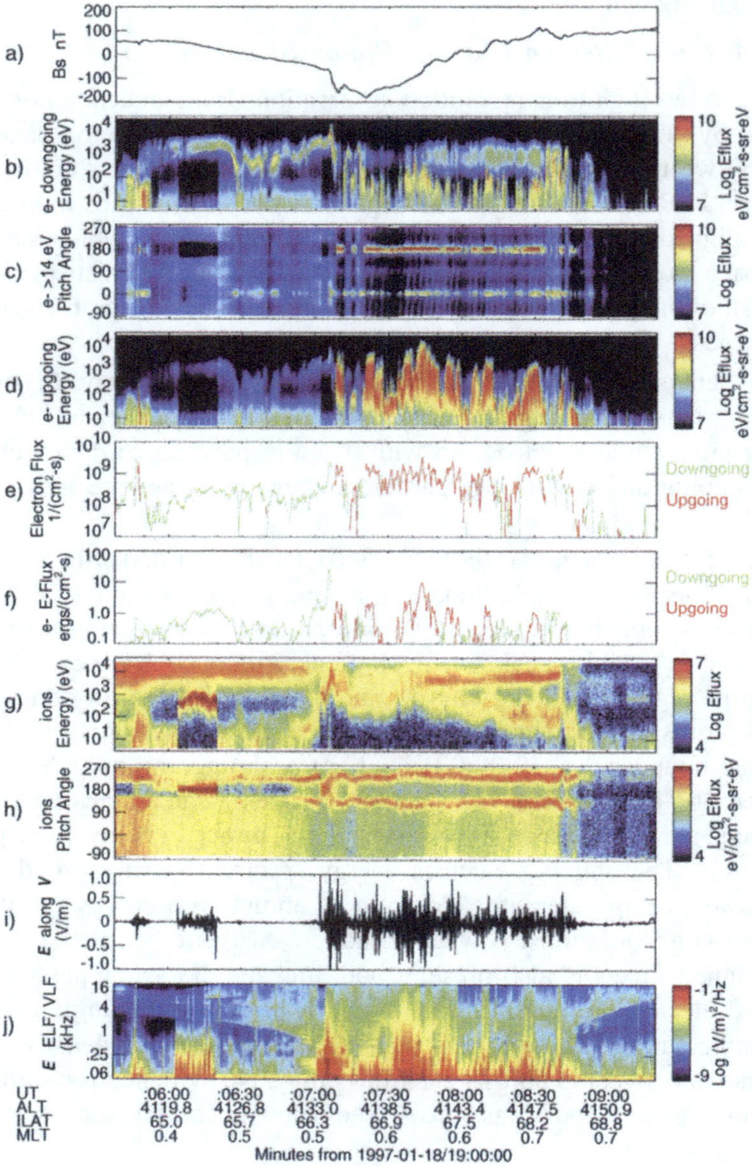

Figure 4.12. FAST observations of an auroral crossing near magnetic midnight, showing an example of a downward field-aligned current region adjacent to an upward current region. From top to bottom: a) The spin axis component (nearly eastward) of the magnetic field. A positive (negative) slope indicates downward (upward) current; b), d) electron energy flux spectrograms versus energy for downgoing and upgoing electrons respectively; c) pitch angle spectrogram for energetic electrons. Fluxes near 180° are upgoing; e), f) total electron number flux and energy flux, color coded for upgoing (green), and downgoing (red); g), h) ion energy flux spectrograms versus energy and pitch angle. The energy spectrogram is averaged over all pitch angles; i) the DC electric field signal perpendicular to B and parallel to the spacecraft velocity; j) power spectral density of low frequency E-field. From Carlson et al. (1998b).

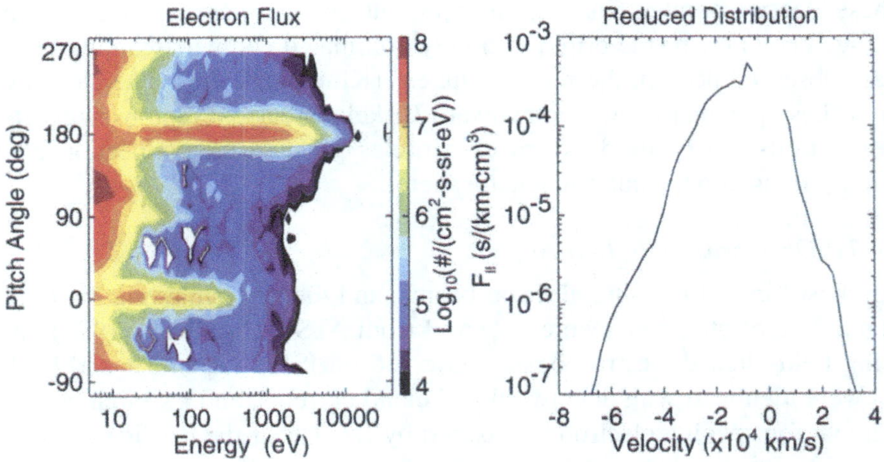

Figure 4.13. Electron energy-pitch angle spectrogram (left panel) and reduced (parallel) distribution function from the event shown in Figure 4.12 at 19:07:43 UT. Adapted from Carlson et al. (1998b).

the narrow pitch angles of the upgoing electrons. More quantitative analysis shows that the perpendicular temperature is no greater than a few eV. The right-hand panel shows the reduced (parallel) field-aligned distribution function $f(v)$, for both upgoing (negative velocity) and downgoing (positive velocity) electrons.

Note that the distribution function plot shows a peak in the upgoing direction (negative velocity) near -1.5x10^4 km s^{-1}, and that the distribution slopes downward on either side of this peak, with an energy width that encompasses the downgoing direction. Above the peak in the upgoing direction, a hot tail extends to over 1 keV. In contrast to this broad distribution, the typical upward current region ion beams (see Figure 4.5 in the upward current region discussion) have a more monoenergetic spectrum.

The apparent 'counterstreaming' seen here is consistent with strong field-aligned energy broadening of an upgoing beam by wave-particle interactions. We will discuss in Section 4.3 the wave-particle interactions which give rise to this effect, turning an upgoing DC-field accelerated beam into an apparently counterstreaming population. Clearly the implications of this picture are very different from one which would give rise to true counterstreaming, i.e., beams travelling in opposite directions. Here, there is only one population, which can be described by an anisotropic temperature ($T_{\parallel}$, $T_{\perp}$), and a thermal velocity that is much larger than its bulk velocity. Note in particular that the downgoing population is here inferred to be (wave-)scattered from the upgoing population of local ionospheric source particles; there is no need for a closed field line nor for activity in the conjugate hemisphere.

These electron populations play an important role in ionosphere-magnetosphere coupling. They may be the origin of highly collimated counterstreaming electrons that are observed deep in the magnetosphere (Klumpar, 1993). The close association of these populations with downward Birkeland current regions suggests that these electrons can be used as 'tracer particles' to map the connection of these current systems into the outer magnetosphere.

4.2.1.3 Observational History

The earliest hints of upgoing electron beams, and their possible relation to downward pointing electric fields were observed from ATS-6 (McIlwain, 1975). Intense upgoing field-aligned electron beams observed on ISIS2 (Klumpar and Heikkila, 1982) were identified with downward Birkeland currents, and a similar conclusion was drawn for upgoing electrons measured by DE 1 near the dayside cusp (Burch et al., 1983). Intense fluxes of upflowing electrons were also reported from Viking observations (Hultqvist et al., 1988). In more recent measurements by the Freja spacecraft, Boehm et al. (1995) identified upgoing magnetic field-aligned electron beams with energies of several hundred eV of clearly ionospheric origin. These beams were directly associated with quasi-static diverging electric field structures with magnitudes of 0.5 to 2 V m^{-1} observed by the Freja electric field instrument in the downward current region, in particular during condition of low plasma density on auroral field lines and in the ionosphere, such as near magnetic midnight and winter solstice (Marklund et al., 1997). Upgoing beams are seen with increasing frequency at altitudes above the Freja apogee (1700 km.). They are much more prevalent in the FAST dataset because of its higher altitude (apogee 4200 km.) and its continuous coverage of field-aligned electron measurements (Carlson et al., 1998a). Based on the FAST dataset, the electrodynamic significance of this accelerated population as the primary carrier of the return current has been firmly established. We will see later in this section that very recent results from the Cluster spacecraft show the temporal evolution of these structures. It is clear now that these highly field-aligned populations are the carriers of the return current Birkeland currents, and we will see that there are many clear parallels between upward and downward current phenomena.

4.2.2 Ion Conics and Ion Precipitation

Returning to Figure 4.12, we now consider the ion signatures in the return current region. In contrast to the ion beams seen within the inverted-V potential drop region, the downward current acceleration region ion signature is a transversely heated distribution, a *conic*. We will see below that the wave particle coupling process which heats these ions is a fundamental part of the control of the return current region electric field structure.

Below the principal electric field region, another ion feature of downward current regions is the appearance at low altitudes of downward ion acceleration. This signature is seen in the return current region only, but its occurrence has different characteristics and requirements than the ion conic regions.

4.2.2.1 Ion Conics and the 'Pressure Cooker'

Ion energization in downward current regions is an integral part of the electrodynamics and quasineutrality definitions of these regions. In the nightside, this is the region of the most energetic ion conics. Figure 4.12 shows an example of a nightside auroral zone crossing with a downward current region. This region is indicated by upward electrons, downward current, energetic ion conics, and strong broadband ELF activity. Conics and weak plasma sheet ion precipitation are the most common ion signatures in the downward current regions (panels g and h of Figure 4.12, beginning at 19:07:10 UT.) The conics are easily identified from the pitch-angle plot as the two symmetric traces that straddle the loss cone, centered at 180°; the plasma sheet ions are isotropic. The ion energy spectrum contains a fairly constant population of plasma sheet ions at an energy of $\approx$5 keV. The higher-energy precipitating ion fluxes which are evident here are always seen in these regions as well (premidnight), making the accompanying ion conics similar to the 'Type 2' conics described in André et al. (1998). Conics are the second population, which varies in energy between 10 eV and several keV. The measured hot ion density varies between 2 and 6 ions cm^{-3} through this region.

The interval from 19:07:20 to 19:08:30 UT is a classic example of the 'pressure cooker' ion heating scenario described briefly in Section 3.5.2. The upgoing electrons, which are strongly field-aligned, are consistent with acceleration of cold ionospheric electrons by a downward-pointing DC electric field. This electric field then retards the upward motion of the ion conics. For ions to have reached the altitude of the observation, they must have acquired sufficient transverse energy to overcome the (altitude dependent) electric field via the mirror force. This energy-filter effect is seen clearly in the ion energy spectrogram from 19:07:20 to 19:08:20 UT; there is a clear minimum energy in the ion spectrum, below which there are very few counts. This distinctive low energy cutoff of the conic energy spectrum visible in this plot is a unique property of the conics that accompany upward electron beams. This cutoff energy level follows the general trend of the electron beam energy. The conic pitch angle also follows the changes in the energy, shifting toward 90° and 270° (see footnote on page 104)with increasing energy. These properties agree with predictions by Gorney et al. (1985) for the behavior of ion conics trapped by downward parallel electric fields.

The ion conic distribution functions patterns seen in these regions have distinct signatures. An example is shown in the bottom right panel of Figure 4.14. The apex of the conic is upward from the $v_{||} = 0$ axis. There is strong velocity-space diffusion as can be seen by the filled-in plateaus between the arms of the conic. The pitch angle roughly increases with energy. There is parallel diffusion caused by the

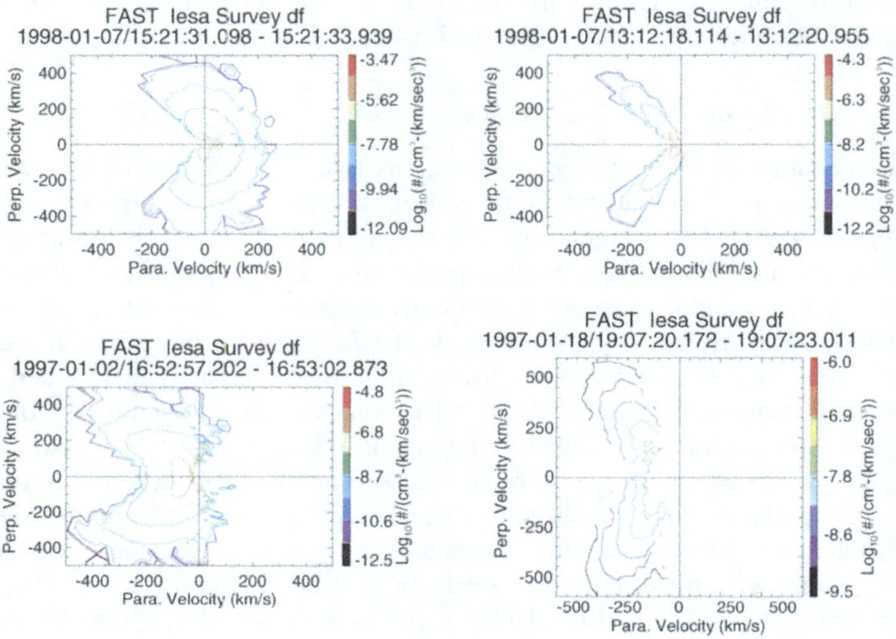

Figure 4.14. Ion distribution functions in $v_{||}$, $v_{\perp}$ space: (upper left) low altitude return current region example; (upper right) inverted-V example; (lower left) Alfvénic arc example; (lower right) pressure-cooker example. Distribution functions are integrated over 8 survey data distribution functions (2.8 or 5.6 s, depending on instrument mode), and a floor of 4 counts per interval is used to cut off the contouring. Mass is assumed to be hydrogen. From Lynch et al. (2002).

electric field. Fluctuations reflecting some of the particles, resulting in a large $T_{||}$ and downgoing ions.

In contrast, and for further illustration of the variety of conics which can be seen, Figure 4.14 also shows distribution patterns from a very low altitude observation (top left) and from an inverted-V region (top right). All of these examples come from midnight or pre-midnight local times, at altitudes within the 2000-4000 km range of FAST observations. The low altitude observation shows the type of conic first described by Gorney's 1985 model, where there is a significant component of the conic population in the downgoing direction. The inverted-V region conic is quite different from all the others, as it has a narrowly defined pitch angle width, and its apex is firmly mounted at the origin. The last example (bottom left) is taken from an Alfvénic polar cap boundary region.

The 'Pressure-Cooker'. Many aspects of the ion behavior in these downward current, downward electric field regions can be understood through a test-particle model of the effects of the Earthward-pointing DC **E** field, the upward pointing mirror force, and the transverse acceleration of ions via waves (Gorney et al., 1985). The fact that the the downward electric field contains the ions in a region of heating, explains why this configuration is known as the *pressure-cooker*. This can be modeled using Eqs. (3.96), and (3.94) with $E_{\|}$ pointing Earthward, and the change in perpendicular energy taken either as a cyclotron-resonant heating (Eq. 3.94) or as a stochastic kick process.

The particle motion thus can be computed, given altitude profiles for $E_{\|}$ and the heating rate, as a closed set of three equations for $v_{||}, v_{\perp}$, and z:

$$\frac{dv_{||}}{dt} = \frac{e}{m}E_{||} + \frac{3}{2}\frac{v_{\perp}^2}{(R_E + z)} \tag{4.2}$$

$$\frac{dv_{\perp}^2}{dt} = -3v_{||}\frac{v_{\perp}^2}{(R_E + z)} + \frac{2}{m}\frac{d\epsilon_{\perp}}{dt} \tag{4.3}$$

$$\frac{dz}{dt} = v_{||} \tag{4.4}$$

An interesting observational feature that is explained nicely by this model is the apparent mass independence of the conic energy. Hydrogen and oxygen are heated to approximately the same energies in these regions, contrary to expectations from cyclotron heating models (Lund et al., 1999). The result of the confining parallel field is to keep all ions within the heating region until they gain sufficient energy such that the mirror force overcomes the parallel electric field force, which removes mass-dependence acceleration effects.

Time Scales. The typical ion conic transit time from 1000 km to 4000 km is about 85 s. The typical electron transit time is a fraction of a second. A typical FAST horizontal transit through the return current region takes about 30 s. Thus the FAST energy-time spectrogram for the electrons shows the *instantaneous* behavior of the potential drop below the spacecraft throughout the crossing. However, the energy-time spectrogram for the ions has a considerable degree of hysteresis. For many examples (Ergun et al., 1998b), the return current potential structures are structured spatially rather than temporally on the time scale of the FAST crossing, as evidenced by matching the electron characteristic energies with the integrated perpendicular electric field. For other examples (Chaston et al., 2002c), the upgoing electrons are part of Alfvénic structures, and the electric field can be assumed to be time varying on the time scale of the FAST crossing. In all cases, the strong perpendicular fields imply strong perpendicular flows, and this can be seen often in the apparent non-gyrotropic nature of many of the ion distributions as observed in the spacecraft frame.

4.2.2.2 Downward Acceleration of Ions

Another aspect of return current region ion data, seen at low altitudes (below 2000 km) is downward acceleration of ions. These downward flowing ions (DFI) appear to be accelerated into the ionosphere by an downward-pointing DC electric field, but there are aspects of the observations which distinguish them from being simply the low-altitude signature (below the potential drop region) of the pressure-cooker regions just described.

The precipitation of keV ions into the atmosphere has been recorded as long as there have been instruments available on satellites for measuring low energy ions. The ESRO-1 satellite (launched in 1968), recorded ion and electron precipitation in generally coinciding regions at auroral latitudes, at altitudes near 1500 km. An unexpected feature was that occasionally field-aligned acceleration of the ions was evident (Hultqvist, 1971) As discussed in previous sections, most measurements at altitudes of one to several $R_E$have observed upgoing ions. These are caused by either direct acceleration by upward-pointing DC fields (see Figure 4.5) or by a combination of transverse acceleration and the mirror force, described just above.

Dynamics Explorer 2, near 900 km altitude, reported field-aligned ion precipitation in the keV and sub-keV ranges (Winningham et al., 1984). The large Freja database (apogee 1750 km) allows a detailed examination of this type of event. Figure 4.15 shows an example of the Freja observations of downgoing ions and plasma sheet electrons at an altitude of 1740 km. The figure shos that all three ion species are precipitated simultaneously, with no velocity dispersion and that all three ion species are precipitated with more or less identical energies. This is typical for all Freja nightside observations of downward accelerated ions (Hultqvist, 2002) Acceleration of these ions in the same potential field is the only possible known mechanism that easily explains these features. Examples of strongly field-aligned pitch-angle distributions can be found, but generally they are closer to isotropic over the downward hemisphere; thus, there are mechanisms that scatter the ions effectively. The events are only seen in return current regions, but are not related to particularly strong current densities. The events are clearly excluded from inverted-V regions. The events are found only in the presence of energetic electron precipitation, generally from the central plasma sheet, with a much wider and more energetic energy spectrum than the ions. The energy of the downgoing ions is often correlated with the energy of the energetic plasma sheet electrons. The events are often seen over wide regions at auroral latitudes, and are seen at all local times.

One mechanism which appears to be consistent with the Freja observations summarized above, is the ambipolar field mechanism (Hultqvist, 1971). The basic accelerator in this model is an ambipolar electric field caused by the large difference in speeds of energetic electrons and ions in the plasma sheet. Note that the downward accelerated ions are not plasma sheet ions, since they are too cold; rather, the source is assumed to be ionized atmospheric neutrals. The ambipolar field is simi-

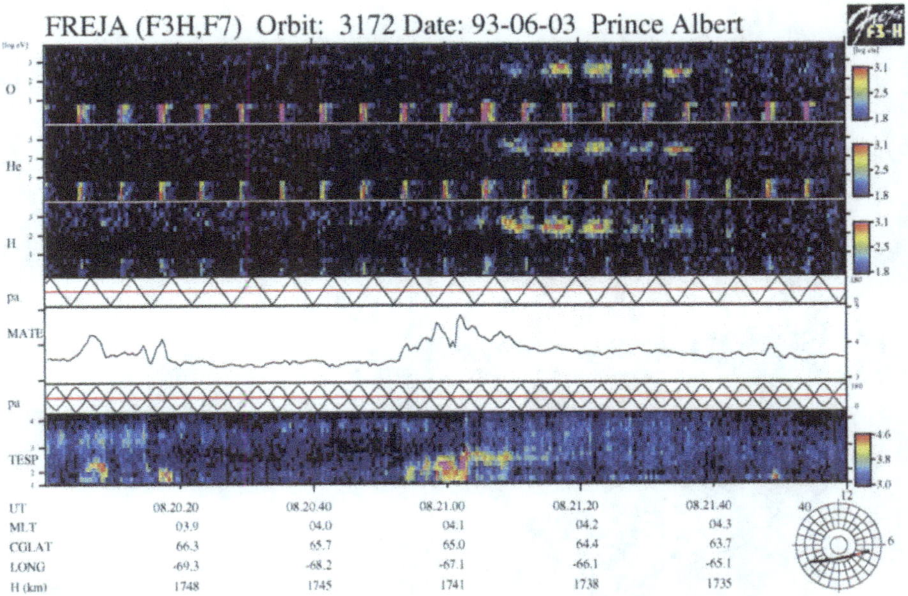

Figure 4.15. Freja data showing the three ion species H^+, He^+, and O^+ with their count-rates color-scale coded, their pitch-angle distributions, the total electron flux in the energy range 1 - 100 keV measured by the electron spectrometer MATE and the electron energy spectrum, color coded as recorded by the TESP spectrometer. The Freja orbit in magnetic local time (MLT) and corrected geomagnetic latitude (CGLAT) is shown in the lower right corner. Between 08:21:08 and 08:21:35 UT all three ion species are precipitated with very similar energies and, in this case, also similar fluxes. After Hultqvist (2002).

lar to the ambipolar field on top of the ionosphere, which helps produce the polar wind outflow of ionospheric ions, but in the present case the particles have much higher energies. For more details see Hultqvist (1971) and Hultqvist (2002).

FAST at higher altitudes, has observed similar phenomena, downgoing ions consistent with acceleration in a parallel electric field. Klumpar et al. (1999) have presented statistical studies (see Section 5.1.3.1), utilizing FAST data at altitudes up to 4300 km, of the occurrence probability of DFI in local time, latitude and altitude, and of their characteristics. They have shown that the ions are predominantly hydrogen, have a structured energy spectrum, and are usually narrowly confined in latitude. Figure 4.16 presents an example of a downflowing ion event. The downflowing ions, observed from 09:10:10 to 09:10:50 UT, had a peak energy increasing from ~10 eV and to ~2 keV, and, although peaked at 0 °, were broad in pitch angle (a horseshoe shaped distribution with a weak beam superimposed, see panel g).

Although the low energy edge of the DFI increased monotonically with latitude, the structure at the high energy edge is more complicated. The electron energy flux was weak until just after the ion event when an intense upflowing electron beam (UFE) occurred with an energies extending up to 1 keV. Klumpar et al. (1999)

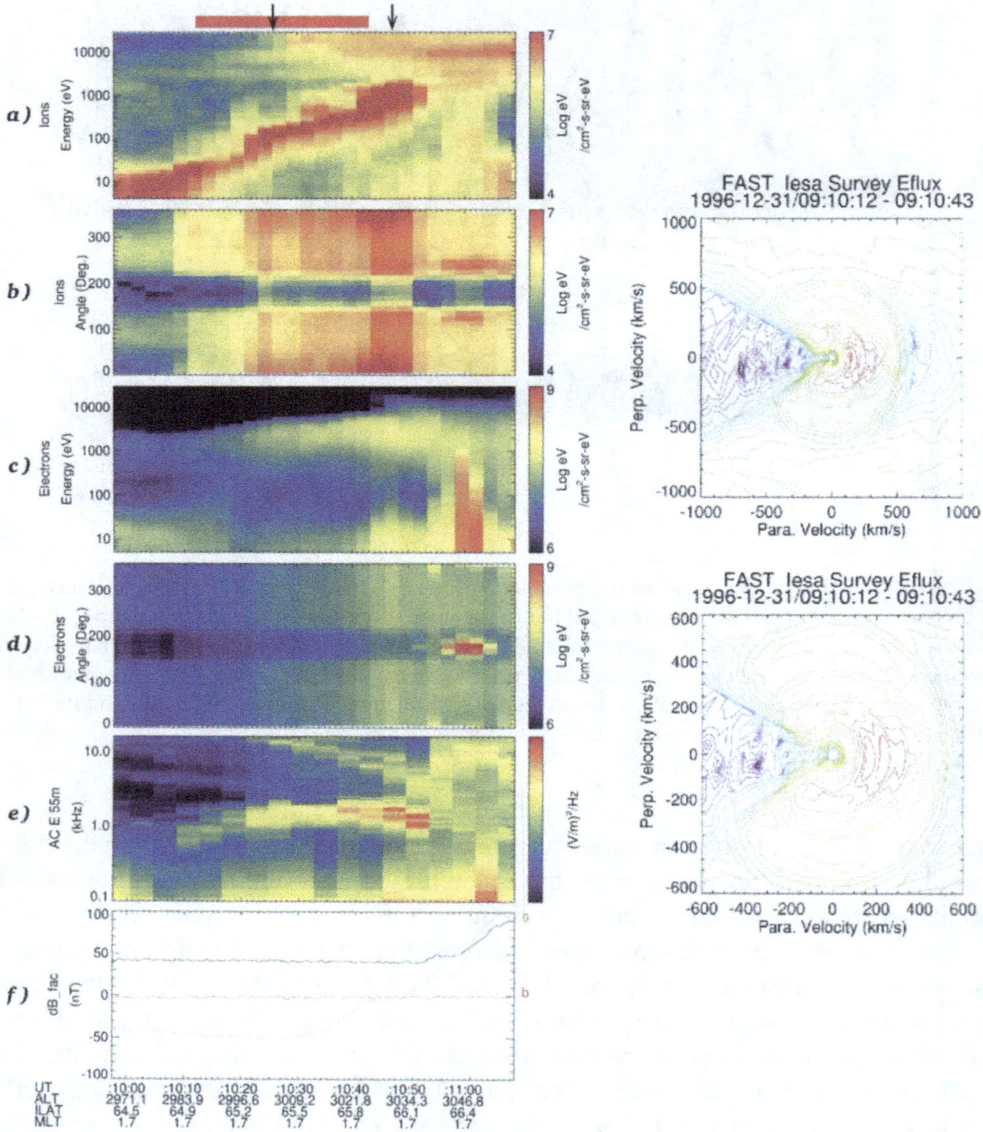

Figure 4.16. Left: FAST particle and magnetic field data for 70s starting at 09:10:00 UT, with a) the ion energy flux (integrated over all pitch angles); b) ion pitch angle spectrogram (integrated over all energies); c) electron energy flux; d) electron pitch angle spectrogram (integrated over all energies); e) the electric field spectral power density as a function of frequency and time between 0.1 and 16 kHz; f) perturbation in the magnetic field, in field-aligned coordinates. Right: two views of the velocity distribution averaged over the interval marked by the red bar above the left top. From Cattell et al. (2002).

have suggested that this commonly observed association of DFI with UFE indicates that the satellite has moved from an altitude below the parallel potential to one above the parallel potential. Ion conics were observed at the time of the upflowing electron beam, as described in Section 4.2.2.1. Plasma sheet ions can also be seen from about 09:10:30 onward. The eastward (green) component of the magnetic field perturbation (panel f) indicated a net field-aligned current flowing into the ionosphere in association with the last half of the DFI and with the upgoing electron beam; however, there is no discernable signature of a current associated with the low latitude, lower energy portion of the DFI event.

The right part of the figure shows the ion energy flux over two different velocity ranges for the detailed shape of the ion distribution for this event. It can be seen that the ion beam has spread out to form a ring in velocity space, as would be expected due to conservation of the first adiabatic invariant as a field-aligned ion beam moved into a region of increased magnetic field strength In a study of 19 events studied by Cattell et al. (2002), the DFI all had the shape of a partial ring in velocity space, some of which had enhancements in the parallel direction.

4.2.3 Electric Fields

Continuing our survey of the return current region features shown in Figure 4.12, we turn now to the characteristics of the electric field measurements. We will focus mostly on the DC signatures, since the wave signatures will be covered in Sections 4.3 and 4.4.

4.2.3.1 Quasistatic Divergent Shocks

As illustrated in the cartoon in Figure 4.1, the return current region has a divergent shock structure which mirrors the convergent structures of the inverted-V. Figure 4.17 shows an example of a pair of these structures seen at 1700 km altitude by the Freja satellite. They are often found together with isolated upgoing electrons. The electric fields accompanying more extended regions of upgoing accelerated electrons are characterized by intense low frequency fluctuations, with amplitudes ranging between 100 and 1000 mV m^{-1}, as seen in the example in Figure 4.12.

Another example of an isolated divergent electric field structure is shown in Figure 4.18. This figure shows an expanded view of FAST measurements in a narrow downward current region on the equatorward edge of the pre-midnight aurora. This isolated electron population lends itself to quantitative comparisons between the measured field and particle measurements. First we compare the current carried by the observed electrons, to the magnetic field deflection observed by the magnetometer. The current shown in panel (b) was computed from the measured electron distribution. The spacecraft trajectory crossed the model auroral oval boundary almost at right angles on this particular pass. The calculated magnetic perturbation expected from a perpendicular transit through a current sheet having the measured intensity is compared with magnetometer measurements in panel (a). The excellent

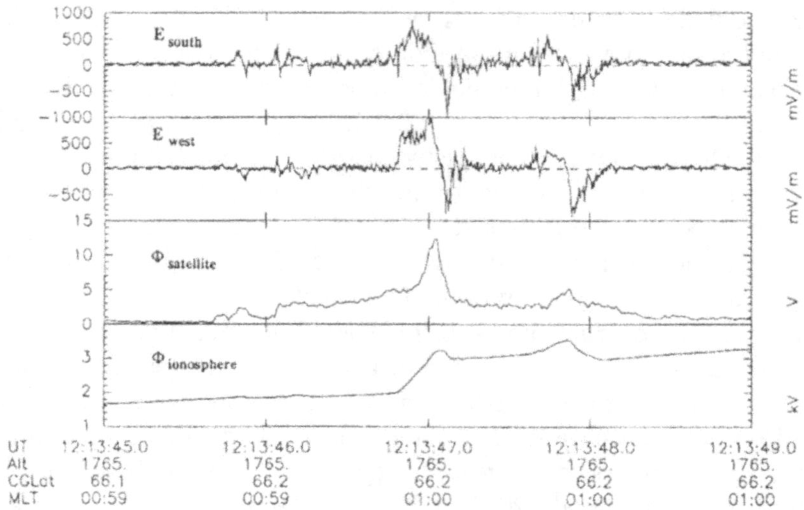

Figure 4.17. Observation of a pair of radially divergent electric field structures seen by Freja at 1700 km altitude on 11 March, 1993. From Marklund et al. (1994).

agreement confirms that the observed suprathermal electrons can account for the total current. We saw a similar result in the upward current region.

Next we compare the electron characteristic energy to the potential found by integrating the perpendicular electric field, as was done earlier with the inverted-V ion beams. The upgoing electron spectrogram shown in panel (e) includes an overlay line plot of its characteristic energy, defined as the ratio of the energy flux to the particle flux of the beam. This quantity represents the average energy that beam electrons gain from passing through a potential. The overlay spans the time when high speed burst data were available. The electric field data in panel (f) shows that a strong diverging electrostatic shock is centered on the beam. The parallel potential displayed in panel (g) was obtained by integrating the observed electric field along the spacecraft trajectory. It is assumed that the parallel field vanished at the edges of the beam and a constant ionospheric electric field is imposed to match these boundaries. The good agreement between the characteristic electron energy and electric potential drop throughout this 15 s event supports the conclusion that the beams are accelerated by quasi-static potential structures. We saw a similar result in the comparisons of ion beam energies with inverted-V region DC electric fields. These results demonstrate a strong similarity between the quasi-static electric field structures that accelerate upgoing electron beams (return current regions) and ion beams (upward current regions), but there are distinctive differences between the resulting ion and electron distribution functions, reflecting the different wave-particle interactions at work in the different regions.

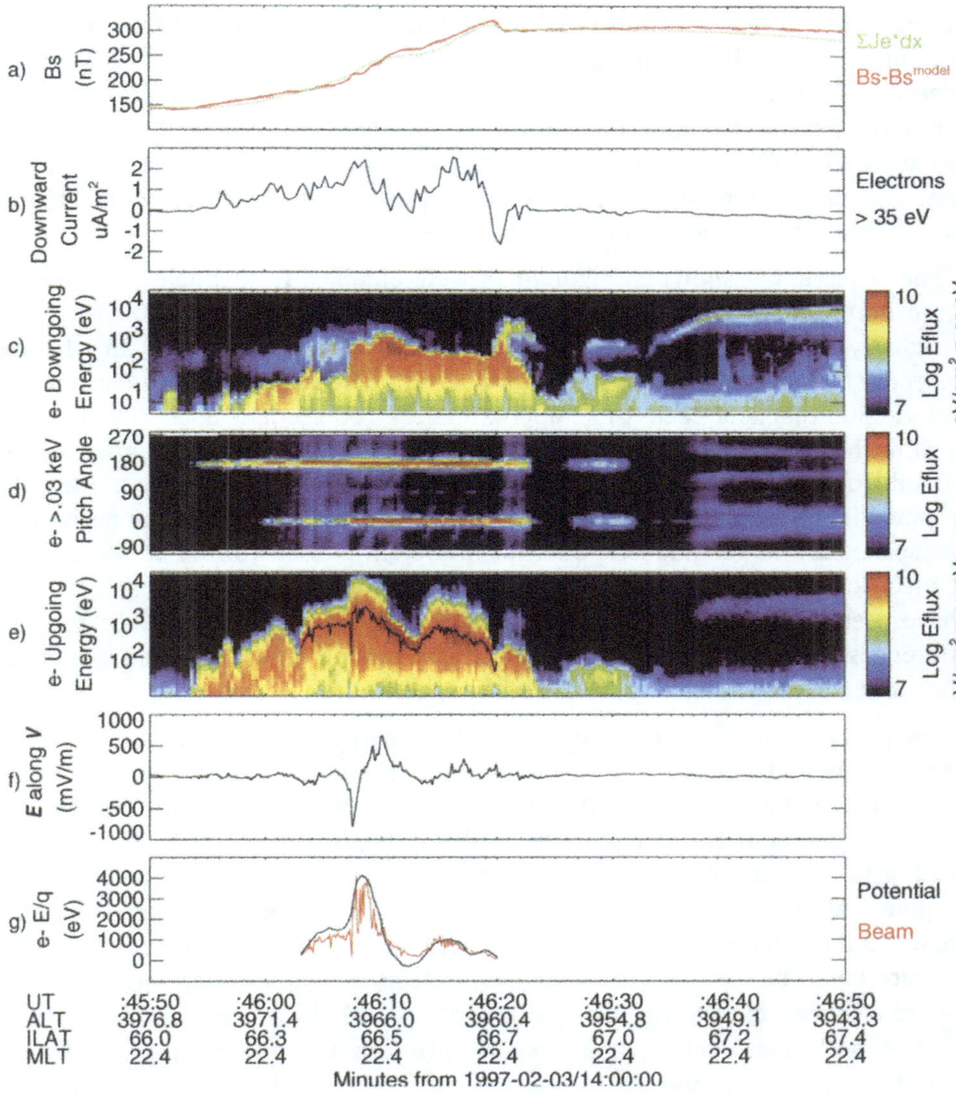

Figure 4.18. Quantitative comparisons of measured magnetic field perturbations and electric potentials with their corresponding electron signatures in an upgoing beam event. Panels: a) measured change of westward **B** field component (red) compared with field expected from a current sheet modeled from the measured electron current (green); b) current carried by electrons having energies between 35 eV and 30 keV; c,d,e) spectrograms of electron energy as functions of energy and pitch angle; in e), the black overlay trace is characteristic energy (energy flux divided by number flux); f) the electric field component perpendicular to **B** and parallel to the spacecraft velocity; g) parallel potential drop (linear scale) inferred from the integrated electric field (black) compared with the characteristic energy of the electron beam (red). From Carlson et al. (1998a).

4.2.3.2 Temporal Evolution of Divergent Structures

The existence of quasi-static divergent electric field structures in the auroral return current region has been demonstrated clearly by the Freja and FAST results. However, there are a number of properties of these structures — altitude extension, growth time, characteristic lifetime — that are not observable by single-point measurements. The four Cluster spacecraft launched in 2000 provide a different view, allowing multiple-point study of these electric field structures with data from auroral oval crossings at a geocentric distance of about 4.3 R_E.

The observations shown in Figure 4.19 were obtained from a poleward crossing of the northern auroral oval by the four Cluster spacecraft around 04:30 UT on 14 January 2001 (Marklund et al., 2001). From an altitude of ~21700 km, the four spacecraft, aligned like a string of pearls, swept across the northern auroral oval around 0300 magnetic local time, approximately 100 seconds apart. The top panels show the dominant northward electric field component, the field-aligned current derived from the eastward magnetic field component, and the negative of the satellite potential (which roughly follows the plasma density variations), for each of the four spacecraft. A geomagnetic coordinate system is used for the presentation of the electric and magnetic field data. Note that the time axis for spacecraft 1, 2, and 4 have been shifted relative to that of spacecraft 3 (reference satellite) by the time differences between the spacecraft as indicated in the figure. Note also that at this time spacecraft 1 was leading, followed by spacecraft 3, 2, and 4.

The panels on the right of Figure 4.19 show energy-time spectrograms for electrons moving in the direction antiparallel to the magnetic field (upward), for crossings of each of the four spacecraft. During this short period, the electric field experiments on spacecraft 1, 3, and 2 observe a diverging, bipolar electric field structure, growing in size and magnitude corresponding to an increase in the positive electric potential of the structure from 500 V to over 2 kV. However, when spacecraft 4 reached the same position, the electric field had vanished and no evidence of a structure could be seen. The corresponding electron data show an increase in the upward electron characteristic energy from about 500 eV to 2 keV. The consistency between the increase in the parallel potential (inferred from the energy increase of the upward electrons) and the increase in the perpendicular potential (integrated from the electric field) indicate that the Cluster spacecraft were crossing the upper reaches of a U-shaped positive potential structure (cf. Figure 4.1), growing in size and intensifying on time scales of a few hundred seconds.

The field-aligned current associated with the structure is downward, as shown in the second panel of Figure 4.19. It can be shown that the total downward current integrated over the structure remains roughly constant during the crossings by the four spacecraft, whereas the width of the current sheet is seen to increase, in particular between the two last crossings by spacecraft 2 and 4. The structure occurs at a plasma density gradient that steepens as the electric field increases, and then

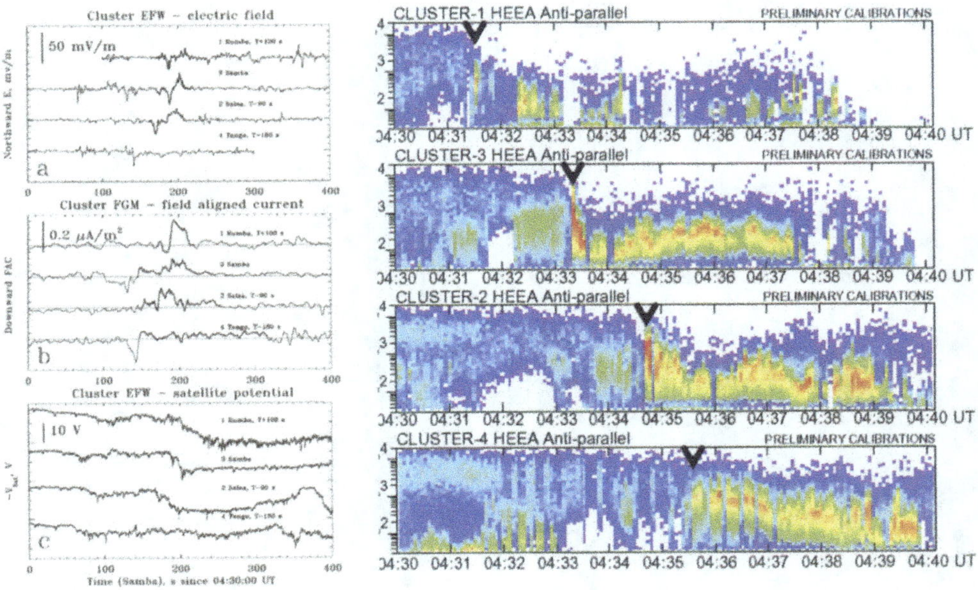

Figure 4.19. Left: northward electric field component (top), field-aligned current (middle), and the negative of the satellite potential (bottom), roughly varying as the plasma density, derived from the EFW and FGM instruments on the four Cluster spacecraft for 400 s after 04:30 UT on January 14, 2001. Right: energy-time spectrograms of electrons moving in the direction antiparallel to the magnetic field as measured by the PEACE instruments on the four Cluster spacecraft between 04:30 UT and 04:40 UT on January 14, 2001. The arrowheads at the top of each panel indicate the times the electric field structures were encountered. From Marklund et al. (2001).

disappears together with the electric field at the time of the crossing by spacecraft 4. Figure 4.20 illustrates these processes schematically.

These data show that the positive potential structures (divergent electric field structures) in the return current region can extend to altitudes greater than 20,000 km and that they can grow in size and intensify over time scales of a few hundred seconds. After this the electric field dies out, the plasma density gradient disappears and the current sheet widens. It is interesting to note that this time period is comparable to the time it takes to evacuate the ionospheric electrons over the narrow region of the downward current sheet. How the evolution of the U-shaped potential structure with its region of a parallel electric field above the ionosphere relates to the plasma evacuation in the ionospheric E- and lower F-region is, however, still a subject for study. A possible explanation for the fading of the potential structure (the disappearance of the electric field and drop in the characteristic energy of the upward electrons) after a few hundred seconds may be found in the sudden widening of the return current sheet seen at this time. Since current carriers become

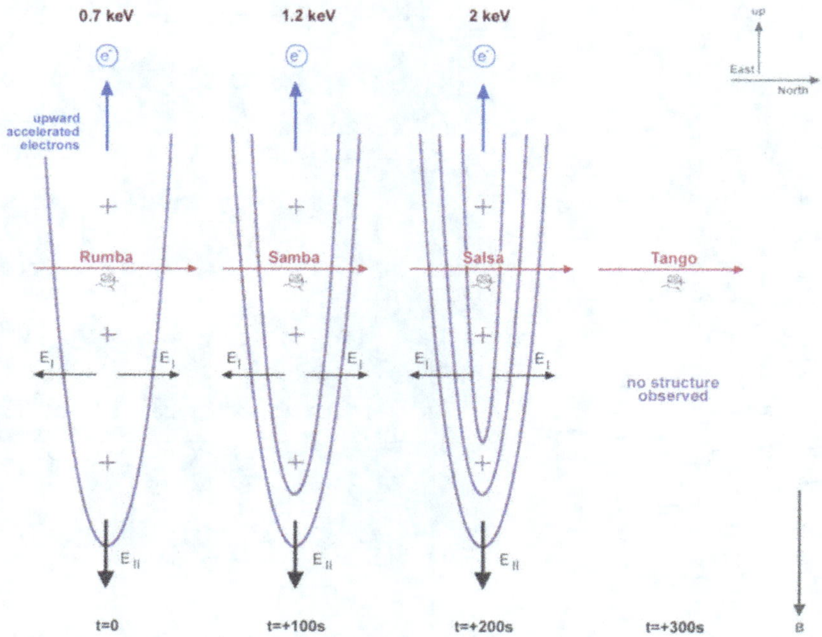

Figure 4.20. A schematic illustration of the sequence of events recorded in Figure 4.19. After Marklund et al. (2001).

available from a much wider region, the return current could be maintained more easily.

4.2.3.3 Wave Fields

Returning now to complete our inspection of Figure 4.12, we consider the intense electric field wave spectrum, extending into the ELF and VLF range. The broadband ELF activity seen here is typical of these regions. In panel (j), the clear signature of a VLF saucer can be seen emanating from each side of the upgoing electron region (before 19:07:00 and 19:08:30 UT). Intense electron solitary waves in **E** and **B** are also associated with the electron beams. They modulate the electron characteristic energy and may account for the large energy spread. These solitary waves also create large perpendicular electric field pulses which can contribute to the observed intense ion heating. We will consider all these varied wave activities in detail in Section 4.3.

4.2.4 Current-Voltage Relations

In the upward current region, we found a clear observational and theoretical relationship between the externally required field-aligned current density, and the resulting parallel electric field (the Knight relation of Sections 4.1.1 and 3.3.) The parallel field is required because the current carriers are limited. In the return current region, at first thought, it seems that the lower ionosphere should be an almost unlimited source of current carriers (the upgoing electrons.) However, as we discussed in Section 3.3.2, *quasineutrality* gets in the way. The current carriers are thus limited here as well. For a particular value of current ($qn_e v_{e,\|}$), electron density can be reduced (and quasineutrality maintained) by accelerating electrons.

The resulting current-voltage relationship is not as straightforward as the Knight relation. A statistical study (Lynch et al., 2002) shows that the amount of parallel field required to meet a certain external current requirement can be parameterized by the local plasma density. Where the density is high (i.e., near the polar cap boundary), relatively weak electric fields will supply the needed current. Where the density is low, stronger and stronger electric fields are required. A second complication is that there is some feedback in the system (Temerin and Carlson, 1998b). The accelerated electrons are accompanied by strong broadband ELF activity, as seen in panel (j) of Figure 4.12. This activity heats the local ion population significantly (see Section 4.2.3 on the pressure cooker), raising the scale height of the ionosphere. This increase in the density thus reduces the required parallel electric field.

The current-voltage relationship here is only just beginning to be understood. Theoretical models of the relationship between an enforced current density and resulting parallel fields include a return current region version of the Alfvén - Fälthammar model (Jasperse and Grossbard, 2000). Statistical studies of particle signatures, and of wave signatures including wavenumber spectra, can provide guidance to these models. We will examine the wave-particle interactions which support this feedback in more detail in Section 4.3 below.

4.2.5 Limitations of a Quasistatic Model

Most of the limitations described in Section 4.1.4 of quasistatic models of the upward current region, apply to the return current region as well. In addition, though, as we will find in the next section, true understanding of the return current region depends strongly on an understanding of the wave-particle processes that support the large-scale structure. In the inverted-V region, the parallel field formation can be seen as a simple response to an externally required current. In the return current region, however, the parallel field strength is a function of strong feedback loops within the acceleration region itself. The wave-particle interactions play an integral part in controlling the large-scale structure, by broadening the upgoing electrons into a population that often appears counterstreaming, by heating the ionospheric ions enough to raise the scale height of the lower ionosphere, and, in raising the scale height, by reducing the required parallel field needed for a given externally required current.

Epilogue: Up/Down Current Region Summary

In Sections 4.1 and 4.2 on the quasistatic aspects of upward and downward current regions, we have found a remarkable symmetry. Limitations on the available supply of current carriers cause the formation of parallel electric fields, accelerating particles up and down the auroral field lines. In the upward current region, the accelerated electron beams create the visible aurora seen in Chapter 2. In the return current region, a similar process (with interestingly different details) causes acceleration of electrons up the field line. Figure 4.21 summarizes some aspects of this symmetry, with a photograph of a classic auroral arc system matching an illustration of the quasistatic potential structures we have discussed.

4.3 Waves and Radiation

The auroral plasma at acceleration region altitudes is collisionless on all relevant time scales. How can momentum and energy exchange take place between different particle populations without collisions? A plasma carrying strong currents or containing non-equilibrium particle distributions, like those caused by 'potential drops', ultimately becomes unstable with respect to various kinds of instabilities. Through wave-particle interactions, these instabilities generate plasma waves, scatter resonant particles in momentum and energy and lead to energy and momentum exchange. Observing a natural collisionless plasma, we are unavoidably confronted with the simultaneous interaction of all particles mediated by field fluctuations. Any observation thus necessarily deals with wave-particle interactions.

The electrodynamic importance of these interactions is determined by the level of coupling between the excited waves and the particles. A parameter suited for classification of wave-particle interactions is the ratio of wave to kinetic energy density in the plasma. This ratio is generally very small, $\delta = W_w/nk_B(T_e + T_i) \ll 1$. A ratio of 10^{-4} typically indicates weak turbulence; of 10^{-2}, strong turbulence. Hence, linear instabilities (of first order in WKB expansions about this ratio) will be the most important among wave-particle interactions (compared to nonlinear terms, of higher order in this ratio.) In the auroral plasma, which is not in thermal equilibrium, one expects that a large number of waves (of different frequencies ω and wave vectors $\mathbf{k}$) and spectra will be generated. These waves carry away energy and momentum to other locations, either distributing it within the plasma or leading to energy and momentum loss from the plasma, depending on whether or not the waves leave the region of excitation faster than they can interact with the particles. It is clear that for interactions of higher order in δ to affect the particle distribution, the waves must stay in resonance with the particles for a longer time. One therefore expects that nonlinear wave particle interactions will be significant only for electrostatic wave modes. These have low phase and group velocities and are confined to the plasma. Electromagnetic modes and in particular free space modes, which propagate at the speed of light, will lead to predominantly linear

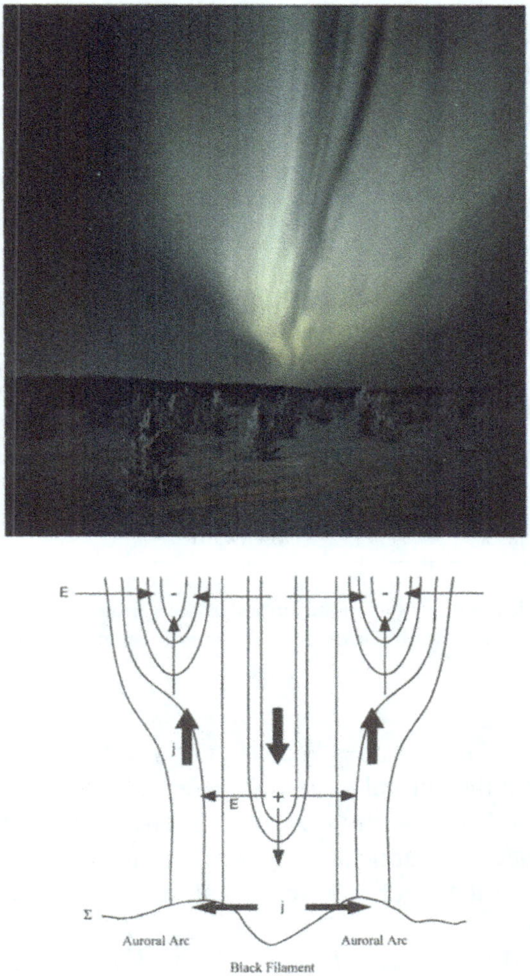

Figure 4.21. Photograph of two parallel auroral arcs, separated by a dark region referred to as black aurora (courtesy Torbjörn Löfgren). The bottom part shows the electric field and current configurations for a double auroral arc system, with the inverse configuration characteristic of black aurora in between. From Marklund et al. (1998).

wave particle interaction. While theory distinguishes between the different levels of wave particle interaction — linear instability, followed by quasilinear reaction of the linearly excited wave field on the resonant particles, and further higher order nonlinear interactions — observations cannot distinguish between these states, seeing only the final state of all interactions. It is thus generally difficult to extract just one of the wave particle interactions from the data.

In the discussions below, we step through the variety of wave-particle interactions seen in the auroral zone. We have delineated them into three separate groups

Electron beam modification		
AKR	Upward	Modification of $f_e(\mathbf{v})$
VLF hiss	Upward	Modification of $f_e(\mathbf{v})$
BBELF	Downward	Broadening of $f_e(\mathbf{v})$
ESW	Downward	Broadening of $f_e(\mathbf{v})$
Ion-electron coupling		
EMIC	Upward	Auroral flicker
IAW, ICW	Upward	Ion beam modification
BBELF pressure cooker	Downward	$E_{\|\|}$ control
Wave diagnostic signatures		
ISW	Upward	
Langmuir waves	Upward	
Roar	Upward	
LHSS	Upward	
VLF Saucers	Downward	

Table 4.3. Wave-particle interactions in the auroral zone: various interactions, their primary location in the upward current/downward current region scenario, and their significance.

roughly according to the ratio $\delta = W_w/nk_B(T_e + T_i)$, as shown in Table 4.3: (i) interactions by which the auroral system transfers energy by modifying the accelerated electron beam; (ii) interactions involving ion effects; and (iii) interactions of diagnostic but not electrodynamic significance. For the first case, δ is large (though still less than one); for the latter, it is very small.

4.3.1 Beam Modification: AKR, VLF, BBELF, ESW

We begin our discussions with *Auroral Kilometric Radiation* (AKR) and *VLF hiss*, the two principal electromagnetic radiative modes in the upward current regions within and below the potential drop, respectively. The AKR emissions are driven by the accelerated electron distributions encountered within the auroral cavity, such as in the second example of Figure 4.4, or in Figure 4.3 near 10:11:26 or 10:11:33 UT. The VLF emissions are generated on the same flux tubes, but at lower altitudes below the potential drop region, such as in Figure 4.3 at 10:11:20 UT.

4.3.1.1 Auroral Kilometric Radiation

The powerful radio emission known as AKR is the outstanding radiative feature of the auroral zone. The very first satellites that carried radio receivers made the surprising discovery (Benediktov et al., 1965) that, as viewed from space, the Earth is an intense radio source. Subsequent work (see, e.g., Gurnett, 1974) established that these emissions are generated nearly permanently at altitudes between 2000-

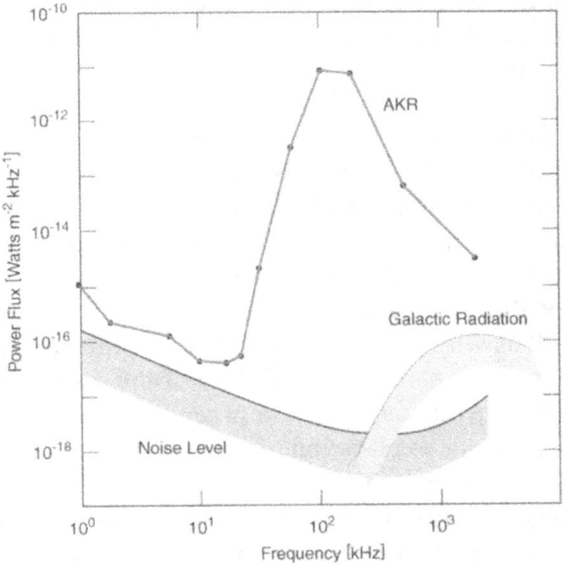

Figure 4.22. Power spectrum of AKR observed by Imp 8 at the remote distance of 25.2 R_E on the evening side of Earth (Gurnett, 1974). The spectrum shows the high peak of AKR at a few hundred kHz far above the galactic background. The 'noise level' curve is instrumental noise. Note the steep decay of the spectrum towards higher frequencies.

10,000 km above the auroral region in association with the appearance of auroral arcs. AKR — so named because at 300 kHz, its free space mode has a wavelength of 1 km — is not detected on the ground because of shielding by the dense ionospheric plasma.

It is now known (see, e.g., Zarka, 1992) that similar emissions come from all of the magnetized planets (see also Chapter 9) and that these auroral emissions are the most intense form of planetary radiation. An example of the frequency spectrum of AKR (measured by IMP 8) is shown in Figure 4.22. The spectrum exhibits a pronounced enhancement above ~ 100 kHz, and peaks at about 100-400 kHz. The typical total power radiated in an AKR event is of the order of 10^7 W, while the peak power level during substorms can be as high as 10^9 W. In comparison, the total electron energy flux into the auroral zone during a substorm has been estimated to be as high as 10^{11} W. Thus the simultaneous wave and particle measurements performed in the AKR source regions indicate an electron-to-radiation energy conversion efficiency of the order of 0.1-1%, a very high fraction for any wave emission process.

AKR originates on auroral field lines (Kurth et al., 1975) in regions of depleted density where down-going, 'inverted-V' electrons are accelerated (Benson and Calvert, 1979). Ray tracing calculations of the radiation, based on the observations far away from Earth, have confirmed the location of the source to lie in the lower auroral magnetosphere in the above range of altitudes (Figure 4.23). The waves propagate primarily in the R-X mode and show strong temporal variations and fine spectral features.

The Electron-cyclotron Maser. Based on the early AKR observations, Wu and Lee (1979) proposed a weakly relativistic linear instability to be responsible for the generation of AKR. Melrose et al. (1982) and Le Quéau et al. (1984) coined the name 'electron-cyclotron maser' (ECM). The ECM is generally acknowledged as being the most likely candidate for generation of AKR (Louarn et al., 1990; Ungstrup et al., 1990). It explains quite well the observed R-X polarization[6] of the radiation as well as why the radiation is generated inside the cold-plasma depleted regions close to the X-mode cutoff. Two of the most significant remaining questions concern the identification of the precise free-energy source that drives the maser instability and the explanation of the fine structure appearing with bandwidths of $\sim$ 1 kHz or even less. The energy source of the cyclotron maser mechanism was originally based on the experimental finding that the auroral electron distribution possesses a pronounced nearly empty loss cone in the upward direction at $\theta \simeq 180°$. Such a loss cone (Figures 4.3 and 4.4 give examples) represents a thermodynamically non-equilibrium distribution with the free energy stored in the transverse velocity derivative $\partial f_e/\partial v_\perp > 0$ of the distribution function $f_e(v_\|, v_\perp)$. The electrons in this gradient near the loss cone boundary shed their excess perpendicular energy, excite waves and fill the loss cone until the gradient is depleted. The loss cone as an energy source was the key element of the original theory of the electron-cyclotron maser.

The particles responsible for the direct generation of the radiation satisfy the relativistic wave-particle resonance condition (e.g., Krall and Trivelpiece, 1973)

$$k_\| v_\| = \omega - \omega_{ce}/\gamma_{\rm rel} \qquad \gamma_{\rm rel} = (1 - v^2/c^2)^{-\frac{1}{2}} \tag{4.5}$$

where $v^2 = v_\|^2 + v_\perp^2$ and ω_{ce} is the non-relativistic electron-cyclotron frequency. It is important to include the relativistic correction factor $\gamma_{\rm rel}$ here. In the non-relativistic case $v_\|$ is a constant, and the loss cone does not contribute to the wave generation. Including $\gamma_{\rm rel}$ transforms the resonance condition into a form that is bi-quadratic in the parallel and transverse velocities. The solution of this form gives an elliptical curve that contours the particles in resonance with the wave. For appropriate parallel wave vectors $k_\|$ and frequencies ω, Wu and Lee (1979) demonstrated that it is possible to shift the resonance ellipse in such a way that its position traces along the loss cone boundary. This requires $k_\| \neq 0$. Thus the wave propagates obliquely to the magnetic field.

[6] R-X is the right-hand polarized free-space electromagnetic mode (e.g., Baumjohann and Treumann, 1996). The R-mode propagates parallel, the X-mode perpendicular to the magnetic field.

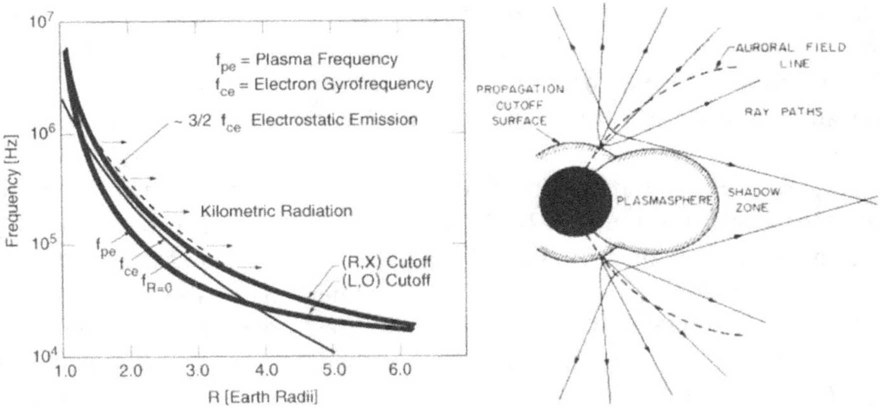

Figure 4.23. Left: rough radial variation of the X-mode cut-off of AKR. Right: qualitative sketch of the ray paths of AKR emitted at low altitudes and fixed frequency along an auroral field line. The refractive index goes to zero at the propagation cut-off. There is a strong tendency for the ray path to be upward away from the propagation cut-off surface. After Gurnett (1974).

In the special case when $k_{\|} = 0$, so propagation is strictly perpendicular, the free space mode is purely in the X-mode and the resonance curve degenerates into a circle. In this case the loss cone does not contribute to the emission. This fact becomes most important when analyzing the measured electron distribution functions in the auroral cavity.

The plasma in the auroral magnetosphere is only weakly relativistic which allows for an expansion of the above resonance relation. If we further assume that the emission is in the R-X mode, which is the free space mode that readily escapes from the generation region, one only needs to consider linear wave-particle interaction theory. The linear growth rate of the R-X mode can be shown to be positive, as long as the major part of the resonance line for $k_{\|} \neq 0$ is in the loss cone. For plasma frequencies corresponding to the auroral cavity, $\omega_{pe} \ll \omega_{ce}$, the maximum growth rate of the emitted oblique free space R-X mode is at frequency $\omega \geq \omega_{ce}$ above the R-X cut off, which is very close to the nonrelativistic cyclotron frequency. Emission just above the electron cyclotron frequency is frequently observed. However, the growth rates calculated from the loss-cone maser are rather small.

Before proceeding we briefly explain why this simple linear instability was named a maser. The connotation implied by the word maser[7] is that the particles occupy an excited state with the ground state being empty of particles. When falling back to the ground state they emit coherent electromagnetic radiation. In the case of the cyclotron maser the situation is analogous. The ground state corresponds to a Maxwellian equilibrium distribution. A non-Maxwellian distribution implies

[7]The acronym stands for Microwave Amplifications by Stimulated Emission of Radiation.

that the electrons are not in the ground state. Resonance allows them to return to the ground state and emit electromagnetic radiation. Viewed macroscopically, this process resembles the macroscopic behavior of the maser, which was the reason for choosing the name electron-cyclotron maser for this particular instability.

Observations. Early remote observations of AKR were made in 1975 by the Alouette spacecraft (Gurnett, 1974; Benson, 1975, 1985). The first spacecraft to perform measurements in situ in the AKR source was the Viking satellite (launched in 1986). It established (Bahnsen et al., 1987; Louarn et al., 1990; Pottelette et al., 1992; Roux et al., 1993) that the radiation is a direct consequence of auroral electron acceleration. Based on these observations, the loss-cone was questioned as the source of AKR by the results of two-dimensional particle simulations (Pritchett, 1984a,b; Pritchett and Strangeway, 1985) using the measured electron distributions. Winglee and Pritchett (1986) suggested that a 'shell' or 'horseshoe' distribution that results from electron acceleration in a parallel electric field could produce the observed strong cyclotron emissions.

Precise measurements of the particle and wave dynamics in the auroral source region came ten years later from the FAST spacecraft (Carlson et al., 1998a). These measurements (Pritchett et al., 1999; Ergun et al., 1998a) confirmed many of the fundamental elements of the electron-cyclotron maser mechanism, but with substantial modifications as already suggested by the early simulations (Pritchett, 1984a,b) and the Viking observations (Louarn et al., 1990). The most important of these modifications is that the emissions do not draw their energy from the loss-cone distribution of the radiating electrons. Rather, the radiation is generated by the unstable horseshoe (or shell) distribution.

We have already discussed the acceleration of electrons by parallel electric fields within the auroral cavity. This acceleration, when combined with the mirror force, generates horseshoe distributions in the electron component as described by the adiabatic model of Chiu and Schulz (1978). Horseshoe or shell distributions are typical in the upward-current acceleration region. The electrons move down the field lines into the increasing magnetic field. Conservation of their magnetic moment (the first adiabatic invariant) causes an increase in pitch angles. Under these circumstances, the energetic electron distribution still remains peaked at 0° but has large $\partial f/\partial v_\perp > 0$ over a wide range of parallel velocities, much wider than in the case of the simple loss cone distribution. The loss cone still remains, giving the distribution the shape of a horseshoe.

The horseshoe distribution makes the $k_\| = 0$ resonance curve attractive. This resonance curve is a circle in the $(v_\perp, v_\|)$-plane which can be brought to surf along the line of largest gradient $\partial f(v_\perp)/\partial v_\perp$ causing a topological change in the integration surface that determines the imaginary part of the X-mode dispersion and thus determining the growth rate. The resulting radio emission is closely perpendicular to the magnetic field. Horseshoe distributions excite nearly pure X-modes.

When one takes into account the weak relativistic effect of the auroral keV-electrons and perpendicular propagation as in Eq. 4.5, the X-mode cutoff shifts to below the non-relativistic local electron-cyclotron frequency. The maximum growth rate is at frequency $\omega < \omega_{ce}$. This fact allows for a definitive test of whether the energy source comes from the loss-cone instability or the horseshoe shell instability.

Verification of the Horseshoe Maser. FAST satellite observations confirmed that the AKR source occurs in a very low density cavity that is dominated by hot electrons and is nearly devoid of low-energy electrons (Strangeway et al., 1998). Examples of this type of cavity are included in Figure 4.3, wherever the ion beams appear. The borders of this cavity contain very intense low-frequency broadband electrostatic noise (BEN). They coincide with the sites of rapid ion acceleration and ion beam formation as well as electron acceleration and heating. All these properties are displayed in Figure 4.24, which collects FAST measurements of waves and particles in the AKR source region.

In the source region the emissions lie within $\sim 2\%$ of ω_{ce} and frequently extend below ω_{ce}, consistent with the horseshoe distribution as the primary energy source. The AKR emissions exhibit strong temporal and/or spatial variations (Ergun et al., 1998a). Of particular significance is the observation of peak electric field amplitudes in the source region which are typically greater than 100 mV m^{-1} and sometimes reach values as high as 300 mV m^{-1}, the saturation level of the FAST instrument.

To within a few degrees the AKR propagation in the source region is consistent with $k_\parallel \approx 0$ as required by the horseshoe cyclotron resonance condition for frequencies extending down to the relativistic cyclotron frequency $\omega_{ce}/\gamma_{\rm rel}$, well below ω_{ce}. In contrast, because a finite value of $k_\parallel$ is required for the loss-cone-driven electron-cyclotron maser instability, emission at a substantial angle, more than 10° away from perpendicular to the magnetic field would be expected. In addition, the pure loss-cone instability produces wave emission at $\omega > \omega_{ce}$ and so is unable to explain the observations made below ω_{ce}. In its present state, however, simple masering still fails to explain how the X-mode, once generated at $k_\parallel = 0$, would be able to escape from the cavity. With regard to this problem, FAST observations seem to provide evidence that, after multiple reflections, the radiation is guided within the density cavity and ultimately is efficiently converted to the R-mode (Ergun et al., 2000). These observations are of interest to the astrophysical community as well as the near-Earth space physics community (Ergun et al., 2000).

AKR Fine Structure. A different problem concerns the explanation of the observed AKR fine structure. Early wideband high-time resolution measurements by ISEE 1 and 2 recorded at large distance from the source (Gurnett and Ander-

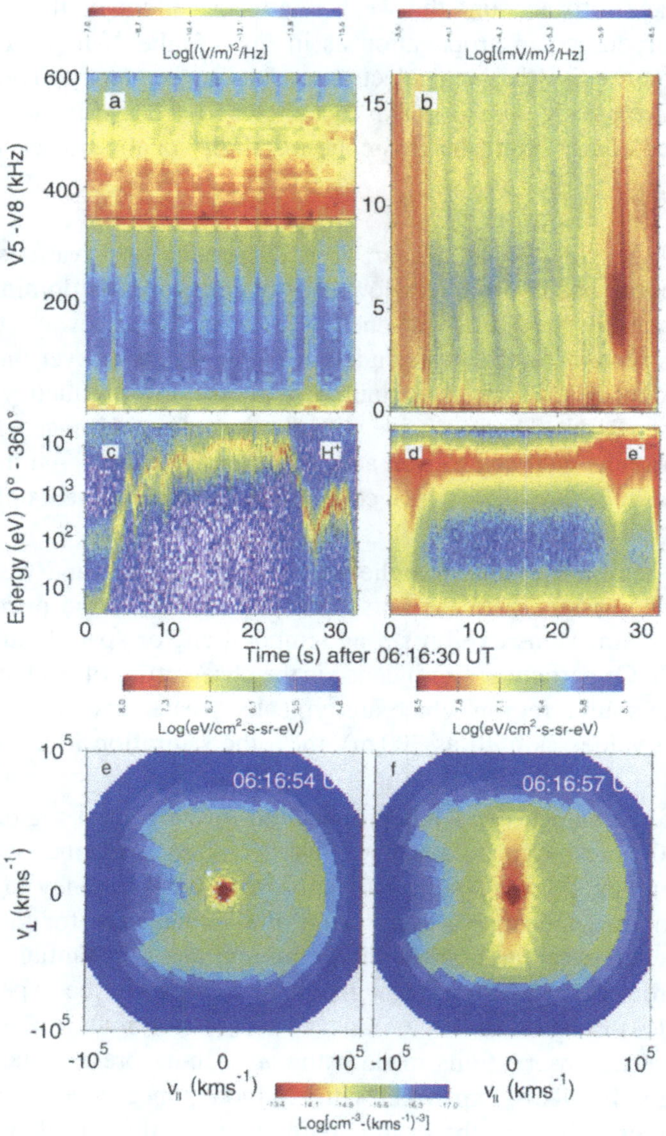

Figure 4.24. FAST wave and particle data in auroral density cavity; (a): HF wave spectrogram showing transversely polarized AKR emission at and below f_{ce} (line near $\sim$ 340 kHz) in X, and at lower frequencise in Z mode as well as VLF waves. Emission at higher frequencies originate at lower altitudes than the spacecraft and contains higher harmonics; (b): VLF wave spectrogram of parallel polarized O mode (above 5 kHz) and electron acoustic noise (below 5 kHz). The intense broadband electrostatic noise at 27 s appears at base of AKR source; (c and d): Ion and electron energy spectra in same time interval, showing ion acceleration and electron heating; (e) and (f): electron pitch-angle distributions in cavity and at base of AKR source with empty upward cone ($v_\parallel < 0$), fast (conical) electron beam ($v_\parallel > 0$), and hot electrons. Note the strong perpendicular heating of the electrons around 06:16:57 UT in coincidence with the broaband noise burst. After Pottelette et al. (1999).

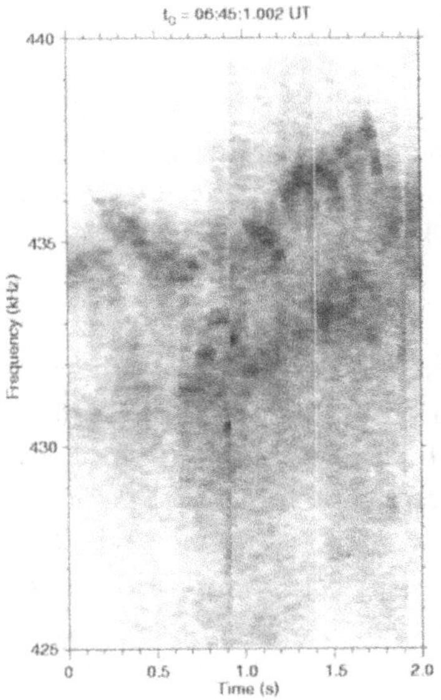

Figure 4.25. High-time and frequency resolution plot of FAST tracker data during the AKR event in Figure 4.24 showing elementary radiation sources moving downward in altitude (upward in frequency). The most intense radiation is emitted from these moving fine structures the nature of which is still uncertain. From Pottelette et al. (1999).

son, 1981) showed that the radiation is not continuous and also is intrinsically not broadband. It rather consists of many discrete narrowband and drifting emissions. A recent high time and frequency resolution example obtained by FAST when it crossed the AKR source is displayed in Figure 4.25.

This figure shows that the intense emission events are built up of several narrowband features. Two of them appear in the frequency range covered by Figure 4.25, which seem to be about equally spaced in frequency. Since the cyclotron frequency changes with altitude, their frequency spacing suggests that they are also about equally spaced in distance along the magnetic field line. Each is composed of the radiative contribution of many narrow-band elementary radiation events. The center frequency of the slow emission features often varies in a systematic manner, sweeping either upward or downward across the spectrum and lasting for several seconds. In contrast, the individual elementary radiation events are typically ≤ 1 s in duration and have a spectral width of ≤ 1 kHz in frequency. So far the nature of these elementary emissions has not been clarified. Their inferred velocity falls into the ion-acoustic wave range; they are possibly related to the ion holes (Tetreault, 1988; Gray et al., 1991) that may evolve in plasmas containing several particle populations of differing temperatures.

4.3.1.2 Auroral VLF Emissions

The AKR emission just discussed is the primary process by which the auroral system radiates wave energy at altitudes within and above the auroral potential drop. *Below* the potential drop, in regions containing significant cold ionospheric plasma densities, this radiation and the accompanying modification of the electron distribution function is taken up by whistler emission in the *Very Low Frequency* (VLF) wave band.

The VLF band in radio science is defined to extend from 3-30 kHz. Ground-based observations from the 1950s onward found a wealth of auroral emissions in this frequency range. In auroral physics the term VLF is used to cover emissions extending from the local lower hybrid frequency, ω_{LH}, up to the plasma frequency, ω_{pe}, or the electron cyclotron frequency, ω_{ce}, whichever is the lowest. This range includes lower-hybrid waves, whistlers and other emissions. We will consider several classes of VLF emissions of particular interest to auroral physics: the ubiquitous broadband waves mostly known as auroral whistler-hiss, lower-hybrid waves forming solitary structures and causing ion heating at low altitudes (Section 4.3.3.2), and the VLF saucers found in the return current region (Section 4.3.3.5).

Auroral Hiss. Auroral hiss is the general name for the persistent broad-band emissions above about 500 Hz seen below the auroral potential drop. For a review see Sazhin et al. (1993) and LaBelle and Treumann (2002). Examples can be seen in the bottom panel of Figure 4.3, at times when the spacecraft is not in the density cavity (i.e., after 10:11:40 UT or before 10:11:20 UT.) The amplitude varies and the lower frequency cutoff moves depending on the ambient plasma properties. Figure 4.26 shows the profile of electron precipitation together with VLF activity during the Topaz3 sounding-rocket northward crossing of a pre-midnight auroral arc. The low-frequency cutoff near the lower hybrid frequency is seen to first decrease and then increase with the local density as the rocket payload rises and then falls in altitude. The electric to magnetic field ratio (the phase velocity) is large near the lower hybrid cutoff, and decreases as the frequency increases to the upper edge of the hiss band. From sounding rockets at altitudes of 800-1000 km, using a quadrupole technique, Ergun et al. (1991) found average wavelengths of 15-50 m, while Vago et al. (1992) interpreted the absence of any phase shift in the double-probe interferometry measurements as the wavelength being at least 50 m.

The VLF-hiss low-frequency cutoff is at the local lower-hybrid (LH) frequency. The peak in the electric field spectrum occurs close to or a few times ω_{LH}, sometimes even higher (Beghin et al., 1989). The E/b ratio in VLF-hiss is generally large such that hiss can be treated in the electrostatic approximation. This fact identifies the waves as belonging to the whistler/lower-hybrid dispersion surface (André, 1985). For $\omega > \omega_{LH}$, this surface exhibits a resonance cone: the constant-frequency cut of the dispersion surface at some $\omega > \omega_{LH}$, for sufficiently high

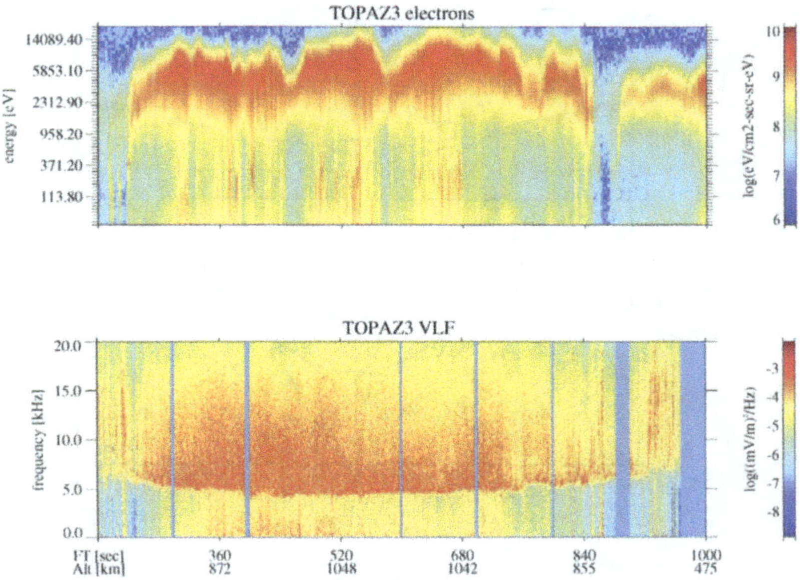

Figure 4.26. VLF activity associated with electron precipitation as seen by the Topaz3 sounding rocket (from Lynch et al., 1994).

values of $k_\perp$, yields a straight line in the $k_\parallel$-$k_\perp$-plane, representing a cone in 3D wave-vector space, on which the waves are quasi-electrostatic and right-hand elliptically polarized. This is shown in Figure 4.27 and agrees with observations.

Coherent Generation: Landau-Cherenkov Resonance. The free energy source of whistler mode VLF hiss is in the parallel deformation of the auroral electron distribution. This distribution has been identified as a horseshoe or shell distribution within the auroral potential drop region by in situ spacecraft observations. While the perpendicular gradients $\partial f/\partial v_\perp > 0$ contribute to AKR, the parallel gradient $\partial f/\partial v_\parallel > 0$, signifying the presence of an electron beam, contributes to hiss. The mechanism of generation depends on the frequency range and the plasma conditions but is based on Landau resonance $k_\parallel v_\parallel = \omega$ with the phase velocity of the growing waves being slightly below beam velocity. The waves are thus riding to some extent on the electron beam, depending on their frequency with slight or long delay.

Figure 4.28 illustrates the emission and convective amplification of the whistler mode hiss for two wave frequencies, one near ω_{ce}, the other well below ω_{ce}. The group velocity for the lower frequency waves is approximately field-aligned, and for higher frequency waves is nearly perpendicular, as shown in the figure. The low frequencies convect downward riding on the beam and thus become amplified. The

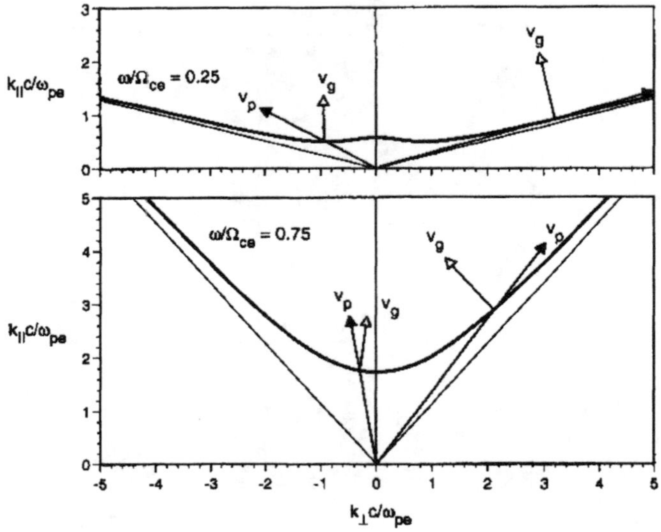

Figure 4.27. Dispersion surfaces for the resonance cone of the whistler mode. After Kivelson and Russell (1995).

high frequency waves escape sideways from the beam and arc region and are lost. Thus it is the low frequency waves where VLF peaks near the LH cut-off, as seen in Figure 4.26. This simple scenario explains most of the properties of the hiss. The amplification does not reach the intensities observed in AKR at higher altitudes. The hiss amplitudes are less than 1% of the AKR power, of order $<10^5$ W and only occasionally $<10^7$ W as measured at sounding rocket altitudes.

Nonetheless, VLF hiss plays an important role in beam modification *below* the parallel potential drop. Note that AKR does not contribute substantially to beam stabilization in the region where it is emitted, because AKR is a free space mode and escapes from the source to places where it is not in resonance anymore and thus does not affect the electron distribution function unless a particular trapping mechanism forces AKR to stay in the source. The distributions measured in the AKR source thus retain their beam-shell like character. Below the potential drop region, however, the nearly complete stabilization of the beam at sounding rocket altitudes is thus probably due to VLF. This would also be in accord with the comparable weakness of VLF indicating that the VLF intensity has been eaten up in the quasilinear plateauing of the beam at the intermediate altitudes. LaBelle and Treumann (2002) recently reviewed the various forms of VLF hiss noting that, in addition, VLF hiss is of interest because of its diagnostic value. Two such diagnostic signatures (see Section 4.3.3) include VLF saucers, and lower hybrid solitary structures, which despite their name are truly localized regions of enhanced VLF hiss.

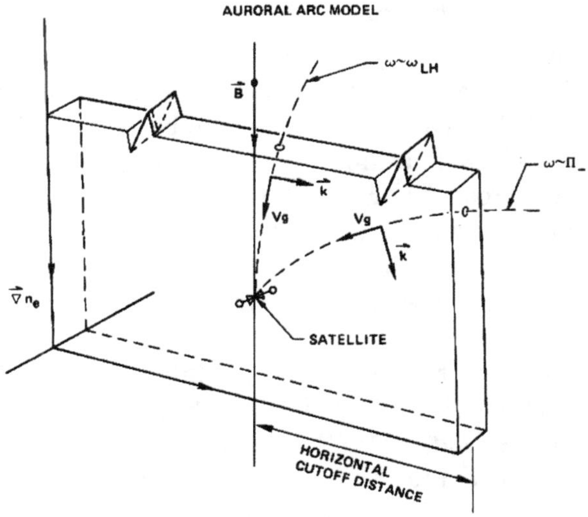

Figure 4.28. Coherent generation of VLF hiss. Phase and group velocity trajectories for wave-growth of whistler noise within a sheet of electron precipitation. The trajectory marked $\omega \sim \omega_{LH}$ is for low frequencies well below ω_{ce}, and the one marked $\omega \sim \Pi_-$ is for frequencies near ω_{ce}. From Maggs (1976).

Lower-hybrid Waves. In principle, the energy flow could go the other way as well: The VLF emissions just described could be an energy source for, rather than a result of, the precipitating electrons. Although lower hybrid waves have been proposed as the main agent for the acceleration of auroral electrons (Bingham et al., 1984; Bryant, 2002), they are no longer thought to provide a viable alternative. The basic problems with this scenario are the following. (i) The relevant LH waves are not commonly observed in the regions where the auroral electrons are accelerated. (ii) A model which produces significant auroral electron acceleration via LH waves must provide an energy source for these LH waves. Such a generation mechanism is not readily available in the auroral zone. (iii) LH waves are very vulnerable to field aligned electron flux. Their high parallel phase velocity lets them resonate with electrons of moderate thermal speed. This velocity-selective interaction weakly heats thermal electrons in the parallel direction and damps the waves. (iv) The 'standard model' of auroral acceleration by quasi-static electric fields explains not only electron acceleration but also the simultaneously observed upflowing ion beams, generation of AKR, and auroral hiss, and associates the aurora with the Poynting energy flux associated with the auroral field-aligned current system of upward and downward currents. One thus concludes that observed lower hybrid waves are rather caused by the unstable electron distribution as part of the VLF (see e.g., LaBelle and Treumann, 2002), and that their effect on the ions is transverse heating.

4.3.1.3 Broadband ELF Waves

We turn now to wave activity and wave-particle interactions that have strong effects on the auroral electron distribution functions in the *return current* regions of the aurora. The two predominant subjects here are Broadband Extremely Low Frequency waves (BBELF), and electron solitary waves (ESW). We begin with the more generic case, the broadband ELF activity.

BBELF is a name given to a number of wave phenomena with similar spectral shape that are found in conjunction with the energization of ion conics and thermalization of parallel electron beams in the downward current region and poleward boundary regions of the auroral zone. The strong link between BBELF and transverse ion acceleration was first noted in lower-altitude (800-1700 km) rocket and Freja satellite observations (Kintner et al., 1996; Bonnell et al., 1996; Lynch et al., 1996; Knudsen and Wahlund, 1998; André et al., 1998; Wahlund et al., 1998). Observations by the higher altitude FAST satellite confirm the statistically significant relationship between BBELF activity and strong ion energization and outflow (Lynch et al., 2002). The BBELF activity is a ubiquitous component of the electrodynamics of the return current region and the polar cap boundary Alfvénic regions. Its intensity is statistically well correlated with the dissipation expected from the coexistence of downward parallel DC electric fields and downward field-aligned current densities. Observationally the relationships of BBELF to the features of the return current region are clear, but its formation and structure, and its microphysical relationship to the upgoing and field-aligned electron distributions it appears to modify, are not as yet understood. The relationship between BBELF and localized electron solitary wave signatures, which appear embedded in a background of BBELF, is an especially interesting open question.

At rocket altitudes, BBELF seems to show a uniform frequency spectrum composed of waves with slow but still finite phase velocities. At the higher altitudes sampled by Freja, FAST, Polar, and S3-3, BBELF exhibits smooth as well as structured spectra with variable magnetic signatures. Indeed, the generic label BBELF is used for a number of different processes that only share similar frequency spectra.

Examples of BBELF spectra can be found in almost every FAST pass shown in this chapter that includes a return current or Alfvénic region. For this discussion, we focus on the examples in Figure 4.29, showing data taken by the FAST satellite at 4100 km altitude and 23 MLT over the northern auroral zone. This example presents several of the different types of BBELF activity. The properties of BBELF seen in these data are similar to the properties observed at lower altitudes by satellites and sounding rockets, but there are also some important distinctions that will be noted below.

As FAST passes from low to high latitudes in Figure 4.29, it passes through several wave and particle regimes. Here we will focus on the regions of up-flowing and counter-streaming electron beams, which are regions of downward and mixed field-aligned current, ion conic energization and ELF wave activity. FAST encoun-

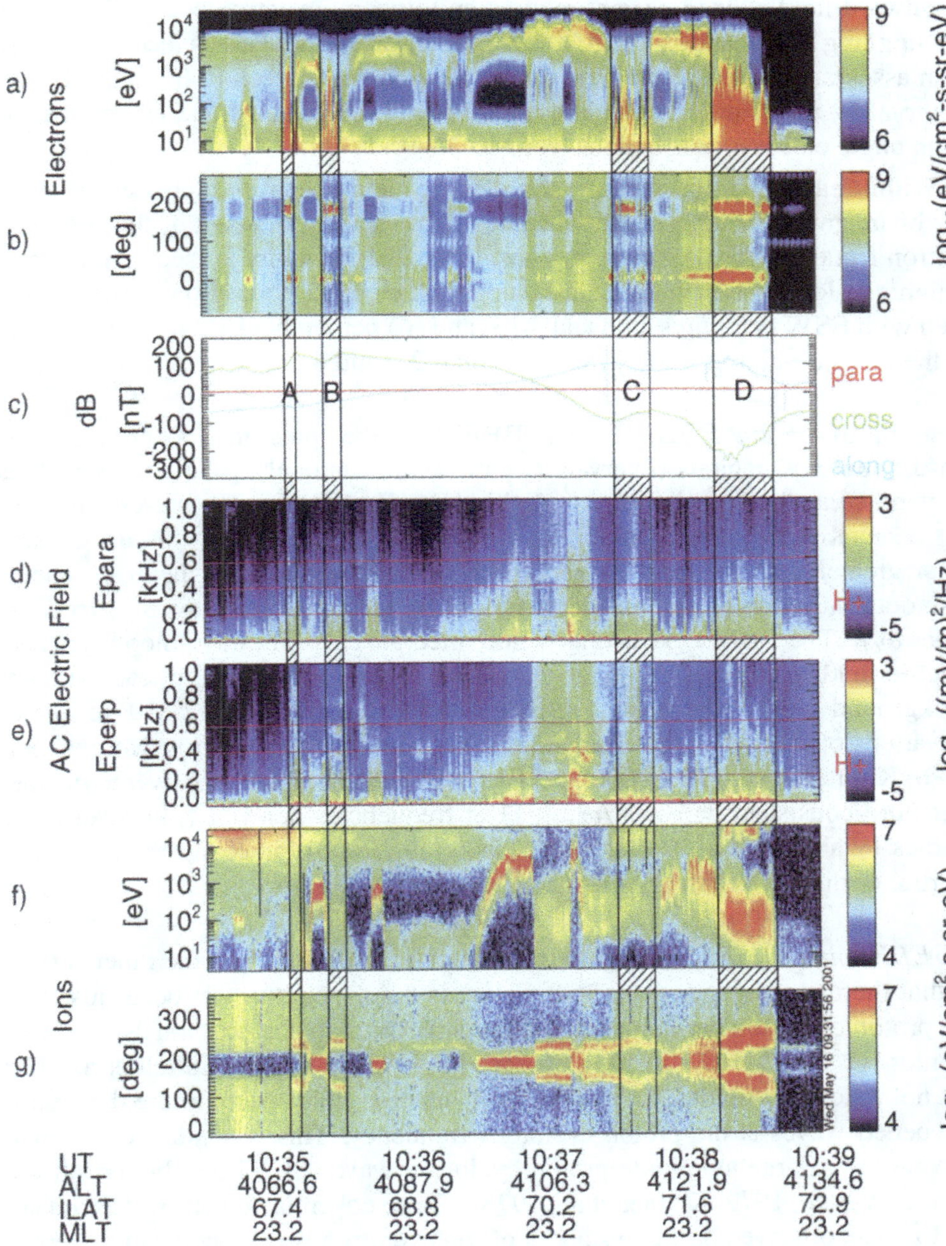

Figure 4.29. An example of BBELF and associated particle and DC field data as observed by the FAST satellite. Panels a) and b): Electron energy and pitch angle spectra. Panel c): Perturbation magnetic field in spacecraft coordinates (green trace is the cross-track component, with positive trend indicating downward field-aligned current). Panels d) and e): Spectral density of parallel and perpendicular components of ELF electric field from 0-1 kHz; the proton cyclotron harmonics are shown in red. Panels f) and g): Ion energy and pitch angle spectra. Figure provided by C. W. Carlsson.

ters four such regions during the pass shown: (intervals A and B) a pair of strong downward field-aligned current regions and upgoing electron beams that bracket the upgoing ion beam at 10:35:20 UT ; (interval C) a weaker electron beam region associated with the 'pressure cooker' ion conic between 10:37-10:38 UT; and (interval D) the poleward boundary region of ion conics and broad-energy, narrow-pitch angle electron distributions from 10:38-10:39 UT.

Within each of these regions, the electric field spectrum increases by an order of magnitude or more at frequencies up to several times ω_{cH}, the proton cyclotron frquency. In all of these regions, the BBELF activity is accompanied by (i) a highly field-aligned, predominantly upgoing, electron population similar to those seen with ESW in Figures 4.33 and 4.13; and (ii) perpendicularly heated ion conics of the pressure-cooker variety (see Section 4.2.2 and 4.3.2)

Smooth Spectrum BBELF. The BBELF in the 'pressure cooker' and 'poleward boundary' regions (intervals C and D) are obliquely polarized, with larger perpendicular than parallel electric fields ($E_\perp/E_\parallel \approx$ 3-10). The waves have small but detectable magnetic components; $E/b \approx 5 \cdot 10^6$-$10^7\,\mathrm{m\,s^{-1}}$ is on the order of or greater than the local Alfvén speed at frequencies up to the ω_{cH}. There is a smooth variation of spectral density with frequency from below ω_{cO} to several times ω_{cH}. The BBELF occurring within interval A also shows a smoothly varying electric field spectrum, but the magnetic component is absent or at least below the background level of the searchcoil magnetometer, with $E/b \geq 7 \cdot 10^7\,\mathrm{m\,s^{-1}}$. Observations of BBELF at lower altitudes found similar $E_\parallel/E_\perp$ ratios (Bonnell et al., 1996; Kintner et al., 1996, 2000). BBELF was found to have an Alfvénic magnetic signature consistent with $k_\perp c/\omega_{pe} \geq 1$ at frequencies below ω_{cO} in several Freja studies (Wahlund et al., 1998; Stasiewicz et al., 2000) indicating the presence of inertial or dispersive effects on Alfvén waves.

BBELF with Structured Spectrum. Interval B differs from the other intervals in that there are enhancements in the electric field spectrum that occur just below the proton cyclotron harmonics. A similar structuring by heavier ion cyclotron harmonics (O^+, He^+) has also been observed at FAST altitudes (Cattell et al., 1998) but not at lower altitudes. The peaks in the electric field spectrum are the result of steepened waves at the proton cyclotron frequency. This is similar to earlier observations of large-amplitude proton cyclotron waves by S3-3 at higher altitudes (Temerin et al., 1979; Kintner et al., 1978). These coherent solitary waves are similar to those observed during instances of ion cyclotron waves (ICW) in the upward current region, although the location and shape of the spectral peaks differ between the two phenomena. The perpendicular cross-spectra during such instances indicate long wavelengths of ≈ 1 km and high phase velocities, establishing that these features are temporal phenomena, rather than spatially-periodic phenomena that are Doppler-shifted to finite frequencies.

BBELF from Doppler-shifted Spatial Irregularities. A third type of commonly observed BBELF results from the Doppler shifting of zero frequency, spatially irregular electric fields into the frequency domain. These 'spatial irregularities' have perpendicular phase velocities much less than the spacecraft velocity, and wavelengths as short as a few tens of meters. The broadband auroral spectral features at frequencies up to a few hundred Hz observed in these cases by Freja and Polar (Angelopoulos et al., 2000; Stasiewicz et al., 2000) can be attributed to such a process. In each of these studies the finite wavelengths of the spatial irregularities was either inferred from the observed wavelength-dependent antenna response, from the frequency dependence of the electric-to-magnetic perturbation ratio, or from estimation of the phase velocity using interferometric techniques. No large-scale survey of the prevalence of spatial irregularities over actual waves has yet been undertaken specific to the downward current and BBELF regions, so it is unclear what proportion of the observed BBELF activity is waves and what is spatial irregularities.

Interferometry: Wave Vector Distribution. An important distinction between these various phenomena with similar frequency spectra is their wave vector characteristics. Interferometeric studies can be used to identify the different wave vector distributions. Figure 4.30 points out the basics of interferometry and coherency measurement techniques. Interferometric techniques estimate (a) how similar the signals at two sensors are, and (b) the time lag or phase shift between two measurements; this information provides a measurement of spatial scale and distinguishes between temporal and spatial effects (LaBelle and Kintner, 1989; Pinçon et al., 1997).

At low altitudes, the wavelengths are found to be well below the estimated local electron skin depth, c/ω_{pe} which indicates that in these cases BBELF is an electrostatic wave mode. The parallel phase velocity $\omega/k_{\parallel}$ of the waves is upward, in the range of 30-50 km/s, with perpendicular phase velocities of around 10 km s^{-1} (Bonnell et al., 1996; Kintner et al., 1996). The waves have a peculiar coherence signature, in that the waves were well-correlated between parallel separated antennas and poorly-correlated between perpendicularly separated antennas. This signature cannot be reproduced by a single plane-wave model, and supports a model of BBELF wherein multiple directions of $k_{\perp}$ with similar values of $k_{\parallel}$ are present over the averaging intervals of 50-100 ms used to measure the wave coherence. Detailed modeling of the perpendicular coherency results (Kintner et al., 2000; Delory et al., 1997) show that the wavelength of BBELF near $\omega_{c\text{O}}$ is on the order of 10-30 m, corresponding to very slow phase velocities.

Models of BBELF: Drift Instabilities. Probably because the term BBELF is such a catch-all for a number of different wave phenomena, there is a similar number of possibilities for formation mechanisms. We examine a few of them now.

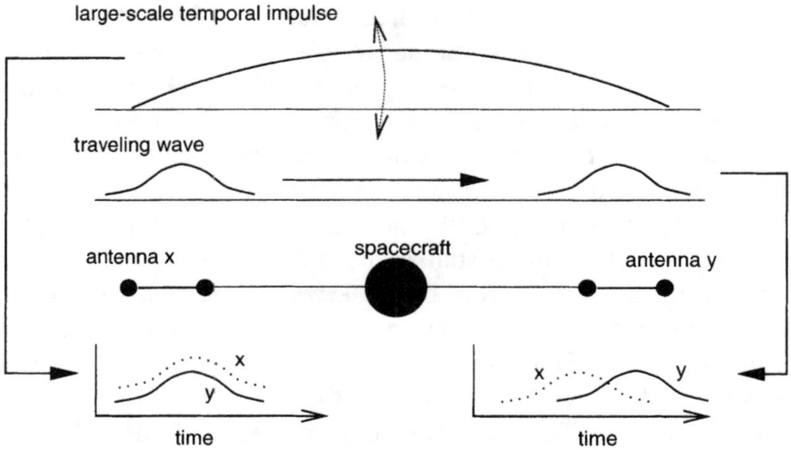

Figure 4.30. A schematic representation of how a large-scale temporal impulse and a traveling wave could produce similar signals on two spatially-separated antennas. Figure provided by J. P. McFadden.

In the downward current region, the upward current drift of accelerated electrons relative to the transversely accelerated ions provides free energy for wave generation. The observed drift speed $v_D \approx 0.1v_e$ and parallel electron to ion temperature ratios $T_{e\parallel}/T_{i\parallel} \approx 10$ suggest that both the observed (Cattell et al., 1998; Chaston et al., 2002c) ion-cyclotron or oblique ion-acoustic waves can be unstably excited (Kindel and Kennel, 1971). The anisotropy of the ion distribution ($T_{i\perp}/T_{i\parallel} > 1$) will drive the most unstable drift-driven modes to larger values of $k_\parallel/k_\perp$ than those predicted in models using an isotropic ion distribution (Lee, 1972). At lower altitudes, BBELF is isotropic in $k_\perp$ and obeys a linear relationship $\omega \propto k_\parallel$ (Bonnell et al., 1996) in agreement with ion cyclotron and ion acoustic waves. However, the small $T_{i\perp}/T_{i\parallel}$ and small v_D make the drift-driven hypotheses less feasible.

Linear models of wave generation in uniform plasma in the BBELF frequency regime predict emission at frequencies between the ion cyclotron frequencies, and damping at and near the cyclotron frequencies, with the widths of the nulls in wave growth dependent upon $T_{i\perp}/T_{i\parallel}$ and $k_\parallel/k_\perp$. This damping may become obscured for phase velocities comparable to or less than the spacecraft velocity, an effect that at low altitudes may explain the broadband nature. At high altitudes, the wave phase velocities are probably much higher, due to the lack of low temperature ionospheric plasma, and the higher electron and ion temperature, limiting the Doppler broadening effect to $< 10\%$.

Models of BBELF: Shear-driven Instabilities. Inhomogeneities in the plasma flow provide an additional source of free energy for wave generation (Bonnell et al., 1996; Amatucci, 1999; Koepke et al., 1999; Hamrin et al., 2000). The attraction of these models that include spatial shears in the flows are three-fold: (i) the resulting waves grow for lower drift speeds than in the current drift models; (ii) positive growth can occur for much lower values of T_e/T_i, and predicting growth for T_e/T_i both much greater than and less than one; and (iii) the growth rate of the waves is positive over a broader range of frequencies, and remains broadband even for small values of $k_{\parallel}/k_{\perp}$. Such models require an inference for the perpendicular scale length of the flows, either by assuming that flow structure has been advected over the spacecraft or that the large scale, lower frequency flows are driven by a phenomenon with a known maximum scale size (such as an inertial Alfvén wave.)

Models of BBELF: Nonlinear Instabilities. Some nonlinear models of BBELF have also been proposed. Freja observations suggest that the breakup of steepened inertial Alfvén waves into short perpendicular wavelength O^+ slow ion-acoustic waves is responsible for observed (Seyler and Wahlund, 1996; Seyler et al., 1998; Wahlund et al., 1998) strongly perpendicularly polarized BBELF. This model assumes a downward inertial Alfvén wave packet, which then steepens and becomes electrostatic until it finally breaks up. At this stage it emits slow ion-acoustic wave packets, converting the initially long-wavelength electromagnetic energy into short-wavelength electrostatic energy that exchanges energy with the ambient ion populations. One apparent difficulty with this model is the requirement that $T_i \gg T_e$ for the breakup to occur. While this condition is met in the Freja observations, it is unclear how the system gets from the initial ionospheric condition of $T_i \approx T_e$ to the hot ion state without another source of ion heating to get it there. Usually it is assumed that ion heating is caused by electrostatic ion cyclotron waves.

Implications of BBELF. We have examined observations and theories for a class of ELF wave activity seen throughout the return current and Alfvénic regions of the auroral zone. Observationally it seems evident that this wave activity plays a crucial role in the electrodynamics of these regions. There is a statistically significant correlation between the intensity of BBELF activity, and the imposed and resulting DC field quantities of the region. The BBELF seems to play a role in broadening the energy spectrum of the upgoing return current electrons, and is clearly related to the ion conic formation which mediates the scale heights of the plasma populations here. We consider next a related wave phenomenon, electron solitary waves. In the ion effects section later, we examine the relationship between the BBELF activity and transverse ion acceleration and conics. While much is known about BBELF activity, much of the microphysics of its generation mechanisms and wave-particle interactions remain an interesting open question.

4.3.1.4 Electron Solitary Waves and Phase-space Holes

Next we study the microphysics of a dramatic wave phenomenon of the return current region, referred to as electron solitary waves or electron phase space holes (see Section 3.5.4). These are two different names for the same type of nonlinear wave structure, emphasizing the electric field signature or the electron response to this structure, respectively. The specific relationship of these structures to the BBELF activity just described is as yet unknown, but the two coexist in the region of the upgoing electrons of the return current region. These strongly localized large-amplitude electrostatic waves (Boström et al., 1988; Mozer et al., 1997; Ergun et al., 1998c,d; Franz et al., 1998, 2000) occur predominantly in the downward (return) current region and at the edges of the AKR source, both the sites of intense localized wave activity. Similar structures have been found in space plasmas outside of the aurora, in the plasma sheet boundary layer (Matsumoto et al., 1994), in the magnetopause, and even in the ramp of the bow shock (Bale et al., 1998). We examine now the details of these structures and their interactions with the upgoing electron population.

Observations. Figure 4.31 shows a very high time resolution display of the electric waveform in the return current region recorded by FAST. The upper panel is ~ 1 min of waveform data exhibiting a sequence of intense bipolar electric spikes in the parallel wave electric field $E_\|$. The frequency spectrum of such impulsive events gives a series of broadband noise pulses. The lower panel shows an expansion of 2 ms of the most intense waves. The impressive fact is that the wave form is bipolar, sinusoidal, strictly symmetric, and localized in spacetime with duration of 0.1 ms. The largest $E_\|$ amplitudes reach about 100 mV m^{-1}. There is no net potential drop across them. The perpendicular amplitudes are of similar size; these are unipolar. Hence, these structures closely resemble electrically neutral solitary waves or phase space holes which arise from the nonlinear dispersive evolution of plasma waves. The origin of this nonlinear process is similar to the Cherenkov emission process of Langmuir waves from an electron beam (see Section 4.3.3).

Electron holes have a net charge close to zero. The peak hole electric field amplitude $|e\Phi_0|$ is most often $<0.2k_BT_{e\|}$. A histogram (see Figure 4.32) of the occurrence of $|e\Phi_0/k_BT_{e\|}|$ seems to obey a power law for large potentials suggesting that large-amplitude electron holes may form through merging of weaker electron holes. There have been very few observations of structures with $e\Phi_0/k_BT_{e\|} > 1$. Thus, on average, electron phase-space holes seem to have no statistically significant net potential drop; $\Delta\Phi/\Phi_0 \approx 0$ with a standard deviation of $\sim 5\%$. Asymmetries could arise from an interaction with background plasma waves, growth or decay of the structure, or random net potentials.

Electron Distributions and Heating. One might consider whether it is these structures which would be responsible for the upward electron acceleration. However, their apparent statistical bipolar symmetry prevents such a conclusion. Only recently have nonsymmetric bipolar structures of amplitude up to 100 mV m^{-1} occasionally been found in the FAST data during passages of the return current re-

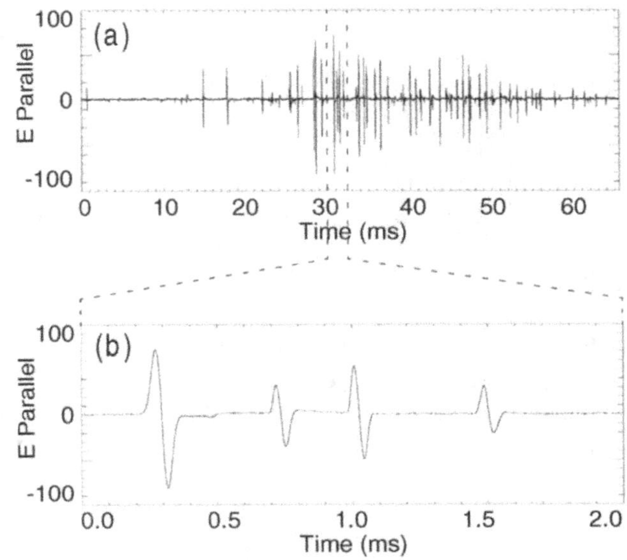

Figure 4.31. FAST observations of the parallel electric field signal of a series of electron phase-space holes. The top panel (a) shows ~65 ms of data at 0.5 ms resolution. The series of bipolar structures are resolved at this resolution. The bottom panel (b) is a further expanded view of four of the electron phase-space holes.

gion. The ESW are observed only in conjunction with upgoing field-aligned ionospheric electrons that are intense, collimated to within $\sim 1°$ pitch angle, and accelerated (10 eV-10 keV). Figure 4.33 shows the type of electron distribution found with the holes. Distributions like this one have been discussed in Section 4.2.1, and are shown in Figure 4.13. The parallel upgoing electron distribution dominates this figure with nonthermal electrons best fitted with a flat-topped steep power law. Interestingly, such distributions are typical generalized Lorentzian distributions which have been shown (e.g., Treumann, 1999) to be thermodynamically stable distributions far from thermal equilibrium arising in the collisionless state. Calculated drift velocities in this example are $v_D \approx -2 \times 10^6$ m s^{-1} while v_e, defined as the second moment of the distribution in the electron drift frame, is $\sim 6\times10^6$ m s^{-1}. The ratio of $|v_D/v_e| \approx \frac{1}{3}$ is not uncommon. Most often $|v_D/v_e| \sim 0.1$-0.2 with only one event out of many hundreds having $|v_D/v_e| \approx 1$. The distribution function is consistent with a DC acceleration of electrons into a beam up the field line which is rapidly plateaued by an interaction with the wave activity. The width of the 'beam' is much larger than the original drift velocity of the accelerated beam. Distribution functions like this are typical for solitary structures and phase space holes.

The parallel and perpendicular scales of the holes in the strong auroral magnetic fields with $\omega_{ce} \gg \omega_{pe}$ are about equal, as supported by the measurements shown in Figure 4.32. In the weaker fields at larger altitude the holes become oblate as

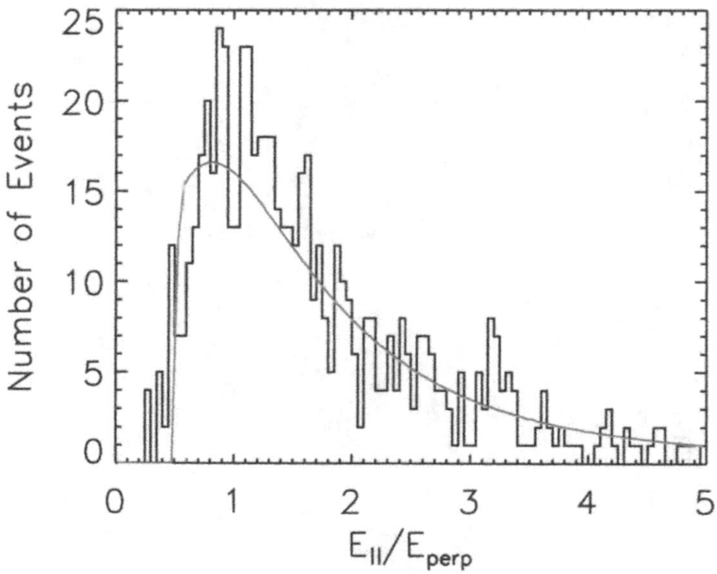

Figure 4.32. Histogram of the electric hole amplitude ratio $E_{\parallel}/E_{\perp}$. The sharp cutoff at ~ 0.5 is due to the selection criteria. With the selection criteria properly accounted for, the histogram fits well to that expected from solitary structures with an oblateness of 1.4. Figure provided by R. E. Ergun.

confirmed by the measurements of the Polar satellite (Franz et al., 2000). The electron distributions display strong modulations that are consistent with the electron phase-space hole interpretation (Muschietti et al., 1999b,a).

Panel (a) in Figure 4.34 has the time axis for $E_{\parallel}$ translated into Debye lengths λ_D assuming a constant parallel velocity of the structures. The measured signal (circles) fits remarkably well to the derivative of a Gaussian (solid line)

$$E(z) = E_0 \frac{z}{z_0} \exp\left[-\frac{1}{2}\left(\frac{z}{z_0}\right)^2\right] \tag{4.6}$$

Here z is the distance along the magnetic field, and the values of $z_0=(0.7\pm0.3)\lambda_D$ and E_0 have been determined by the fit. The charge density of a Gaussian structure lies between the two traces in Panel (b) of Figure 4.34 which represent spherical and planar geometry. The structure, if assumed to be spherical, has a positive core such that the deviation from charge neutrality is roughly $\delta n/n_0 \approx 8\%$. The positive core is surrounded by a negative halo thus, in total, making them electrically neutral. Whether or not they can substantially contribute to the formation of field-aligned potential drops remains theoretically unclear though in the FAST data asymmetric hole-electric field signatures have been seen occasionally (Ergun et al., 1998d; Pottelette et al., 2002).

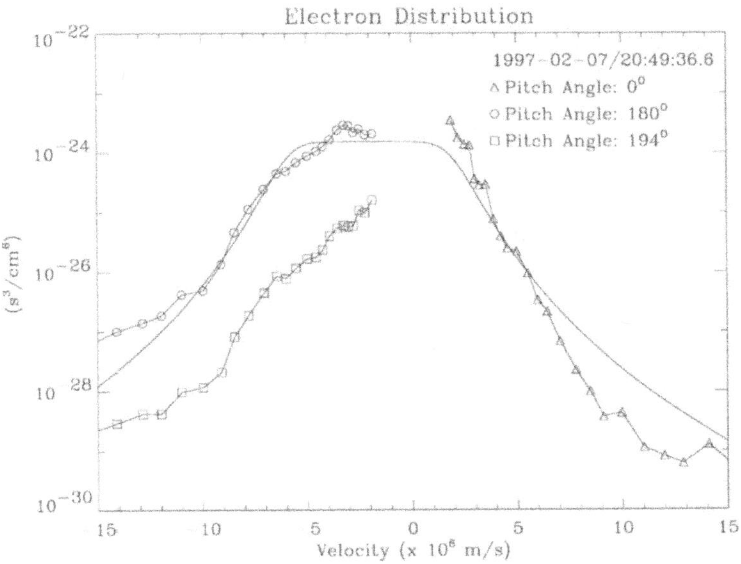

Figure 4.33. A return current region electron distribution at three pitch angles. The anti-parallel distribution (180°) dominates over all other angles, indicating the electron fluxes were nearly field-aligned. This distribution had $v_D \approx -2\times10^6$ and $T_{e\parallel} \approx 30$ eV ($v_e \approx 6\times10^6$ m s^{-1}). Measurements below 1×10^6 m s^{-1} could be affected by spacecraft charging. The solid line shows a model flat-top distribution. Figure provided by R. E. Ergun.

Comparison with Theory. The theory of solitons and electron holes has been given in Section 3.5. In the case of the auroral cavity with no cold plasma it can be based on the BGK theory (Bernstein et al., 1957; Saeki et al., 1994; Rasmussen, 1982; Dupree, 1982; Schamel, 1982b; Turikov, 1984; Muschietti et al., 1999b,a,c). When small amounts of cool plasma are present, it can be based on fluid theory (Berthomier et al., 2000), taking into account two-component two-temperature electron-beam plasmas (Pottelette and Treumann, 1998), in which case they are nonlinear electron-acoustic structures. Simulations have shown (Omura et al., 1994; Goldman et al., 1999; Oppenheim et al., 1999) that the minimum parallel size of electron phase-space holes increases with increasing potential. The relationship between the amplitude ($e\Phi_0/k_BT_{e\parallel}$) and size ($z_0$) has been displayed in Figure 3.14 of Chapter 3 which is in perfect agreement with the prediction, showing the diagnostic value of the observation of electron holes. They evolve in beam-plasma systems and have speeds far greater than the ion thermal speed and can undergo various kinds of instability thereby contributing to other wave types.

Simulation studies of the evolution of electron holes in 1D and 2D (Goldman et al., 1999, eg.,) have shown that the holes evolve through several stages. The structures initially appear to be nearly 1D, that is, with the perpendicular size much larger than the parallel size. In the early stages, a large number of moderate-

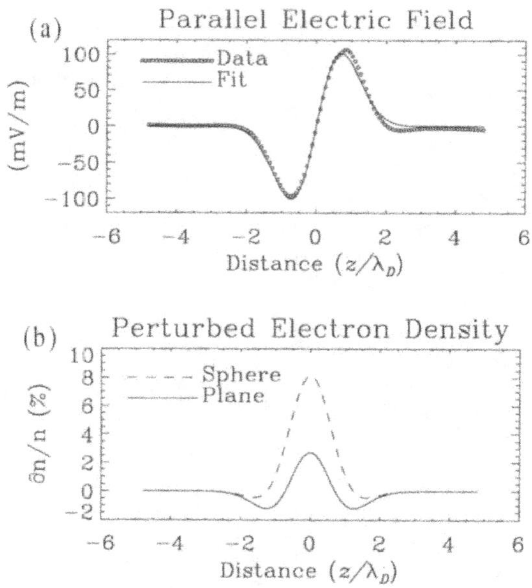

Figure 4.34. (a): The parallel electric field $E_{\parallel}$ in an observed electron hole. The circles are the measured waveform data at 5 ms time resolution, translated into Debye lengths assuming a constant parallel velocity of the structure. The smooth trace is the fit to Eq. 4.6. (b): Calculated charge densities assuming spherical and planar geometry. From Ergun et al. (1998d).

amplitude structures rapidly coalesce into a small number of large-amplitude structures. After ~100 plasma periods, the hole emits electrostatic whistler waves, and the amplitudes and perpendicular scale-sizes decrease (Oppenheim et al., 1999). In the final stages, ion-cyclotron waves dominate the electromagnetic spectrum (Goldman et al., 2000; Newman et al., 2001). Results like these suggest a very close relation between holes and other wave emissions such as hiss, whistlers, and saucers in the auroral zone.

Ion Heating by ESW. Unlike the field-aligned electrons, ion distributions in these regions are dominated by heated ions at near-perpendicular pitch angles when electron phase-space holes are present. The ion heating is a direct consequence of the formation of 2D or 3D electron holes. In the frame of the solitary structure, ions are incident at very high velocity compared to their thermal velocity, and will scatter from the positively charged core. The scattering angle is small. Assuming the perturbation to the ion trajectory is small and that the ions are unmagnetized, when the ion transit time is far less than the gyro-period, the change in perpendicular momentum of a proton passing through the solitary structure with impact parameter r is

$$m_i \Delta v_{i\perp}(r) = e \int_{-\infty}^{\infty} E[r, z(t)] \mathrm{d}t \tag{4.7}$$

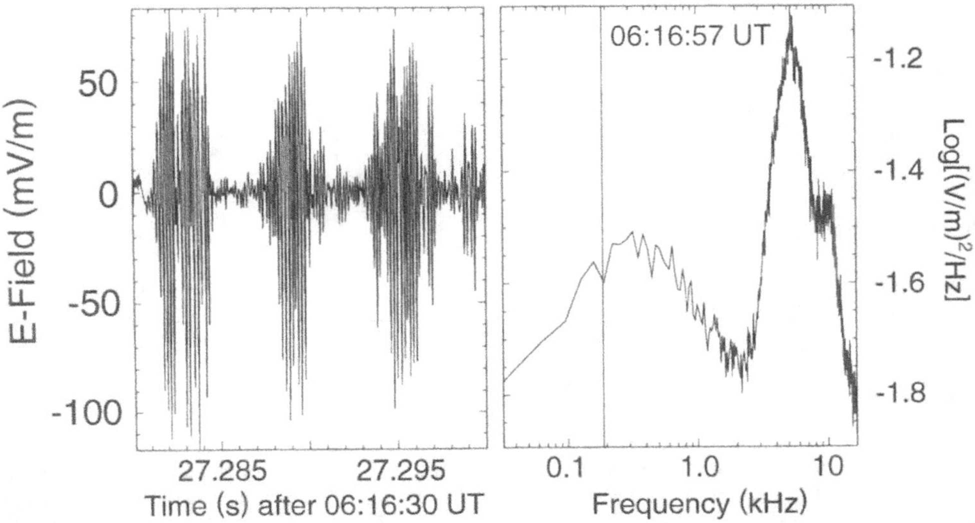

Figure 4.35. Left: FAST waveform data of broadband noise at edges of an AKR source. The signal consists of highly coherent nearly monochromatic burst, indicating wave trapping. Right: frequency spectrum of broadband noise showing peaks corresponding to the electron acoustic wave (at $\sim$ 5 kHz) and the plasma frequency (at $\sim$ 10 kHz). From Pottelette et al. (1999).

In the ion frame, the particles experience many such perpendicular impulses. For ion momenta randomly distributed in time or direction, the ions heat stochastically with heating rate proportional to $\langle E \rangle^2$. Heating rates can exceed $100\,\mathrm{eV\,s^{-1}}$, in extreme cases leading to keV ion energies. This interaction between ions and holes contributes to the dissipation of the holes; most of the dissipation goes into wave generation.

Impulsive Broadband Noise. So far we have dealt with single electron holes only. However, the observations suggest that they occur in multitudes. An example is given in Panel b of Figure 4.2 where the strong events in the waveform associated with electron signatures indicate that highly space-time modulated structures have been crossed by the spacecraft. A higher time resolved example is shown in Figure 4.35 together with its spectrum. Being localized in space while moving at high speeds, the holes appear in the spectrum as impulsive broadband noise features. Figure 4.36 shows how well a statistical ensemble of electron-acoustic solitons fits the spectrum of some broadband electric pulses in the auroral region (Dubouloz et al., 1993).

This broadband electric noise may be involved in electron heating. Dubouloz et al. (1993) have shown that the evolution of an initial electron distribution under

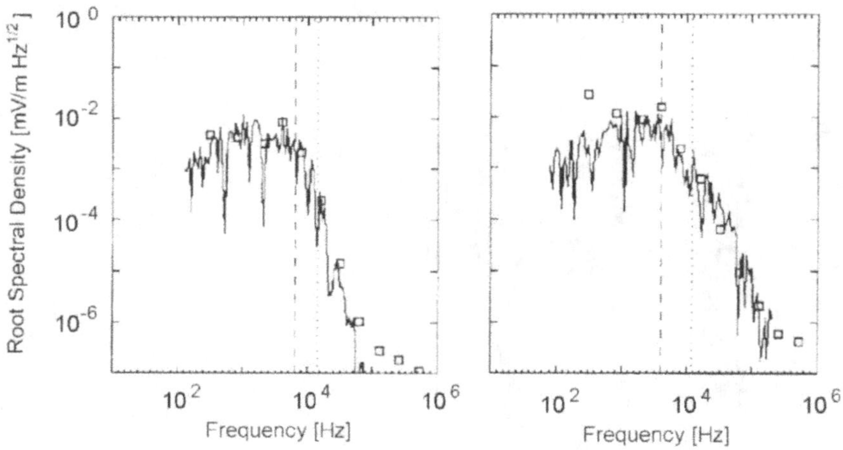

Figure 4.36. Viking spectral observations of two bursts (squares) of broadband electric noise in the auroral zone fitted by the spectrum of a statistical ensemble of electron-acoustic solitons in a two electron component two temperature plasma (solid lines). After Dubouloz et al. (1993).

quasilinear energy diffusion in presence of such noise ends up forming an extended electron tail by transit-time damping of the solitary waves. Heating can be as large as a factor of >10 in energy. However, as mentioned above, the mechanism requires the presence of a substantial fraction of cold electrons for which there is little experimental evidence in most cases. Moreover, the energy spectrum of the accelerated electrons rather points to direct acceleration in an electric potential without stochastic acceleration.

4.3.2 Ion Effects: EMIC, EIC, IAW, ELF

Now we shift to wave-particle interactions involving ion effects and ion-electron coupling. In the upward current region, such interactions include the electromagnetic ion cyclotron and electrostatic ion cyclotron activity associated with auroral flicker, and ion acoustic waves which grow from the upgoing ion beam populations within the auroral density cavity. Waves below the lower hybrid frequency and above the frequencies of Alfvén waves in the upward current region can be broadly categorized as electromagnetic ion cyclotron (EMIC), electrostatic ion cyclotron (EIC), ion acoustic and beam plasma modes. These waves are largely thought to be driven by current or beam instabilities growing from the accelerated electron and ion distributions characteristic of the auroral acceleration region. The following review of wave observations and possible source mechanisms is by no means exhaustive and is intended as an overview. Within the upward current region accelerated distributions and multiple ion and electron components present a plethora of possible instabilities for not only the homogeneous cases, as discussed below, but also non-homogeneous waves as discussed by Seyler et al. (1998) and Ganguli et al. (1994).

In the downward current region, the dominant ion interaction is the transverse acceleration of conics from BBELF and ESW. We consider here some of the wave-particle interaction details of this 'pressure-cooker' region that we have already seen in Section 4.2. We have seen that the ion heating here raises the scale height of the entire ionospheric plasma, and is a significant player in moderating the electrodynamics of the region.

4.3.2.1 Electromagnetic Ion-cyclotron Waves

EMIC waves were first identified in the auroral oval by Temerin et al. (1984). These waves are oblique, with $k_\perp/k_\parallel > 10$ and frequency $\omega < \omega_{ci}$ less than the local proton gyro-frequency and also below the He^+ and O^+ frequencies. They are observed at altitudes below the primary acceleration potential and are thought to be L-mode waves growing through Landau resonance with the downgoing monoenergetic electron beam. The fifth panel of Figure 4.29 shows an observation of this mode near 10:37 UT. The wave in this case occurs as a sharp peak in the electric field spectrum at 0.75 ω_{ci}. Coincident with this observation the bottom panel of this figure shows a heated ion distribution, or conic. These waves have been shown to heat ion conics through gyro-resonance (Lund et al., 1998; André et al., 1998), and are associated with flickering aurora (see the next section).

4.3.2.2 Electrostatic Ion-cyclotron Waves

EIC waves occur largely within the primary acceleration potential where upgoing ion beams are observed and are in essence ion-acoustic waves coupled to ion gyromotion. In the current-driven case, Kindel and Kennel (1971) have shown that the threshold electron drift to drive EIC waves unstable is less than that for the ion-acoustic instability and so these waves should be more commonly observed. They occur above ω_{ci} and often have a harmonic structure separated roughly by multiples of ω_{ci} (Kintner et al., 1979; Cattell, 1981). The waves are oblique with $k_\perp/k_\parallel > 5$ and occur on the same dispersion surface as ion-Bernstein modes but with finite $k_\parallel$. Several studies have shown that these waves are unstable to growth through inverse Landau resonance with the electron beam (Berger and Field, 1984), a Doppler shifted gyro-resonance with the ion beam (Kaufmann and Kintner, 1982) and for extremely oblique waves through cyclotron resonance on the plasma sheet ion loss cone (André, 1986). Recent observations from FAST show that although these waves have been traditionally identified as electrostatic, they often carry a measurable magnetic component (Cattell et al., 1998; Chaston et al., 1998) with phase velocities E/b as low as c. These waves are nearly always observed in coincidence with ion beams, as shown in Figure 4.29 at 10:37:15 UT, which may be transversely heated in the wavefield to temperatures approaching that of the plasma sheet ions. The generation of these wave has been shown to occur through an absolute instability of the field-aligned electron beam (Chaston et al., 2002a), leading to narrow bandwidths, amplitudes as large as 1 V m^{-1} and keV modulation of the downgoing electron energy at the wave frequency.

4.3.2.3 Flicker

EMIC wave activity below the potential drop, and EIC within the acceleration region, are both associated with a modulation of the downward electron energy flux of the auroral precipitation causing flickering aurora (Temerin et al., 1986; McFadden et al., 1987), a modulation of the visible aurora discussed in Chapter 2. Figure 4.37 illustrates the strongly modulated electron flux as seen at sounding rocket altitudes. The modulation involves both the primary auroral inverted-V population (top two panels) and also accelerated cold ionospheric electrons (bottom panel.) The modulation in the inverted-V population can be seen both as a fluctuation in the total count rate, and as dispersion in the pitch angle signature. Similar data from the FAST spacecraft show that this modulation is consistent with a modulation of the energy, rather than the number flux, of this population (McFadden et al., 1999b).

In addition to this modulation of the precipitating main electron population, EMIC is also associated with wave-particle interactions with the colder ambient, back scattered, and secondary populations below the acceleration zone. Evidence for this interaction with the cold ambient population can be seen in the lower panel of Figure 4.37, where dispersed bursts of electrons are seen at the cyclotron frequency. Depending upon instrument resolution, altitude of the spacecraft, and wave activity, narrow or broad spectral features in the electron spectrum can be produced with energy less than the inverted-V spectral peak. The electrons below the spectral peak are part of a circulation pattern of up-going cold and down-going accelerated electrons that carry no net current but extract energy from waves and transfer the energy to the atmosphere. The waves which play the most important role in this process are EMIC waves. Another example of these accelerated populations is seen in Figure 4.3, at 10:11:40 UT.

The accelerated cold electrons seen in Figure 4.37 further support the concept of strong sheath effects in the acceleration region. The cold populations are accelerated up to the inverted-V potential energy; in order for the cold ionospheric population to diffuse upward into the potential drop region during one cycle of the modulation, the spatial distance they must travel is quite limited (Arnoldy et al., 1999). This acceleration is related to the Alfvénic acceleration discussed in Sections 4.4.2 and 7.6.

Another complication to the quasistatic arc picture, related to these wave-accelerated populations, is presented by the accelerated field-aligned electrons which are common at the edges of inverted-V arcs (Whalen and McDiarmid, 1972; McFadden et al., 1986) or within narrow arcs (McFadden et al., 1990). This precipitation

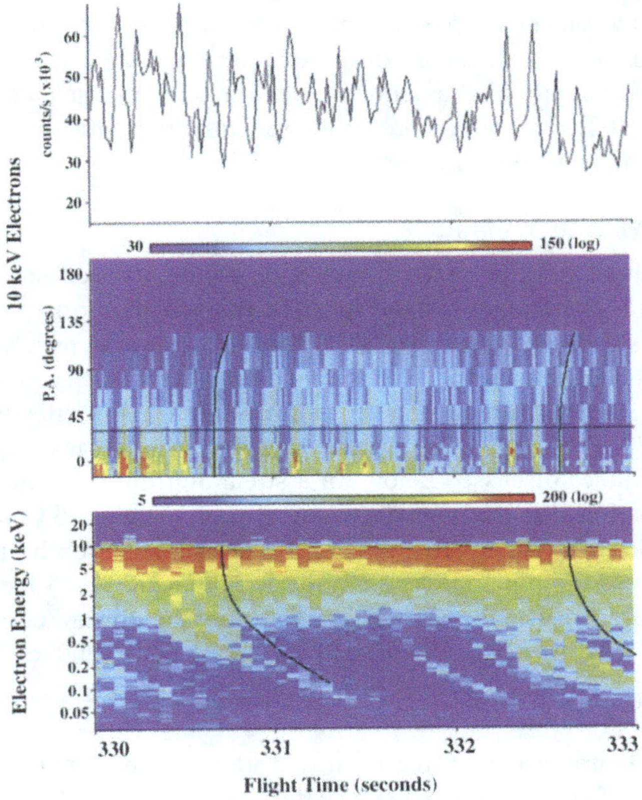

Figure 4.37. Modulation of electron precipitation associated with flicker. The principal inverted-V population is modulated at the oxygen cyclotron frequency, and dispersed bursts of accelerated ionospheric electrons are seen at the same frequency. From Arnoldy et al. (1999).

does not show the modulations seen in EMIC wave generated flux. It has been proposed that these features are due to convection of cold plasma into the side of the inverted-V acceleration region (McFadden et al., 1986). Boehm et al. (1995) argued that cold plasma convection into the edge of arcs is inconsistent with Freja observations of field-aligned electrons on both sides of narrow arcs, where it is clear that the arc motion through the plasma is in a single direction rather than the arc spreading on both edges. The cold electrons should rather be accelerated away only on the leading edge of the arc and not on the trailing edge. However, field-aligned electrons are observed throughout these narrow arcs. The quasistatic model of the auroral arc potential structure, with strong perpendicular electrostatic shocks at the edges, a strong double layer at the bottom, and a turbulent acceleration region with very weak parallel electric fields, may be consistent with several $\mathrm{km\,s^{-1}}$ drifts of cold plasma into these structures before being scattered and accelerated away (see Lotko (1986)). This model (McFadden et al., 1990) explains the presence of the secondary field-aligned inverted-V peak of about one third of the primary inverted-V peak energy observed within narrow arcs. However, these speculations have not provided a definitive answer on the subject of edge precipitation. A 2D simulation of auroral arc structure, with appropriate wave turbulence, may be required to solve this problem.

4.3.2.4 Ion-acoustic Waves

Within the upgoing ion beams of the acceleration region, ion-acoustic waves (IAW) may grow from a two-stream instability due to the relative drift of the multiple ionospheric ion species (Bergmann et al., 1988). These waves have been observed as parallel propagating waves in the primary potential drop with a broadband spectrum extending from zero frequency up to ion plasma frequencies (Cattell et al., 1998) and satisfy $\omega/k_{\parallel} = c_s \pm v_D$. These waves are the linear counterpart to the ion phase space holes discussed below. In a streaming multi-electron-component plasma such as is found in the upward current region, ion acoustic waves may also become destabilized by the relative drift of the ion and electron beams. Since the electron beam speed is much larger than the ion acoustic speed these waves are usually termed beam-plasma modes. These waves in the upward current region as observed from FAST have been discussed by Chaston et al. (2002c).

4.3.2.5 ELF Ion Conic Acceleration

In the downward current region, the most important wave-particle interaction involving ions is the feedback loop known as the 'pressure cooker' model, which we have seen in Sections 3.5.2 and 4.2.2. Here we consider some of the details of this wave-particle interaction.

In the original application of this model, to high altitude dayside heating from cyclotron wave activity, the perpendicular heating rate was parameterized as a single constant value. In BBELF and ESW regions, the perpendicular energization can be modeled as a series of random perpendicular velocity 'kicks' from the interaction of the ions with the broadband ELF wave activity. Similar calculations have been made in other models (Temerin, 1986; Hultqvist, 1996). Including these random perpendicular velocity 'kicks' in the model is important for reproducing the observed reflected population, and for reproducing the observed temperature ratios.

Reproducing the ion conic distribution function data with a model is a matter of choosing electric field and ion energization altitude profiles. Constraints on these choices are found from (a) the electron data, which indicate the value of the integrated parallel electric field up to the observation altitude; (b) the ion data, both energy and pitch angle distribution, which constrain the allowed values of the electric field; and (c) reasonable assumptions about the possible efficiency of the wave-particle interaction between the ions and the BBELF.

The 'pressure-cooker' model is useful for understanding the ion distribution function patterns, but it is a highly simplified model. The underlying physics of this region is a tightly coupled system involving the ion and electron populations and the evolution of the parallel field in response to current and quasineutrality requirements. While the pressure-cooker model tells us what the ion distributions should look like in response to $E_{\parallel}$-morphologies, the more important question is what determines $E_{\parallel}$. Statistical observations show that both strong and weak electric field downward current regions can be parameterized by the 'pressure-cooker' model, and the relationships between the particles, fields, and wave parameters give clues to the formation of the parallel fields here. These more complete current-voltage models are considered later in this book.

4.3.3 Diagnostic Wave Signatures

Finally we consider wave-particle interactions which are classified as *diagnostic*, as they can be useful probes of the plasma environment, but are electrodynamically insignificant, to the extent that the ratio of wave to kinetic energy density in the plasma, $\delta = W_w/nk_B(T_e + T_i)$, is extremely small. In the upward current regions, these diagnostic signatures include ion solitary waves within the acceleration region, and lower hybrid solitary structures below the acceleration region, as well as Langmuir wave, upper hybrid, and 'roar' signatures. In the downward current region the most visible diagnostic signature is that of the dramatic VLF saucers. We examine each of these signatures in turn, demonstrating for each how they can serve as diagnostics of plasma processes.

4.3.3.1 Ion Solitary Waves

Ion solitary structures, and their asymmetric versions known as weak double layers, are a typical signature of the upward current acceleration region. We discuss their observations, interpretations of these observations, and possible implications.

Observations. Figure 4.38 shows an example of *ion solitary waves* (ISW) measured by the Polar spacecraft. At the altitudes of Polar, the structures are primarily one dimensional, and have bipolar parallel electric field signatures of 10-100 mV m^{-1} (very similar to that observed for the electron holes of Section 4.3.1). Observations suggest that these are ion holes which grow out of an ion-acoustic instability driven by the upward current/downward electron drift velocity or ion-two stream instability. Their characteristic periods vary between 1 ms and 10 ms. They have scale sizes the order of about $\sim 10\,\lambda_D$, and travel at velocities near the ion-acoustic and upward proton beam speed. Typical measurements do not provide the necessary information to determine scale size and speed; they can be interpreted in many ways if one does not know the speed at which the pulse is passing the spacecraft. Polar spacecraft observations near perigee, as shown in Figure 4.38 provide some of the best velocity estimates of ISW using four antennae operated in voltage-

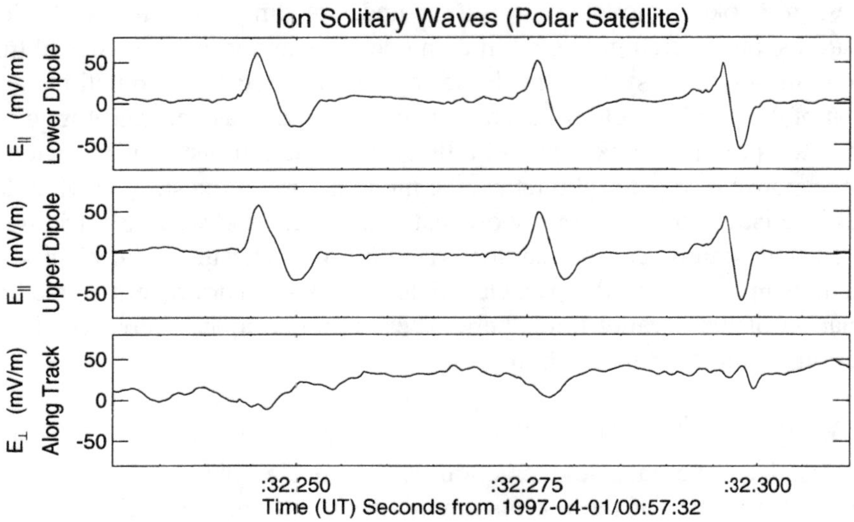

Figure 4.38. Ion solitary wave (ISW) signatures from the Polar spacecraft, showing one-dimensional bipolar parallel electric field signatures. Adapted from Dombeck et al. (2001).

mode. The measurable delay between two different baselines (two top panels) separated by either 50 m or 65 m along **B** gives the parallel electric field signal. The small 0.15 ms delay (not observable from the figure) between the baselines implies a ~200 km s^{-1} upward propagation velocity. Such delay measurements find ISW velocities in the range ~75-400 km s^{-1} for events where high resolution burst data were available. These velocities require solitary wave scale sizes of ~10-15 λ_D and potentials of $< 0.1 k_B T_e$. Dombeck et al. (2001) report that the solitary structures observed by the Polar spacecraft propagate at velocities intermediate between the H^+ and O^+ beam speeds, which supports a model that these structures are generated from an ion two-stream instability.

Further insight into ISW structure is provided by high-time-resolution particle measurements available on the FAST satellite. The upper panels of Figure 4.39 show a group of large amplitude ISW while the lower panels show the accompanying flux of 2 keV and 4 keV electrons. With the passage of each solitary wave, the 2 keV electron flux increases and the 4 keV flux decreases. These flux modulations are caused by the wave potential, shifting the electrons to lower energy as they pass through the structures. At 2 and 4 keV the distribution function has positive and negative slopes, so energy shifts result in positive and negative changes in counts, respectively. The measured shifts in the electron distributions showed that these structures had associated potentials of 300-900 V, scale sizes of 3-12 km (6-20 Debye lengths), and parallel phase velocities of 800 km s^{-1} (McFadden et al., 2002). This velocity is greater than the highest velocity H^+ ions in the beam, where

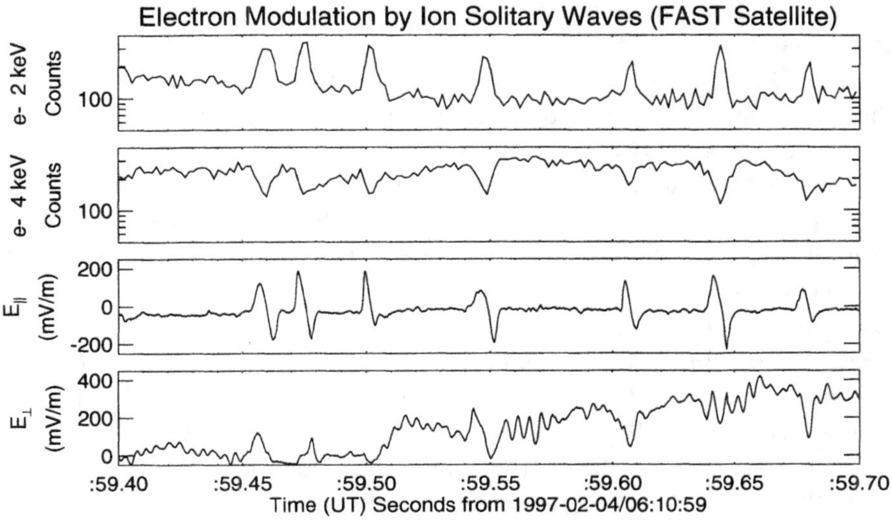

Figure 4.39. ISW signatures at lower altitudes from the FAST spacecraft, showing the accompanying modulation of the electron distribution function. From McFadden et al. (2002).

the H+ beam is generally observed to have broad distribution below a characteristic beam energy. This differs from the Polar observations where the ISWs had velocities between the H^+ and O^+ velocities and may indicate different generation mechanisms at the two different altitudes.

From the similar amplitude of parallel and perpendicular signals seen in Figures 4.39c and d, one infers that the parallel and perpendicular scale lengths of these lower altitude structures are comparable. These FAST observations at ~4000 km differ from the higher altitude Polar observations of Figure 4.38, and from those made by the S3-3 (6000-8000 km) and Viking (5000-13,500 km), where it was found that solitary waves had little or no perpendicular electric field component (Temerin et al., 1982; Boström et al., 1989). At the higher altitudes, it has been concluded that the structures are much more oblate with larger extension in the perpendicular direction.

The asymmetric electric field waveforms in Figure 4.39 would imply a potential double layer. However, as has been observed in electron structures, ion fluxes return to approximately the same level after passage of the wave, indicating no net potential drops (McFadden et al., 2002). Reasons for an observed apparent asymmetry are: (i) a non-ideal response of the electric field experiment, (ii) changes in the solitary wave structure on time scales of their passage, and (iii) non-Gaussian shape potentials. Observationally, smooth nearly symmetrical bipolar electric fields such as those seen in Figure 4.38 are the exception.

Interpretations. The importance of ion solitary waves within the auroral acceleration region has been a subject of speculation since the time when they were initially observed. Do they represent an essential element for maintaining the parallel electric fields? Or are they a consequence of the acceleration processes, forming through energy exchange between various unstable particle distributions? If a large number of asymmetric structures of same polarity exist along a magnetic field line, they might represent a coalescence of current driven turbulence that could maintain the parallel electric field which accelerates electrons and ions in inverted-V arcs (Hudson et al., 1983).

Early analyses of the S3-3 observations indicated that ISW could be described as BGK modes, the one-dimensional, time-stationary solutions of the Vlasov-Poisson equations that were observed in particle simulations (Hudson et al., 1983). This analysis suggested that small potential $e\Delta\Phi/k_B T_e \sim 1$ double layers arise non-linearly from *ion holes* which form from inhomogeneities in current driven ion-acoustic turbulence. In the simulations, as the symmetric ISW or ion holes grow, they reflect a portion of the electron current, resulting in a net potential drop typical for 'weak double layers' (see Section 3.3.3). Comparison with early spacecraft observations seemed consistent with the simulations, with inferred ion solitary wave velocities of $\sim 80\,\mathrm{km\,s^{-1}}$ and scale lengths of $\sim 32\lambda_D$. However, S3-3 estimates of the plasma density and temperature ($5\,\mathrm{cm^{-3}}$, $10\,\mathrm{eV}$) are inconsistent with more recent observations of the density cavity ($n < 1\,\mathrm{cm^{-3}}$, $k_B T_e \sim 1\,\mathrm{keV}$) within the acceleration region (McFadden et al., 1999b), where the Debye length must be more than an order of magnitude larger. In addition, the wave phase velocity could not be resolved by S3-3, and the measurements set only a lower limit ($> 50\,\mathrm{km\,s^{-1}}$) on the propagation speeds.

An alternate picture says that ion solitary waves result from the non-linear evolution of ion acoustic and ion cyclotron waves generated by a two-stream instability between the different-mass ion beams (Bergmann et al., 1988). Evidence for this instability is observed within ion beams, where protons have lower average energies than the heavier mass ions, as discussed earlier in this chapter. This would indicate energy exchange between the ions. Since the initial observations of these structures, other bipolar structures have been observed in association with electron beams. Resolution of the role that ISW play in auroral acceleration depends critically on reliable observations of the physical properties of these structures (their size, shape, and velocity) and precise measurements of the plasma distribution functions.

Implications. The idea that a large number of aligned ISW, through asymmetries in their potential drop structure, could cumulatively support a large potential drop in the auroral density cavity, has had an extensive impact on auroral literature. These asymmetric ISW, initially identified by the S3-3 satellite, termed *weak double layers* (Temerin et al., 1982) indicated net voltage drops across some of these structures. As described above, however, not many ISW are asymmetric, and it is not known that statistically they are cumulative. Many of the early results which seemed to show a preferred direction to the asymmetry can be shown to be a remnant of the measurement technique.

Observations by the Viking satellite (Boström et al., 1988) attempted to improve on S3-3. Viking's four-probe electric field experiment could be operated in either 'voltage mode', which measured electric fields, or in 'current mode', which was used to estimate the plasma density. Viking electric field observations of solitary waves were similar to those by S3-3, and showed that these structures were present at altitudes up to ~13,500 km (Mälkki et al., 1993).

The current-mode operation was used to estimate the ISW speed by measuring the delay time between current spikes to a pair of Langmuir probes aligned along the magnetic field. The measured delays indicated that the parallel scale lengths were about the length of the antenna (~80 meters), that the phase velocities and wave potentials were small (typically $< 50\,\mathrm{km\,s^{-1}}$ and $< 2\,\mathrm{V}$), and that the density depletion within the structures was large ($10\% < \delta n/n < 50\%$). However, problems with current-mode operation are indicating since >90% of the Langmuir probe current was from spacecraft- or antenna-produced photoelectrons when the probes were aligned along the magnetic field (Hilgers et al., 1992). The current spikes were due to interaction of the wave with the photo electrons. This problem renders the Viking estimate of phase velocity and other physical properties questionable. Similar problems arise on FAST when their electric field instruments are run in current mode (McFadden et al., 2002).

Finally, the more recent FAST observations show that asymmetric electric field signals do not necessarily imply a net potential drop. This suggests that the net potential drops across the ISW might be artificial, in which case the ISW contribute negligibly to the overall potential along the magnetic field. If this is the case, then they are secondary structures of diagnostic value for the state of the plasma in the upward current region, only.

4.3.3.2 Lower Hybrid Solitary Structures

Lower hybrid solitary structures (LHSS) are a ubiquitous signature within VLF hiss seen throughout the upward current regions of the aurora at all local times. On the nightside, these structures are associated with transverse ion acceleration. When evolving to large amplitudes, VLF whistler hiss with nearly perpendicular electric fields can form localized enhancements. These were first observed on the MARIE rocket flight (LaBelle et al., 1986) from Fort Churchill, which detected pulses of localized LH waves. Figure 4.40 shows an example of these measurements, taken at an altitude of ~600 km.

Numerous LHSS have been observed throughout the auroral zone (Kintner et al., 1992; Norqvist et al., 1996; Lynch et al., 1999b) and are associated on the nightside with ion-conic formation. The formation of these solitary structures was originally

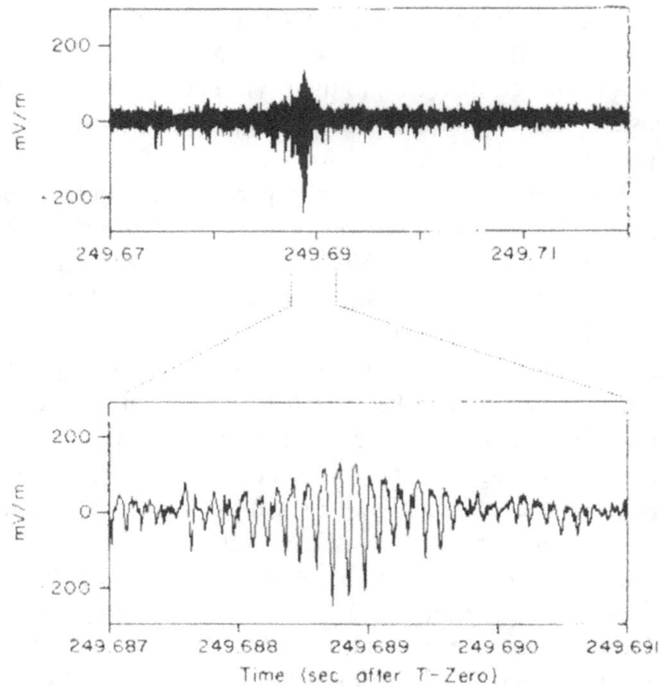

Figure 4.40. The original lower-hybrid 'spikelet' observations from the MARIE rocket at Fort Churchill. From LaBelle et al. (1986).

believed to be similar to Langmuir soliton formation, but more recent results have suggested that a linear scattering mechanism may be operative instead. Figure 4.41 illustrates a comparison of theory and model for one such event (Pinçon et al., 1997; Schuck et al., 1998).

LHSS are extended along the field line, while their transverse dimensions are the size of the ion gyroradius. Little is known so far about the dynamical consequences of LHSS though it is reasonable to assume that they predominantly lose their energy in resonant transverse ion heating and ion-conic formation. The ion acceleration associated with LHSS is based on the fact that the LH waves and LHSS have nearly perpendicular wave electric fields. An ion gyrating around the magnetic field and meeting for a while such a structure, the perpendicular size of which is less than the ion gyroradius, would be kicked in an unpredictable way in phase during each orbit when meeting the wave electric field. This causes transverse acceleration and conic formation.

The ion acceleration observed with nightside LHSS is narrowly confined to 90° pitch angle, in spatially confined regions of up to a few hundred meters across **B**. The acceleration process does not affect the thermal core of the ambient distribution and does not directly create a measurable effect on the ambient ion population

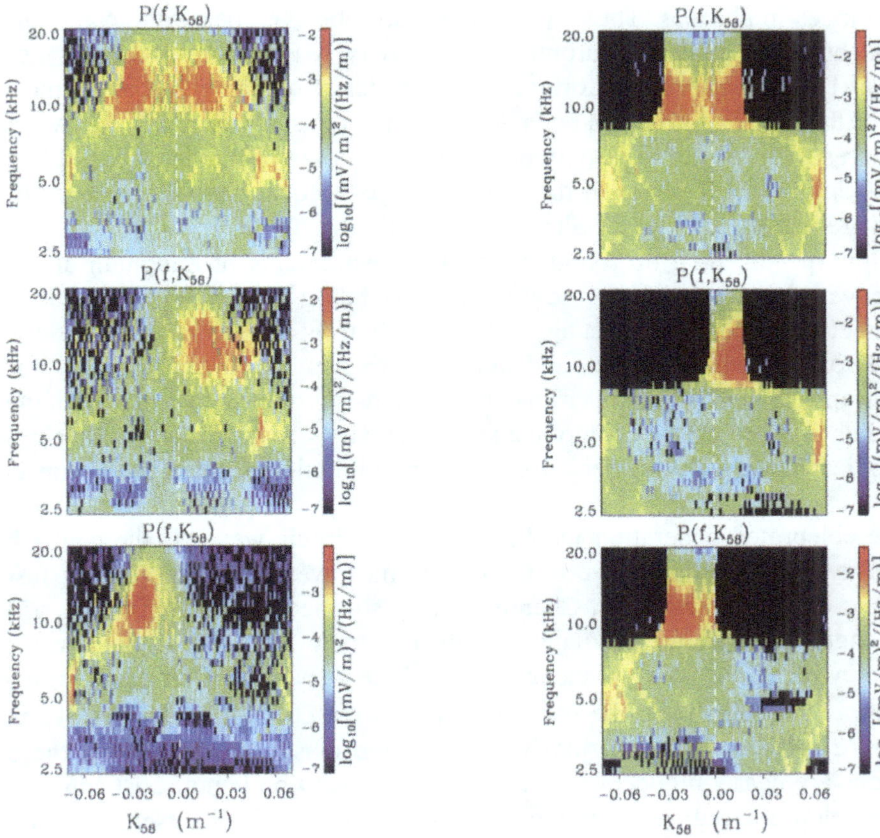

Figure 4.41. Observation (left) and simulation (right) of the dispersion relation (frequency versus wave number) of a lower hybrid solitary structure propagating state. In each case the top panel shows data from the entire event, the second panel from the first half-crossing of the event, and the third from the second half-crossing of the event. From Pinçon et al. (1997).

outside the LHSS themselves. This precludes observation of any nonlinear feedback between the ion acceleration and the existence or evolution of the density irregularities on which LHSS events grow. Within the LHSS region the acceleration process creates perpendicular heating for ions a few times the thermal ion speed. The cumulative ion outflow from the LHSS is consistent with Freja statistics for VLF-type pre-midnight auroral upflow events (Lynch et al., 1999b). It is not known why this transverse ion acceleration process is not seen on the dayside in otherwise similar LHSS wave events.

4.3.3.3 Langmuir and Upper Hybrid Waves

Other 'diagnostic' signatures in the upward current region include Langmuir (or in a magnetized plasma, upper hybrid) waves, the canonical electron-beam excited wave modes in plasmas. These wave modes are observed below the potential drop region, but not within the auroral density cavity where there is little or no cold plasma. This is somewhat unfortunate as their detection can give a direct measurement of the otherwise barely known local plasma density. The Langmuir waves are poorly resolved within the cavity because of the low plasma density, and because of masking by VLF waves in the same range of frequencies, while upper hybrid waves are masked by the prevalent AKR emission.

Below the potential drop region, many measurements of Langmuir and upper hybrid waves are provided by low altitude sounding rockets (Ergun et al., 1991; Boehm et al., 1990; LaBelle et al., 1999; McAdams et al., 1999). Here, these waves couple to VLF, upper hybrid, and plasma waves, producing weak radiation and mode trapping in existing upper ionospheric density irregularities. Their diagnostic value lies in the reconstruction of the spatial density variations. They also can serve as proxies for velocity-dispersed electron precipitation signatures (Ergun et al., 1991).

The celebrated generation mechanism of Langmuir waves is the gentle beam instability. It is most effective in dense plasma traversed by a relatively fast, dilute, and cool electron beam. Cherenkov emission leads to amplification of short wavelength electrostatic fluctuations with frequency close to the electron plasma frequency $\omega \sim \omega_{pe}$ and wavelengths $\lambda \geq \lambda_{De}$. Landau damping is unimportant as long as the velocity of the beam v_b satisfies the conditions $v_b > \sqrt{3}v_e$ where $v_e = \sqrt{2k_B T_e/m_e}$ is the thermal velocity of the plasma electrons, and the beam electrons are nearly monochromatic. The latter condition is not satisfied in the auroral acceleration region where the beam is hot and the plasma itself is dilute. One therefore expects a very low level of Langmuir turbulence. However, in the VLF region below the acceleration region, these waves are seen and can be used to determine plasma density.

4.3.3.4 Upper Hybrid Waves and Auroral Roar

Within the auroral cavity the upper hybrid frequency cannot be distinguished experimentally from the electron cyclotron frequency. Outside the cavity and at lower latitudes, where the plasma frequency exceeds the cyclotron frequency, a type of weak radio emission is observed which has been given the name of auroral roar. These emissions have first been detected from space (Kellogg and Monson, 1979) and verified by the ISIS 1 spacecraft (Benson and Wong, 1987). Later on they have been given much attention from ground based observations (Weatherwax et al., 1993, 1995) which could resolve them in frequency and time and show that they contain a very pronounced frequency-time fine structure.

Auroral roar has been always found in connection with aurora and substorm conditions. Energetically auroral roar is entirely unimportant. However, its detection can contribute to the diagnostics of the auroral region because it provides information on the plasma density. Auroral roar is emitted at frequencies close to harmonics of the electron cyclotron frequency and in the vicinity of the upper hybrid frequency. Ground based stations detect only those emissions which have high enough frequencies to pass across the dense ionosphere or leak through ionospheric density holes to reach the ground. The interpretation of these emissions is still unclear. Two possibilities arise. The first one is based on the cyclotron-maser mechanism acting now at relatively low altitudes in ionospheric density depletions (Weatherwax et al., 1995), the latter having been detected from ground based density measurements (Shepherd et al., 1998). It can act only in local plasma cavities. The second explanation is based on the double resonance mechanism when the upper hybrid frequency matches with the second or any higher harmonic of the cyclotron frequency. It thus works under the opposite conditions of high plasma densities. In this case the emission is narrowband and located in a very definite altitude depending on the ionospheric plasma density. It is possible that both mechanisms are at work in different places.

4.3.3.5 VLF Saucers

Perhaps the most conspicuous of the plasma waves in the downward current region is a type of electrostatic whistler emission known as a VLF saucer. VLF saucer emissions were among the first radio-wave features observed in the auroral zone (Smith et al., 1966; Gurnett, 1966) and since have been established as a common characteristic (James, 1976). The phenomenon is identified by a V-shaped or saucer-shaped appearance in a time-frequency-power spectrogram of wave electric fields in the kilohertz frequency range. Such V-shaped structures are seen over a range of time (or distance) scales ranging from less than a second (< 5 km) to hundreds of seconds (> 1000 km).

The saucer- or V-shape comes from propagation properties of quasi-electrostatic whistler waves. The lowest frequency emissions have a group velocity nearly parallel to the ambient magnetic field. Higher frequency, quasi-electrostatic whistler emissions have an oblique group velocity; the higher the frequency, the larger the angle from **B**. (We saw this already in our discussion of VLF whistler hiss; see Figure 4.28.) Thus, as a satellite passes over a small source region, it detects the higher frequencies farther away from flux tube of the source and the lowest frequencies as it passes over the flux tube of the source.

One of the most intriguing features of VLF saucers is the small physical size of the source regions that are inferred from the observations. Source sizes are as small as 0.5 km in latitude and less than 10 km in altitude (James, 1976). Such small source regions require strong, rapid wave growth. James (1976) postulated an instability from cold (< 5 eV), dense, upgoing ionospheric electrons carrying the downward or return current in the auroral zone impinging upon a warm magnetospheric population. At the time of this theory, no electrons had been detected with VLF saucers; the cold ionosphere electrons were thought to be too low in energy to be detectable by the satellite instruments. Over a decade later, the Viking satellite

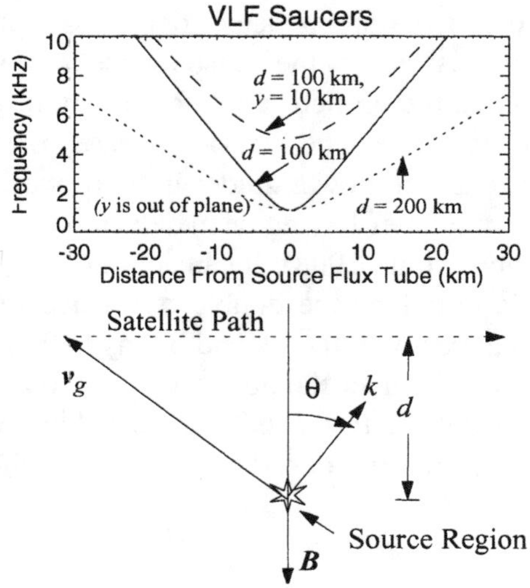

Figure 4.42. The top box shows the relation between the frequency and horizontal distance between the satellite and the source flux tube. The solid trace is the peak in the wave spectra from a point source as measured by a satellite 100 km above the source region. The dotted trace is from a point source 200 km below the satellite, and the dashed trace is 100 km below the satellite and 10 km out of the plane defined by the satellites's path and the magnetic field. The V- or saucer-shaped time-frequency-power spectrograms of electric field waves comes from the propagation characteristics of quasi-electrostatic whistlers. The bottom box shows the relation between the wave vector and group velocity for a saucer. From Ergun et al. (2001).

detected upgoing electrons associated with VLF saucers in several events, and verified that the majority of events were in the downward current region (Lonnqvist et al., 1993). Observations from the FAST satellite subsequently demonstrated that a majority of the VLF saucer sources were on flux tubes carrying intense, upgoing energetic electron fluxes.

A close examination of the VLF saucer wave spectra shows that intense broadband bursts are a frequent phenomenon at the vertex (or source flux tube) of VLF saucers (Lonnqvist et al., 1993). Subsequently, FAST observations revealed that these broadband bursts represent a series of solitary structures that have been identified as electron phase-space holes (Ergun et al., 1998c,d) indicating that electron phase-space holes are traveling upward on the same flux tubes as the VLF saucer source.

Properties. Figure 4.42 demonstrates how wave spectra vary with source location and size. The traces are examples of ideal power-frequency-distance (or time) spectrograms from point source locations. The solid trace is from a point source 100 km in altitude below the spacecraft as measured at the vertex. The plasma conditions for all the examples have an electron density n_e of 20 cm^{-3}, a constant magnetic field B at 10^4 nT, and a homogenous H^+ plasma. The trace forms a 'V' in the spectrogram with a low-frequency cutoff at the lower hybrid frequency. The dotted trace displays the expected wave spectrogram from a point source 200 km below the spacecraft. The 'V' is less steep but otherwise shows the same low-frequency cutoff. Comparing these two traces demonstrates how the source size can be determined. The extent of the source in altitude determines the bandwidth of the spectra at the higher frequencies (far from the vertex), but does not dramatically change the spectra at the vertex. Thus, one determines the location and extent in altitude of the source region by examining spectra at higher frequencies far from the saucer vertex.

The dashed trace in Figure 4.42 illustrates the effect of an out-of-plane displacement of the source region. The source is 100 km below the spacecraft but 10 km out of the plane defined by the spacecraft path and magnetic field. In the case of an auroral satellite traveling in the north-south direction, the displacement would be in longitude. When compared to the solid trace, the dashed trace shows the greatest change near the vertex. It has a cutoff at a significantly higher frequency because the minimum propagation angle is non-zero. Farther from the source, that is, in horizontal distance along the spacecraft track, the effect of the out-of-plane displacement of the source is less pronounced. Thus, one determines the longitudinal position and size by examining spectra near the saucer vertex. Unfortunately, broadband emissions from electron phase-space holes at the vertex often complicate such an analysis.

Observations. Figure 4.43 shows a FAST observations of a VLF saucer. The top panel displays the DC electric field signal along the path of the spacecraft (nearly southward) and perpendicular to the ambient magnetic field. The signal is negative beginning at 00:03:25 UT and then shifts to positive at 00:03:30 UT. This signature is that of a diverging electric field structure which implies a downward parallel electric field is on the flux tube of the spacecraft at a lower altitude (see Section 4.2). Panel (b) displays the mostly east-west perturbation magnetic field signal (δB); a model magnetic field has been subtracted. The slope of the magnetic field signal indicates a current, assuming a sheet-like structure. A negative slope implies an upward current and a positive slope implies a downward current. The region of interest (00:03:25 UT to 00:03:35 UT) has a strong positive slope indicating a downward current of $\sim$2 μA m^{-2}.

Panel (c) displays the electric field spectral power density in the 0-16 kHz range. The dominant feature is a complex, multiple-arm, VLF saucer with the vertex at 00:03:30 UT. The vertical features in the plot (e.g., 00:03:25 UT, 00:03:29.5 UT, 00:03:45.5 UT) are bursts of broadband electrostatic noise. Panel (d) displays the electron differential energy flux (color) as a function of energy (vertical axis) and panel (e) displays the electron differential energy flux as a function of pitch angle.

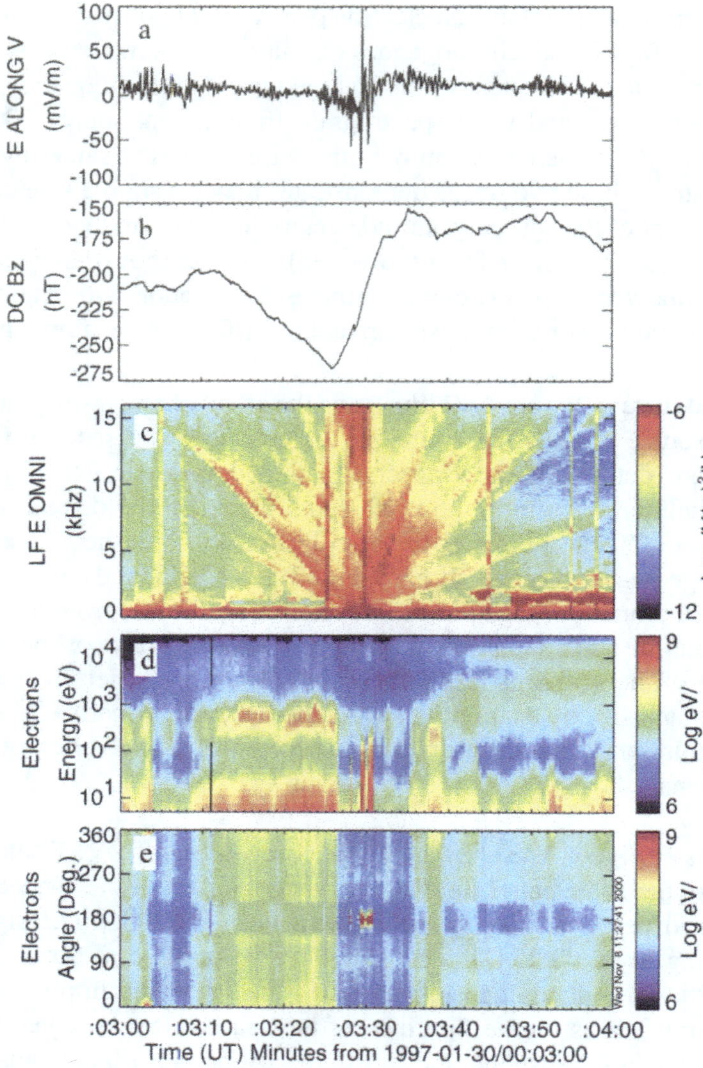

Figure 4.43. (a) The perpendicular electric field in the spin plane and nearly along the spacecraft path (mostly south). (b) ΔB along the spacecraft spin axis (mostly west) and normal to the spacecraft velocity. A negative slope indicates upward current; a positive slope indicates downward current. (c) The electric field spectral power density summed over all directions. (d) Electron differential energy flux versus energy, and (e) versus pitch angle. From (Ergun et al., 2001).

Downgoing electrons are at 0° and 360° and upgoing electrons are at 180°. During most of the 60 s period, the electron fluxes were traveling predominantly earthward. At the vertex of the VLF saucer, however, intense fluxes of highly field-aligned, ~ 100 eV electrons were traveling upward. Figure 4.43 displays a clear example in which the VLF saucer source is associated with upgoing electron fluxes that were accelerated by a parallel electric field.

The electric field spectral power density reveals multiple VLF sources. There appear to be at least 16 distinct arms in the VLF saucer. The slopes and bandwidths of the arms provides information on the individual source sizes and locations. The narrow bandwidth ($\sim 40\,\text{Hz}$) of the arms far from the source indicates that the sources are limited to at most $\sim 10\,\text{km}$ extent in altitude and in latitude.

The slopes of the arms indicate the distance of source from the spacecraft. Distances range from ~ 100 km to over 1000 km. The vertices of the arms map from 00:03:25 UT to 00:03:35 UT, during which time the spacecraft traversed $\sim$50 km, so the individual sources may be dispersed over this region in latitude. The size of the nearest sources are less than $\sim$2 km in latitude. The bandwidth of the arms increases near the vertex (at lower frequencies), indicating a finite longitudinal size. Because of the multiple arms and the broadband electrostatic noise (vertical features), the longitudinal size cannot be determined for all of the arms.

The above observations support many of the conclusions drawn from the ISIS 2 data (James, 1976) and Viking measurements (Lonnqvist et al., 1993). The energy source appears to be from upgoing electron beams. The source regions are confined in both latitude and altitude to less than $\sim$10 km.

4.4 Time-Dependent Alfvénic Processes

Electromagnetic energy is transmitted along magnetic field lines by shear Alfvén waves when the time variability of the coupled magnetosphere ionosphere system contains power at frequencies matching or exceeding the fundamental frequency of the field line eigenmode – the frequency of a standing Alfvén wave with one-half parallel wavelength extending between the northern and southern ionosphere. Basic properties of shear Alfvén waves are described in Section 3.4, including their interaction with the ionosphere. Here we consider in situ observations of Alfvén waves and associated auroral phenomena in the ionosphere and lower magnetosphere. Observed properties depend on the frequency spectrum and transverse structure of the Alfvénic fluctuations, and on whether their source is transient in nature or quasi-harmonic. The observational signatures are also influenced by measurement effects in the lower magnetosphere, as space-time aliasing (Doppler shifting) resulting from the rapid motion of the (satellite) measurement platform across field lines is always present to some degree.

The effect of a simple Alfvén wave on the ambient medium is to induce a reversible sloshing motion of ions which carry the wave polarization current and electrons which carry the wave field-aligned current. The density perturbation accompanying the Alfvén wave is generally weak. When an Alfvén wave develops a parallel electric field, which occurs at small transverse wavelengths, it exchanges momentum and energy with the medium and can deposit substantial energy in the form of plasma heating and field-aligned electron beams. This irreversible conversion of the electromagnetic wave power to plasma kinetic energy is one of the more interesting aspects of the plasma-Alfvén wave interaction. It also produces dynamic, luminous auroral phenomena.

Below we consider in situ measurements of both travelling and standing Alfvén waves and their interactions with the ambient plasma. We begin with a discussion of the conductivity measure $\mu_0 E/b$, as a manner of distinguishing between the quasistatic arcs described in previous sections, and the strongly time-dependent phenomena considered here. Then we consider various observations of travelling dispersive Alfvén waves. We see that Alfvénic activity can bring significant amounts of energy into the auroral zone through its Poynting flux, and that many of the travelling Alfvén waves exhibit dispersive phenomena which accelerate electrons into the lower ionosphere creating narrow, dynamic visible arc structures. Finally we discuss standing Alfvén wave resonators, both in the lower ionosphere, and along the entire auroral field line. In this section on time-dependent Alfvénic processes, we emphasize the specific Alfvénic processes at work, illustrated by examples from various observations. We will revisit many of these processes in their proper context later in the book, including the discussion of the energy flow through the nightside polar cap boundary in Section 4.5, and discussions of specific magnetosphere-ionosphere coupling models, resonant arc formation models, and propagating arc models, in Chapter 7.

4.4.1 Wave Impedance

Interpreting Low-frequency Electromagnetic Fluctuations. The magnetosphere and ionosphere are coupled together by particle flows and by electric fields and currents. This coupling varies strongly in space and time. In fact, strong, low-frequency electromagnetic fields that fluctuate in the spacecraft frame are good indicators of auroral-zone crossings. Interpreting these auroral electromagnetic fluctuations is complicated by the fact that typical spatial structures (0.1-100 km) are Doppler-shifted by spacecraft motion into the same frequency range as typical temporal fluctuations (0.001-1 Hz). Thus, it can be very difficult to separate time and space when observing auroral electromagnetic fields. The distinction is important since static fields and field-aligned currents interact with the ionosphere much differently than do time-varying ones. Specifically, time-varying fields (Alfvén waves) are subject to wave phenomena including reflection, interference, mode conversion and trapping, as discussed in Section 3.4. Here we compare two limiting cases: the static case $\omega = 0$, $\lambda_\perp =$ finite, and the temporal case $\omega \neq 0$ with $\lambda_\perp \to \infty$. The ratio E/b between the electric and magnetic field perturbations will have measurably distinct values for these two cases.

In the static limit, treated in Section 3.3.4, $E/b = \pm 1/\mu_0 \Sigma_P$ (see Eq. 3.45). Here it is assumed that E_x is constant throughout the Pedersen layer for horizontal spatial scales λ_x of order 1 km or larger (Forget et al., 1991). Thus E_x and b_y in structured, quasi-static current systems are related by the ionospheric conductivity.

When the scale of the structure decreases to a few km or less, the ratio E_x/b_y increases as static electric fields map through the ionosphere less efficiently. While it is not always easy to determine Σ_P for the ionosphere below a satellite, typical values range from several to tens of Siemens.

For the case of an ideal Alfvén wave propagating in a single direction, treated in Section 3.4.1, $E_x/b_y = \pm v_A = \pm 1/\mu_0\Sigma_A$ (Eq. 3.55). As Σ_A is usually much less than Σ_P, the magnitude of E_x/b_y is generally larger for Alfvén waves than it is in the static case. If the Alfvén wave has a small perpendicular wavelength, this ratio is increased even further by the effects of electron inertia and ion gyroradius (Eq. 3.76). Lysak (1998) has plotted this ratio as a function of the plasma parameters for the fully kinetic model of the Alfvén wave.

Observations near the ionosphere are complicated by the fact that incident and reflected Alfvén waves will interfere with each other. In this limit, treated in Section 3.4.2, the ratio $\tilde{E}_x(\omega)/\tilde{b}_y(\omega)$ of the Fourier transformed field components, for non-zero reflection coefficient R (Eq. 3.64), is a complex function of frequency and position. It is bounded in magnitude by the values $(\mu_0\Sigma_P)^{-1}$ and $\Sigma_P/\mu_0\Sigma_A^2$. This can be seen by dividing Eq. (3.59) by Eq. (3.60) and setting $t = 0$. For $R < 0$, E_x is minimum and b_y is at a maximum at the reflection point $z = 0$, with ratio $(\mu_0\Sigma_P)^{-1}$. The upper bound is found by evaluating the ratio at $k_z z = \pi/2$.

In general, the complex nature of the ratio $\tilde{E}_x(\omega)/\tilde{b}_y(\omega)$ implies a phase shift between E_x and b_y. Thus the relative amplitudes and phases of E_x and b_y vary in a predictable way when shear Alfvén waves reflect off the ionosphere. Figure 4.44 compares amplitude and phase of electromagnetic fields measured by a sounding rocket (boxed points) in the morning-side auroral zone. Also plotted are (dashed horizontal line) the $1/\Sigma_P$ value prediction for the impedance in the static case, and (thin line) the predictions of a full-wave model of Alfvén waves reflecting from the auroral ionosphere. (This model accounts for the fact that, for wave periods less than 10 s, the ionospheric reflection coefficient is more complicated than that given by Eq. 3.64.)

The phase relation between E_x and b_y matches quite well that expected from large-λ_x Alfvén waves, indicating that for this one flight, electromagnetic fields were likely dominated by temporal rather than spatial fluctuations. The match between measured and modeled E_x/b_y is not as good, possibly because of at least some contribution from static current systems.

4.4.2 Dispersive Travelling Alfvén Waves

4.4.2.1 Poynting Flux Energy Input

Poynting flux from Alfvénic activity can bring significant amounts of energy into the auroral zone from the outer magnetosphere. High-altitude auroral observations (Wygant et al., 2000) demonstrate that Alfvén waves mediate and often dominate the energy transport from tail reconnection processes to the auroral acceleration region.

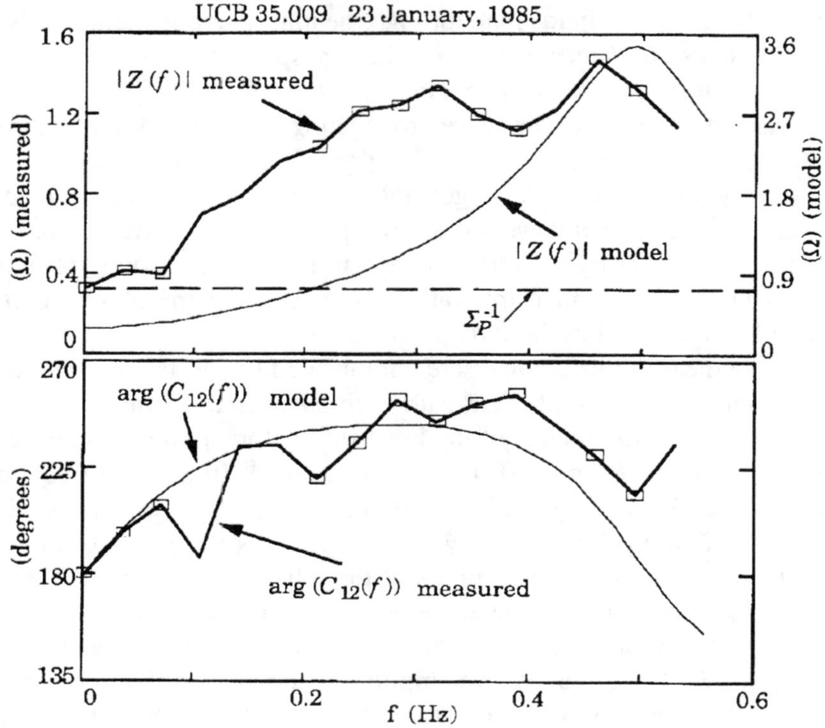

Figure 4.44. Modeled and measured (top) wave impedance $|\mu_0 \tilde{E}_x(\omega)/\tilde{b}_y(\omega)|$ and (bottom) phase shift between E_x and b_y for a morning-side sounding rocket flight. Square data points indicate higher coherency (and therefore confidence) for the phase measurements. From Knudsen et al. (1992).

Figure 4.45 shows in situ data from the Polar spacecraft at 4-6 $R_E$together with a UV image of the corresponding nightside aurora overlaid with the projected footpoint of the Polar trajectory. During the interval shown, Polar moved from the lobe into the plasma sheet at 22 MLT (L = 9.4 - 4.8). The particle data show that Polar entered the plasma sheet boundary at 0537 UT at a geocentric distance of about 4.9 R_E. The entry into the plasma sheet is most clearly indicated by the order of magnitude increases in the flux of 2-20 keV electrons. The electric field enhancements are located on the high-density side of a jump in density from about 0.1 cm^{-3} on the lobe side to 0.4 cm^{-3} on the plasma sheet side.

Large amplitude electric and magnetic field fluctuations (110-140 mV m^{-1} and 20 nT, respectively) perpendicular to the background magnetic field are embedded in the high-density side of the plasma sheet boundary, together with an upward current sheet as evidenced by the large-scale gradient in B_y. The ratio of the amplitudes of the electric and magnetic fields, $E_z/b_y = 4000$ to $10,000$ km s^{-1}, agrees reasonably well with the local Alfvén speed, and is one or two orders of magni-

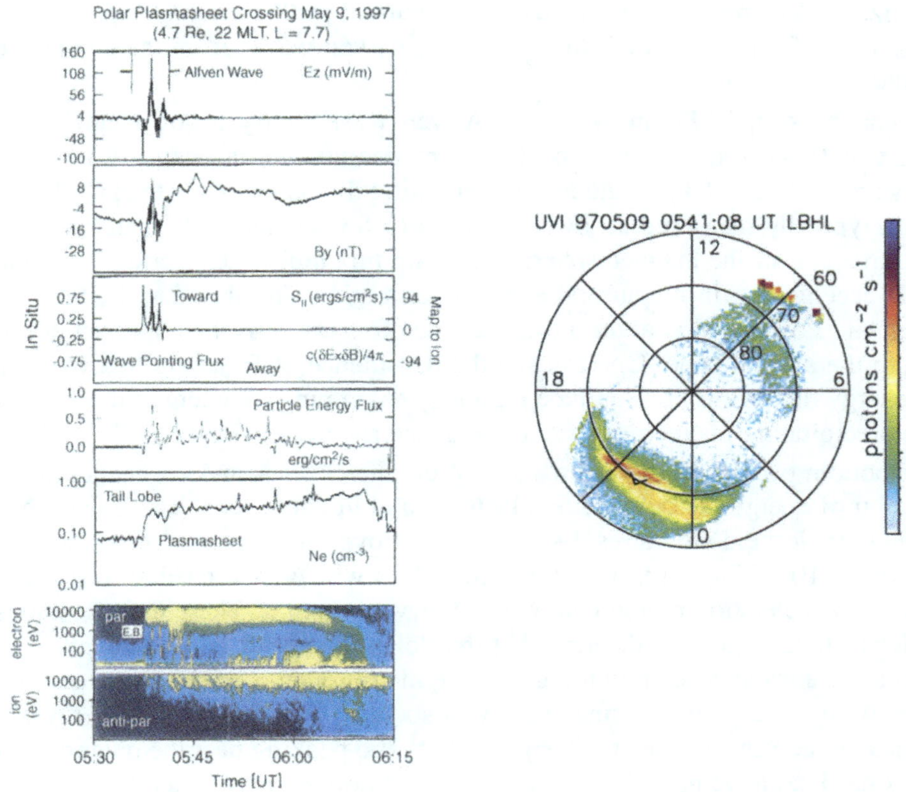

Figure 4.45. Polar observations (left) of in situ fields and particles, together with (right) a remote UV image of the corresponding auroral luminosity. The top two panels show the dominant components of the electric and magnetic field perturbations during this pass. The third panel displays the Poynting fluxes along the ambient average magnetic field, calculated from the full three-dimensional electric and magnetic field measurements. The fourth panel shows the sum of the electron and ion energy flux along the magnetic field (positive is downward) and the fifth panel shows the electron density. The bottom two panels are energy-time spectrograms of upgoing ion and downgoing electron differential energy flux. The vertical brackets in the top panel delineate time periods when the foot of the magnetic field line mapped from the spacecraft to within one degree of the regions of intense (greater than 16 erg cm^{-2} s^{-1}) auroral emission determined from the UVI imaging. From Wygant et al. (2000).

tude greater than the $1/\mu_0 \Sigma_P$ expected for a static field aligned current system. In addition, the amplitude ($\approx$ 1 nT) of magnetic fluctuations parallel to the ambient magnetic field is far smaller than those perpendicular to the field, supporting the conclusion that these electromagnetic fluctuations are shear Alfvén waves, rather than compressional magnetosonic waves. Higher time resolution analysis of the electric and magnetic field fluctuations (Wygant et al., 2000) shows that there is a phase shift between the electric and magnetic field fluctuations (electric field pulses maximize on leading edge of magnetic field pulses) in this event, consistent with that expected for reflection of slowly-varying Alfvén waves from the conducting ionosphere.

As can be seen in Figure 4.45, the Alfvén waves carry a downward Poynting flux of 0.6-1.0 erg $cm^{-2}s^{-1}$ locally, corresponding to an energy flux of 24-100 erg $cm^{-2}s^{-1}$ at 100 km altitude. This Poynting flux is far larger than the Poynting flux typically measured at lower altitudes (a few erg $cm^{-2}s^{-1}$), and suggests that a large part of the magnetospheric Poynting flux input is converted to parallel particle acceleration in the auroral acceleration region. In this example, the electromagnetic Poynting flux at the altitude of the observation is the dominant energy flux into the auroral zone. One can see the dominance of Poynting flux over any other energy flux in this high-altitude region by examining the energy flux provided by the precipitating magnetospheric ion and electron populations.

The bottom two panels of Figure 4.45 present the electron and ion energy flux as a function of energy for downgoing electrons and upgoing ions. These data show that the electric field structures and earthward Poynting flux coincide almost exactly with 1-10 keV upgoing ions (labeled 'UFI') which have been accelerated in the auroral acceleration region below the observation point. These ions are almost completely field aligned, indicating that the Polar spacecraft is magnetically conjugate to the auroral acceleration region. The measurements of the particle energy flux show that the entire plasma sheet is associated with earthward energy flux with an average value of about 0.3 erg $cm^{-2}s^{-1}$ and peaking near the outer edge of the plasma sheet at values of 0.8 erg $cm^{-2}s^{-1}$. Although only the total energy flux is plotted in Figure 4.45, the electron contribution dominates over the ion by 1-2 orders of magnitude. The particle flux dominates over Poynting flux everywhere except at the outer edge of the plasma sheet, where both the Poynting and particle energy fluxes increase, and Poynting flux dominates over particle energy flux.

The energy being deposited in the lower ionosphere can be estimated from an examination of ultraviolet images of the auroral emissions. By comparing the energy flux consistent with the images to the energy fluxes measured at high altitudes, one can infer the presence of acceleration processes occurring along the field line and powered by the high-altitude Poynting flux. Figure 4.45 includes one of a sequence of UVI images (LBH-long) taken during the interval corresponding to the large-amplitude Alfvén waves. The image is color coded in photon flux and in energy flux with a spatial resolution of $\approx$50 km. The image indicates an auroral arc that persisted throughout Polar's transit of this region. The highest amplitudes of Poynting flux, near the poleward edge, correspond to the brightest region of the aurora. The energy flux associated with the region of bright auroral emission conjugate to the spacecraft is typically 20 erg $cm^{-2}s^{-1}$, and is far less than the mapped Poynting flux of 100 erg $cm^{-2}s^{-1}$. The apparent overabundance of Poynting flux relative to the observed energy deposition is caused by the relatively small dimensions of discrete auroral arcs (0.1-50 km) compared to the spatial resolution of the images. The rough parity between the Poynting flux input at high altitudes and the energy deposition at the ionosphere demonstrates the essential role that Alfvén waves have in providing the energy to power auroral acceleration in this region. A statistical comparison of Alfvénic Poynting flux in the plasma sheet to the intensity of the magnetically conjugate auroral emissions was done for 40 plasma sheet crossing by the Polar spacecraft including both, cases with large in situ values of Poynting flux and cases with low values (Keiling et al., 2001). It was found that the energy fluxes at high and low altitudes are correlated: large (small) Poynting fluxes observed by Polar is associated with large (small) auroral emission. The large Alfvénic Poynting fluxes were generally found near the outer boundary layer of the plasma sheet and mapped to the poleward border of simultaneous auroral emissions. This suggest that the aurora near the polar cap boundary is associated with Alfvén wave activity. However, how exactly do those Alfvén waves accelerate the electrons that produce the aurora? Next, we discuss how Alfvénic activity can result in particle acceleration.

4.4.2.2 Inertial Effects

Alfvén waves can exhibit dispersion – different frequencies or wavelengths travelling at different velocities – resulting from finite ion gyroradius and/or electron inertia effects. Such finite-$\lambda_\perp$ effects lead to the development of a parallel electric field $\mathbf{E}_\parallel$ that is slowly-varying on the time scale of electron transit along the field line, allowing for the acceleration of electrons. An example of such small scale Alfvén waves can be found within the Poynting flux input example discussed above. The event contains small-scale, large-amplitude dispersive Alfvén waves/spikes at the plasma sheet boundary layer (Wygant et al., 2000). These observations suggest that large scale Alfvén waves can striate into intense small scale electric field spikes.

Figure 4.46 shows an expanded view of the fields observed during a 25 minute subinterval of Figure 4.45, at the time of the Alfvén waves seen near the outer boundary of the plasma sheet. The top left panel shows the x-component of the electric field data at the full resolution of 20 samples s^{-1}. The bottom left panel shows the y component of the magnetic field at the full resolution of ~9 samples per second. The adjacent two panels show the electric and magnetic field data after averaging to produce a time resolution with a Nyquist frequency of 4 Hz. These data have also been detrended to remove variations slower than the spin period of 6 s. The electric field panels show that embedded within the large-scale Alfvén wave are individual electric field spikes with amplitudes up to 400 mV m^{-1} peak-to-peak, coincident with magnetic field fluctuations with amplitudes of 1-10 nT peak-to-peak.

The ratio of the electric to the magnetic field amplitude for the higher-frequency fluctuations is 20-50 mV $m^{-1}nT^{-1}$ = 20,000-50,000 km s^{-1}. This is a factor of 2-5 greater than that expected for an Alfvén wave, but consistent with dispersive

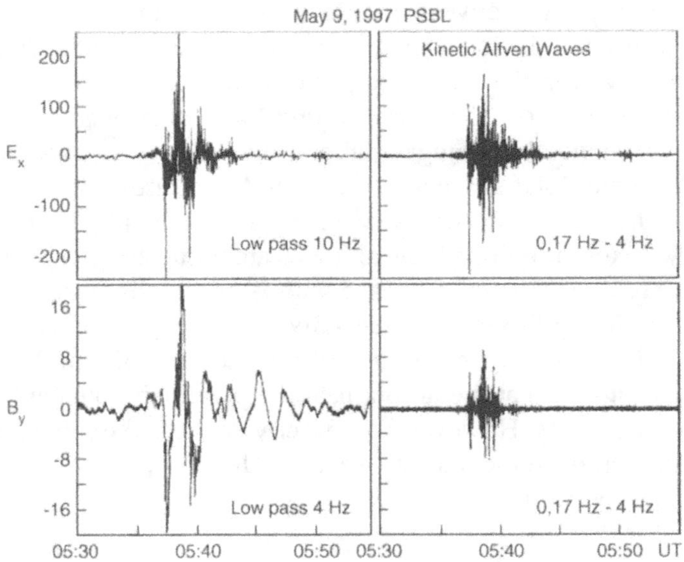

Figure 4.46. Left: Polar measurements taken on May 9, 1997 showing E_x at a time resolution of 10 Hz and B_y at 4 Hz. Right: the detrended E_x and B_y, bandpass filtered at frequencies between 0.17 and 4 Hz. From Wygant et al. (2000).

Alfvén waves having a finite perpendicular scale size (i.e., $k_\perp \neq 0$) (see Eq. 3.76.) While these single-point field measurements cannot confirm that the lower- and higher-frequency Alfvénic fluctuations represent long- and short-perpendicular wavelength structures, the divergence between the field ratios in the two frequency regimes is strong evidence for such an interpretation. Using Eq. 3.67, one can see that the enhanced field ratio of the higher-frequency fluctuations corresponds to a perpendicular scale size on the order of an electron inertial length ($c/\omega_{pe} \sim 20\,\text{km}$) or about 0.5 ion gyroradii.

Theoretical analysis of the observed spikes suggests that these waves very efficiently accelerate electrons along **B**. Dispersive Alfvén waves possess small but significant parallel electric fields extending over long distances parallel to **B**, and can efficiently accelerate charged particles. Simultaneously measured electron distribution functions provide strong circumstantial evidence that parallel acceleration occurred, as both electron parallel heating and earthward beams were observed. The characteristic parallel energy of the electrons is on the order of ~1 keV, consistent with estimates of the integrated parallel electric field associated with small scale dispersive Alfvén wave structures.

Small scale Alfvén waves are unlikely to be generated directly by reconnection processes in the tail, and so one needs to consider mechanisms by which larger scale Alfvén waves can be converted into the observed smaller scale features. Theoretical and laboratory studies of Alfvén waves have shown that small scale Alfvénic structures can be generated from larger scale waves and density gradients through a variety of linear and nonlinear processes. Both the amplitude and direction of the electric field associated with the observed small scale spikes is similar to that of the large scale Alfvén wave. Since the large scale waves carry more Poynting flux than the small scale waves, it is reasonable to suppose that the large scale waves generate the smaller scale waves. These considerations lead to the possibility that small scale density variations are effective scattering surfaces which interact with either the large scale Alfvén wave or other smaller scale dispersive Alfvén waves propagating through the medium. These surfaces exist in addition to the ambient, steady-state density jump at the tail lobe and plasma sheet boundary. Such interactions with density gradients could result in the production of small perpendicular scale size structures through processes which invoke reflection and refraction of incident wave fronts off the variations in the index of refraction (Lysak and Song, 2000) or which invoke linear mode conversion (Hasegawa, 1976).

4.4.2.3 Electron Acceleration

Next we consider lower altitude signatures of these travelling waves, focusing in particular on the resulting field-aligned electron acceleration. Low altitude spacecraft observe Alfvén waves throughout the auroral oval but most commonly in the vicinity of the polar cusps and the polar cap boundary near midnight (Louarn et al., 1994; Chaston et al., 2002c). These waves obey the dispersion relation

$$v_{ia} = \omega / k_{\parallel} = v_A (1 + k_{\perp}^2 \lambda_e^2)^{-1/2} \tag{4.8}$$

where λ_e is the electron inertial length or electron skin depth. This is Eq. (3.80) without the gyroradius effect. The perpendicular scale size of these waves is such that $k_{\perp}\lambda_e \approx 1$ (Chaston et al., 1999). Correlated deviations in $E_{\perp}$ and $b_{\perp}$ time series, with $E_{\perp}/b_{\perp}$ of the order of the local Alfvén speed (but often larger) observed from the FAST spacecraft are invariably accompanied by coincident observations of field-aligned electron beams with energies extending up to about one keV in the dayside cusp (Chaston et al., 2002c), and to more than 10 keV in the nightside polar cap boundary. These distributions are predominantly directed Earthward, although upward accelerated electron distributions are observed, particularly at higher altitudes. Figure 4.47 (left) shows observations from FAST recorded on the edge of the dayside cusp. The first two panels show the electric and magnetic field components perpendicular to $\mathbf{B}_0$. These time series contain a broadband spectrum of Alfvénic turbulence. Throughout the period of turbulence the electron spectra contained in the next two panels show suprathermal fluxes extending up to 1 keV at pitch angles collimated along the field line and directed towards the ionosphere. The black trace in the lowest panel shows that these electron fluxes provide energy densities of up to 3 erg cm^{-2} s^{-1} into the ionosphere and are correlated with similarly directed wave Poynting flux (red trace).

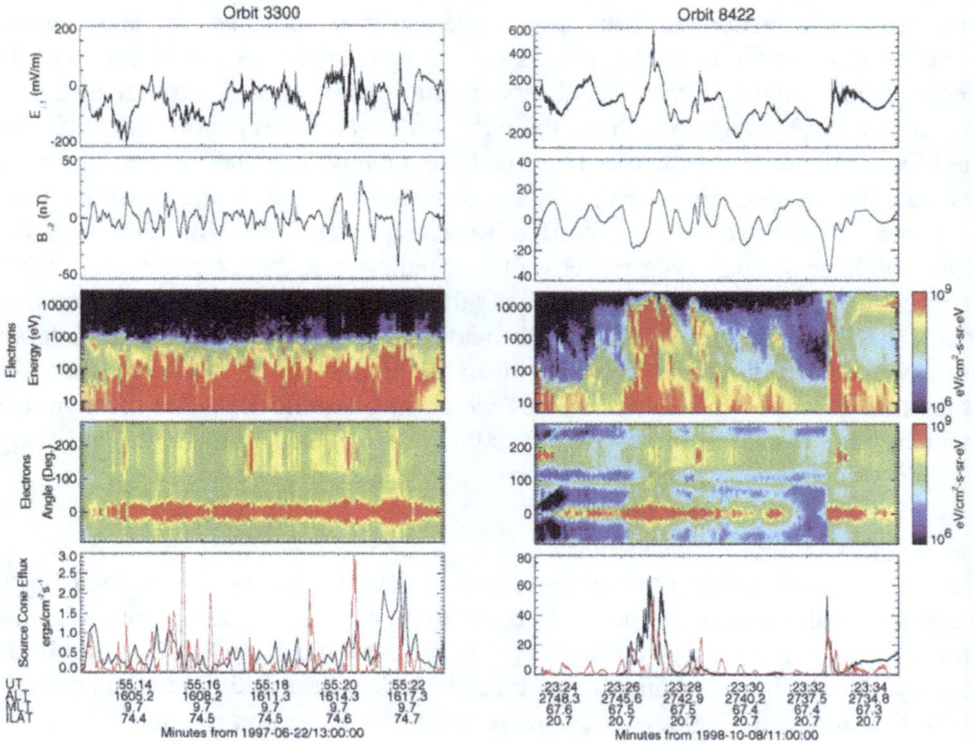

Figure 4.47. Left: observations from FAST of Alfvénic activity on the edge of the polar cusp. The bottom panel shows the integrated energy flux in the field-aligned electron beam (black trace) mapped to the ionosphere and the downgoing wave Poynting flux (red trace) also mapped to the ionosphere but divided by 5 (from Chaston et al., 1999). Right: observations from the nightside polar cap boundary. As in (left), but in this case the Poynting flux has instead been multiplied by 5.

Figure 4.47 (right) shows a similar observation in the nightside polar cap boundary. The wave in this case is accompanied by fluxes of electrons with energies exceeding 10 keV distributed as a downgoing field-aligned beam. The integrated energy flux shown in the next panel peaks at 60 erg cm^{-2} s^{-1} and is again correlated with the wave Poynting flux. Simulations suggest that this flux in the ionosphere will give rise to an auroral arc with width of about one km. In some cases the Poynting fluxes and integrated energy fluxes in the source cone for such observations in the nightside polar cap boundary can exceed 100 erg cm^{-2} s^{-1} when mapped to ionospheric altitudes (Chaston et al., 2002c). Transport code results and empirically derived emission efficiencies indicate that such distributions can yield auroral arcs of more than 100 kR intensity and so power the very brightest aurora (Chaston et al., 2002c).

Electron acceleration in the Alfvén wavefield is caused by electron inertial effects which requires the shear Alfvén wave to carry an $E_\parallel$ of magnitude given by

$$E_\parallel/E_\perp = k_\parallel k_\perp \lambda_e^2/(1 + k_\perp^2 \lambda_e^2) \tag{4.9}$$

This is Eq. (3.77) in the cold plasma limit. This can yield parallel electric fields up to 1 mV/m over wavelengths of more than 1000 km for typical altitude dependent composition and density profiles and observed perpendicular electric fields. Consequently the potential drop in the wavefront of the Alfvén wave can be of the order of kilovolts. If $\frac{1}{2} m_e v_{ia}$ (see Eq. 4.8) is less than this potential, a cool electron can be picked up by the wave and accelerated towards the ionosphere. As the wave approaches the ionosphere it speeds up (since the magnetic field increases more rapidly than the density) and so the resonant electron continuously gains energy as it travels with the wave down the fieldline. In some cases FAST observations have revealed densities below 1 cm^{-3} at altitudes as low as 3500 km (Strangeway et al., 1998), providing inertial Alfvén wave phase speeds and parallel wavefields sufficient to accelerate electrons to energies approaching 10 keV, as observed.

The observed electron distribution can be reproduced in detail (Chaston et al., 2000; Thompson and Lysak, 1996; Chaston et al., 2002b) by simulating the electron acceleration process from an altitude of 5 $R_E$in an inertial Alfvén wave field. The acceleration turns off at altitudes below 3500 km due to the rapid increase in ionospheric density below this altitude and the associated diminishing inertial effect. Electrons then overtake the wave to form time dispersed field-aligned electron bursts as observed particularly from rockets but also from low altitude spacecraft (Stasiewicz et al., 2000). The most energetic electrons typically will come from higher altitudes (up to a few R_E) since they gain energy continuously after being picked up, while those electrons picked up by the wave at lower altitudes encounter the wave with a smaller electric field and phase speed and so gain less energy. Because of this, calculation of the source height for dispersive electrons under the assumption of a localized acceleration region may not yield meaningful results and often there may be no dispersion apparent at all.

Other low-altitude examples of electron acceleration from inertial effects have been observed by sounding rockets and by the Freja spacecraft. Later we will return to this subject in the discussion of energy flow at the nightside polar cap boundary, integrating many of the above processes (Section 4.5): Poynting flux input, development of inertial effects, electron acceleration, all of which result in wave generation and energy transfer to ambient ions. We will see that it is not coincidental that many of these examples are drawn from crossings of the polar cap boundary regions, where gradients in the magnetosphere provide source regions for these Alfvén waves.

Sounding Rocket Observations of Electron Acceleration. EMIC waves generated near 1 $R_E$become Alfvénic as they propagate downward into the ionosphere.

The parallel field of these Alfvén waves snowplows electrons into the upper atmosphere (Temerin et al., 1986), as we have seen in various examples so far. Thus there is a close relationship to the flickering aurora effects seen in Figure 4.37, with its bursts of accelerated cold electrons at cyclotron frequencies, and the Alfvénic electron acceleration described just above in Figure 4.47. Examples of these wave-accelerated electrons are found on many auroral sounding rockets (Boehm et al., 1990; Lynch et al., 1999a; Arnoldy et al., 1999), including the Topaz3 rocket shown as an example for VLF hiss in Figure 4.26. We will see below yet another related effect, where the Alfvén waves become trapped in the auroral resonator and accelerate bursts of electrons at resonator frequencies.

Freja. The Freja database provides an excellent source of data for studying the low altitude effects of Alfvén waves on particle acceleration and heating. A collection of these data and their interpretations, particularly regarding dispersive Alfvén waves and solitary kinetic Alfvén waves (SKAW) is found in Stasiewicz et al. (2000).

4.4.3 Standing Alfvén Waves

4.4.3.1 The Auroral Resonator

We have seen in Section 3.4.2 that Alfvén wave reflections between the lower ionosphere, and the Alfvén speed maximum, can set up a resonator cavity effect on low-altitude auroral field lines. Figure 4.48 shows an example of such an observation (Ivchenko et al., 1999; Lynch et al., 1999b). The electric and magnetic field oscillations were observed, by two payloads separated by about one km, just inside the poleward edge of a pre-midnight inverted-V arc. During the second half of the event, when the electric field variations are smaller, bursts of field-aligned precipitating electrons were also seen at the same frequency. The electrons are thought to have been accelerated by the parallel field of the wave at about 10,000 km altitude. We will examine further aspects of reflected dispersive Alfvén waves in Section 7.3.

4.4.3.2 Field Line Resonances

The period of the auroral resonator oscillations described above scales as the field-aligned integral, $\int ds_{\parallel}/v_A$, with limits extending from the ionospheric E-layer to the altitude of the peak in Alfvén speed, typically 1-2 R_E. The period of torsional oscillations of the field line may be similarly estimated with limits extending along the entire field line between the northern and southern ionosphere. Typical periods of fundamental field line eigenmodes occurring at auroral latitudes are in the so-called Pc 5 range of 100-1000 s. These ultra-low-frequency torsional waves are called field line resonances (FLRs). Classical FLRs are torsional because they correspond to displacements of the geomagnetic field lines in the azimuthal or LT

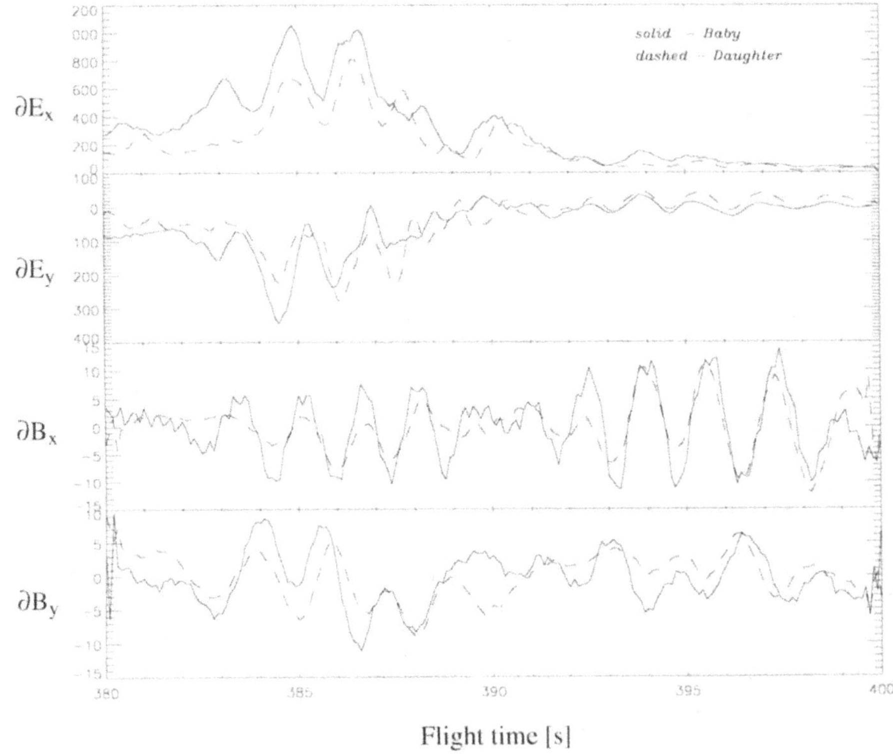

Figure 4.48. Electric and magnetic fields from the AT II sounding rocket observation of the auroral resonator. Solid and dashed lines are from two separated payloads. The event was seen inside the poleward edge of an inverted-V arc. From Ivchenko et al. (1999).

direction. The strength of a dipole geomagnetic field or, more generally, an axisymmetric geomagnetic field, is constant in the azimuthal direction. Therefore, displacements of the magnetic field in the azimuthal direction change the direction of the field but not its intensity. The FLRs are approximately magnetic shell-aligned with a variable north-south width extending from about 200 km down to 10 km or less referenced to the ionospheric height. Because the meridional velocity of a low-altitude, polar orbiting satellite is on the order of 5 km s^{-1}, FLRs appear as quasi-static, Doppler-shifted spatial structures in low-altitude satellite data. In contrast, auroral resonator oscillations have periods of 1-10 s and appear primarily as temporal variations in low-altitude satellite observations.

The structure of a field line resonance measured in situ at an altitude of 4100 km is shown in Figure 4.49. Based on the center frequency of the resonance at 1.3 mHz, determined independently from ground-based magnetometer measurements, this FLR is a fundamental mode with essentially one half-wavelength spanning the field line between the northern and southern ionosphere. Its east-west magnetic

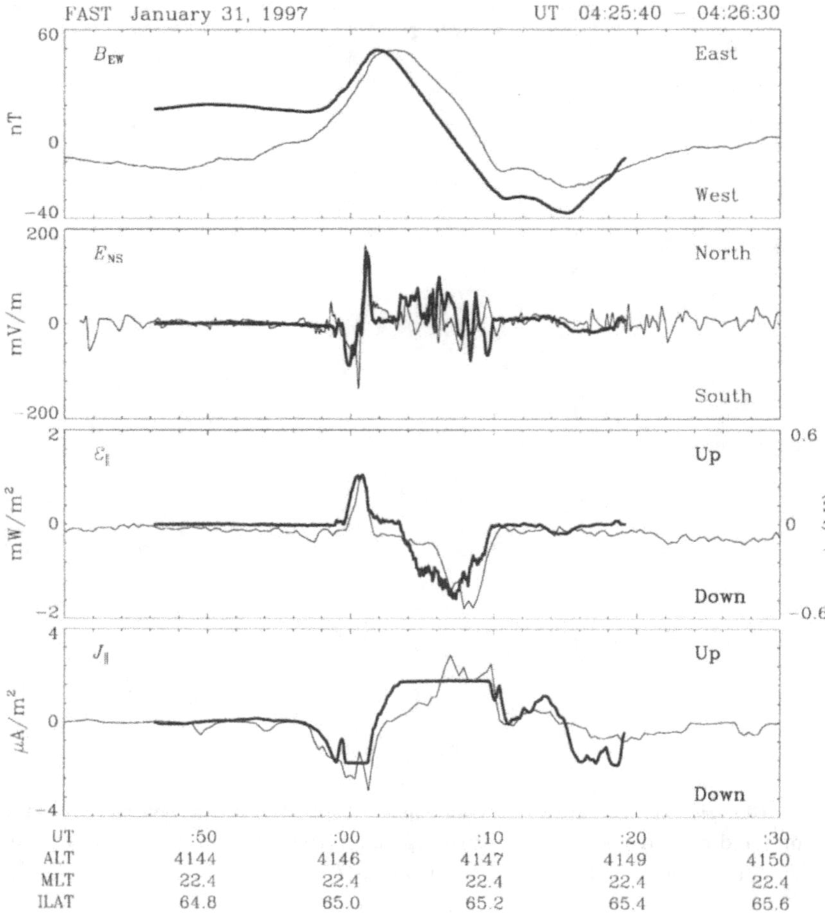

Figure 4.49. Light curves: reduced FAST data from 04:25:40 - 04:26:40 UT of orbit 1760 on 31 January 1997. From top to bottom: spin-axis magnetic field detrended over the one-minute time interval; detrended electric field projected on the satellite velocity vector; electron field-aligned energy flux; electron field-aligned current density, both from electron distribution functions integrated from 0.01 - 20 keV and projected along the IGRF to a reference altitude of 100 km with positive indicating upward. Heavy curves: computer simulation results. From Lotko et al. (1998).

perturbation at 4100 km altitude is plotted in the top panel of the figure. The magnetic deflection is produced by the pair of oppositely directed field-aligned currents shown in the bottom panel, with the lower latitude sheet flowing toward the ionosphere while the higher latitude sheet flows away from it.

The north-south electric field plotted in the second panel exhibits fine-scale structure on transverse length scales smaller than a few kilometers. Some of this structure, especially the intense electrostatic ‘shock’ structure appearing near 04:26 UT, is evident in the field-aligned current but not in the magnetic signal. Insights

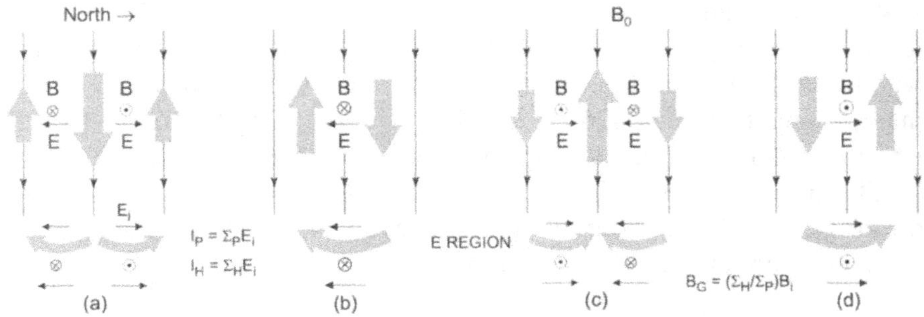

Figure 4.50. Four quarter cycles of the electric and magnetic field and field-aligned current density expected near the ionosphere in a field line resonance. Adapted from Greenwald and Walker (1980).

into its origins can be found in the work by Vogt and Haerendel (1998), which demonstrates that small-scale, dispersive Alfvén waves are efficiently reflected by collisionless dissipation layers in which $E_{\parallel} \neq 0$. More specifically, if the FLR spectrum contains power at transverse wavelengths satisfying $k_{\perp}^2 \Sigma_A / K \gg 1$, then Eq. 3.66 indicates that these fluctuations will be strongly reflected, with a reflection coefficient approaching +1. As discussed in Section 3.4, the electric field of the reflected signal is strongly enhanced under these circumstances while the small-scale magnetic field is suppressed. The reflected wave propagates as a dispersive, inertial Alfvén wave. Numerical simulations of this FLR event, based on a two-fluid MHD model, largely confirm this behavior (Streltsov and Lotko, 1999). The heavy lines in each panel of Figure 4.49 are from the simulations.

The third panel of Figure 4.49 shows the field-aligned electron energy flux measured by the satellite. Both the energy flux and field-aligned current in the figure have been projected to a reference altitude of 100 km. We saw in Figure 2.2 that an electron energy flux exceeding a few mW m^{-2} at 100 km altitude will produce auroral light. We should therefore expect the FLR-powered, electron energy flux flowing toward the ionosphere to form an auroral arc with a north-south thickness of about 10 km. Simultaneous ground-based observations using all-sky cameras and meridian scanning photometers indeed confirm this expectation and, together with magnetometers and Doppler radars, show that the FLR actually persisted for several hours while pulsating at 1.3 mHz.

The large-scale electric field signature of the resonance is masked in Figure 4.49 by the large-amplitude, small-scale fluctuations. The Fourier spectrum of the magnetic signal is found to roll off rapidly at 0.15 Hz, corresponding to a wavelength of 33 km. When a low-pass filter with roll-off at 0.15 Hz is applied to E_{NS}, B_{EW}, and $J_{\parallel}$, the resulting waveforms closely resemble the schematic waveforms illustrated in panel (d) of Figure 4.50. This diagram shows the relation between E_{NS}, B_{EW}, and $J_{\parallel}$ during four quarter cycles of oscillation of an ideal FLR observed above the ionosphere or in the low-altitude magnetosphere where the measurements shown in

Figure 4.49 were made. Had the satellite traversed the resonance 3 minutes earlier or later (one-quarter period), it would have recorded one of the triple current sheet configurations illustrated in panels (a) and (c). The weak downward field-aligned current registered between 0426:15-25 UT in Figure 4.49 appears to be the remnant of the third current sheet in panel (c) in Figure 4.50 that formed one quarter-cycle earlier. The similarity between the configuration shown in panel (a) and that illustrated in Figure 4.21 may be more than coincidental. Theoretical aspects of field line resonances are discussed in Chapter 7.

4.5 Specific Regions

We have ordered most of this observational chapter according to electrodynamic processes: upward current regions, downward current regions, Alfvénic regions. However, there are also characteristics of the aurora which are strong functions of their geophysical location. To a large extent the aurora is the footprint of magnetospheric boundaries which serve as sources of particles and/or energy. The magnetopause and cusp region on the dayside is one example of such a boundary. The nightside polar cap boundary is another. A characteristic observational signature of these regions, because of their spatial constraints, is velocity, mass, or energy dispersion in the particle data that in turn allow remote sensing of these boundaries. In this sense, investigations of the aurora also contribute to the understanding of magnetospheric processes at these boundaries. In this section, we will discuss the nightside and dayside polar cap boundaries (Sections 4.5.1 and 4.5.2). Similar dispersion features also occur elsewhere in the auroral zone (Section 4.5.3). Finally, the region equatorward of the oval of discrete auroral forms is the site of diffuse aurora and some special types of aurora that we will address in Sections 4.5.4 and 4.5.5.

The most characteristic spatial boundary of the auroral zone is the polar cap boundary, the boundary between open and closed field lines. In this context, 'open field lines' indicate magnetic field lines that connect to the IMF, and 'closed field lines' are those with both footpoints on the Earth. In addition to being a separation between two different magnetic topologies, the open-closed boundary also marks the difference between fundamentally different magnetospheric plasma domains (see Section 5.2.3) and particle precipitation mechanisms.

Although the magnetic separatrix definition looks rather straightforward, locating the boundary on the basis of in situ observations often proves difficult, in particular when there is flow across the boundary (e.g., when magnetic reconnection is occurring). The boundary therefore may not necessarily coincide with changes in optical emissions or particle or wave signatures. Also, the actual location of the boundary depends strongly on the geomagnetic activity level and on the direction of the interplanetary magnetic field.

4.5.1 Nightside Polar Cap Boundary

A study of energy flow at the nightside polar cap boundary allows us to integrate many of the processes described in Sections 4.1 through 4.3, and to see how these processes are related in a causal energy flow. The nightside boundary region is a junction point for magnetospheric-ionospheric coupling, and an entry point into the ionosphere for energy from the magnetosphere. Irregularities and events in this region provide an energy source for Poynting flux from the magnetosphere into the ionosphere. Events such as the example we shall discuss in Figure 4.51 are associated with substorm activity, suggesting the low altitude signature of bursty flows and magnetic reconnection regions in the tail (Tung et al., 2001).

Figure 4.51 illustrates a poleward pass of the FAST spacecraft through the auroral zone near midnight local time. It provides a textbook example of the processes outlined in the introductory illustration of Figure 4.1. The upward current region and its inverted-V electron structure (Section 4.1) can be seen from 17:23 to 17:24 UT. Then the magnetic field deflection reverses, and the downward current region stretches from 17:24 to 17:25:20 UT. This region shows upgoing electron beams and 'pressure cooker' ion conics (Section 4.2). Next the spacecraft passes through another inverted-V region (note the changed magnetic deflection from 17:25:20 to 17:25:40), this time passing within the acceleration region altitude as indicated by the upgoing ion beam. For our discussion here, we concentrate on the region poleward of these, from 17:25:40 through 17:27, and the entry to the polar cap.

This boundary region is characterized by (a) filamentary Alfvénic currents with no preferred sense, (b) intense Alfvén wave accelerated field-aligned electrons which produce some of the brightest midnight aurora, and (c) intense low-energy ion outflows which are the dominant source of nightside ion outflow. This region is located at the nightside open-closed field line boundary, and consists of recently reconnected field lines in which the magnetospheric plasma has not come into convective equilibrium with the ionosphere. Until a convective equilibrium is reached, Alfvén waves are exchanged between the magnetosphere and the ionosphere. Part of the energy of these waves accelerates electrons, which in turn generate higher frequency waves that heat the ions producing ion conic outflow.

Particle Signatures of the Nightside Polar Cap Boundary. The boundary identification can be made by examining two of the particle populations: (a) high-energy velocity-dispersed ions, and (b) polar rain. During the interval from 17:26 to 17:27, the high energy ions (above 1 keV) show the signature commonly referred to as velocity-dispersed ion signature, type 2 (VDIS-2). This dispersion signature, first reported by AUREOL-3 (Bosqued, 1987; Kovrazhkin et al., 1987; Zelenyi et al., 1990), is a signature of the open-closed field line boundary. Many VDIS-2 events can be reasonably approximated by a velocity-filter dispersion, resulting from a superposition of downward particle motion along magnetic field lines and

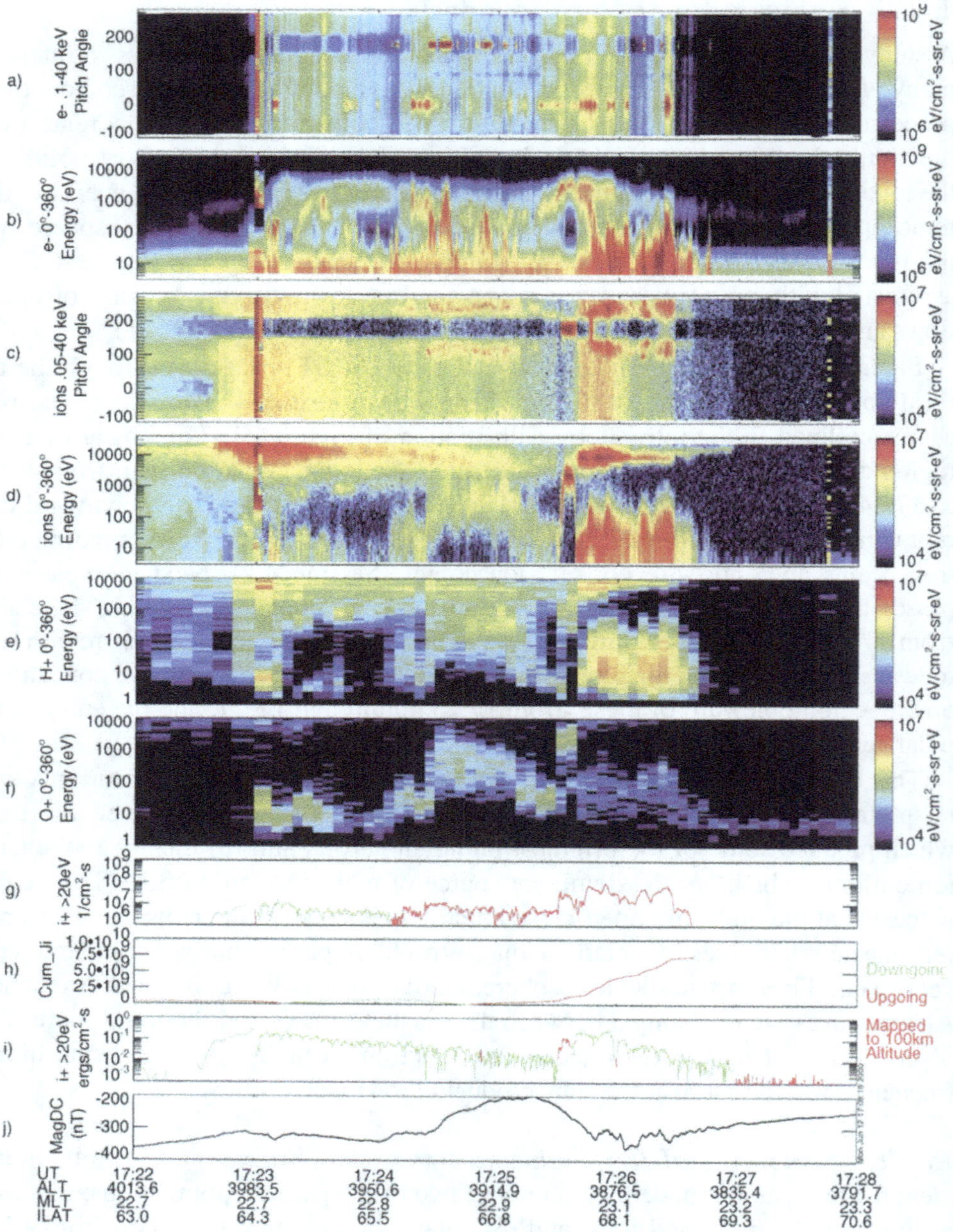

Figure 4.51. A nightside crossing of the auroral zone by the FAST spacecraft, illustrating upward current, downward current, and poleward boundary regions of the aurora. Panels from top to bottom show: (a-d) electron and ion pitch angle and energy spectrograms, (e-f) hydrogen and oxygen ion energy spectrograms, (g-i) ion number flux, integrated number flux, and energy flux both upward (red) and downward (green), and (j) DC magnetic field deflections indicating field-aligned current. From Tung et al. (2001).

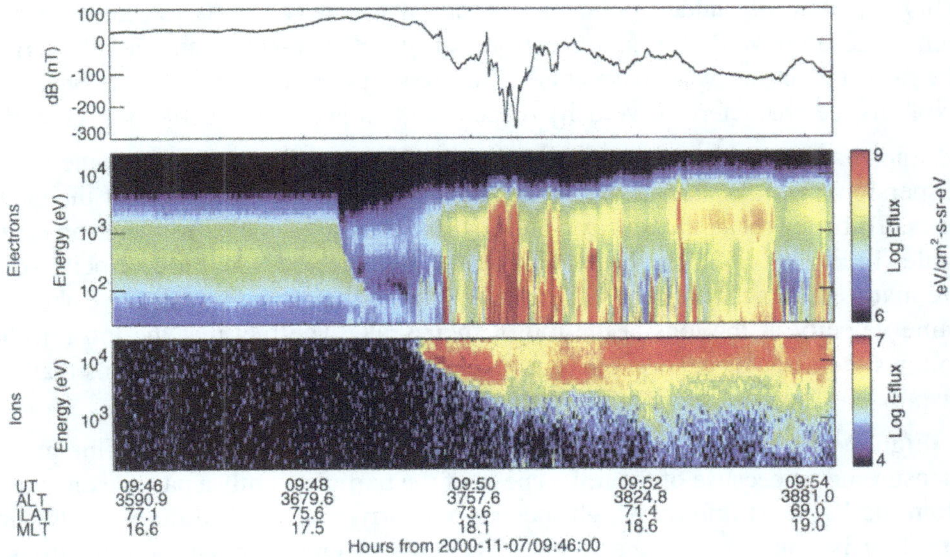

Figure 4.52. FAST observations of magnetic field perturbation, electron and ion spectrograms for crossing of the nightside polar cap boundary. Figure provided by C. W. Carlson.

convective motion of these field lines. The ion dispersion observed is consistent with a plasma sheet boundary layer source at distances from ~30-100 $R_E$in the tail. Modeling (Onsager and Mukai, 1995) allows a close reproduction of observed VDIS-2 events, given an origin 5 $R_E$earthward of a 'distant neutral line' source at $X = -60$ R_E. The colocation of the boundary region indicated in Figure 4.51 by accelerated field-aligned electrons and ion outflows, with this VDIS-2 signature, is interpreted to mean that the region is magnetically connected to the nightside reconnection X line (Tung et al., 2001). Another indication of the polar cap boundary is found from examination of the polar rain signature. This population is visible (weakly) as a several-hundred eV peak in the electrons in the polar cap region poleward of 17:26:40 UT.

The relationship between the velocity-dispersed ions and polar rain electrons becomes clearer in another example, shown in Figure 4.52. Here the FAST satellite was moving equatorward, from the polar cap into the plasma sheet. The polar rain is much more intense than in Figure 4.51 and thus is clearly visible as the distinct electron population between 100 and 1000 eV that extends up to about 09:48:30 UT. Plasma sheet electrons appear immediately at the polar rain cutoff.

To understand this behavior of the polar rain electrons, let us consider what happens to open polar cap field lines as they convect towards the plasma sheet. Because they are open, they are populated with fast solar wind ('strahl') electrons that are the source of the polar rain. The moment such a field line undergoes reconnection at a distant X-line, it abruptly becomes disconnected from the solar wind source.

This explains the sudden polar rain cutoff in Figure 4.52. As the field lines are convected earthward after reconnection, this cutoff is not the same at all energies: the fastest electrons will arrive at FAST altitudes poleward of the slower ones. This explains the characteristic velocity dispersion that is clearly visible in Figure 4.52.

Ions that are accelerated and heated at the X-line will undergo the same type of dispersion, with the fastest ones reaching FAST altitudes poleward of the slower ones. To better relate the electron and ion dispersions, the electron and ion energy scales in Figure 4.52 have been chosen such that electrons and ion velocity scales are matched, i.e., the electrons at the lower end of their energy scale have about the same velocity as the ions at the top of their scale. The fact that the ion injection follows the same dispersion curve as the polar rain electrons thus means that both dispersions have the same origin, namely the distant X-line.

Figure 4.52 illustrates that the polar cap boundary is smeared out in the particle measurements because of the finite speed of the particles. Only a particle emanating from the X-line at infinitely high speed would arrive at FAST altitudes on the field line that is connected to the X-line. The highest energy polar rain electrons in Figure 4.52 move at about 40000 km s^{-1}, and thus their cutoff nearly coincides with the location of the field line connecting to the X-line.

Energy Flow: From Poynting Flux to Ion Outflow. Having established the interval from 17:25:40 to 17:27 UT in Figure 4.51 as a boundary region associated with the nightside open-closed field line, we can now consider the energy flow in this region. This allows us to consider the Alfvénic processes of Section 4.4, and the wave-particle interactions of Section 4.3, in context.

Our example here is a low-altitude counterpart of the lobe – plasma sheet interface seen in Figure 4.45. That observation from the Polar spacecraft showed energy coming into the auroral ionosphere in the form of Alfvénic Poynting flux. This is occurring on flux tubes connected to reconnection sites in the magnetosphere. As the energy moves toward the lower ionosphere, it is converted to particle acceleration by the dispersive Alfvén wave processes described in Section 4.4.2, and is visible here in Figure 4.51 as intense field-aligned electrons. Although not shown in this particular example, often these field-aligned populations are variable in direction or counterstreaming, consistent with the irregular current sense shown by the DC magnetic field deflections.

The end result of this energy flow is (a) extremely bright auroral forms and (b) the strongest flux nightside ion outflows. Panels (d), (e), and (f) in Figure 4.51 show ion energy spectrograms, indicating hydrogen-dominated low-energy ion outflows exceeding 10^7 $cm^{-2}s^{-1}$. The field-aligned electron populations generate BBELF waves very similar to those described in the downward-current region, and this wave activity heats the ambient ions and causes outflow. Unlike the downward current region, there is not a strong downward electric field to cause a strong ‘pressure-cooker' signature, and the heated ions are freely evacuated from the region by the mirror force.

The resulting copious ion outflow found here during a substorm expansion phase is one of the striking features of the poleward boundary of the nightside aurora. This outflow is also seen during generally active auroral conditions without clear onset and expansion phase signatures. A statistical study (Tung et al., 2001) has demonstrated that this outflow region is located near midnight, on field lines that most probably map to the reconnection site in the tail, and covers an extent in MLT delimited by the region of enhanced auroral activity as determined from Polar UVI images.

The ion outflow itself is primarily H^+ ion conics with energies from a few tens to a few hundred of eV. The upgoing ion fluxes typically exceed 10^8 cm^{-2} s^{-1}, and are an order of magnitude larger than the fluxes observed at lower latitudes in the auroral zone. These fluxes represent a source of typically 10^{23} ions s^{-1} that, while not a dominant portion of the total estimated ion outflow rate of several times 10^{25} ions s^{-1} (Moore et al., 1999), is certainly a dominant portion of the nightside sector outflows. The nightside ion outflows also have a direct path to the plasma sheet, unlike the dayside outflows that must pass through the lobe and mantle prior to being incorporated into the plasma sheet (if they are incorporated at all).

4.5.2 Polar Cusp

Typical nightside features, such as low-energy inverted Vs, counterstreaming electrons, and ion heating, are also found on the dayside. There are, however, some unique features associated with the dayside polar cap boundary that should be mentioned, such as the cusp ion dispersion, stepped ion injections, and the cleft ion fountain. With regard to the region often referred to as cleft, we note that for the purposes of this book, we will not distinguish between cusp and cleft. A full description of this distinction can be found in Hultqvist et al. (1999).

4.5.2.1 Dispersions and Steps

Low altitude satellites traversing the auroral zone often observe a clear layering of the electron and ion distributions as a function of latitude (see Figure 5.21). An example of a low altitude satellite pass over the (southern) dayside polar region is given in Figure 4.53. Since we are interested in the dayside features, we we will discuss it in reverse order. Starting from the right, the spacecraft is initially in a region of closed field lines which maps to the plasma sheet, recognizable by the presence of high-energy particles. The sudden intensification of the electron and ion fluxes marks the entry into plasma having entered through the cusp. The equatorward edge of the cusp maps into the reconnection region of the dayside magnetopause. Poleward of this boundary are recently reconnected field lines that that are convecting tailwards. Ions moving downward along these field lines reach satellite

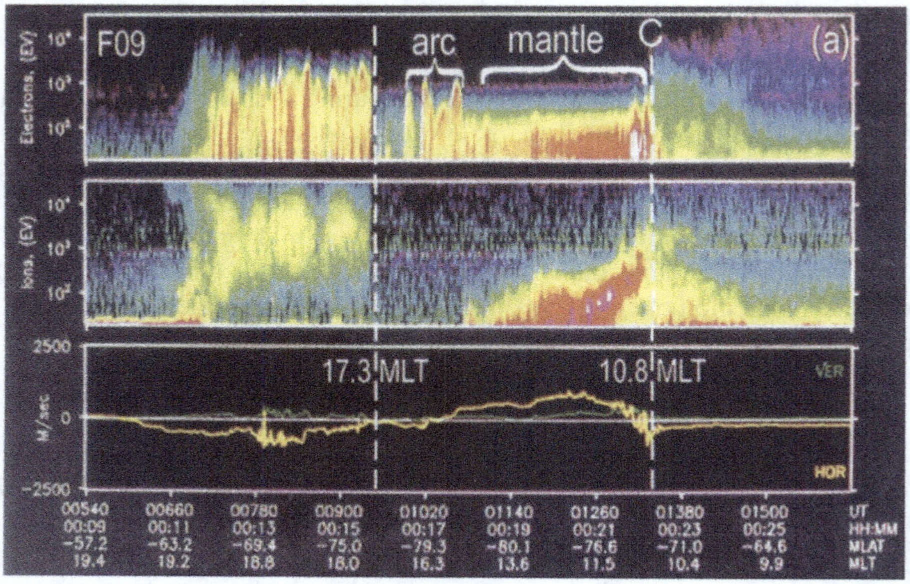

Figure 4.53. Energy-time spectrograms of electrons and ions, and the plasma convection velocity, for a southern polar cap traversal by the low-altitude polar-orbiting DMSP 9 spacecraft. The two dashed lines mark the position of the polar cap boundary, with the line denoted C being the dayside boundary, i.e., the equatorward edge of the cusp. From Lyons et al. (1996).

altitudes more poleward the lower their energy. Thus the ions in Figure 4.53 show the energy dispersion characteristic for conditions of southward IMF.

Often the cusp ion dispersion is not smooth but contains a number of almost discontinuous steps (Escoubet et al., 1992). In optical data these correspond to poleward-moving auroral forms (PMAFs) (Sandholt et al., 1990). Both features are thought to arise from pulsed reconnection at the magnetopause (Lockwood and Smith, 1992).

While these features are interesting, they shed more light on the processes happening at the magnetopause than on auroral acceleration, and we thus will not elaborate on them further. Instead we will discuss a very localized heating mechanism that causes large ion outflows from the cusp region.

4.5.2.2 Ion Heating Wall

A well-defined localized region which produces distinct signatures is the ion heating wall in the cusp/cleft region. Low-energy (tens of eV) ion measurements from the Dynamics Explorer 1 satellite in the early 1980's showed upflows of ionospheric ions at all local times in the auroral zone (Yau et al., 1984). However, the most intense outflows were seen to emanate from the dayside cusp/cleft region. These outflows included both light and heavy ions (H^+, He^+, N^+, O^+, O^{++}). The ions in this cleft ion fountain (Lockwood et al., 1985) are carried back over the

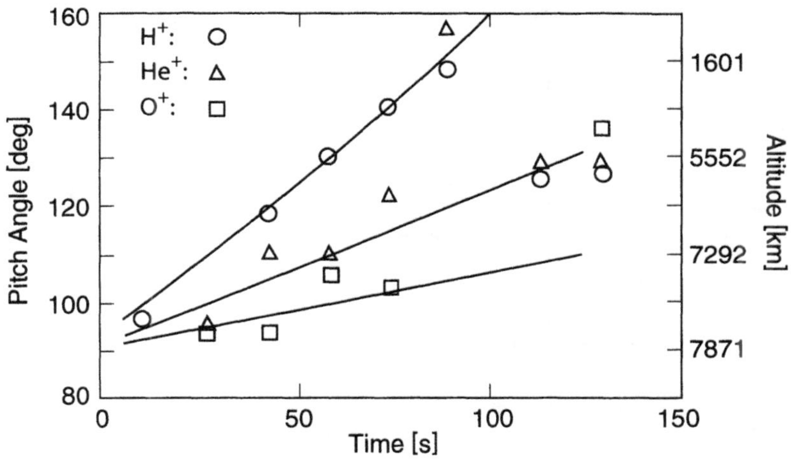

Figure 4.54. Pitch angle dispersion in cusp ions of ionospheric origin. From Knudsen et al. (1994a).

polar cap during periods of anti-sunward convection (IMF $B_z < 0$), with heavier ions tending to fall back into the nightside polar cap due to gravity.

While the ions comprising the cleft ion fountain are observed throughout the cusp/cleft and polar cap, the specific region in which they are heated can be quite localized. The direct signature of an ion heating region is an ion conic distribution with a pitch angle near 90°, coincident with plasma waves capable of such heating. Akebono satellite observations (Whalen et al., 1991) show ion populations with 90° pitch angles near the equatorward boundary of the cusp. Northward of this boundary, ion distributions were folded upward to form conics; this folding was observed to evolve more rapidly (in latitude) for lighter ions. Figure 4.54 shows observed pitch angles for H^+, He^+, and O^+ as a function of latitude, illustrating this effect.

Ion pitch-angle dispersion allows one to infer the morphology of the cleft ion heating region. The observations in Figure 4.54 are consistent with a 'heating wall' that is narrow (~30 km) in latitude, extended in longitude, and spans altitudes from ~2,000-10,000 km or higher. Ions are heated transversely to **B** at a rate of a few eV s^{-1} as they convect northward through the wall. This scenario is illustrated in Figure 4.55.

While the broad features of the ion heating region are fairly consistent with the simple morphological model shown in Figure 4.55, recent studies demonstrate the need for more sophisticated models. For example, heated ion energies depend on the IMF B_y component (Miyake et al., 2000). Such a dependence could result either through increased energy input into the cusp/cleft for large $|B_y|$, or through changes in ion convection patterns that can in turn affect ion residence time inside the heating wall. Significant variations in convection direction make it necessary

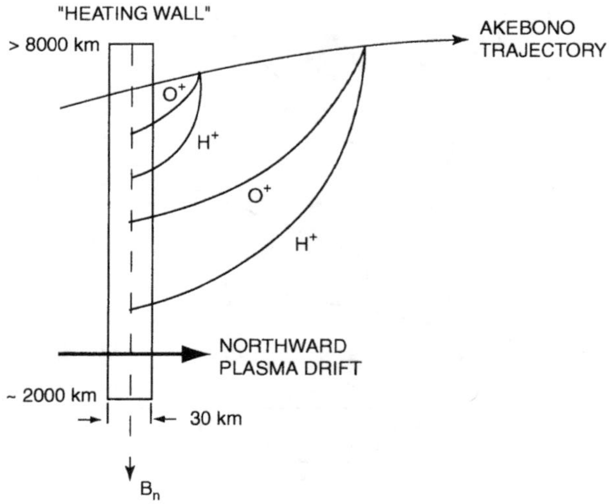

Figure 4.55. Morphology of the cusp ion heating wall. The vertical axis is field-aligned, and the horizontal axis represents a poleward cut through the auroral zone. From Knudsen et al. (1994a).

to include inertial forces in addition to wave-particle interactions when calculating ion trajectories (Dubouloz et al., 2001).

Structure within the cleft ion heating region extends down to very small spatial scales. The SCIFER sounding rocket observed that at 1000 km altitude, transverse ion heating and BBELF plasma waves turned on and off simultaneously as plasma density changed abruptly on scales of order 100 m (Kintner et al., 1996). Bursts of low-energy (0-20 eV) ion flux over similar spatial scales were observed as Freja flew through the the cusp/cleft near 1700 km altitude (Knudsen et al., 1994b); these bursts were shown to be consistent with localized ion heating associated with small-scale Alfvén waves (Knudsen and Wahlund, 1998).

Another population seen in this region indicates that wave particle interactions associated with the cleft heating region appear to extend well above altitudes at which heating has been observed directly (1,000-15,000 km). Interball Auroral Probe observations (Vovchenko et al., 2000) show a descending superthermal proton population inside the cusp-mantle region having a V-shaped pitch-angle dispersion. These particles must have originated at mid-altitudes within the cusp plasma flux tubes, possibly accelerated as a result of velocity diffusion of the magnetosheath protons due to beam-beam interaction with ascending ion conics.

4.5.3 Other Auroral Dispersion Signatures

There are other particle motions in the magnetosphere which lead to time-of-flight and path dispersion effects, especially when the source is localized in space and/or time. Many of these motions have signatures that are seen in or near auroral zone observations. We describe briefly two examples of such dispersion signatures.

4.5.3.1 VDIS-1 Events Within the Auroral Oval

The upgoing ion beams seen above auroral inverted-V structures (Section 4.1.2) form a localized source for a dispersion signature known as VDIS-1 (Velocity-Dispersed Ion Structure of type 1). Here ionospheric ions of different masses are upward accelerated to the same energy by field-aligned potential differences and then dispersed during their bouncing in the magnetosphere. They were first identified and studied with DE 2 data (Winningham et al., 1984) and AUREOL-3 data (Bosqued et al., 1986) at low altitudes, and then most thoroughly from AKEBONO at altitudes mostly above the field-aligned potential drop (Hirahara et al., 1997a,b). The VDIS-1 ion beam, having been accelerated up and out of the potential drop region, moves along the closed magnetic field lines of the near-Earth central plasma sheet. During its bouncing between the conjugate ionosphere and back, it is convected both towards lower latitudes and in azimuth. Thus it becomes both velocity and mass dispersed, as lower velocity particles are displaced further because of their longer bounce times. VDIS-1 events are observed from low-altitude satellites such as AUREOL-3, DE 2, DMSP (i.e., below the acceleration region) only when the bouncing beams are displaced outside the conjugate (decelerating) potential drop. At altitudes above the potential drop they are seen much more often.

A schematic explanation of different ion drift paths is given by Hirahara et al. (1997a), with a kinematic interpretation of their observed energy-mass spectra. Figure 4.56 demonstrates an example of a VDIS-1 event in the auroral oval and diffuse zone. The VDIS-1 events show conclusively that the nightside auroral oval is located on closed plasma flux tubes in the tail plasma sheet, where ion bouncing is possible.

4.5.3.2 Time-of-flight Dispersion During Substorm Expansion

Another example of dispersion signatures seen in the auroral zone is caused by substorm expansion. During the expansion phase of a substorm the auroral dynamics in the expanding auroral bulge (see page 251) are rapid and still not well understood. The velocity of a mid-altitude satellite trajectory across the respective magnetic flux tubes during this short time is slow and often may be neglected. Thus nearly pure time-of-flight effects for the injected ions may be seen in the plasma flux tubes of the poleward expanding auroral bulge (Sauvaud et al., 1999). These measurements are important for remotely delineating the realistic particle trajectories in the changing near tail during substorms. The energy-time spectrogram and the ion trajectory tracings back to the source shown in Figure 4.57 indicate a localized burst injection at 1247:30 UT. These tracings together with simultaneous UV images from POLAR consistently indicate that the bursty ion source was at

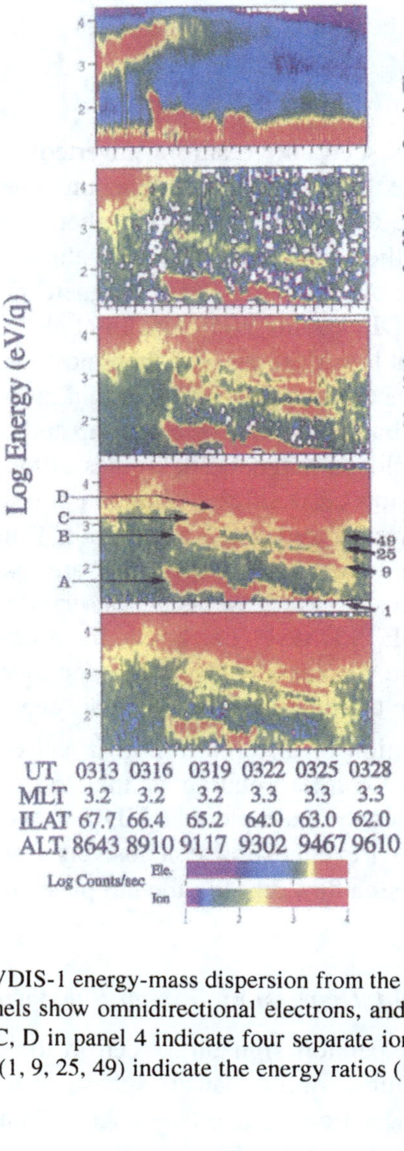

Figure 4.56. An example of VDIS-1 energy-mass dispersion from the AKEBONO satellite above the acceleration region. The panels show omnidirectional electrons, and ions sorted into four pitch angle ranges. The labels A, B, C, D in panel 4 indicate four separate ion dispersion signatures; the corresponding numerical labels (1, 9, 25, 49) indicate the energy ratios ($1{:}3^2{:}5^2{:}7^2$). From Hirahara et al. (1997a).

the equatorial plane at a distance of 7-8 R_E within the magnetic flux tubes of the expanding auroral bulge during substorms.

4.5.4 Diffuse Aurora

The existence of a wide band of relatively structureless (on scales less than tens of km), or diffuse, electron precipitation and auroral emission adjacent to the night-side auroral oval from equatorial side was noted both from ground-based auroral photometry and from early particle measurements from satellites in the late 1960s

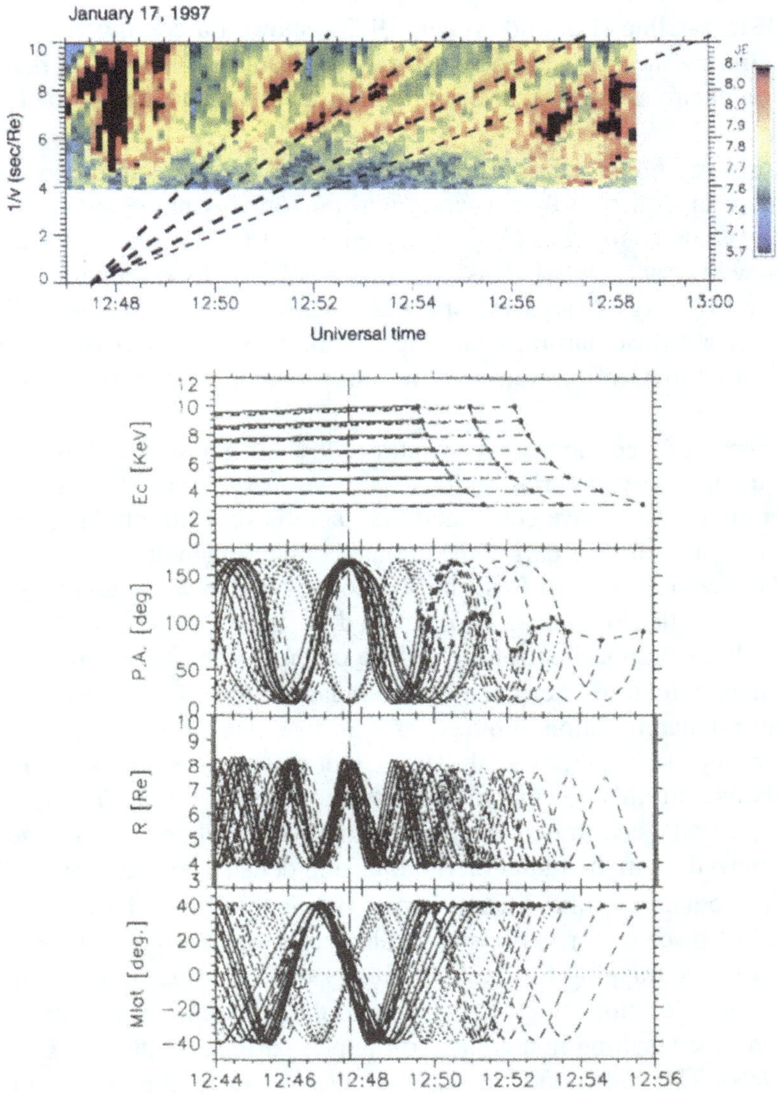

Figure 4.57. Upper panel: An example of pure time-of-flight ion dispersion from an expanding auroral bulge as seen from the INTERBALL-2 satellite. The spectrogram has been inverted so that the vertical axis shows inverse velocity (sec/R_E); the x-axis is time. The converging straight lines, fitting the ion traces, correspond to the expected TOF dispersion effects for bouncing ions. Lower part: Results of backward trajectory tracing calculations for the bouncing ions shown in the upper panel. The source (burst) is shown to be located at 7 $R_E$in the equatorial plane. Twenty-four energy and pitch angle points are traced backward in time starting from the dots. From top to bottom the panels show ion energy, pitch angle, radial distance, and magnetic latitude. Note the focal point at 12:47:45 UT as indicated by the vertical line. From Sauvaud et al. (1999).

(Galperin et al., 1970). The first global patterns of auroral luminosity, obtained from the ISIS satellite (Lui and Anger, 1973), showed a continuous auroral oval pattern encircling the magnetic pole with an adjacent diffuse auroral belt. Diffuse precipitation encircles the oval of discrete aurorae on its equatorial side (Feldstein, 1974).

An useful operational definition of this diffuse auroral precipitation, as distinct from the auroral oval of bright discrete auroral forms, can be found by computing the correlation coefficient C between sequential ($\sim$ 1 s apart) electron energy spectra (Newell et al., 1996a). This coefficient is found by correlating particle intensities at each energy in adjacent spectra. A value of $C \geq 0.6$ is a good defining parameter for a diffuse aurora, while significantly lower values of C are usually found within the oval of discrete aurorae indicating more small-scale structuring there.

Comparisons of electron energy spectra at low altitudes with those at magnetically conjugate near-equatorial regions (geostationary orbit) (Meng et al., 1979; Schumaker et al., 1989) have confirmed that there is very little field-aligned acceleration in this quasi-dipolar part of the nightside inner magnetosphere. The precipitation of the electrons in the diffuse auroral zone is due to weak angular scattering, or diffusion, into the loss cone. As discussed by Blanc et al. (1999), the precise mechanism that is responsible for the diffuse precipitation is not known at present.

In stationary conditions the equatorial boundary of the diffuse electron precipitation is a natural manifestation of the equatorial boundary of the large-scale convection on the nightside (Khalipov et al., 1977). Hot plasma from the tail is transported across L-shells until this boundary, where the convection turns. Thus in stationary conditions the soft electron boundary on the nightside collocates with the plasmapause – a sharp drop in the radial thermal plasma density profile at high altitudes.

The polar boundary of the diffuse auroral belt is the equatorial boundary of the auroral oval of discrete auroral forms. It also coincides on the nightside with the boundary of stable trapping for the high-energy outer belt particles (Feldstein and Starkov, 1970; Feldstein, 1974). This is due to the stretching of the magnetic field to the tail and the resulting non-adiabatic trapped particle scattering (Büchner and Zelenyi, 1989). The transition from quasi-dipole to stretched magnetic field in the near-Earth's tail usually occurs somewhere between the distances $\sim 10-15$ $R_E$and the geostationary orbit. The physical reason for the tail magnetic field stretching is the effect of the cross-tail current flowing from dawn to dusk in the form of a current sheet embedded in the plasma sheet. That is why this transition coincides with the boundary of stable trapping, or of the particles' isotropization.

Field-aligned currents generated within the quasi-dipole part, in absence of particle injections during a magnetic storm, are large-scale and locally weak. Respectively, such currents (within the Region 2) lead to a widespread, relatively structureless diffuse precipitation, but not to bright localized auroral forms closely related to concentrated large field-aligned currents (except during large magnetic storms).

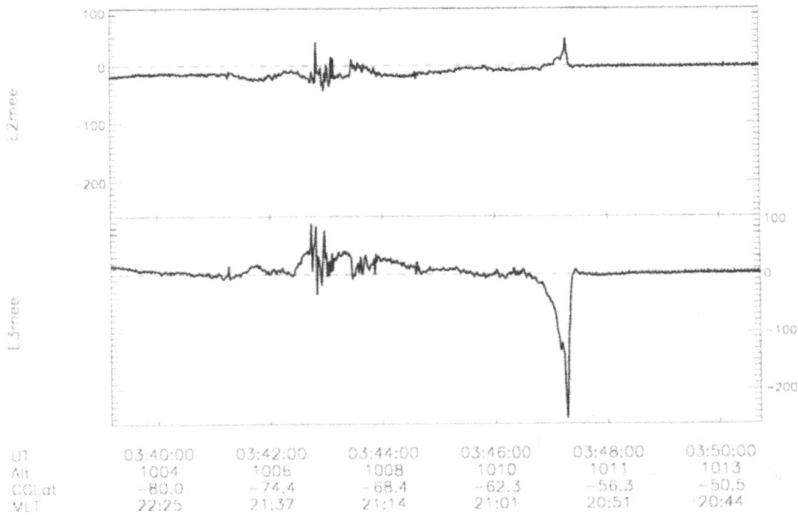

Figure 4.58. Astrid 2 electric field data of a SAID structure (shortly after 03:47 UT). The top panel shows the eastward component of the electric field, while the second panel shows the equatorward component. The SAID electric field is almost exactly poleward directed, and much larger than the auroral electric fields observed near 03:43 UT. This is an observation from the southern hemisphere; SAID events seem to occur equally often on both hemispheres, and indeed can be observed simultaneously in conjugate hemispheres. Figure provided by T. Karlsson.

At the same time large-scale convection still transports low-energy plasma sheet particle populations to this region of the outer radiation belt. Their slow scattering to the loss-cone is the origin of the diffuse auroral luminosity.

4.5.5 Subauroral Ion Drifts

Normally the region equatorward of the oval is associated with weak convection plasma drifts and electric fields. However, during the substorm recovery phase sometimes a band of enhanced westward plasma drift (or equivalently, poleward electric field) is encountered at subauroral latitudes. The electric field can be several hundreds of mV m^{-1}, quite comparable to the largest ionospheric electric fields in the auroral oval. This phenomenon was first discovered by the Kosmos-184 satellite (Galperin et al., 1974) and is usually referred to as subauroral ion drift (SAID) or sometimes as polarization jet. An example of a measurement of a SAID-related electric field is shown in Figure 4.58.

The region of enhanced plasma drift is almost exclusively encountered on the evening side (Spiro et al., 1979; Karlsson et al., 1998), as shown in Figure 4.59, and at any one moment probably extends over many hours in local time (Anderson et al., 1993). The latitudinal width is of the order of 100 km, at the ionospheric level. These morphological features suggest a relation to the low plasma density

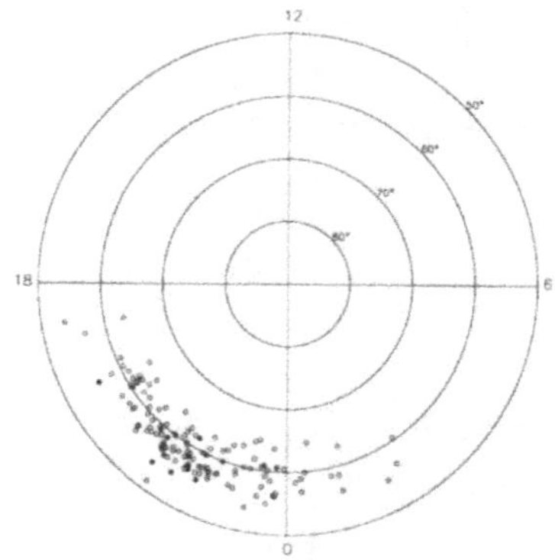

Figure 4.59. Polar plot (in magnetic local time and corrected geomagnetic coordinates) of the locations of SAID observed by the Freja satellite. Solid circles represent SAID where the maximum poleward component of the electric field is greater than 200 mV m^{-1}. It can be seen that SAID are almost exclusively an eveningside phenomenon, and that the largest fields seem to be centered around 2200 MLT. From Karlsson et al. (1998).

region of the mid-latitude trough, and SAID are indeed observed to be associated with deep density troughs (Anderson et al., 1991).

There is no generally accepted physical model for the SAID generation mechanism. There are two main theories. One explains the SAID in terms of substorm-associated Region 1 currents moving down to very low latitudes and closing through (and modifying) the mid-latitude trough, where the low conductivity of the trough causes large electric fields to be set up to ensure current continuity (Anderson et al., 1991). The other is in terms of shielding of substorm-enhanced electric fields, produced by the increased separation of the inner edges of the ion and electron populations of the plasma sheet (Southwood and Wolf, 1978; Galperin, 2002).

The large electric fields associated with SAID cause a substantial Ohmic heating of the ionospheric plasma, which even at times may cause excitation of atomic oxygen and red line emissions at 630 nm, similar to the aurora (Khalipov et al., 2001). These emissions constitute a special case of subauroral red arcs (SAR; the acronym is sometimes taken to mean 'stable red arcs'). These are in contrast to the classical SAR, associated with the decay phase of magnetic storms, where the excitations are caused by energetic ring current ions (Kozyra et al., 1997, and references therein).

Chapter 5

Statistics and Mapping of Auroral Features

Statistical studies of the large-scale convection electric field, field-aligned currents, auroral luminosity, accelerated electron and ion distributions, small-scale electric field structures ('electrostatic shocks'), plasma density, and waves, such as auroral kilometric radiation, can provide important constraints on interpretations of both the drivers and the mechanisms of auroral acceleration. The microphysics of acceleration and wave processes has been discussed in Chapter 4. In the first section of this chapter, we describe statistical properties of these phenomena and their dependence on solar wind conditions, the IMF, magnetic activity, solar illumination ('season'), and solar cycle, as well as altitude, magnetic local time and magnetic latitude. In the second section we discuss the mapping of the low altitude auroral features to their source region in the magnetosphere.

5.1 Statistical and Empirical Studies of Auroral Distributions

Observations of ionospheric convection or the large-scale electric field, auroral precipitation, and currents can be made from both ground- and space-based instruments. However, a complete instantaneous MLT and invariant latitude picture is difficult, if not impossible, to achieve. Studies of ionospheric electrodynamics often rely on statistical information under similar geophysical conditions but obtained over a long period of time. Despite these limitations, statistical and empirical distributions of ionospheric electrodynamic fields provide useful information on the properties of the solar wind-magnetosphere-ionosphere coupling processes.

Statistical studies of accelerated electron and ion distributions can be utilized to examine the distribution of the parallel potential and to provide constraints on driving mechanisms. In the upward current region, measurements of the down-going electrons provide information on the size of the potential drops above the satellite, while observations of ion beams yield information on the parallel potential below the satellite, and on the altitude where parallel potentials occur. In the downward current region, studies of upflowing electrons can be used to examine the occurrence of downward directed electric fields, as described in Chapter 4, while downgoing ion beams are an indicator for the potential above the satellite.

Very large amplitude (100's of mV m^{-1}), small-scale electric fields, oriented primarily perpendicular to the earth's magnetic field ('electrostatic shocks'), were first observed by the S3-3 satellite (Mozer et al., 1977) and were shown to be associated with upward parallel electric fields as indicated by downward accelerated electrons and upflowing ion beams (Mozer et al., 1980), and with visible aurora (Kletzing et al., 1983). More recently, such fields (with diverging perpendicular components) have been associated with the upward acceleration of electrons (Marklund et al., 1997; Carlson et al., 1998a). Because both the converging shocks and the diverging shocks are associated with parallel acceleration, statistical studies of these shocks can provide information on auroral acceleration that is complementary to studies of accelerated particles.

Density 'cavities' – large depletions in the ambient plasma density – are observed over a range of length scales in the auroral acceleration region. Because many microphysical processes are dependent on the density structure, an understanding of the characteristics of these structures, their altitude dependence and formation is critical for understanding acceleration. Processes involved in cavity formation have been discussed in Chapter 3. For observations of cavities, see Chapter 4.

As discussed in Section 3.5.2 and Section 4.2.2, many spacecraft observations show that ions in the ionosphere and magnetosphere can be transversely energized. These ions are then subjected to the upward mirror force in the inhomogeneous geomagnetic field, thereby forming so-called 'conics' in velocity space. The upflowing ions constitute an important source of magnetospheric plasma (e.g., Moore et al., 1999). Most of the energization takes place above the collisional ionosphere, and thus depends on interaction between the charged ions and some (often time-varying) electric field. Statistical studies of conics and the associated waves can help elucidate the plasma physics associated with the extraction of these ions and the dependence of this important plasma source on solar and magnetic activity.

The most complete statistical studies to date of accelerated electrons utilize data from the DMSP and the NOAA-TIROS satellites. The statistical occurrence of ion beams and conics has been derived from many satellites, including S3-3, DE, Viking, ISEE, Aureol-3, Freja, Polar, and FAST. Studies on the net fluence of various ion species will not be specifically addressed since they have been discussed in Moore et al. (1999), and are not as directly related to the auroral acceleration process. Numerous statistical studies of large-amplitude fields have been conducted using data from S3-3, DE, ISEE, Akebono, Aureol-3, Freja, Polar, and FAST, including the association with ion beams, ion conics, currents and electrons. Auroral luminosity studies are primarily obtained from DMSP, Viking, DE 1 and Polar, and, more recently, IMAGE.

5.1.1 Ionospheric Observations and Empirical Models

5.1.1.1 Ionospheric Convection

The solar wind-magnetosphere interaction forces plasmas to circulate or convect in the magnetosphere and in the magnetically connected ionosphere. When the interplanetary magnetic field (IMF) is southward the typical plasma convection pattern has two cells, with anti-sunward flow in the polar cap and return sunward flow in the auroral zone in the frame corotating with the Earth. The IMF B_y controls the asymmetry of the dawn-dusk convection cells with respect to the noon-midnight meridian. In the northern polar cap, the dayside plasma convection shifts toward dawn when B_y is positive and toward dusk when B_y is negative. The Earth's rotation in the non-rotating (magnetospheric) frame adds the so-called co-rotation component. As a result, the dawn (dusk) convection cell is crescent shaped and the dusk (dawn) cell is round shaped for $B_y > 0$ ($B_y < 0$). In the southern hemisphere, the convection patterns are roughly the mirror images of those in the northern hemisphere.

When the IMF points northward the convection patterns are often very complicated, consisting of either a distorted two-cell pattern (Friis-Christensen et al., 1985; Heppner and Maynard, 1987; Marklund et al., 1991) or three or more cells (Burke et al., 1979; Potemra et al., 1984; Reiff and Burch, 1985). Furthermore, the convection patterns in the two hemispheres can be drastically different from one another during northward IMF. Transition of the convection configuration from 'northward-like' to 'southward-like' occurs when the ratio of $|B_y|/B_z$ becomes greater than one (Knipp et al., 1993; Lu et al., 1994).

A number of statistical and empirical models of ionospheric convection based on satellite observations (Heppner and Maynard, 1987; Weimer, 1995, 2001; Rich and Hairston, 1994), radar measurements (Foster et al., 1986; Ruohoniemi and Greenwald, 1996; de la Beaujardière et al., 1991), or ground magnetometer data (Friis-Christensen et al., 1985; Papitashvili et al., 1994) have been developed. For southward IMF conditions, all the statistical models have a similar two cell configuration although the cross-polar-cap potential drops may differ from one model to another. For northward IMF, however, there are dramatic differences in the convection patterns given by the different models. Figure 5.1 shows the statistical convection patterns of the Weimer model (Weimer, 1995) for different IMF orientations. For southward IMF, the convection has two cells that are roughly symmetric about the noon-midnight meridian. When the IMF points northward, the convection pattern consists of 4 cells, with two primary, reversed-polarity cells in the central polar cap and two secondary cells of normal polarity at lower latitudes on the dawn and dusk sides. The figure illustrates how ionospheric convection evolves from two cells to four cells as the IMF rotates from southward to northward.

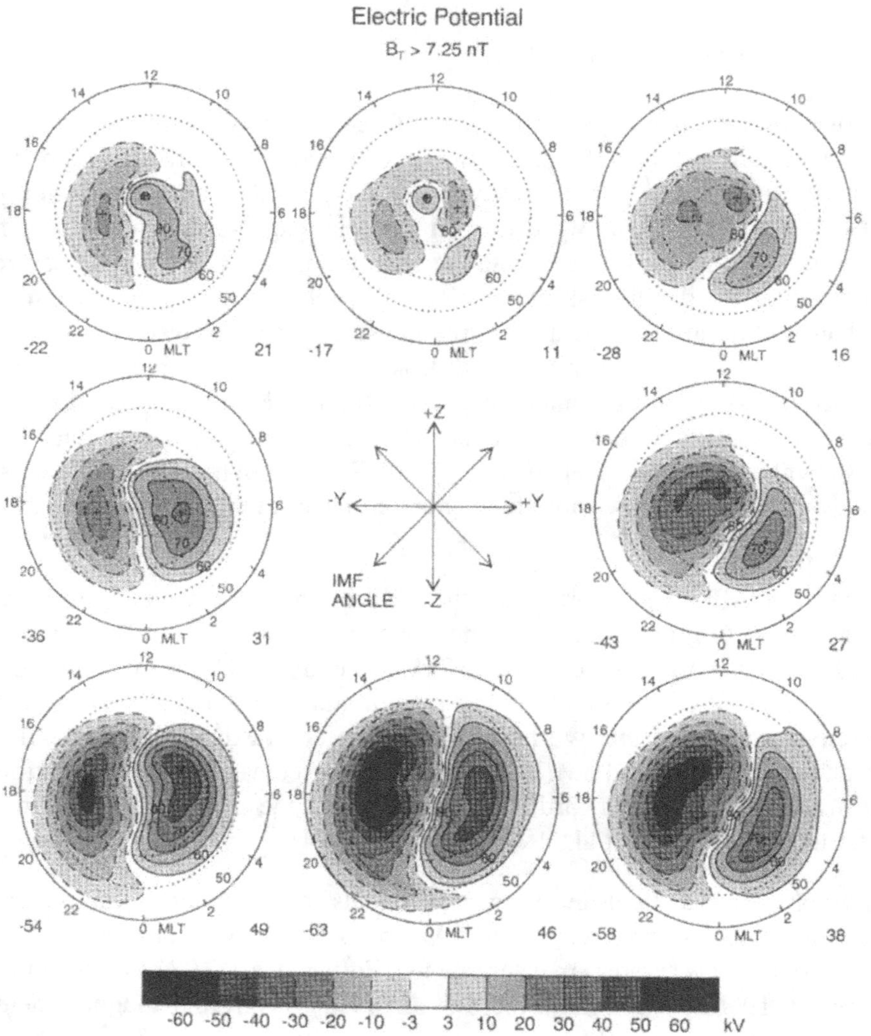

Figure 5.1. Statistical ionospheric convection patterns of the Weimer model for different IMF orientations. From Weimer (1995).

5.1.1.2 Field-Aligned Currents

Magnetospheric convection is generated by the solar wind-magnetosphere 'dynamo'. The momentum and energy transfer from the outer magnetosphere to the ionosphere is accomplished by magnetic field-aligned currents. The presence of field-aligned currents has been confirmed from both particle and magnetic field observations. Field-aligned currents are most commonly inferred from measurements of the resulting magnetic field perturbations. Statistical analysis of satellite

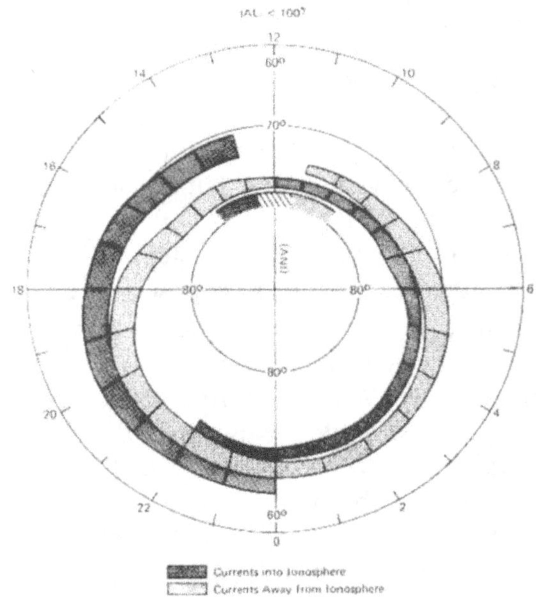

Figure 5.2. Statistical patterns of the field-aligned currents. From Iijima and Potemra (1978).

magnetometer data reveals that large-scale field-aligned currents are concentrated in two ring-shaped zones encircling the geomagnetic poles as shown in Figure 5.2, and already introduced in Section 1.4.

During quiet geomagnetic conditions the Region 1 current intensities peak on the dayside near 13 MLT and 10-11 MLT (Iijima and Potemra, 1978). During disturbed conditions, the prenoon Region 1 peak shifts to 07-08 MLT, while the postnoon Region 1 peak shifts to 14-15 MLT. The equatorward Region 2 currents exhibit no distinct peak during quiet times but show a peak in the midnight sector during active times. The dayside Region 1 peaks are associated with 'hot spots' in auroral emissions, described in Section 5.1.2.

On the nightside, the Iijima and Potemra (1978) model shows a triple current sheet pattern located between 22 and 24 MLT, a region often referred to as the Harang discontinuity (Harang, 1946; Heppner, 1972; Koskinen and Pulkkinen, 1995). The basic pattern of the Region 1/2 system does not change very much with geomagnetic activity, but the regions become wider and also tend to shift to lower latitudes during more disturbed periods. More detailed discussions of substorm-related nightside current systems are given in Chapter 6.

On the dayside and poleward of the Region 1 currents, an additional pair of field-aligned currents flow in the opposite direction to the Region 1 currents. These currents have been variously identified as 'cusp currents' by Iijima and Potemra (1976), 'mantle currents' by Erlandson et al. (1988) and Bythrow et al. (1988), and 'Region 0 currents' by Heikkila (1984). Although the dayside cusp currents are a pronounced feature during southward IMF conditions, they exist even during the so-called 'magnetospheric ground state' when the IMF $B_z \approx B_y \approx 0$ (Rich and Gussenhoven, 1987; Hoffman et al., 1988; Watanabe et al., 1998). When the IMF is northward, the Region 1 and Region 2 field-aligned currents become very weak and highly structured on the nightside, but remain nearly the same as for the southward IMF conditions on the dayside. A single third current sheet, poleward of Region 1, rather than an additional pair, consistent with a boundary layer source (Cattell et al., 1996), is often seen in the morning sector (Cattell et al., 1979).

The distributions of field-aligned currents, like ionospheric convection patterns, also show a strong dependence on the IMF orientation, especially those of the dayside Region 1 and cusp currents. When the IMF is southward or weakly northward, the Region 1 currents shift duskward (dawnward) and the dominant cusp currents flow out of (into) the ionosphere for $B_y > 0$ ($B_y < 0$) (D'Angelo, 1980; Erlandson et al., 1988). When the IMF becomes more northward, a pair of field-aligned currents with polarity similar to the cusp currents starts to evolve over the polar regions. Unlike the cusp currents, which are generally confined to a latitudinally narrow region near the cusp, this current system extends to most of the polar cap. This broadly distributed current system has been dubbed the 'NBZ' (standing for northward B_z) currents by Iijima et al. (1984). The NBZ currents intensify and become more stable as the value of B_z becomes larger. For strongly northward IMF, the NBZ currents can exceed both the Region 1 and Region 2 currents and become the dominant current system in the high-latitude ionosphere.

While there is general agreement regarding the large-scale distribution of the Region 1 and Region 2 currents, controversy over the cusp/mantle/Region 0 currents on the dayside has yet to be settled. Some studies suggest that cusp currents form a unique current system that is topologically separate from the Region 1 current system (Iijima et al., 1984; Potemra et al., 1980; Burch et al., 1985; Erlandson et al., 1988; Taguchi et al., 1993; Ohtani et al., 1995), while others suggest that the cusp currents are an extension of the Region 1 currents from either dawn or dusk depending on IMF B_y (McDiarmid et al., 1979; Rich and Maynard, 1989; Cowley et al., 1991; Saunders, 1992; de la Beaujardière et al., 1993; Lu et al., 1995; Weimer, 1999).

The intensity of field-aligned currents also exhibits a strong seasonal dependence, with the average currents into or out of the sunlit ionosphere being larger than those flowing into the darker winter ionosphere (Fujii et al., 1981).

Empirical models of field-aligned currents have also been derived without using space-based magnetic-field measurements. Utilizing statistical ionospheric convection patterns and an ionospheric conductivity model, Weimer (1999) derived a qualitative model of field-aligned currents for both substorm and non-substorm conditions for various IMF orientations. The upward currents form a continuous oval for $B_y > 0$ and link the dawnside Region 2 currents with the duskside Region

1 currents in both the local noon and midnight sectors. For $B_y > 0$, the upward currents near local noon are interrupted by a downward current that links the duskside Region 2 current to the dawnside Region 1 current. One of the main findings is that the 'Region 0' or cusp currents near local noon appear to be an extension of the dawn and dusk Region 1 currents, rather than a separate entity. The difference between substorm and non-substorm field-aligned current distributions lies mainly in the midnight sector where currents are enhanced during substorms in association with the enhancement of westward electrojets. Empirical models of field-aligned currents derived from ground magnetometer observations (Friis-Christensen et al., 1985) exhibit large-scale features similar to those of the Iijima and Potemra (1976) statistical pattern but with differences in some smaller-scale features. One of the main features of the Friis-Christensen et al. (1985) model is the apparent separation of the dayside and nightside Region 1 currents, implying the existence of separate magnetospheric source regions for the dayside and nightside parts of the Region 1 currents.

5.1.1.3 Global Auroral Precipitation and Conductances

Fuller-Rowell and Evans (1987) obtained statistical global patterns of auroral precipitation and height-integrated Pedersen and Hall conductivities based on particle measurements made by the NOAA-TIROS spacecraft. Their model is parameterized according to the hemispheric power index (Foster et al., 1986), corresponding to an estimated total energy deposition into a single hemisphere. The model shows an expanded auroral oval and harder (more energetic) precipitation with increasing hemispheric power index. A clear asymmetry of auroral precipitation with respect to the magnetic midnight meridian is also found. For small index levels, precipitating electrons in the dawn and late morning sector tend to be harder than those in the afternoon and nightside regions; for high index levels, the hardest precipitation is found in the pre-midnight sector. Electron precipitation is discussed in more detail below.

Auroral precipitation and conductivity models, sorted by the Kp[1] index, have also been derived from the DMSP observations (Hardy et al., 1985; Hardy et al., 1987). These models show that auroral conductivities increase and the auroral oval moves equatorward with increasing geomagnetic activity. Two conductivity peaks are found, one near midnight from 23.5 to 01.5 MLT and the other in the early morning sector between 03 and 05 MLT. Because the precipitation is typically harder in the morning than the evening, the Hall to Pedersen conductivity ratio is typically higher in the morning. The midnight conductivity peak is generally slightly larger than the morning peak in magnitude. Overall, average Pedersen and Hall conductivities vary smoothly with both latitude and local time.

[1] Kp is a planetary 3-hour index in the range 0–9, describing the average level of magnetic activity.

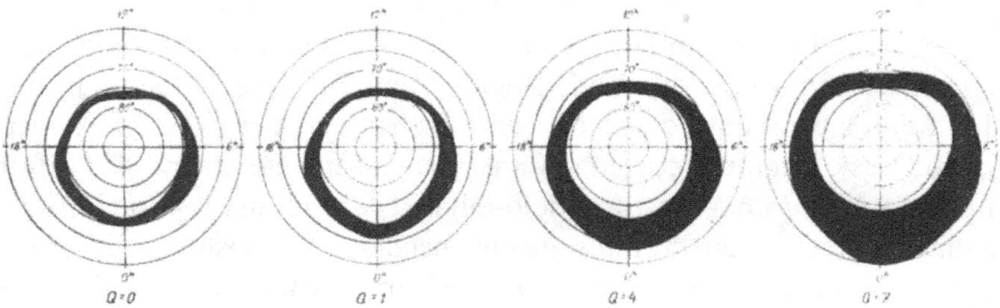

Figure 5.3. The Feldstein auroral oval in magnetic local time and corrected geomagnetic coordinates and its variation with increasing auroral activity in the northern hemisphere. From Feldstein and Starkov (1967).

5.1.2 Auroral Luminosity

5.1.2.1 Distribution of Various Types of Auroral Forms

The intensity of visual auroral features closely tracks the precipitated electron and proton energy flux as measured from rockets and satellites (Stenbaek-Nielsen et al., 1998; Hallinan et al., 2001). Thus statistical averages of the auroral intensity distribution around the auroral oval should match closely those of electron and proton energy flux. Nonetheless, imaging of aurorae, whether from satellites or from the ground, provides a somewhat different perspective than that obtained from particle measurements.

Whereas particle data represent a series of point measurements along a single line through the aurora, imaging offers an instantaneous two-dimensional distribution. Sequential images can provide temporal information as well, with resolutions varying from fractions of a second to minutes depending on the imager. Because of these differences, auroral imaging tends to emphasize the shapes, motions, and temporal variations of auroral forms rather than broad statistical averages, although these also are sometimes obtained from satellite images.

The various types of auroral forms are described more fully in Chapter 2. Our concern here is mainly with the distribution of these forms around the oval in space and time. Although this distribution shows great variety, a general pattern of the latitudinal extremes of its average boundaries first emerged from a synthesis of all-sky camera images obtained during the International Geophysical Year (1957-8). The average extrema sorted by quiet, moderate and active auroral conditions are shown in Figure 5.3. The auroral oval is typically thinner during quiet periods and expands both poleward and equatorward on the nightside with increasing activity.

The dynamical progression from the ground state represented by the quiet and relatively thin oval configuration in Figure 5.3 to the much broader and disturbed nightside oval was first deduced by Akasofu (1964). The evolution of this pattern is based on a sequence of events called an auroral substorm as illustrated in Figure

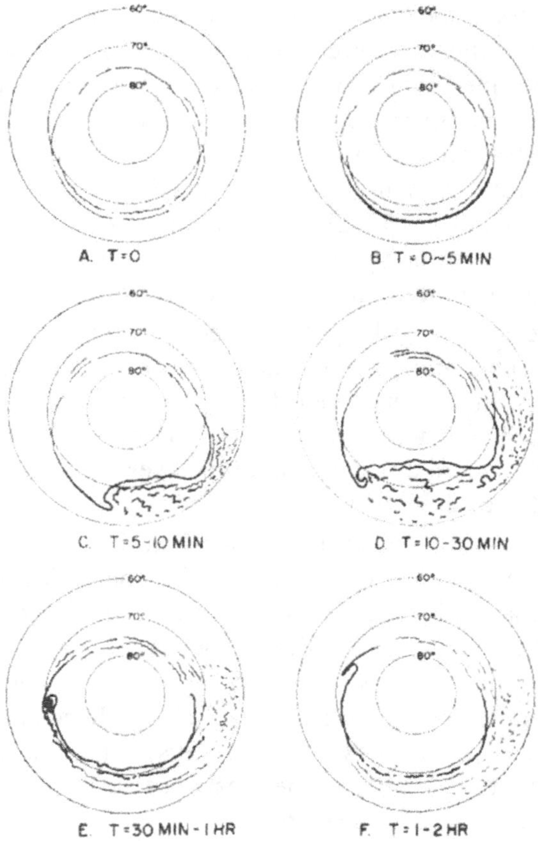

Figure 5.4. The development of an auroral substorm illustrating expansion of the auroral oval from its quiet ground state and eventual recovery over a period of 1 - 2 hours. After Akasofu (1964).

5.4. This sequence continues to provide a basis for organizing observational data and for development of more detailed models.

The pattern starts with a quiet phase, characterized by a diffuse aurora on the equatorward side of the auroral oval and discrete auroral arcs along the poleward edge of the oval. The diffuse aurora in the evening, although caused mostly by precipitating electrons, also includes the maximum proton energy flux. This quiet phase is interrupted by a growth phase during which both the diffuse aurora and the discrete arcs brighten slowly and move equatorward. During the growth phase, auroral field lines become progressively stretched.

The substorm onset is very sudden and starts with the most equatorward discrete arc suddenly brightening and then moving or expanding poleward (expansive phase). This onset occurs near magnetic midnight (Akasofu, 1964) or more often about an hour prior to magnetic midnight (Craven and Frank, 1991). Within a few minutes, the brightness of the discrete aurora increases by orders of magnitude and the arc evolves into a gigantic spiral (Hallinan, 1976) up to a thousand kilometers in diameter. The spiral typically drifts rapidly westward. A general region of brightening and activation may proceed westward faster than the individual spiral and other spirals may form within this westward traveling surge (Rostoker et al., 1987a, cf. Section 6.2). The spirals and associated arcs are also characterized by small flickering patches (4 – 15 Hz) (Beach et al., 1968), folds, and ray motions. An observer in the early evening sector will see this activity start on the eastern horizon and propagate overhead and then off to the West.

In the morning sector, along the poleward edge of the oval, a similar pattern of activity occurs with the region of bright active aurora, including spirals and curls, expanding eastward from the onset meridian. In the morning sector, the difference between the drifts of individual forms and the rate of expansion of activity (eastward surge velocity) is more obvious than in the evening. The main proton energy flux is poleward of the oval in this sector.

After the discrete aurora located in the meridian of substorm onset reaches its maximum poleward extent, it begins to retreat in what is known as the recovery phase. During this phase, the discrete aurora becomes less intense. In the morning sector, and also in the region between the westward surge and the midnight meridian, the equatorward aurora is characterized by pulsating patches. In this region there are also (intermittently) eastward drifting omega bands (cf. Section 6.3) that are bright and include rapidly moving curls (rays). The pulsations typically last 20-30 minutes in the region between the westward surge and midnight, but may persist much longer in the morning sector.

Auroral forms on the dayside consist of a latitudinally narrow (< 200km) zone of arcs and bands not unlike the discrete forms on the night side (Deehr et al., 1980). They are 20 km higher in the atmosphere, reflecting the generally less energetic electron flux (Sigernes et al., 1996). The spirals and curls are smaller and weaker. The main differences between the dayside and the nightside aurora are that on the dayside the discrete auroral forms are embedded in the main proton energy flux, their latitudinal movement is poleward instead of equatorward, there are no dayside substorm effects (Lorentzen et al., 1996), and their general occurrence is controlled directly by the interplanetary magnetic field direction and magnitude (Sandholt et al., 1998). Equatorward of this region of discrete arcs in the prenoon sector is the dayside diffuse auroral region in which pulsating auroral forms are observed. The pulsation frequencies range from 1 through 0.01 Hz and are sometimes coherent over a large part of the sky for a single observer. Such large coherency allows simpler correlation of particle, optical and magnetic signatures. It appears that such diverse phenomena as magnetic impulse events, magnetic pulsations, X-ray microbursts, magnetic vortices, and pulsating aurora are all related in this region (Lorentzen et al., 1996).

Synoptic patterns of auroral luminosity and related features in particles and fields based on averages in both space and time are described in the next sections. When considering such averages, it is important to keep in mind the temporal variability that aurorae almost always exhibit.

5.1.2.2 Synoptic Global Imaging

Global images of the aurora have provided a new tool for examining the time evolution of aurora and their dependence on magnetic activity, conductivity and IMF. Synoptic studies of ultraviolet luminosity (Liou et al., 1999; Shue et al., 2000) have provided new insights into large-scale stable features together with their variability. A synopsis of over 17,000 POLAR UVI images (Figure 5.5) statistically delineates the primary nightside, pre-midnight feature, as well as the two dayside features in terms of LBH (165-175 nm) emission intensities (Liou et al., 1997). As discussed in the next section, enhancements in precipitating particle fluxes also accompany these features.

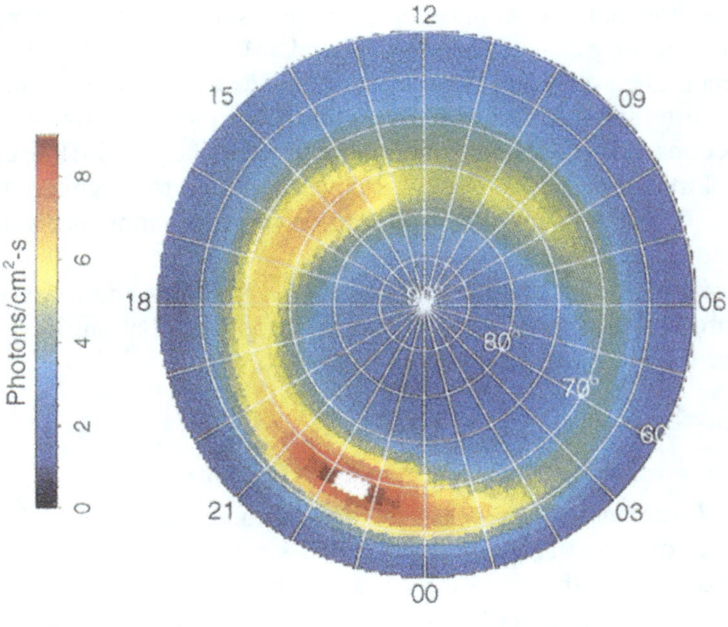

Figure 5.5. Map of average northern hemisphere auroral intensity in the LBH-long band ($\lambda \sim$ 165 – 175 nm) superposed on a geomagnetic grid. A total of 17372 images recorded by the Polar UVI imager during the period April 1 to July 28, 1996 was used to construct this average distribution. From Liou et al. (1997).

The stronger of these two dayside regions (a 'hot spot') appears statistically from 1400 hours to 1600 hours MLT in both the particle precipitation and emission maps (Cogger et al., 1977; Evans, 1985; Newell et al., 1996a). It is roughly coincident with the peak in the dayside Region 1 field-aligned current. This afternoon hot spot is relatively more prominent (because of weaker oval emissions) when B_z is northward than when B_z is southward (Evans, 1985; Newell et al., 1996a). A weaker region of emissions (a 'warm spot') appears in the pre-noon sector from 0600 hours to 0900 hours MLT (Meng and Lundin, 1986; Newell et al., 1996a), coinciding with the pre-noon peak in the Region 1 current.

Like the strong dependence of the distributions of ionospheric convection and field-aligned current on the IMF, auroral morphology also changes as the IMF varies. Generally speaking, the auroral oval tends to be a circular ring during southward IMF. During northward IMF the oval is distorted into a horseshoe shape and sun-aligned auroral arcs, such as 'theta' aurorae (Frank et al., 1986), form over the polar caps. The influence of interplanetary magnetic field on global auroral distributions has recently been reported by Jankowska et al. (1990) and Shue et al. (2000). Some representative auroral patterns are shown in Figure 5.6. By averaging the global auroral images obtained from the Polar UVI instrument, they found that the dependence of global auroral patterns on the IMF orientation changes with season. The afternoon bright spot becomes more distinct from the nightside aurora for $B_y < 0$ than for $B_y > 0$ under northward IMF or small IMF B_z conditions during summer. The average position of the duskside aurora varies with the sign of the IMF B_y during summer; but its location remains almost the same during winter. The B_y effect on auroral position is more pronounced when the IMF becomes more southward. Northern hemisphere auroral power is found to be greater for $B_y < 0$ than for $B_y > 0$, and the B_y effect on auroral power is also more pronounced when the IMF is more southward.

Because of their lower occurrence frequency, variability and lower luminosity, polar cap auroral features do not show up in synoptic global auroral maps. They will therefore be discussed separately in Chapter 6.

5.1.3 Auroral In Situ Observations

5.1.3.1 Accelerated Electron and Ion Distributions

Downgoing Electrons. The local time and latitude dependence of accelerated electron events observed at an altitude of 800 km (below the altitude where the acceleration occurs) is shown in Figure 5.7. The criteria utilized for identifying accelerated distributions, described in detail in Newell et al. (1996b), is based on the existence of a sharp drop in flux above a spectral peak and selects events with an acceleration potential that is larger than the source thermal energy by a factor of 3 to 4. The time resolution of each individual spectrum is 1 second which corresponds to a spatial resolution $\sim 0.068°$ in latitude or $\sim$7.5 km. The region of enhanced occurrence matches the statistical oval. The maximum probability for events with energy fluxes greater than 0.25 erg cm^{-2}s^{-1} (which is below the visual threshold of $\sim$1 erg cm^{-2} s^{-1}) occurred from 18-24 MLT near 70° and near 15 MLT and 72°. There is also a weaker peak near 6-9 MLT, and accelerated electrons are least

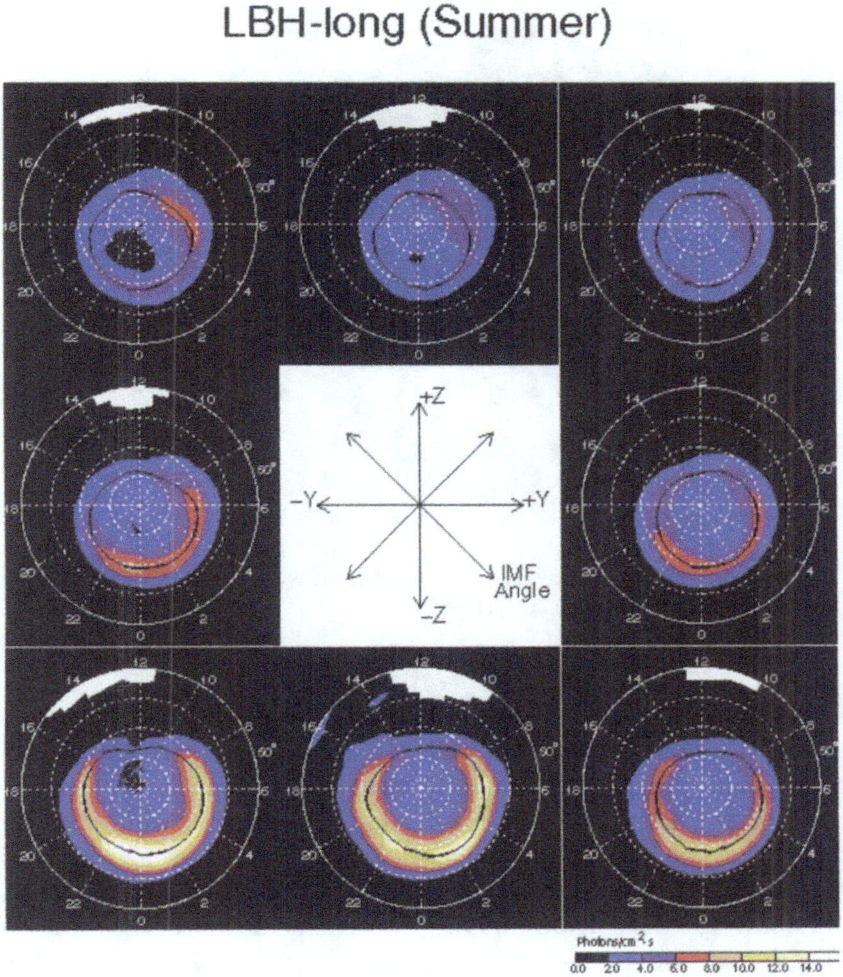

Figure 5.6. Distributions of mean auroral emissions from the Polar UVI LBH-long band for different IMF orientations during summer. From Shue et al. (2000).

likely between 11 and 12 MLT. Note that the pre-noon peak and the peak at 15 MLT correspond to the previously described peaks in auroral emissions and in Region 1 currents. The peak energy and the width of the accelerated distributions was also determined. The inferred potential drop has a narrow peak at 700 eV for the 1500 UT hot spot, and potentials above 3 kV are almost never observed anywhere on the dayside. For both the weak and intense events, the distribution of inferred potentials is quite broad for the nightside and extends to the highest value observable with the DMSP detector (~30 keV). The observed distribution of

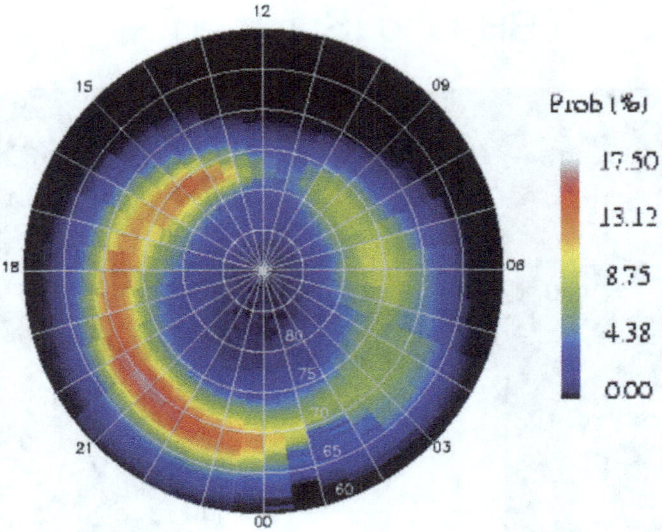

Figure 5.7. Probability of observing accelerated electrons with energy flux > 0.25 erg $cm^{-2}s^{-1}$ for all IMF conditions. From Newell et al. (1996b).

scale sizes for the accelerated electrons for all local times and IMF conditions is exponential with a mean width of ~30 km.

In contrast (see Figure 5.8), the most intense (>60 erg $cm^{-2}s^{-1}$) electron acceleration events occur almost exclusively in the dusk sector with the maximum probability at ~ 65° and 2100 MLT, as shown by a study of over 12,000 events observed during all phases of the solar cycle and levels of magnetic activity by the NOAA/TIROS satellites (Evans, 1994). The characteristic energy was typically ~10-15 keV, and distributions are usually strongly peaked in energy. The usual widths were less than ~ 0.3 ° in latitude (30 km) and may sometimes be less than the ~11 km resolution of the instrument.

The magnetic activity dependence of electron acceleration has been examined in two different ways: (1) sorting data by Kp; and (2) sorting by IMF B_z. The IMF B_z dependence is shown Figure 5.9. The occurrence probabilities are higher for southward IMF, and the '1500 hot spot' is not clearly distinct from the nightside events. For northward IMF, there are three distinct regions of accelerated electrons, centered at ~15, ~08, and ~22 MLT. For the most intense events (> 60 erg cm^2s^{-1} - see Evans, 1994), the probability of observing an event increases with increasing magnetic activity, as indicated by Kp. An intense event is 12 (35) times as likely when Kp is 4 (7) compared to when Kp is 1.

In addition to the dependence on location, interplanetary magnetic field and magnetic activity, the average probability of observing accelerated particles depends on whether the foot of the field line is sunlit or dark conditions, and on solar

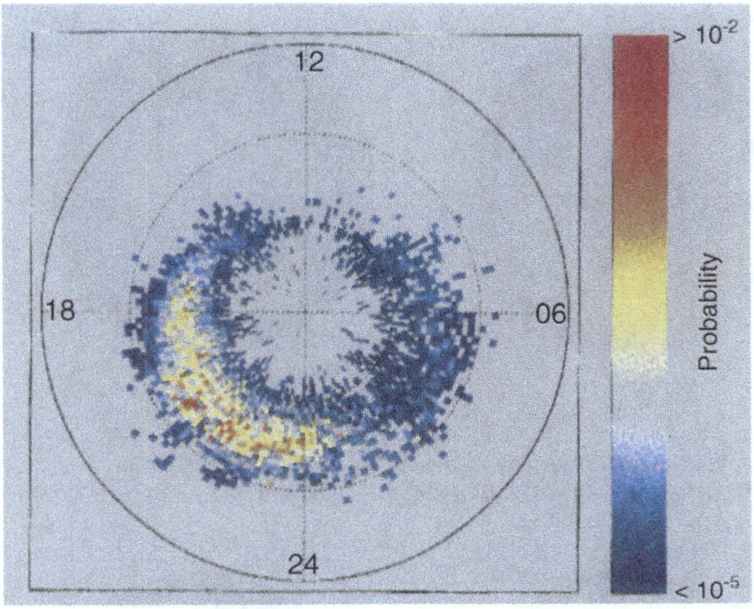

Figure 5.8. Map of absolute probability of a satellite encountering a large auroral electron energy flux event as a function of corrected magnetic latitude and magnetic local time of satellite location. From Evans (1994).

cycle. Newell et al. (1996c) produced maps of the occurrence probability of intense electron events (defined as having energy flux >5 erg $cm^{-2}s^{-1}$) for sunlit and dark conditions, using electron distribution function data from the DMSP satellites (see Figure 5.10). They found that intense events, which were primarily restricted to the dusk sector (~19-23 MLT) at invariant latitudes of $\sim 68 - 72°$, were a factor of ~ 3 more probable in dark compared to sunlit conditions. Intense electron flux events were also observed, but with a much smaller occurrence probability and latitudinal extent, near 15 MLT and in the morning.

The solar cycle dependence of auroral electron acceleration was examined by Newell et al. (1998), using the same DMSP data base of intense (> 5 erg cm^2s^{-1}) events. They concluded that the number of intense events in darkness was not correlated with solar cycle, whereas the number in sunlight was inversely correlated with solar activity. Note that studies of auroral image data are not in agreement with this result (Nevanlinna and Pulkkinen, 1998). A study of extremely intense events, (Evans, 1994) which did not distinguish between events occurring in light and dark, reached a different conclusion, finding a strong solar cycle variation. The likelihood of an intense event is a factor of 4 lower than the average at solar minimum and a factor of 4 higher in the declining phase after solar maximum (see Figure 5.11).

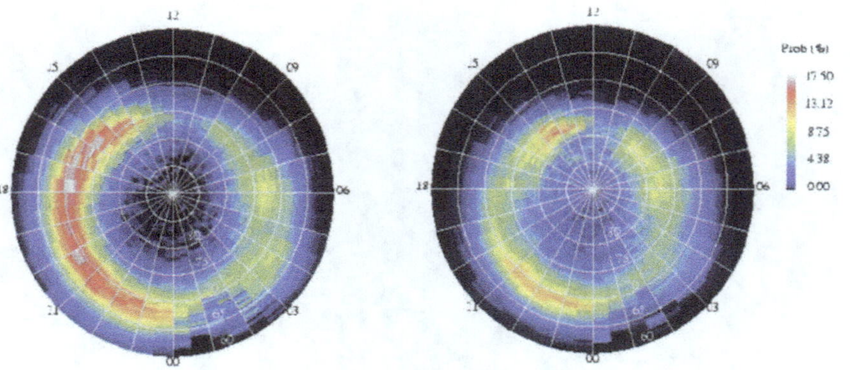

Figure 5.9. Probability of observing accelerated electrons with energy flux > 0.25 erg/cm^{2}s for southward IMF (left) and northward IMF (right). From Newell et al. (1996b).

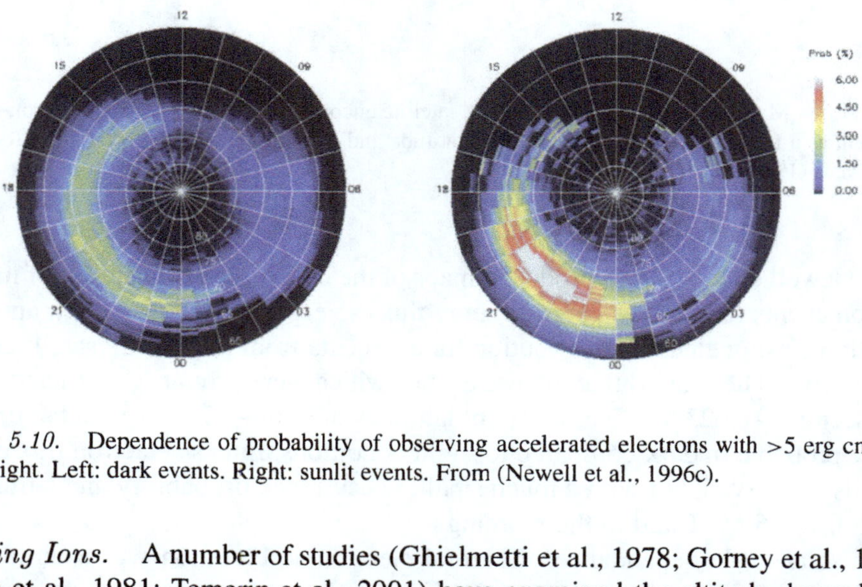

Figure 5.10. Dependence of probability of observing accelerated electrons with >5 erg cm^{2}s^{-1} on sunlight. Left: dark events. Right: sunlit events. From (Newell et al., 1996c).

Upgoing Ions. A number of studies (Ghielmetti et al., 1978; Gorney et al., 1981; Collin et al., 1981; Temerin et al., 2001) have examined the altitude dependence of ion beams, in addition to their magnetic local time and latitude dependence. Ion beams occur primarily in the dusk sector at latitudes where the upward Region 1 currents are observed. Their occurrence frequency increases rapidly with altitude, and most ion beams occur above 4000-5000 km.

The distribution of ion beams at energies of 90 eV to 3.9 keV observed by the S3-3 satellite at altitudes of 200-8000 km was described by Gorney et al. (1981). This study, which distinguished between quiet (Kp < 3) and active periods (Kp >3), found that the altitude distribution of ion beams did not depend on the activity level and that most beams occurred above 4000-5000 km with increasing

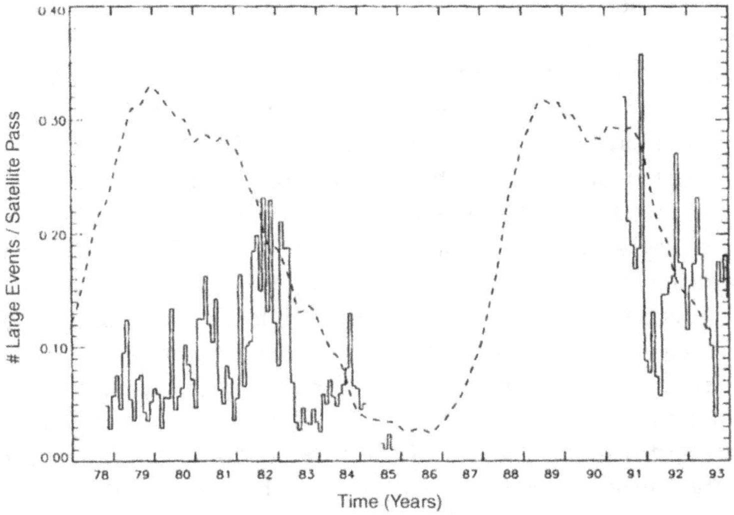

Figure 5.11. Dependence of probability of observing intense auroral energy flux on solar cycle from NOAA/TIROS data. The probability is calculated as the number of instances of an energy flux exceeding >60 erg cm^{2}s^{-1} per satellite transit over the polar regions, averaged over 10 days. The dashed line is the sunspot number. From Evans (1994).

occurrence frequency to the satellite apogee (8000 km). Ion beams are a dusk sector phenomenon with a stronger peak near 18 MLT for active times. During active periods, ion beams were more common (almost every orbit) and more energetic than during quiet periods. In particular, during quiet times, most beams had energies less than 400 eV, whereas >2 keV beams were as common as <400 eV beams during active times.

Strong ion outflow in the poleward part of the midnight auroral oval, i.e., near the polar cap boundary was discussed in Chapter 4. The occurrence of intense ion conic outflow (> 10^8 cm^{-2}s^{-1}) at the polar cap boundary was examined by Tung et al. (2001) utilizing FAST data obtained during January and February, 1997 (passes were near midnight and ~ 4000 km altitude). Tung et al. (2001) concluded that the strong outflow maximized near midnight and in association with substorms and periods of active aurora. Wygant et al. (2000) showed a case of strong ion outflow observed at the polar cap boundary in association with intense Alfvénic Poynting fluxes into the ionosphere and bright aurora. The associated Alfvén waves are correlated with intense downward electron fluxes at energies up to a few keV (Chaston et al., 2002b). Complete statistical studies to examine the relative importance of this nightside polar cap boundary outflow to outflow at other local times and latitudes, during quiet and active times, have not yet been performed.

The increase of energetic ion outflow from the dayside ionosphere during substorms has also been investigated using Viking ion data and one-minute values of the AE index[2] (Øieroset et al., 1998, 1999). The data in these studies cover altitudes between 6000 and 13500 km, 06-18 MLT and 65° - 90° invariant latitude. A considerable amount of energetic ion outflow from the dayside ionosphere is observed even for very quiet geomagnetic conditions (AE < 100 nT) and the observed increase in average number flux for 300 < AE < 400 nT is only a factor of two. The highest average number flux values are found in the cusp region, and the outflow prenoon is higher than postnoon. However, the postnoon outflow is more affected by changes in AE. The outflow intensity and the region of energetic ion outflow are different for positive and negative IMF B_z. However, this difference is not as strong as the difference in ion outflow intensity and location observed for AE < 50 nT and AE > 200 nT. The results imply that the dayside energetic ion outflow is controlled partly by nightside (driven by the release of stored energy in the magnetotail, as observed in the AE index) and partly by dayside (directly driven, as observed in the polarity of the IMF B_z component) activity. The effect of the directly driven component is strongest in the cusp and the postnoon region.

Data from the FAST satellite have provided the most complete survey of the altitude, magnetic local time, invariant latitude, and solar illumination dependence of upgoing ion beams in the altitude range sampled by FAST, from 1000 - 4000 km (Temerin et al., 2001). Consistent with previous studies, the probability of observing an ion beam increases dramatically with altitude; however, the distribution is strongly dependent on solar illumination. Ion beams are rarely observed, even at 4000 km, when the foot of the field line is illuminated. Although the occurrence frequency peaks in the 20-24 MLT sector, beams are also common on the dayside (~9-15 MLT) when the foot of the field line is in darkness. The energy of the ion beams is strongly dependent on solar illumination and local time. For the same solar illumination conditions, ion beams observed near noon are much less energetic (approximately 15 %) than those observed in the pre-midnight sector.

Collin et al. (1981) examined ion beams with energies of 500 eV to 16 keV for MLT=18-24 and altitudes from 6000–8000 km for both O^+ and H^+. The time resolution was 64s for a full distribution. They found that the energy of the O^+ and H^+ beams were very well correlated, with the O^+ energy on average ~1.7 times that of H^+. The pitch angle distribution of the O^+ was also substantially wider than that of the H^+, consistent with H^+ gaining energy from a parallel potential and O^+ from the parallel potential plus a mass-dependent wave interaction (see Chapter 4). The average energy of the downgoing electrons was well correlated with the ion beam energy and approximately equal, implying that at these altitudes approximately half the potential is above the satellite and half below.

---

[2]The AE index is a measure of the strength of the auroral electrojet. It is derived from the horizontal component traces from a chain of magnetic observatories in the auroral zone. The amplitude of the upper (lower) envelope of these traces is called AU (AL). The range between them defines AE.

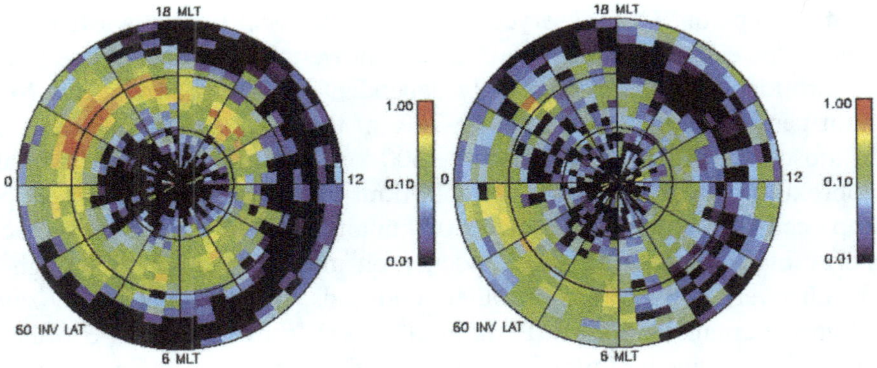

Figure 5.12. Left: The frequency of occurrence of upflowing beams of any ion species as a function of invariant latitude and magnetic local time over the southern polar region during the local winter, April through September 1997. Note that midnight is to the left. Occurrence frequency is color coded on a logarithmic scale running from 1% to 100%. Invariant latitude runs from 60 to 90 in 1.5° steps and magnetic local time is in steps of 0.5 hours. Right : same for the local summer, October 1996 through March 1997. From Collin et al. (1998).

Collin et al. (1998), using TIMAS ion composition data from Polar perigee passes (altitude of $\sim$ 1 R_E) obtained from October, 1996 through September, 1997, examined the seasonal variation in upflowing ion beams. Figure 5.12 shows that the regions of high occurrence probability of upflowing ion beams are $\sim$18-24 MLT and $\sim$15 MLT during winter time. During summer time (right hand side of Figure 5.12), the probability decreases by a factor of 2-3, with no clear MLT peaks. Collin et al. (1998) found no seasonal dependence in the occurrence of ion conics (which have their largest occurrence frequency between $\sim$9 and 12 MLT). All three ion species (H^+, O^+, and He^+) showed the same trends.

The solar cycle dependence of ion beams has been investigated by Collin et al. (1987) and by Cattell et al. (1991, 1993). Statistics obtained from the S3-3 satellite at radial distances below 2.2 $R_E$suggest that ion beams (i.e., parallel potential drops below the satellite) are much less common during solar maximum than during solar minimum, consistent with the results of Newell et al. (1998). Comparison with ISEE 1 observations of ion beams implies that the potential occurs primarily between $\sim$3 and 4 $R_E$at solar maximum, and that the potential drop is usually less than $\sim$1 keV since ion beams above that energy are rarely observed. In addition, during solar maximum, oxygen beams were more common than hydrogen beams. Details of the composition, solar cycle, and magnetic activity dependence of the net ion fluence are given in Moore et al. (1999).

Upflowing Electron Beams. Statistical studies of upflowing electrons (UFE) (Peria et al., 2000; Elphic et al., 2000) provide information on the potential drop below the satellite in downward current (downward parallel electric field) regions.

In a study comparing UFE observed near midnight local time at solstice to those observed at the equinox, Elphic et al. (2000) showed that the occurrence probability and altitude dependence is strongly dependent on solar illumination. None of the sunlit passes had beams, whereas ~71% of the passes in darkness did. There were almost no UFE observed below ~1600 km, and the occurrence probability was approximately constant above ~2500 km, implying that the parallel potential drop occurred between 1600 and 2500 km altitude. Neither the occurrence nor the energy of UFE had a clear dependence on magnetic activity, although energetic beams were rare during very quiet conditions. In a complementary study of field-aligned currents and UFE, Peria et al. (2000) showed that UFE occurrence increases with the length of time the foot of the field-line has been in darkness (i.e., inversely with the density at the footpoint).

Downflowing Ion Beams. High time resolution from FAST have shown that downflowing ion beams (DFI) are a common occurrence in the auroral oval (e.g., Klumpar et al., 1999), as shown in Figure 5.13 and as previously discussed in 4.2.2. These intense downflowing ion beams usually are observed in the downward current region. The existence of the DFI thus provide an indicator of regions where the potential drop is above the satellite altitude, just as the upflowing electrons indicate that it is below the satellite. This interpretation is consistent with the fact that the high latitude edge of DFI is often associated with an upflowing electron beam. The UFE/DFI elucidate the downward current region parallel potential in the same way as the upward flowing ion beams (UFI)/auroral electrons have traditionally been used to determine the potential in the upward current region. The occurrence frequency indicates that the peak occurrences of DFI and upflowing ion beams (and downgoing auroral electrons) are symmetric about midnight. The DFI occurrence peaks about two hours after midnight, whereas auroral electrons peak about two hours before midnight.

5.1.3.2 Large-amplitude Electric Fields

As discussed in Section 4.1, large-amplitude, small-scale perpendicular electric fields are often called 'electrostatic shocks' in the literature. These large amplitude electric field structures are strongly associated with auroral particle acceleration in upward current regions, and upflowing electrons in downward current regions. A detailed study of electrostatic shocks, currents, and particles, as well as the simultaneously observed waves, for a set of events was made by Temerin et al. (1984). Several specific events for the upward current region with converging electric fields and downward current region with diverging electric field were presented in Chapter 4.

Studies of large-amplitude (defined as >120 mV m^{-1}), small-scale electric fields observed by S3-3 near radial distances of 2 R_E(Bennett et al., 1983; Redsun et al., 1985) show that the magnetic local time and latitude occurrence of these fields is

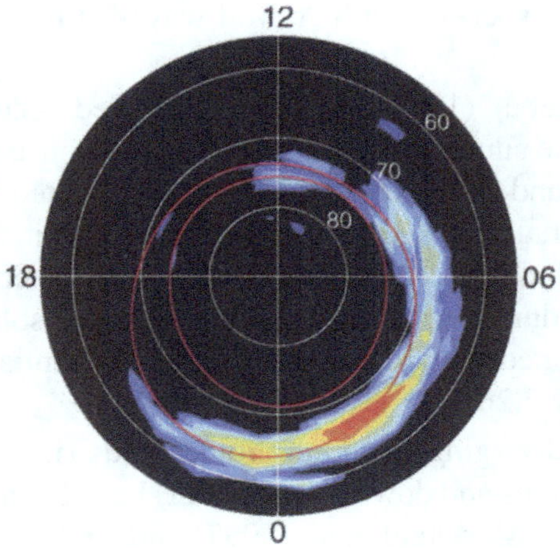

Figure 5.13. The occurrence frequency in magnetic local time/invariant latitude of downflowing ion beams observed by FAST. From Klumpar et al. (1999).

consistent with the location of the statistical auroral oval. Shocks with ion beams occur primarily in the dusk sector, while shocks with ion conics occur uniformly in local time. The potential drop across a shock is correlated with the energy of the associated ion beam (Redsun et al., 1985), as seen in the FAST data (see Chapter 4).

By combining the results of studies from S3-3 (up to 2 R_E), DE 1 (up to 4.3 R_E), ISEE-1 and Polar, the altitude dependence of such fields over the range from a few thousand km to >8 $R_E$can be examined. Large-amplitude fields (> 100 mV m^{-1}) are more rarely observed below 1.3 R_E, and the occurrence frequency increases rapidly with radial distance up to ~1.6 R_E and more slowly up to 2-3 R_E, and decreases above that altitude (see e.g., Bennett et al., 1983; Weimer and Gurnett, 1993). By examining the variation in the average magnitude of the fields in comparison to that expected for mapping along equipotential field lines, it can be inferred that parallel electric fields occur in the altitude range around 2 R_E. Analysis of the occurrence frequency of electrostatic shocks in DE 1 data (Weimer and Gurnett, 1993) suggests that there may be a second region of parallel potential above 2.7 R_E.

A study of electric fields from Polar (Keiling et al., 2001) at radial distances of 4-7 $R_E$shows a distribution of events limited to the auroral oval similar to the low-altitude observations. At high altitudes, a majority of the events occur in the nightside centered on midnight with a smaller enhancement near 15 MLT. The MLT distribution is consistent with the auroral electron and luminosity studies described above. The altitude distribution of the fields is consistent with either Alfvén waves or with a small additional parallel potential drop at these altitudes. However, specific event studies (Wygant et al., 2000) have shown that the large-amplitude electric fields are often associated with Alfvén waves at these altitudes (see Section 4.2.2).

The study of Bennett et al. (1983) implied a decreased occurrence frequency of electrostatic shocks in the summer compared to winter, consistent with the studies of Newell et al. (1996c) and Collin et al. (1998) of accelerated particles. This study was limited by the seasonally-locked nature of the S3-3 orbit, however, so detailed local time and invariant latitude dependence could not be determined. Electrostatic shocks are less common during solar maximum than during solar minimum, and, in particular, shocks associated with ion beams (parallel potential below the satellite altitude) were rare below 2.2 R_E.

Statistical studies of diverging large amplitude fields (i.e., associated with upward flowing electron beams and downward current) have been discussed by Karlsson and Marklund (1996), Marklund et al. (1997) and Elphic et al. (2000). In contrast to the upward current shocks, shocks in downward current regions are often observed at low altitudes, with many events occurring at altitudes of 1500-3000 km. A few cases have been observed below 1000 km. The magnetic local time, latitude, and seasonal dependence of such intense electric fields, as determined from Freja data at altitudes 600 to 1760 km (Karlsson and Marklund, 1996; Marklund et al., 1997), is shown in Figure 5.14. The latitude distribution of intense fields follows that of the statistical auroral oval and events are most common from midnight to early morning. The largest amplitude events have the smallest scale sizes. There is no clear dependence on magnetic activity (as indicted by Kp). There is, however, a strong dependence on solar illumination, as indicated by the seasonal dependence shown in panel (d), and in Figure 8 of Marklund et al. (1997). The largest amplitude events occur when the conductivity is lowest, i.e. near winter solstice and magnetic midnight.

Parallel Electric Fields. Examining parallel electric fields measured at altitudes of 2000 to 14000 km on the Viking satellite, Marklund (1993) has shown that, on average, the parallel electric field at high altitudes (above ~8000 km) is directed upward, whereas, at low altitudes (below ~5000 km), the field was usually downward. This observation is consistent with the studies described above on the accelerated particle distributions and intense electric fields. Evidence for downward parallel electric fields below the satellite (i.e., upflowing electron beams and diverging electric fields) was often observed at lower altitudes of ~1000-4000 km. Evidence for upward parallel electric fields below the satellite (i.e., upflowing ion beams) was most often observed at altitudes above 4000 km. Several studies have determined the parallel electric field by comparing the perpendicular field at different altitudes. A statistical study of the perpendicular electric field versus altitude was interpreted as indicating that there was, on average, a downward parallel electric field at latitudes near ~70° (Mozer and Torbert, 1980). The occurrence of parallel fields has also been examined by comparing the perpendicular field at two different

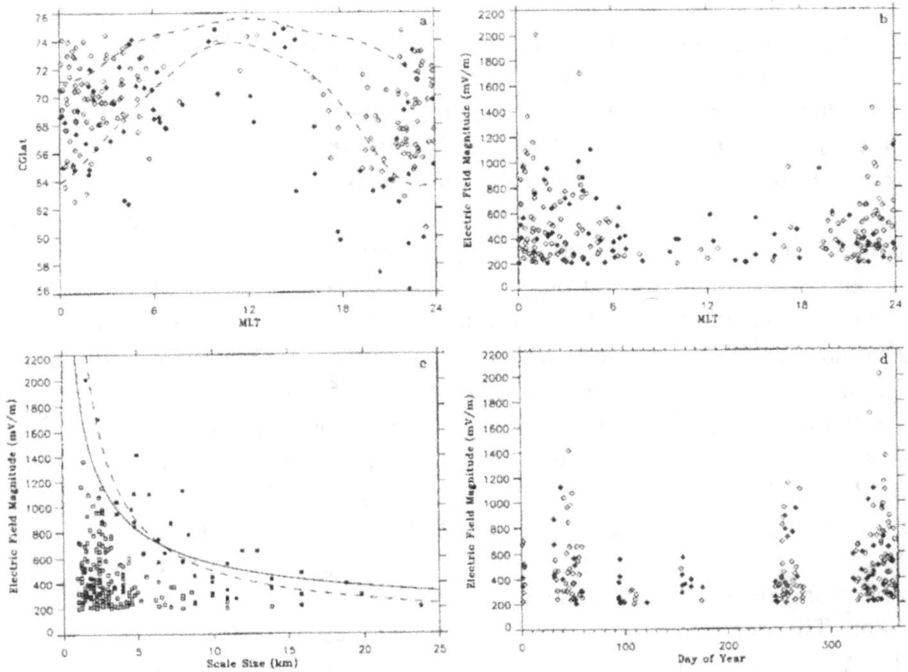

Figure 5.14. Statistical properties of intense electric field events. a) Latitude and MLT distribution; b) Magnitude of field versus MLT; c) Scale size, and (d) Day of the year. Solid and open squares refer to large and small potential drops, respectively. Solid diamonds refer to Kp > 3 events and open diamonds to Kp < 3 events. From Marklund et al. (1997).

altitudes. For example, Weimer et al. (1985) showed, in a comparison of DE 1 and DE 2 data, that parallel potential drops were associated primarily with small-scale perpendicular electric fields.

5.1.3.3 Density Structure and Density Cavities

Initial studies of the density structure in the auroral acceleration region were made using Langmuir probe and wave data from the S3-3 satellite (Mozer et al., 1979). It was shown that the average density decreased with altitude (Lysak and Hudson, 1979) and that, in the region of the parallel potential drop, there was a region of reduced density, i.e., a 'cavity' (Mozer et al., 1980). Statistical studies, over a limited local time and latitude range, of the occurrence, altitude dependence and structure of density cavities were performed by Calvert (1981) and Persoon et al. (1988), utilizing DE 1 wave cutoffs. Mesoscale density cavities were observed in the nightside auroral oval.

Spacecraft potential measurements from the electric field instrument (EFI) onboard the Polar spacecraft during one year of perigee passes from April 15, 1996 to April 15, 1997, were used to make the first synoptic maps of the thermal plasma density throughout the auroral zone and polar cap at an altitude of ~ 1 R_E, at all local times (Johnson et al., 2001). To determine the effects of variations in ionospheric illumination, the data were sorted by whether the measurement was made when the footpoint of the field-line was in sunlight or in darkness. The resulting maps are shown in Figure 5.15b (non-illuminated) and 5.15d (illuminated). These maps show that there is a large scale density cavity in the auroral zone which extends to all local times except noon. The depth of this cavity and the density in the polar cap both depend very strongly on the solar illumination of the ionosphere. The average plasma density for illuminated (summer) conditions is a factor of 5 larger than for non-illuminated (winter) conditions. The average density in the auroral zone under non-illuminated conditions is $\sim$ 5-10 cm^{-3}; whereas under illuminated conditions, it is $\sim$ 20-40 cm^{-3}. During winter, the lowest average plasma density is a few tenths cm^{-3} near 19 MLT and 72° invariant latitude. The density is depleted below the average cavity density from 15-24 MLT, and 3-5 MLT. In the polar cap, the average density in darkness is $\sim$20 cm^{-3}; in sunlight, it is $\sim$100 cm^{-3}. In addition, when all of the polar cap is illuminated, there is a region of strong density enhancement poleward of 78-80° sunward of the dawn-dusk meridian. This region disappears when the polar cap is in darkness.

Mäkelä et al. (1998) presented results of a statistical study of meso-scale density cavities utilizing data from the Freja Langmuir probe, at altitudes of 600-1760 km. To be included in their study, a cavity had to have a rapid (faster than a few seconds) decrease in the plasma density by at least 50%, to be clearly within the auroral zone, to be associated with clearly identifiable high frequency (kHz) wave activity, and to be the only such event within a given auroral zone crossing. No information on the absolute density within each cavity or the adjacent region was included in the data base. After correcting for effects due to the low orbit inclination, they found that the occurrence probability of a cavity had: (1) a significant maximum in near $\sim$02 MLT and a minimum near $\sim$19 MLT; (2) no dependence on magnetic activity as indicated by Kp; and (3) an even distribution around $\sim$68° in the dawn and dusk sectors. Although they found no seasonal dependence, they did find that most cavities (with the exception of those in the cusp) occurred when the Freja footpoint was in darkness.

5.1.3.4 Waves and Radiation

There have been many statistical studies of waves in the auroral zone. In this section, a few studies which are directly related to acceleration processes will be described. Section 4.3 presents a more complete discussion of the wave modes observed in the auroral zone.

Auroral Kilometric Radiation. Kumamoto and Oya (1998), in a study restricted to the 18-24 MLT sector, showed a seasonal dependence of auroral kilometric radi-

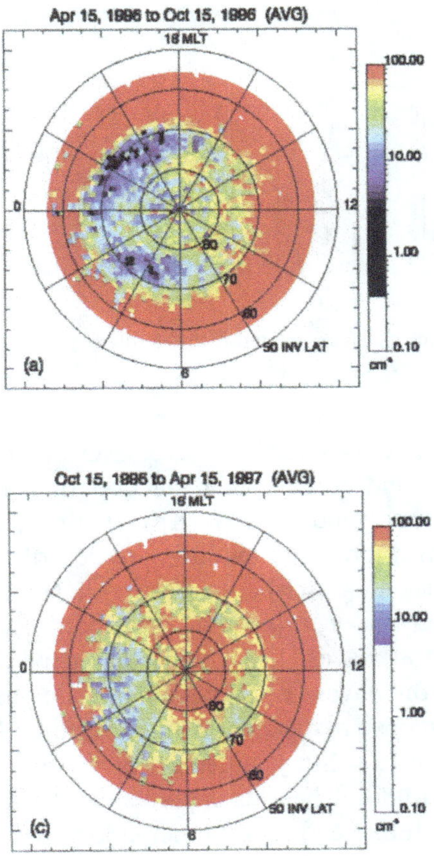

Figure 5.15. Dependence of density structure in the low-altitude auroral zone on sunlight. Synoptic maps of the southern hemisphere using Polar spacecraft potential data for 1 year starting April 15, 1996; (a) winter months from 4/15/1996 to 10/15/1996; (b) the entire year selecting data for dark ionospheric conditions; (c) the summer months from 10/15/1996 to 4/15/1997; and (d) the year selecting for sunlit ionospheric conditions. Note that noon is to the right. From Johnson et al. (2001).

ation (AKR) with emissions in the winter hemisphere being more intense and more common. Their findings are consistent with the Newell et al. (1996b) results on intense electron beams. In addition, they showed that the seasonal dependence was stronger for the higher frequency emissions. This is consistent with observations of the seasonal dependence of the densities within the cavity. AKR is produced at the electron cyclotron frequency so that higher frequencies are emitted at lower altitudes. Because the ratio of the electron gyrofrequency to the plasma frequency must be large for the generation of auroral kilometric radiation by the relativistic Doppler shifted cyclotron maser instability, AKR is more likely to be generated at higher frequencies in the winter when the density cavities are at lower altitudes.

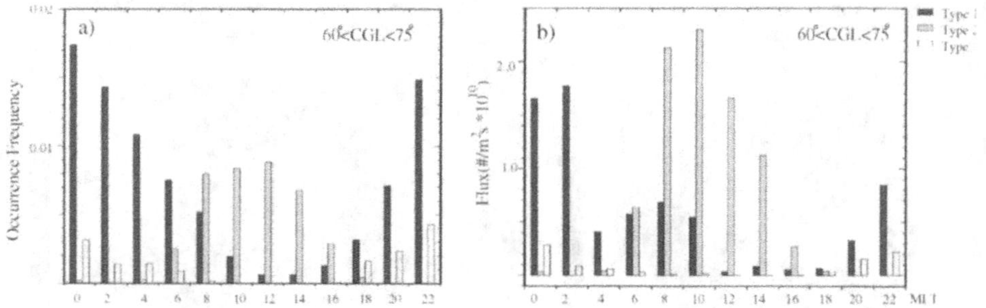

Figure 5.16. Occurrence frequency (left) and fluence (right) associated with the three type of ion heating events described in the text. After Figure 5 of Norqvist et al. (1998).

Ion Conics and Associated Waves. The previous sections have described the occurrence and rates of ion outflow (beams and conics) as a function of position (invariant latitude and magnetic local time) and activity (Kp). In this section, statistical studies of waves correlated with ion conic outflows are described. Three types of waves (the specific wave modes are described in Section 4.3) have commonly been associated with perpendicular ion energization – broadband ELF waves (BBELF), electromagnetic ion cyclotron waves (EMIC) and lower hybrid waves (auroral hiss). Electrostatic ion cyclotron waves (EIC) and small-scale quasistatic electric fields (electrostatic shocks) are also associated with ion heating. In addition, it has been suggested that lower hybrid solitary structures (LHSS) may provide some initial heating of ions at low altitudes. Theories of perpendicular ion energization are discussed in Section 3.5.2. Satellite and rocket observations of ion heating and outflow invariably find that heating is associated with plasma wave activity (see Section 4.3.2). Several recent studies have quantified this relationship (Yau and André, 1997; André et al., 1998; Norqvist et al., 1998; Lund et al., 2000). Figure 5.16 shows the occurrence frequency and fluence of O^+ ion outflow/heating events as a function of MLT as observed in 20 months of observations by the Freja satellite.

Each MLT bin (2 hours wide) is subdivided by type of wave activity and particle precipitation observed during the event. Types 1 and 2 denote ion heating associated with BBELF wave activity (without and with concomitant magnetospheric ion and soft electron precipitation), while type 3 denotes ion heating associated with auroral electron precipitation and either lower hybrid or electromagnetic ion cyclotron wave activity. As can be readily seen, BBELF waves (types 1 and 2) are associated with most of the ion outflow events on both the dayside and the nightside, as well as the bulk of the ion fluence, with additional outflow in the midnight sector associated with EMIC waves.

BBELF energization gives the highest average ion energies (hundreds of eV on the nightside), and the highest outflows (more than 10^{13} ions $m^{-2}\,s^{-1}$ in the prenoon sector) observed by Freja. During specific events, typically in the premid-

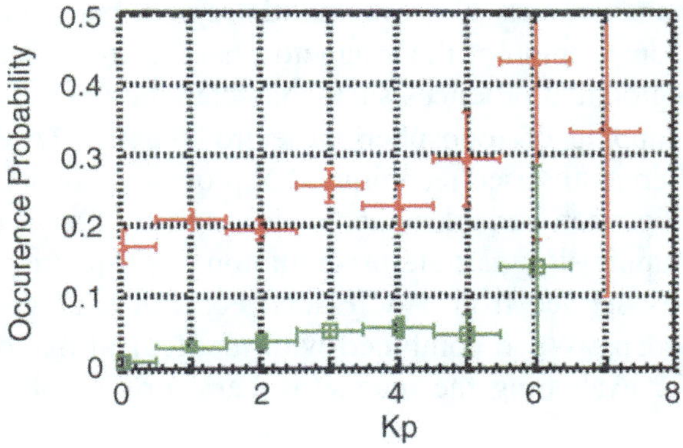

Figure 5.17. Normalized occurrence probabilities of BBELF (red) and EMIC (green) ion conic events as a function of Kp. From Lund et al. (2000).

night auroral region and associated with downgoing keV auroral electrons, lower hybrid or electromagnetic ion cyclotron (EMIC) waves may dominate the ion energization.

Figure 5.17 shows the results of a study using FAST data of the dependence of the different types of outflow events on magnetic activity. The occurrence probability roughly doubles from lowest to highest activity levels for the BBELF events, and increasing by an order of magnitude for the EMIC events. While the cause of the differing dependencies of the two classes of events upon magnetic activity is unclear, it is thought that the clear dependence of EMIC wave activity itself upon Kp is responsible for the EMIC trend; the weaker trend of the BBELF events reflects a weaker control by Kp (Lund et al., 2000).

The seasonal dependence of the broadband waves and their association with ions has been examined by Kasahara et al. (2001). The wave activity is limited to the cusp in the summer, but intense waves are observed over a broader local time region in the winter.

5.2 Mapping Auroral Features to the Magnetosphere

Observed distributions of auroral precipitation and luminosity are usually interpreted as the low-altitude signatures of magnetospheric processes. Two different techniques have been used to connect these low-altitude features to their source regions in the magnetosphere. Both are based on the concept of mapping between low and high altitude.

The first of these techniques, called topological or magnetic field mapping, connects low- and high-altitude regions by following the set of magnetic field lines threading the two regions. The accuracy of topological mapping obviously depends on the quality of the magnetic field model used to implement it. The second technique is a form of morphological mapping. It connects the two regions by identifying particle characteristics in a low-altitude region that are known to be more or less unique to some region of the magnetosphere or to a particular process in a magnetospheric region that produces such characteristics.

Assimilative Mapping of Ionospheric Electrodynamics (AMIE - see Section 5.2.2) is another frequently used technique that provides global information on the interrelationships between variables such as ionospheric convection, field-aligned currents, auroral luminosity, particle precipitation, ionospheric Joule heating and so on. The patterns derived from AMIE are geospace analogues of the familiar synoptic weather maps. When combined with topological mapping, AMIE offers a powerful tool for evaluating the magnetospheric origins of large-scale auroral electrodynamics.

Topological and morphological mapping concepts are discussed in more detail below. The use of AMIE for event studies is also illustrated together with an application that combines AMIE with magnetic mapping. The results reveal some interesting insights into the complex interaction between the magnetosphere and ionosphere. Mappings are most reliable for quiet or slowly evolving magnetospheric conditions. Complicating factors influencing their interpretation are discussed in the final section of this chapter.

5.2.1 Topological Magnetic Mapping

Chapter 1 outlined the most important regions of the magnetosphere and also gave their morphological connectivity to the auroral regions at ionospheric altitudes. Such mappings are based on models for the magnetic field and the fact that the auroral particles move almost along the magnetic field lines. Apart from this large-scale picture, the magnetic mapping between various regions of the magnetosphere is still very much an open issue, which makes it difficult to determine definite locations of the source regions of particular auroral forms. The issue is further complicated by the fact that the mapping changes substantially as a function of time, and that the mapping at any particular moment is a complex function of the previous history of dynamic events. This subsection outlines three methods to obtain magnetic mappings: empirical magnetic field models, isotropic particle boundaries measured at low altitudes, and global magnetohydrodynamic (MHD) simulations.

For magnetospheric modeling purposes, a large data set has been collected from available high-altitude magnetic field measurements in the region ranging from about 4 $R_E$out to 40 R_E. The data have been tagged with spacecraft location information, geomagnetic activity indices, and solar wind and interplanetary magnetic field values. Presently, the data set consists of about 100,000 half-hour or half-$R_E$averages of the magnetic field values (Fairfield et al., 1994).

These data together with modeling of the extraterrestrial current systems have been utilized to build a series of statistical, empirical magnetic field models for the external part of the geomagnetic field (Tsyganenko and Usmanov, 1982; Tsyganenko, 1987, 1989, 1995; Hilmer and Voigt, 1995; Koskik, 1998). These models are based on parameterized representations for the major magnetospheric current systems, the ring current in the inner magnetosphere, the cross-tail current extending from dayside to the distant magnetotail, and the currents at the magnetospheric boundaries. Each of the current systems was parametrized by several free parameters that depended on location, dipole tilt angle relative to the Sun-Earth line, level of geomagnetic activity, solar wind pressure, and interplanetary magnetic field values. The parameter values were then evaluated by a least-squares fitting technique to minimize the errors between the model and the observed values (Tsyganenko, 1990). The resulting models then give the external magnetic field as a function of location, dipole tilt angle, and the level of geomagnetic activity. The early versions of the Tsyganenko models (Tsyganenko and Usmanov, 1982; Tsyganenko, 1987, 1989) are divided into six subsets according to the value of the 3-hour planetary Kp index, whereas the more recent versions (Tsyganenko, 1995, 2000) give the field values as continuous functions of the 1-hour ring current index Dst (see Section 8.2.3), the solar wind pressure, and two components of the interplanetary magnetic field (B_y and B_z). The Hilmer-Voigt model (Hilmer and Voigt, 1995) gives the field values as functions of the dipole tilt angle, the Kp index, and the equatorward boundary of the diffuse auroral precipitation.

The internal geomagnetic field can be represented either by the dipole approximation or the International Geomagnetic Reference Field (IGRF), depending on the desired level of accuracy. Near Earth ($R < 6\ R_E$), the dipole approximation can lead to quite substantial errors, whereas further out the higher order multipole terms are small and the dipole term gives the internal field contribution quite accurately.

Once the magnetic field values are defined for each point in space, magnetic field lines can be traced from the Earth's surface to the magnetospheric boundaries. Such mappings then give a one-to-one correspondence between various regions of the magnetosphere-ionosphere system. Figure 5.18 illustrates the mappings from the auroral ionosphere to the magnetosphere. The upper left panel shows concentric circles in the ionosphere and their mapping to the magnetotail in the upper right panel. Note how the typical latitudes of the auroral oval (about 65-73°) map to the plasma sheet, whereas the polar cap at higher latitudes is on the open lobe field lines. The lower panels show a more realistic configuration of the auroral regions and their mapping into the magnetosphere. In this case, a reverse mapping has been made from the regions shown in the magnetospheric panel on the right to the polar ionosphere. For this case the regions close to magnetospheric boundaries all map to a very small region near the dayside cusp near the poleward edge of the auroral oval. Furthermore, the inner and mid-magnetotail occupy field lines that cover most of the auroral zone, whereas the distant magnetotail from 40 R_E outward maps to a very narrow arc-like structure near the poleward boundary of the auroral oval.

The statistical models give average mappings as functions of magnetic activity conditions, but fail to include the history of the dynamics in the mapping. These models have been further developed to represent the dynamics during particular

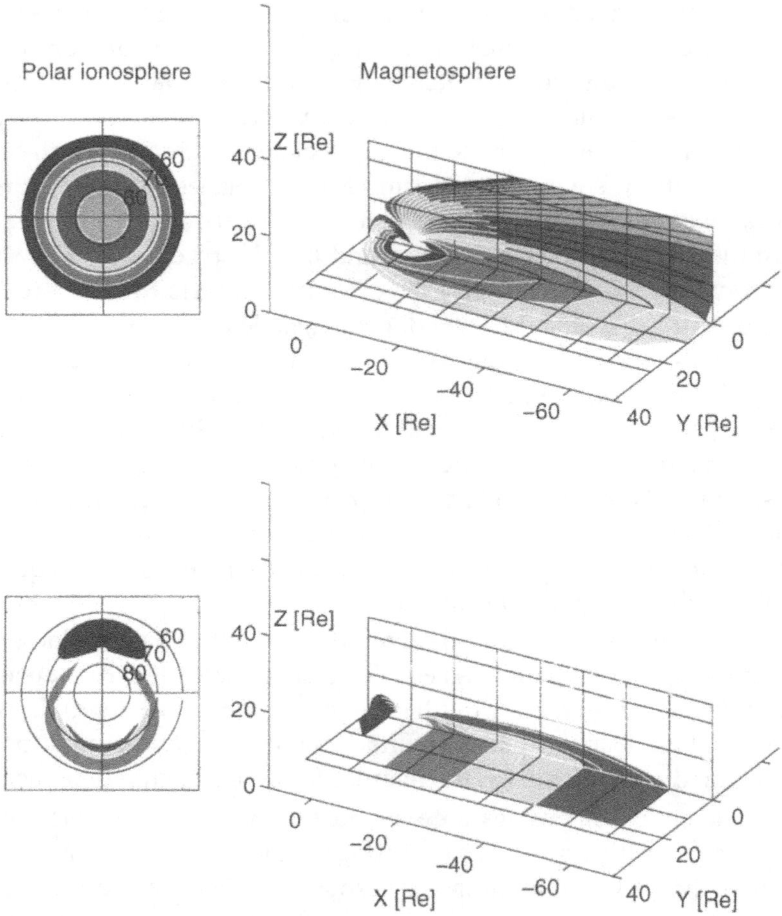

Figure 5.18. Upper panels: Concentric circles in the ionosphere and their corresponding mapping to magnetospheric regions. Lower panels: Reverse mapping from the specified regions in the magnetosphere to the ionosphere. Figure provided by T. Pulkkinen.

events by fitting the model to in situ measurements (Pulkkinen et al., 1992, 2000; Lu et al., 1997). This event-oriented approach utilizes the statistical field model as a starting point, but adds free parameters to the model that are evaluated based on the specific conditions during a particular period of time. These models have been developed mostly for slowly evolving configurations when the parameters can be assumed to vary linearly with time and hence one set of parameters is sufficient to represent a longer period of time. A further requirement of the geophysical situation is that the field-aligned currents are small enough that they do not dominate the field configuration. Especially during the substorm growth phase (see Chapter 8), the magnetotail evolves slowly from a quiet state (see top panel of Figure 5.20 for the field-line configuration in the noon-midnight meridian plane) toward a more stretched and high-energy state (bottom panel of Figure 5.20). Both observational and modeling studies addressing the field line configuration changes during the growth phase (e.g., Pulkkinen et al., 2000) show very consistently the tailward stretching of the field lines, which strongly modifies the magnetic mapping between the ionosphere and the magnetosphere. This is illustrated also in Figure 5.18, which shows that a quite broad area in the magnetotail results in a narrow, almost arc-like structure in the ionosphere. Assuming that the arc source region in the magnetotail does not move, the temporal evolution of the magnetic mapping leads to southward motion of the auroral arc by several hundred km, or several degrees in latitude during the growth phase that typically lasts from 30 min to about 1 hour.

In addition to high-altitude magnetic field measurements, particle measurements made by low-altitude polar-orbiting spacecraft can also be used to create magnetic mappings. Plasma sheet particles trapped on closed field lines bounce continuously between the two strong-field mirror points at off-equatorial latitudes. For a small portion of the particles the mirror points are at ionospheric altitudes, and these particles are precipitated to the ionosphere. If the particle distributions are isotropized in the magnetotail, the loss-cone is continuously filled and the flux levels at ionospheric altitudes (around 1000 km) remain high. However, if the particle motion in the plasma sheet is adiabatic, the loss cone remains empty and consequently the low-altitude fluxes are low. In the midtail and distant magnetotail the particle distributions are effectively isotropized by non-guiding center motion in weak magnetic field regions and scattering by plasma waves. In the inner magnetosphere the particle motion is adiabatic and the loss of particles to the ionosphere leads to anisotropic distributions as there is no continuous refilling of the loss cone. The characteristics of the particle motion within the current sheet depends on the ratio κ^2 of the minimum field line curvature radius to the maximum particle gyroradius. Strong pitch-angle scattering, which fills the loss cones, has been found to occur for κ^2 values between ~ 0.1 and 8 (see Section 8.1).

The low-altitude measurements usually show a clear boundary, where the particle distribution for a given energy range changes from a loss-cone distribution to an isotropic distribution as the spacecraft moves from lower to higher latitudes. These boundary locations give B_n as a function of latitude by considering a number of energy ranges and assuming that strong scattering begins when $\kappa^2 = 8$. Here, B_n is the component of $\mathbf{B}$ normal to the neutral sheet. Using this information to adjust empirical models such as those described above, one can obtain a magnetic field model which corresponds to the observed radial distribution of the magnetic field component normal to the current sheet.

A third, and the most self-consistent, way to obtain a magnetic mapping is to use global magnetohydrodynamic (MHD) simulations of the solar wind–magnetosphere–ionosphere system (Walker et al., 1993; Fedder and Lyon, 1995; Raeder et al., 1995; Janhunen and Koskinen, 1997). These simulations take the solar wind and interplanetary magnetic field as input, and compute the self-consistent dynamics of the magnetosphere–ionosphere system under these boundary conditions. At any given time, the global MHD simulations give the magnetic field, the flow velocity field, and the plasma density and temperature in all regions of geospace. Thus, by interpolation, the magnetic connectivity between the magnetosphere and ionosphere can be found. The problem with this approach is that the representation of the inner magnetosphere, which strongly controls the mapping, is still not very accurately included in the models, which cannot account for the non-MHD ring current in the inner magnetosphere. However, results of such mappings indicate a close correspondence between flow bursts in the magnetotail and auroral activations (regions of enhanced precipitation) in the ionosphere.

5.2.2 Assimilative Mapping and Synoptic Maps

As already mentioned in the introduction to Section 5.2, synoptic maps of auroral features can be derived from the data assimilation technique known as Assimilative Mapping of Ionospheric Electrodynamics, often abbreviated AMIE[3] (Richmond and Kamide, 1988). The resulting maps overlay various fields of interest on the same ionospheric projection and ideally represent a synoptic configuration as a snapshot in time. When combined with a global magnetic field model, AMIE patterns can be projected into the outer magnetosphere using the topological mapping technique discussed above. The first set of examples described below illustrates the use of AMIE in developing a synoptic map. The second example describes results from an event study in which AMIE is combined with magnetic field mapping to infer the magnetospheric source regions of field-aligned currents and auroral precipitation.

5.2.2.1 Example 1: Steady IMF and Non-Substorm Conditions

Figure 5.19 shows the patterns of ionospheric convection and field-aligned currents derived from AMIE under steady IMF conditions, plotted on top of the corresponding Polar UV auroral emissions. The patterns on the left side correspond to 07:01 UT on 12 January 1997 when the IMF B_x, B_y, and B_z were about 2, 3, and -5 nT, respectively. The patterns on the right side correspond to 16:55 UT on 25 June 1998 when the three components of the IMF were -5,-2, and 15 nT, respectively. The convection and field-aligned current patterns were derived based on magnetic perturbations recorded by ground magnetometers, ion drifts observed by the SuperDARN radars and DMSP satellites, along with the ionospheric conductances inferred from Polar UV auroral images, using the AMIE procedure.

[3]AMIE is a mathematical procedure that performs an optimally constrained, weighted, least squares fit of a set of coefficients to the relevant data obtained from spacecraft and ground-based instruments.

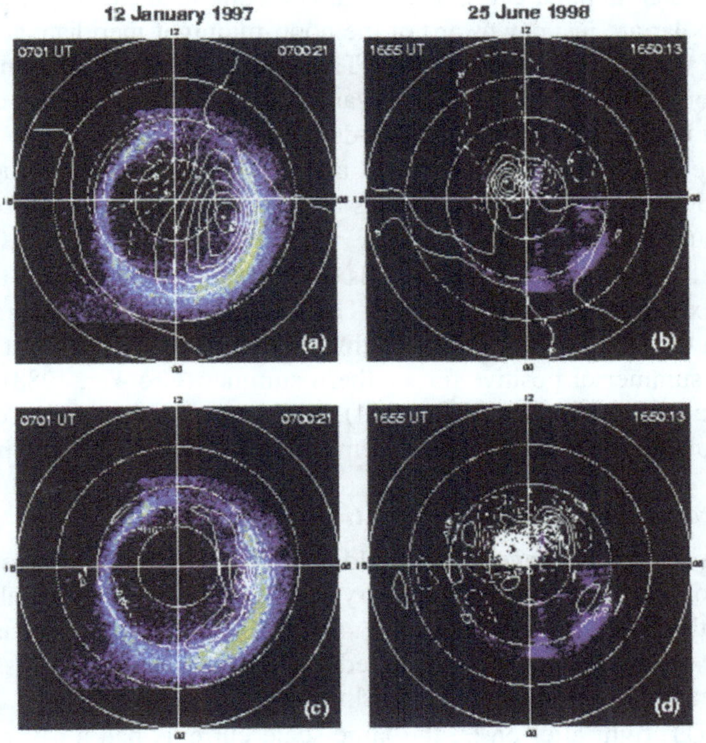

Figure 5.19. Distributions of ionospheric convection (patterns at top) and field-aligned currents (patterns at bottom) in the northern hemisphere above 50° magnetic latitude, plotted on top of the corresponding Polar UV images. The patterns on the left-hand side are for 12 January 1997 (IMF southward case), and the patterns on the right-hand side are for 25 June 1998 (IMF northward case). Solid contours represent positive electric potentials or downward currents, and dashed contours for negative electric potential or upward currents. The contour interval is 5 kV for electric potentials and 0.3 $\mu A/m^2$ for field-aligned currents. Figure provided by G. Lu.

For the southward IMF case on 12 January 1997, the ionospheric convection pattern consists of a positive potential cell on the dawn side and a negative potential cell on the dusk side. The slight asymmetry between the dawn and dusk cells with respect to the noon-midnight meridian is consistent with the well-known IMF B_y effect. The auroral oval, as indicated by the UV image, resides at the equatorward edge of the dawnside convection cell, but nearly coincides with the convection reversal on the nightside and is slightly equatorward of the duskside reversal boundary. The distribution of field-aligned currents closely resembles the traditional Region 1-2 pattern. The upward currents on both dawn and dusk coincide with the auroral oval.

For the strongly northward IMF case observed on 25 June 1998, the northern polar cap is dominated by a pair of reversed convection cells, with sunward convection in the central polar cap. The auroral UV emissions are very weak. There is a faint transpolar arc just dawnward of the noon-midnight meridian, and it crosses the center of the negative potential cell. The distribution of field-aligned currents is less well organized than in the southward IMF case. Field-aligned currents are concentrated on the dayside of the dawn-dusk meridian, with a pair of 'NBZ' currents over the polar cap and the Region 1 and 2 currents at lower latitudes. On the nightside, however, field-aligned currents are too weak to be recognized in terms of any particular current system.

Note that the reversed convection cells over the polar cap seen in this case do not always exist during northward IMF. They usually occur in the summer hemisphere when the IMF B_z is strongly northward and the B_x component is negative for northern summer or positive for southern summer (Crooker, 1988). As shown by Knipp et al. (1993) and Lu et al. (1994), while reversed convection exists in the sunlit hemisphere, weak and disturbed convection is found in the dark hemisphere.

5.2.2.2 *Example 2: Substorm Conditions*

Figure 5.20 shows the changes in the field-aligned currents and auroral emissions during an isolated substorm on 9 January 1997. The onset of the substorm was detected by the Polar UVI camera at 07:47:16 UT. During the expansion phase, the nightside auroral emissions intensified rapidly and the main emission region shifted gradually poleward as well as duskward.

At 07:48 UT, right after onset, the large-scale currents had a pair of Region 1 and 2 currents on the dawn side, but only the upward region 1 current on the dusk side. Downward field-aligned currents are also seen in the post midnight sector. By 07:51 UT (3 minutes into the substorm expansion phase) the dayside field-aligned currents have diminished, whereas the pair of field-aligned currents with the nightside region-1 polarity has intensified significantly. This pair of field-aligned currents is reminiscent of the substorm wedge currents (McPherron et al., 1973). The nightside field-aligned currents continued to grow during the expansion phase, not only in intensity but also in spatial extent. The large-scale upward currents generally coincided with the bright aurora in the central auroral oval. The majority of the downward field-aligned currents were found at the poleward edge of the auroral oval, but much weaker downward currents were also seen at the equatorward edge where auroral emissions are considerably lower. However, the brightest and most localized auroral emissions tend to reside at the boundary between upward and downward currents. This is probably due to the inability of AMIE to capture the fine-scale current structures associated with these discrete aurorae.

The mapping of auroral images and field-aligned currents to the equatorial plane was carried out using an improved time-dependent magnetic field model specified for this event (Lu et al., 1999). The equatorial maps show that the initial auroral brightening at substorm onset mapped to the inner plasma sheet at X = -5 to -7 R_E. During the expansion phase, the corresponding magnetospheric source re-

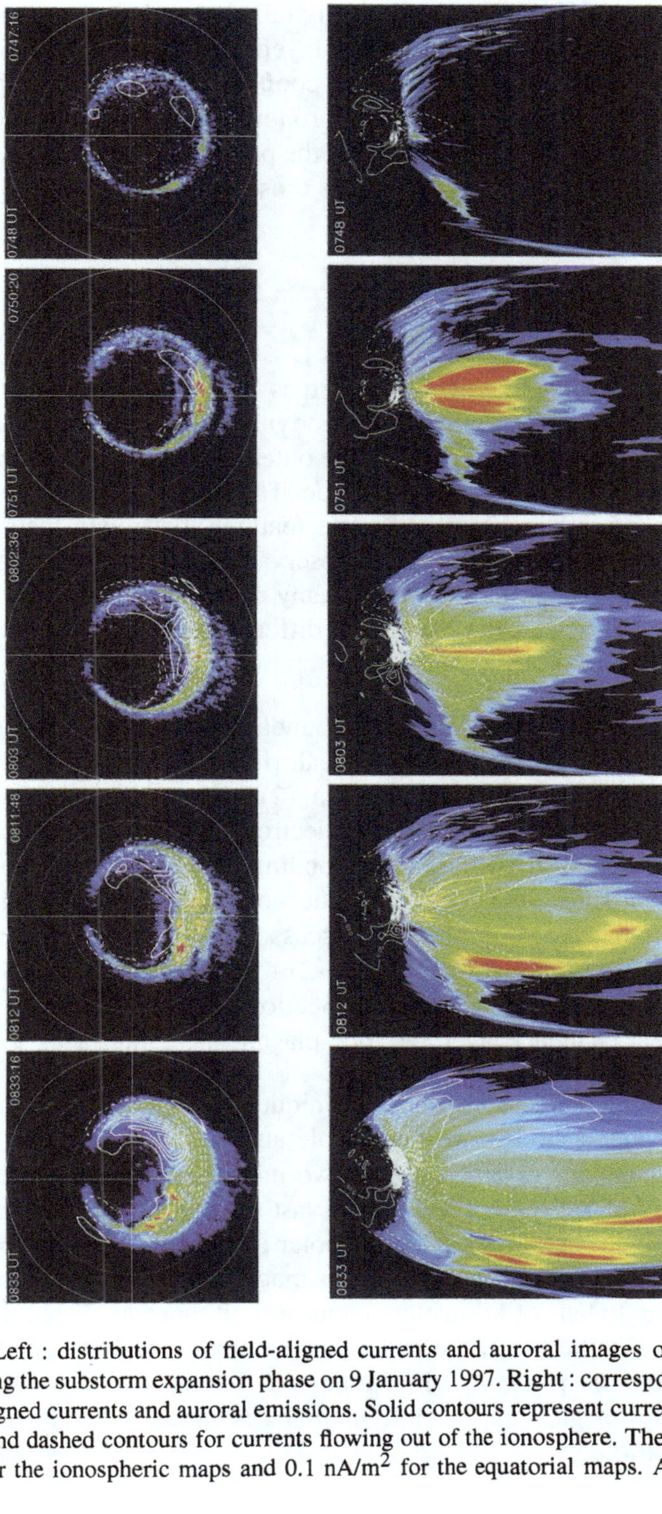

Figure 5.20. Left : distributions of field-aligned currents and auroral images over the northern hemisphere during the substorm expansion phase on 9 January 1997. Right : corresponding equatorial maps of field-aligned currents and auroral emissions. Solid contours represent currents that flow into the ionosphere and dashed contours for currents flowing out of the ionosphere. The contour interval is 0.3 μA/m^2 for the ionospheric maps and 0.1 nA/m^2 for the equatorial maps. Adapted from Lu et al. (2000).

gion of auroral precipitation moved slightly tailward and also gradually toward the dusk flank near the low-latitude boundary layer. The substorm-related field-aligned currents, on the other hand, mapped to a confined region close to the Earth, with the current density peaks around geosynchronous altitude. The apparently different magnetospheric source regions imply that the processes that produce energetic precipitating particles may not be the same as those generating the strongest magnetic perturbations in the magnetosphere.

5.2.3 Morphological Mapping

Morphological mapping identifies the sources of auroral plasmas and, by association, the dynamos that drive auroral electrodynamics by connecting robust boundaries and large-scale plasma regions in the outer magnetosphere to specific particle features commonly observed at low altitude. The 'morphological' approach makes use of 'natural tracers' – specific plasma features that retain their characteristic signatures regardless of the altitude of observation. Its validity is based on statistical inference and the accumulation of many observations showing the similarity of these characteristics when observed at different altitudes along a magnetic flux tube.

Examples of natural tracers and their boundaries include: (1) the steady plasmapause corresponding to both the large-scale inner convection boundary and the equatorial boundary of diffuse precipitation; (2) the stable trapping boundary where the pitch-angle distributions of energetic electrons undergo a transition from a loss-cone to an isotropic distribution, corresponding also to the inner boundary of the cross-tail current sheet where the magnetic field changes from a dipole-like to a more highly stretched configuration as discussed in Section 5.2.1; and (3) the type 2 velocity-dispersed ion signature (VDIS-2) that is observed at low altitude near the polar cap boundary, as described in Section 4.5, and that projects outward to the plasma sheet boundary layer and lobe-plasma sheet interface.

Although morphological mapping techniques were developed long before reliable magnetospheric models were available and before topological mapping was possible, the mappings obtained by the two methods are roughly consistent. Due to the geometry of the geomagnetic field, vast regions in the outer magnetosphere project to relatively small regions in the polar ionosphere. Regions of hundreds or thousands of kilometers in the near-Earth magnetotail, and up to many $R_E$in the distant tail, map to tens of kilometers at auroral altitudes as illustrated graphically in Figure 5.18. Furthermore, the entire magnetopause region effectively maps to a relatively small cusp segment in the dayside ionosphere. As will be discussed below, the morphological approach to mapping is valid for slowly varying magnetospheric conditions.

5.2.3.1 Nightside Auroral Features

The boundaries delineating the gross plasma domains near the midnight meridian ($Y_{GSM} \cong 0$) are described in detail by Feldstein and Galperin (1985) and Galperin and Feldstein (1991, 1996). The main results are summarized schematically in Figure 5.21, which illustrates the morphological relationship between characteristics of electron and ion precipitation observed at low altitude and their source regions in the plasma sheet and more distant magnetotail. Some of the low-altitude particle characteristics have been described already in Section 4.5.

The upper panel of the figure (not to scale) shows the following main magnetospheric plasma domains:

- *Plasmasphere.* The region of nearly equilibrated and relatively dense thermal plasma and stably trapped high-energy particles (blue);
- *Outer plasmasphere.* The region of decreased thermal plasma density immediately tailward of the plasmasphere containing some admixture of hot plasma remaining from previous magnetotail plasma injections (light green);
- *Inner magnetosphere.* The region of quasi-dipolar magnetic flux tubes projecting to the diffuse auroral zone at low altitude (green);
- *Plasma sheet.* The region of hot dense plasma in the tail consisting of an outer boundary layer region (green), the central plasma sheet (yellow) and a high plasma beta ($\beta \gg 1$) neutral sheet (red) near the tail midplane where the intense cross-tail current flows.
- *Stable trapping boundary.* Under the effect of the cross-tail current, starting from its earthward edge, the magnetic field stretches, and a stable magnetic trapping becomes impossible. Its exact location depends on the particle Larmor radius.
- *Equatorial boundary of the auroral oval.* For steady conditions this boundary coincides with the stable trapping boundary, also the boundary of discrete auroral forms, which form within strong localized field-aligned currents diverging from the cross-tail current.
- *Polar cap.* The region of plasma flux tubes that extend into the tail lobes where the plasma is very diluted (violet). The small red arrows near the plasma sheet lobe interface represent ion beams flowing earthward in the plasma sheet boundary layer and returning beams at somewhat lower latitudes that are convected toward the tail midplane following reflection of the earthward beams by the magnetic mirror force.

The panel below the magnetospheric domains in Fig. 5.21 shows an idealized radial profile of the electron density n_e in cm^{-3}. Two apparent plasmapauses or sharp

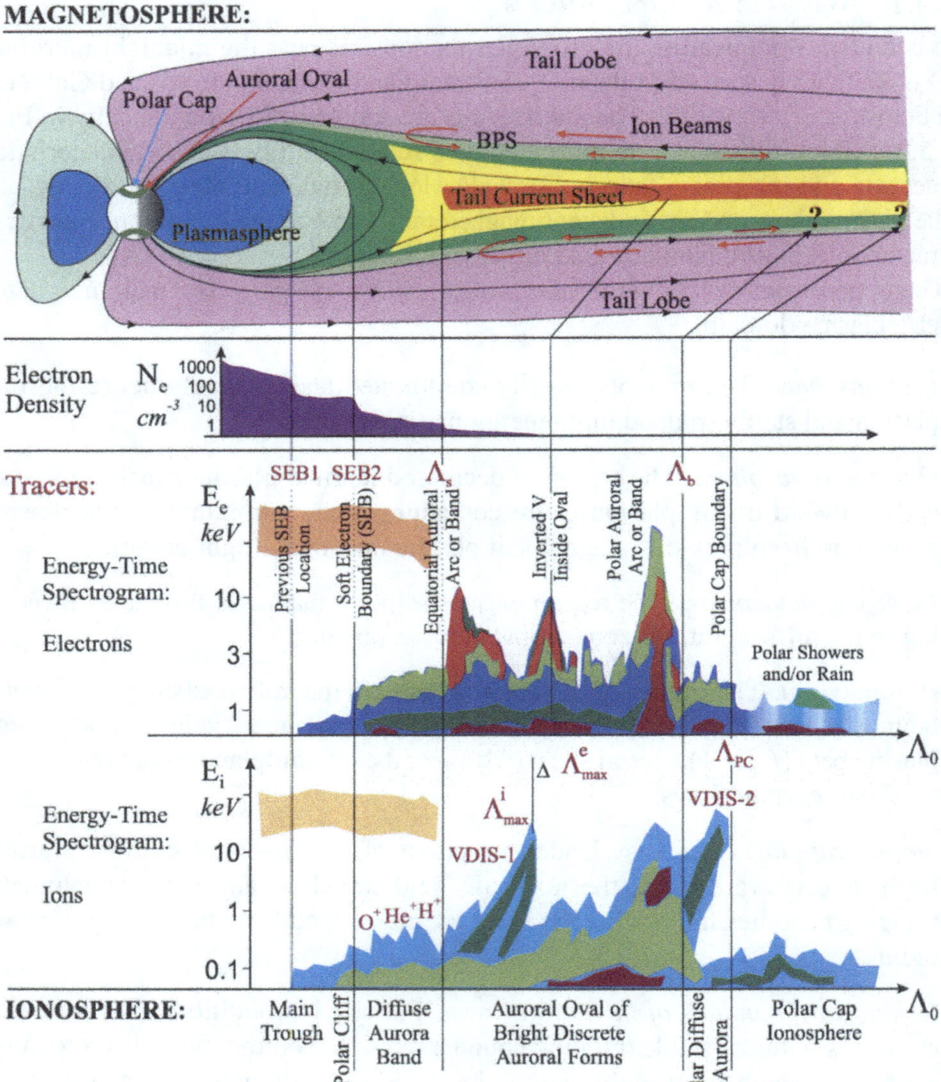

Figure 5.21. Mapping of nightside particle precipitation and ionospheric features to the magnetosphere. For details, see text. Figure provided by Y. Galperin.

transitions in electron density are sometimes observed depending on the past history of the magnetotail. The inner one occurs at the transition between convection- and corotation-dominated plasma flow established during a previous magnetotail convection surge. Upon relaxation to a weaker convection state, the convection boundary retreats tailward, while the plasma-depleted flux tubes between the new transition and the previous one begin to refill. This intermediate-density plasma region will eventually disappear for steady magnetotail conditions.

The first sharp plasma density gradient corresponds to a boundary of weak, soft electron precipitation (SEB1) – the remnant of a recent convection surge that brought hot plasma from the tail inward to this location. The outer plasma density gradient corresponds to a sharp precipitation boundary (SEB2). It indicates the current location of the large-scale convection boundary and is accompanied by a gradient in the intensity of diffuse electron precipitation (next panel).

The bottom two panels in Figure 5.21 show idealized energy versus invariant latitude spectrograms for precipitating electrons and ions that a low altitude satellite would record during a nightside auroral pass. Typical structures are labelled, although not all of them are usually observed during a single satellite pass. Some of these features such as VDIS-1 and VDIS-2 have been described in Section 4.5. Characteristic features that can be used as *tracers* are marked in brown color, and their mapping to the magnetosphere is indicated by straight lines. From left to right, these include:

- *SEB1 and SEB2 boundaries.* They delineate a weak mass-dispersed low-energy ion precipitation zone (marked as O^+, He^+, H^+) sometimes seen at subauroral latitudes during the storm recovery phase.

- *Equatorward boundary of the auroral oval.* This boundary is often (but not always) marked by a homogeneous auroral arc. It projects to the quasi-dipole/stretched tail-like magnetic field transition.

- *Auroral oval of discrete forms.* Poleward of the boundary just described, a more structured and intense discrete auroral precipitation is present forming the auroral oval of discrete forms. It projects to the main body of the tail plasma sheet.

- *Polar boundary of auroral oval.* A bright auroral band which usually consists of a series of arc-like features is found at the poleward border of the auroral oval of discrete forms. It projects to the distant tail, presumably to the region where reconnection processes are taking place.

- *Polar diffuse auroral zone (PDAZ).* Further poleward low-energy electrons of low and varying intensity produce a weak auroral glow dominated by the 630-nm red line in the F-layer. Due to the long lifetime of this metastable emission ($\sim$ 100 s), the red glow is diffuse, from which the designation 'Polar Diffuse Auroral Zone' originates. It has a distinct poleward border which is identified with the Polar Cap Boundary. An important feature of the PDAZ is that the class of ion dispersion events VDIS-2 (see lower energy-time spectrogram) often appear to originate at or near the polar cap boundary and can extend equatorward through the Polar Diffuse Auroral Zone and even in the polar part of the auroral oval of discrete auroral forms. Thus the PDAZ projects to the Plasma Sheet

Boundary Layer (PSBL) at midtail distances, but its ultimate origin in the distant tail and mapping remain debatable.

- *Polar cap.* Very low electron and ion intensities at energies of tens to hundreds of eV are usually observed in the polar cap. During extended, very quiet conditions the PDAZ extends poleward and can cover most of the polar cap. Conversely, during active periods, intense auroral features are sometimes observed even at very high latitudes in what would otherwise be considered as the polar cap, especially during substorm expansion. No reliable models of the distant magnetospheric tail structure at such extreme conditions are currently available.

5.2.3.2 Dayside Auroral Mapping

It should be possible, in principle, to develop a computer search algorithm that can identify well-defined and distinctive characteristics of particle precipitation in a satellite data base. Such an algorithm has been developed by Newell and Meng (1994) and applied to dayside particle precipitation originating in the polar cap/lobe, cusp, mantle, low-latitude boundary layer, and central and boundary plasma sheet. The statistically determined spatial dependence of precipitation originating from these regions constitutes another type of morphological map as shown in Figure 5.22. The results have been derived from a large data base of particle measurements carried out under the Defense Meteorological Satellite Program (DMSP).

The plot shows the global morphology of dayside auroral particle precipitation in magnetic local time and magnetic latitude coordinates. The auroral oval – the region in which aurora are statistically likely to be observed – encircles the magnetic pole in an essentially continuous band of emissions as shown in Figure 5.3. In Figure 5.22, the dayside portion of the auroral oval is associated with the 'traditional CPS'. The figure emphasizes the dayside regions that couple magnetically to various magnetospheric regions. A synoptic map overlaying these particle precipitation signatures onto polar region field-aligned currents and convection boundaries has also been constructed by Feldstein et al. (2001).

5.2.4 Mapping Complications

It has been mentioned already that both topological and morphological mapping are most reliable when the magnetosphere is quiet or changes slowly with time. Inductive electric fields become important when the ambient magnetic field varies more rapidly (Pellinen and Heikkila, 1984). These fields are always present at some level during storms, substorms and other dynamic magnetospheric processes. In weak field regions of the magnetotail, even moderate inductive fields are large enough to locally cancel convection fields mapped from the ionosphere (Toivanen et al., 1998), and they can cause significant deviations in the particle orbits deduced from

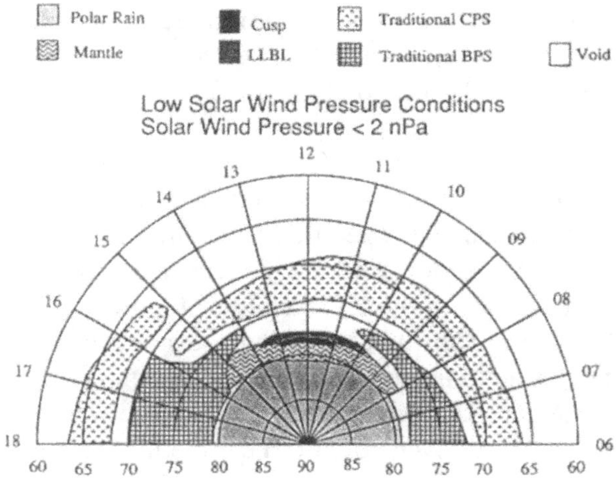

Figure 5.22. Global morphology of the dayside aurora observed from above the magnetic pole. Dayside aurora can be separated into several regions based on their plasma populations: central and boundary plasma sheets at low magnetic latitudes, the low latitude boundary layer (LLBL), the cusp or cleft at higher latitudes, and the plasma mantle at the highest latitudes. Shown here for conditions of low solar wind pressure, the morphology is based on precipitating particle observations from thousands of passes of the DMSP satellite. In the words of the authors, the map is 'less than reality but more than a cartoon'. From Newell and Meng (1994).

otherwise static convection and magnetic field models (Toivanen, 1997). Because inductive fields influence magnetically mapped features like convection as well as the dynamics of particle tracers, care must be exercised in interpreting both topological and morphological mappings for active magnetospheric conditions.

Other factors can also influence the mapping of particles and fields in the vicinity of auroral arcs, even for quiet magnetospheric conditions. These include the effects of parallel electric fields, field-aligned currents and non-ideal Alfvén wave characteristics.

Topological mapping usually assumes that magnetic field lines are equipotentials and that moving fluid elements are tied to magnetic field lines. This assumption is implicit in the magnetic mappings illustrated in Figure 5.20. However, the development of a parallel electric field in some region and the resulting field-aligned potential drop between mapping points permits an otherwise 'frozen-in' $\mathbf{E} \times \mathbf{B}$ fluid element to leave its anchoring field line. The maximum traversal or transverse slippage may be estimated as the net field-aligned potential drop divided by the perpendicular electric field magnitude above the field-aligned potential. Typical parameters for an inverted-V precipitation structure imply a maximum slippage on the order of one degree of invariant latitude. One consequence of this effect is that the location of convection reversal observed in the ionosphere will map to locations in the equatorial magnetosphere that may differ from equipotential mapping by as much as 1 R_E. This effect is especially important when mapping features in the dayside auroral ionosphere to boundary regions such as the low-latitude boundary layer (Kan and Lee, 1980; Lotko et al., 1987) or to the magnetotail (Toivanen et al., 1998) .

The mapping illustrated in Figure 5.20 projects field-aligned currents observed in the ionosphere to source regions in the equatorial magnetosphere by tracing along field lines derived from a prespecified magnetic field model (a modified Tsyganenko model in this example). The magnetic fields produced by the observed field-aligned currents shown in the left panels of Figure 5.20 are not included in the mapping of the patterns shown in the right panels. While the effects of these neglected magnetic fields are relatively minor when mapping to strong ambient field regions of the inner and middle magnetosphere, their inclusion can yield surprising results when mapping to the magnetotail or near the magnetopause. Depending on the strength of the field-aligned currents, the actual mapping is very nonisotropic and can be displaced from the inferred position by many R_E. Nonisotropic mapping causes circular structures in the ionosphere, for example, to appear very stretched and twisted in the equatorial magnetosphere (Kaufmann and Larson, 1989; Donovan, 1993).

The auroral processes observed at low-altitude are powered by magnetospheric dynamos that are usually located far from the low-altitude sites of energy dissipation, e.g., the ionospheric E layer and auroral acceleration regions found near 1 $R_E$altitude. In the intervening medium between the dynamo and dissipation regions, electromagnetic power is transmitted with minimal loss along magnetic field lines. For very quiet conditions, this energy is transported quasi-statically. More typically, however, the energy transport is mediated by MHD waves. Compressional MHD waves propagate energy isotropically away from magnetospheric source regions and are not effective in forming discrete auroral structures. Alternatively, energy transport by ideal shear Alfvén waves is perfectly guided along the ambient magnetic field. Shear Alfvén waves are therefore closely connected with the formation of discrete auroral structure. The magnetic guidance of shear Alfvén waves implies that simple magnetic mappings between magnetospheric dynamo regions and low-altitude sites of dissipation continues to be valid, in first approximation, even in dynamic situations. Deviations arise from the two-fluid effects discussed in Section 3.4.3, which modify the ideal shear Alfvén characteristics. In particular, dispersive effects introduced by the finite ion gyroradius and electron inertia enable perpendicular group propagation. These effects are important on the transverse length scales of discrete auroral arcs–less than 10 km (projected to the ionosphere). As a consequence, discrete arcs tend to project upward along Alfvén wave characteristics rather than magnetic field lines.

Chapter 6

Electrodynamics of Auroral Forms

This chapter provides a description of some major night side auroral forms and of the polar cap aurora, with particular emphasis on the ionospheric electrodynamics of these forms. By this we mean the local characteristics of the electric field, current, and conductivity patterns, and how these are interrelated. This relationship is governed by Ohm's law, the requirement of current continuity, and use of field-aligned currents as the connecting agents between the ionosphere and the magnetosphere. This chapter provides a representative view of the characteristics and electrodynamics of the selected auroral forms for quasi-static conditions, leaving much of the discussion on the dynamical aspects and magnetospheric coupling to Chapter 8. The electrodynamic models presented in this chapter integrate ground-based observations, low-altitude satellite and rocket observations, and numerical modeling results into a synoptic picute of several major auroral forms.

The stable eveningside discrete arc, described in Section 6.1, represents the basic auroral form and has been studied extensively. Its occurrence is by no means limited to the evening sector of the auroral oval, but can appear at any local time. The electric field pattern associated with this discrete arc can take on many different forms, depending upon the magnetic local time sector, the intensity of precipitation, and the way by which current continuity is maintained across the arc. However, each of these patterns may be understood using the model described in the next section.

The nightside auroral bulge is the common name for the extensive region of auroral activity that develops and expands in the midnight sector due to a massive energy input from the magnetospheric tail during the main phase of a substorm (see Figure 5.4). In Sections 6.2 to 6.4 of this chapter, some major auroral forms within different regions of the auroral bulge are described and subjected to detailed two-dimensional analysis of the ionospheric electric fields and current systems. In each case, the following questions are addressed: how are the field-aligned currents (FAC's) associated with the various auroral forms closed in the ionosphere; what is the relation between FAC's, the particle-induced conductivity, and the aurora;

and what is the relevance of the Cowling current slab model (see Section 3.1, for describing the auroral electrodynamics).

The northwestern part of the substorm auroral bulge often develops into a large-scale surge, called the Westward Traveling Surge (WTS), which is the most spectacular of the large-scale auroral displays and represents the optical signature of the westward part of the substorm expansion. Models of the electrodynamics of the WTS are described in Section 6.2. The poleward boundary of the post-midnight bulge, characterized by diffuse aurora, may evolve into a periodic pattern called Omega Bands during the recovery phase of substorms. These are treated in Section 6.3. At the poleward boundary of the auroral bulge yet another spectacular auroral phenomenon, named Auroral Streamers, may evolve in the form of finger-like, north-south-aligned auroral forms. These streamers are believed to be the ionospheric signatures of bursty bulk flows in the central plasma sheet, and are discussed in Section 6.4.

Whereas Sections 6.1 - 6.4 are focussed on aurora within the night-side auroral oval, Section 6.5 discusses the rich variety of auroral forms that appear poleward of the auroral oval during certain conditions, in what is conventionally called the polar cap. A summary is given of the observed properties of various polar cap aurora forms, their electrodynamic features, and how these fit into a more global perspective.

6.1 Eveningside Discrete Arcs

The electrodynamics of a stable, discrete auroral arc has mainly been described from two slightly different viewpoints. The first one considers the arc and its associated current system as a quasi-static system, neglecting any time variations. The electrodynamics is then developed from the assumption that all currents must close in the ionosphere. A requirement for these types of models to be valid is that the arc is stable on time scales larger than the Alfvén travel times between the ionosphere and the acceleration and equatorial source regions, which are of the order of seconds and minutes, respectively. This section will describe the meso-scale features of the arc-associated electric field.

6.1.1 The Arc as Part of a Quasi-Static Current Circuit

The ionospheric electric field around auroral arcs has been studied extensively by means of rocket, satellite and radar measurements. In spite of the wealth of data, there are several factors that make the interpretation rather difficult: rocket and satellite measurements suffer from the inability to distinguish between spatial and temporal variations, and they are measurements in only one dimension. Radars can provide a two-dimensional measurement of one electric field component, but the spatial and temporal resolution is much lower than for in situ measurements, and in order to get both components of the horizontal field, two radar beams at an angle to each other are required.

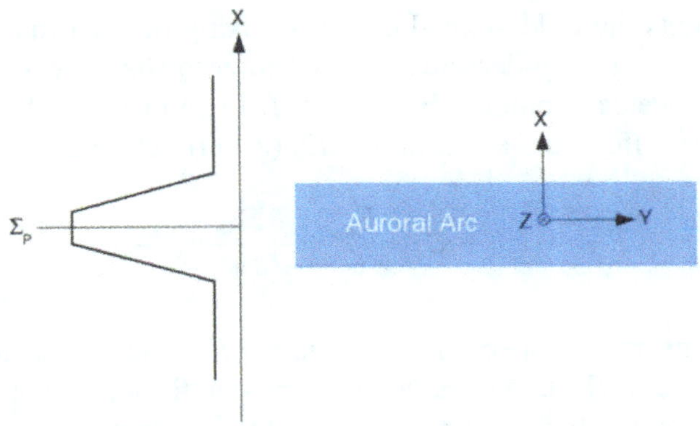

Figure 6.1. Definition of the coordinate system (right) used for a nightside auroral arc in the northern hemisphere, together with (left) a corresponding conductivity distribution. Figure provided by T. Karlsson.

In order to overcome these difficulties, much research has focused on stable auroral arcs with a large east-west extent. In this way all time variations and variations in the east-west direction can be neglected, at least to a first approximation. This type of discrete auroral arc is most common in the evening side, but may occur at any local time sector.

To first approximation, the ionospheric electric field in the vicinity of such an auroral arc is mainly determined by three factors. Together with the assumption of a stationary state, and that of quasi-neutrality, they determine what the electric field must be to fulfill the requirement of current continuity. These three factors are: (i) the background convection electric field, which is of magnetospheric origin (cf Section 1.4). (ii) the relationship between the ionospheric conductivities inside and outside of the arc, and (iii) magnetic-field aligned currents associated with the arc. We also neglect the effects of the velocity of the neutral background (the neutral wind). Such effects are discussed in Section 8.4.3.

The way to quantify this into a mathematical formulation is to invoke the current continuity for the case that the arc is quasi-stationary, i.e., $\frac{\partial n}{\partial t} = 0$ in the current continuity equation.

Figure 6.1 shows the orientation of the arc and defines the coordinate system we will use. Now, for the case where the neutral wind velocity is zero, the height-integrated, horizontal current vector is given by

$$\mathbf{J}_\perp = \Sigma_P \mathbf{E} + \Sigma_H \hat{\mathbf{b}} \times \mathbf{E} \tag{6.1}$$

where Σ_P and Σ_H are the height-integrated Pedersen and Hall conductivities, respectively, $\hat{\mathbf{b}}$ is the unit vector along the magnetic field (cf. footnote on page 43). In a stationary situation, this current must be divergence-free (cf Section 3.1):

$$\nabla \cdot \mathbf{J}_\perp = j_z \tag{6.2}$$

where j_z represents the field aligned currents. Using the fact that we neglect time variations, and setting all y-derivatives equal to zero (because we ignore all east-west variations), we can combine Equations (6.1) and (6.2) to form a one-dimensional, first-order differential equation for E_x (as a function of x):

$$\frac{\partial \Sigma_P}{\partial x} E_x + \Sigma_P \frac{\partial E_x}{\partial x} = \frac{\partial \Sigma_H}{\partial x} E_y + j_z, \tag{6.3}$$

where E_y is a constant. By specifying a value for E_y, and a value for E_x outside of the arc (which we call E_{x0}), to act as a boundary condition, the solution to Eq. (6.3) is uniquely determined for a given j_z, if also the conductivities are specified. E_y and E_{x0} together represent the background convection electric field.

Case $j_z = 0$. We now begin by considering the case of $j_z = 0$ and study the ionospheric electric field pattern that results from introducing a highly conducting region into the nightside background convection pattern. (For the auroral arc, the high conductivity of course comes about as an effect of the ionization by the precipitating particles causing the aurora.)

Figure 6.2 shows a number of solutions for the conductivity profile of Figure 6.1, and for background electric fields of the same magnitude (50 mV m^{-1}) but varying directions. For simplicity, the Hall conductivity was set equal to the Pedersen conductivity. The background field is varied from a purely northward field (representative of the background field in the evening sector) via a westward field (midnight), to a purely southward one (representative of the morning sector), which thus can be interpreted as moving in local time from the evening sector to the morning sector (cf Section 1.5). Except for the 'westward electrojet' case (cf Section 3.1.2), these are taken from a classification scheme introduced by Marklund (1984), which is shown in Figure 6.3. This scheme is based on a large number of measurements of auroral arc-associated electric fields and classifies them according to the physical mechanism that maintains current continuity across the arc, i.e., by polarization electric fields or by field-aligned currents. The polarization electric field is defined as the difference between the background field, and the electric field inside the auroral arc, i.e., $E_{x0} - E_x$.

It is easy to understand how this polarization field comes about. Consider, for example, the evening anticorrelation case, where the background field is northward. Here, anti-correlation means that the variation in electric field magnitude is anti-correlated with the conductivity variation across the arc. Due to the high conductivity inside the arc the background electric field drives a much larger northward current inside the arc, than outside it. If no field-aligned currents are present this leads to an accumulation of charge on the arc edges, positive on the northern edge, and negative on the southern one. This accumulation will continue until the resulting southward field due to the accumulated charge, the polarization field, is

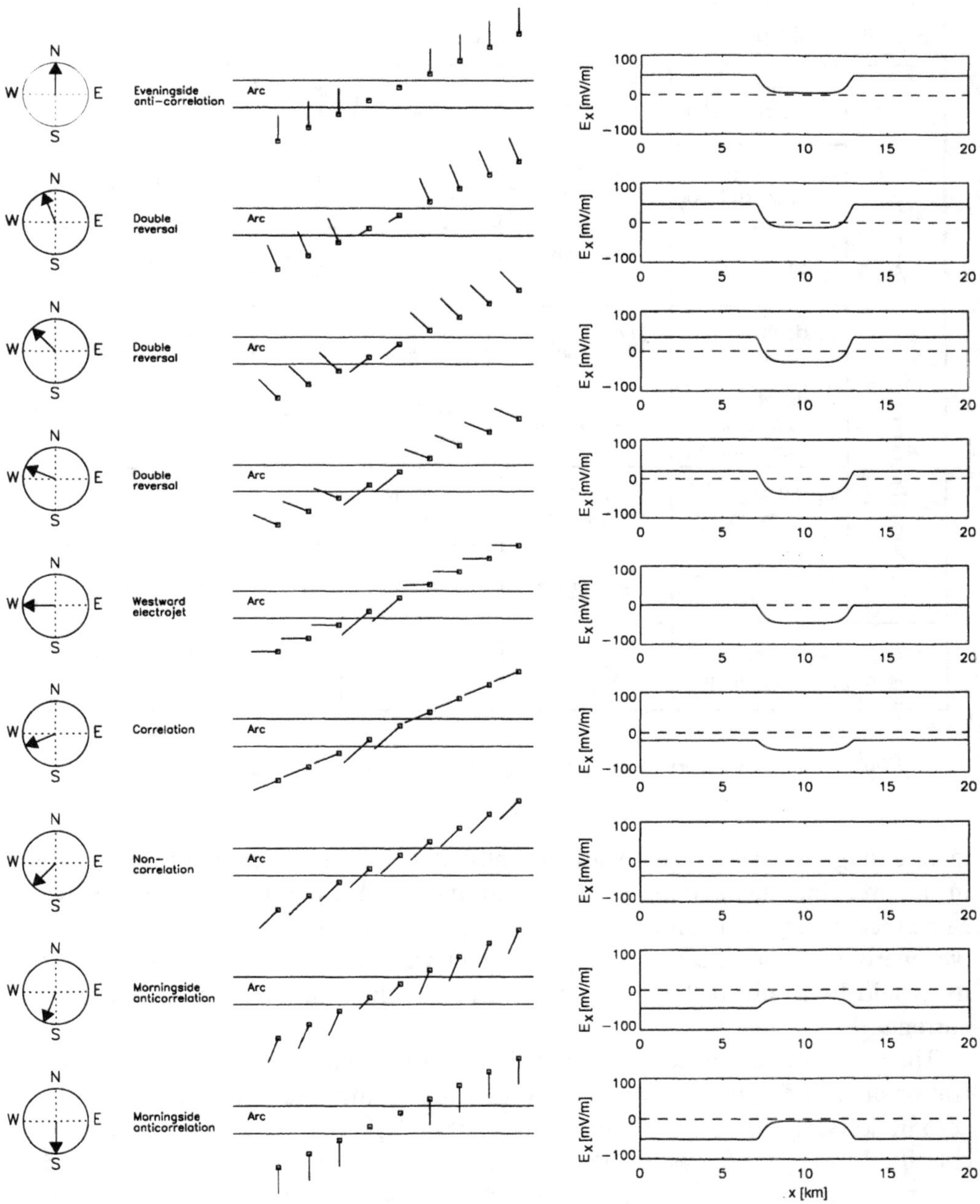

Figure 6.2. Calculations of arc-associated electric fields, given a background electric field and a conductivity distribution (no FAC's). From left to right is shown the direction of the background field, a designation, a vector plot of the electric field along an imagined rocket trajectory, and the poleward component of the electric field, E_x. The terms 'correlation' and 'anti-correlation' are used to represent the variation of the electric field magnitude with respect to the conductivity variation across the arc. Figure provided by T. Karlsson.

Arc Cate-gory	Subtype	E_x - conf.	E_x - (amb.)	E_y - (amb.)	E polarisation
I. Polarisation arcs	I_a: evening-anticorrelation		large N-ward	small, E- or W-ward	S-ward
	I_b: morning-anticorrelation		S-ward	small, E- or W-ward	N-ward
	I_c: correlation		S-ward	W-ward	S-ward
	I_d: double-reversal I	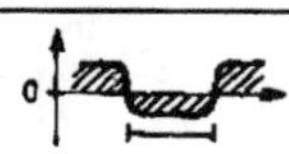	N-ward	W-ward	S-ward
II. Birkeland current arcs	II_a: Inverted V (single or tri-ple reversal)		small, N- and S-ward	small, typic. W-ward	0
	II_b : double-reversal II		small, N-ward	small, E- or W-ward	0
	II_c : non-corre-lation	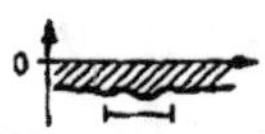	S-ward	typic. W-ward	0
III. Combin. arcs	III_a: evening asymmetric		large N-ward	small, E- or W-ward	probably S-ward at equatorward arc edge, small at the poleward

Figure 6.3. Summary of auroral electric field observations. After Marklund (1984).

large enough to drive a southward current, balancing the excess northward current in the arc, thus maintaining current continuity across the arc edges. (This is the same mechanism which is discussed in the section on the Cowling channel and the westward electrojet, Section 3.1.2) The other way of maintaining current continuity would be to divert the excess current out of the ionosphere, via field-aligned currents.

The solutions presented in Figure 6.2 then correspond to what are called 'polarization arcs' in Figure 6.3, which is consistent with the assumption that the FAC be zero in deriving Eq. (6.3). (We can also see that the 'non-correlation arc', although classified in Figure 6.3 as a Birkeland current arc, may equally well be classified as a polarization arc, since we can reproduce it as a solution to Eq. (6.3), without any FAC's.)

Figure 6.4 gives two examples of measurements of an arc-associated electric field. The left part shows rocket observations from the evening sector (Marklund et al., 1982). The auroral arc can be identified with the highly increased electron energy flux, and extends from approximately 160 s flight time to $\sim$ 380 s. The rocket was moving approximately northward. The background field, as measured at

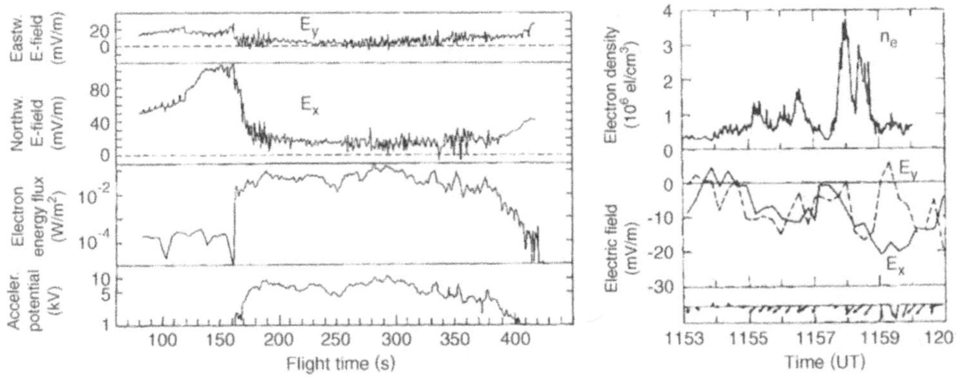

Figure 6.4. Left: Rocket observations of an evening anticorrelation arc (Marklund et al., 1982). Right: Radar observations of a correlation arc (de la Beaujardière et al., 1977).

around 100 s, is mainly northward (for the moment we neglect the intensification of the northward electric field component at the southern edge of the arc). Comparing with Figure 6.2, we should thus expect to see a depression of the northward electric field component (or expressed in another way: an arc-associated southward-directed polarization field). This is exactly what we see in the data; the northward component decreases from around 50 $\mathrm{mVm^{-1}}$ south of the arc, to approximately 10 $\mathrm{mVm^{-1}}$ inside it. At the end of the rocket passage, the electric field increases again, but the measurement is then terminated by the rocket re-entry into the atmosphere. Several similar measurements of an evening anticorrelation arc have been reported (e.g., de la Beaujardière et al., 1977; Carlson and Kelley, 1977).

The right part of Figure 6.4 shows an example of a 'correlation' arc (de la Beaujardière et al., 1977). The southward component of the electric field is enhanced within the arc. The fact that the peak value is observed outside of the arc is probably due to the long integration time of the radar measurement. The background electric field outside of the arc is approximately southwestward, in good agreement with the correlation case in Figures 6.2 and 6.3.

Case $j_z \neq 0$. The second type of electric field configuration around auroral arcs comes about when the horizontal current is not continuous across the arc edges, i.e., $\nabla \cdot \mathbf{J}_\perp \neq 0$. One way to think about this is to say that the polarization electric field does not have the right value to maintain current continuity. The excess current must then be diverted into, or away from the ionosphere as a field-aligned current. More formally the northward electric field is modified from the above case by a new source term, since j_z is now not zero.

It is clear that by specifying the field-aligned current j_z in various ways we can reproduce a number of different electric field configurations. In Figure 6.5 we show two examples of simple current systems that reproduce the arc types called

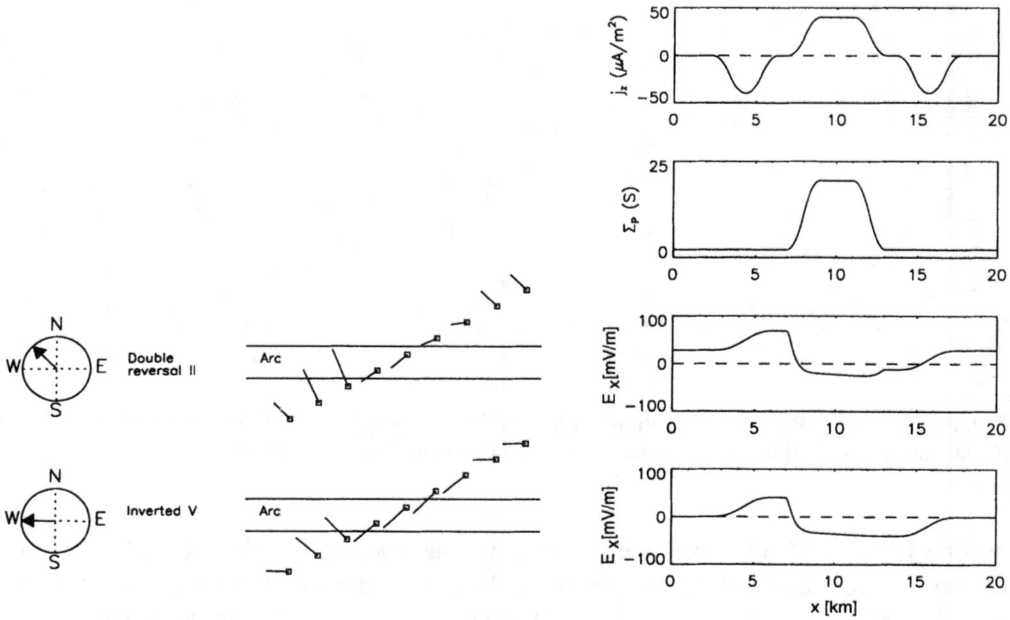

Figure 6.5. Electric fields, similar to those of Figure 6.2, but with a field-aligned current distribution as shown in the top panel on the right. The second panel on the rights shows the conductivity distribution. Figure provided by T. Karlsson.

'inverted-V' and 'double reversal II' in Figure 6.3. In both these cases the FAC is chosen so that the upward current is colocated with the auroral arc, and closed both northward and southward of it. This is a common current configuration for auroral arcs, and can be thought of as a signature of a U-shaped acceleration potential above the arc (see Section 4.1.3). It is worth noting that in order for the Birkeland currents to be dominant in determining the electric field pattern, they have to be quite large. Thus it is clear that the background electric field and the conductivity distribution around the arc usually are important factors in determining the ionospheric arc-associated electric field, just outside of the precipitation region defining the optical arc. An example can be seen in Figure 6.4a, from approximately $t = 120$ s to $t = 160$ s. This signature is sometimes called 'radar arc', since it has been studied mainly by radar measurements (e.g., Stiles et al., 1980; Aikio et al., 1993). The model put forward by Aikio et al. (1993) was that this signature is associated with a current system such that the upward current associated with the arc is closed on only one side. This intense return current creates a region of low ionospheric conductivity due to the outflow of cold ionospheric electrons carrying the current (Doe et al., 1995; Karlsson and Marklund, 1998). The low conductivity there leads to an enhancement in the electric field, in order to drive the Pedersen current closing the upward and downward currents (see Figure 2.7). If the upward

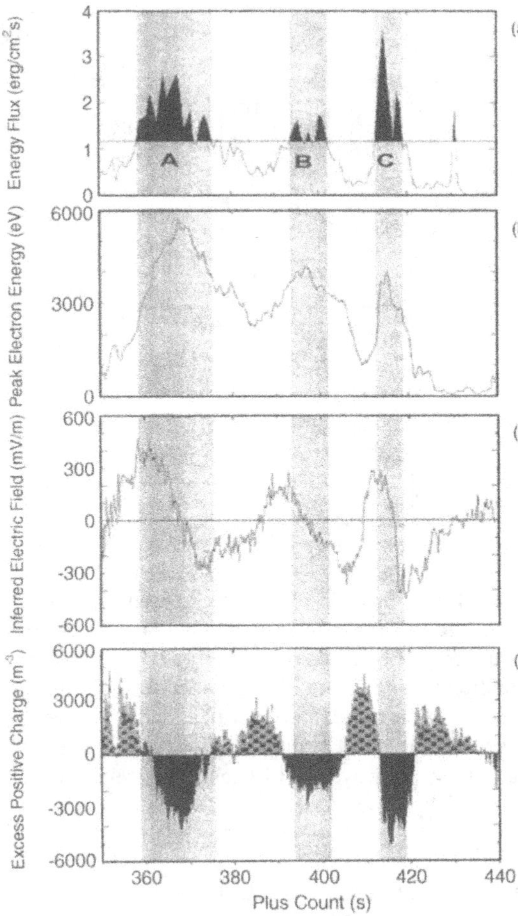

Figure 6.6. The first and second panels show measurements of the electron energy flux, and the peak electron energy for three adjacent auroral arcs, the third and fourth panels show the inferred electric field and space charge, respectively. From Hallinan et al. (2001).

and downward currents are oriented in the same way as the large-scale Region 1 and 2 currents, this theory then predicts that the enhancement should be located equatorward of the arc in the pre-midnight region, and poleward of it at local times after midnight, which is precisely what is observed.

6.1.2 Relation to the High Altitude Electric Field

As mentioned above, the high-altitude electric field above an auroral arc is typically much larger than the corresponding ionospheric electric field, even though it should be smaller if one only considers the geometric mapping due to the divergence of the geomagnetic field lines, under the assumption that the field lines be equipotentials. In reality, of course, the field lines above the aurora are not equipotentials,

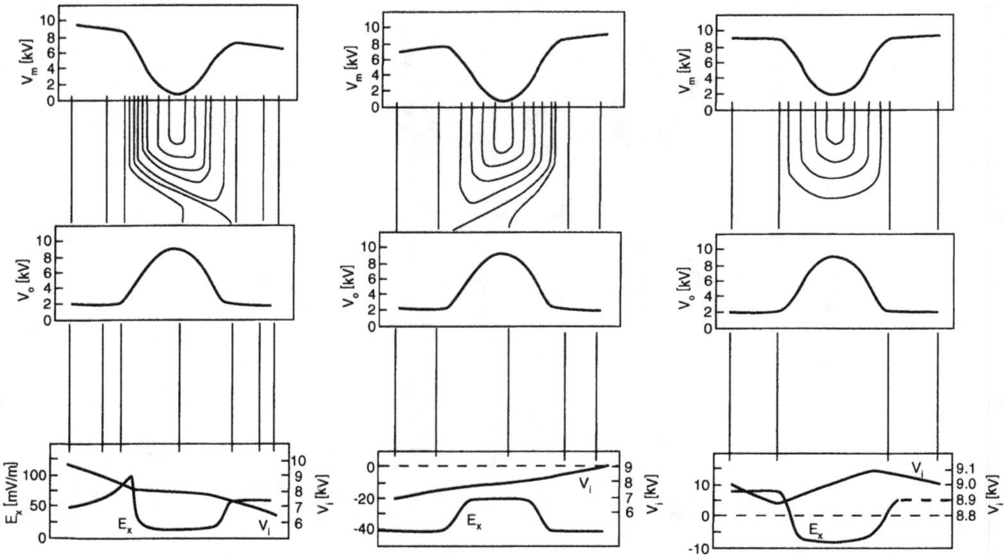

Figure 6.7. Three examples of the connection between the high-altitude potential distribution, the ionospheric potential across the arc, and the acceleration potential. The bottom panels show the ionospheric electric field, and the potential obtained by integrating this across the auroral arc. The middle panels show a typical acceleration potential, and the top panel the high-altitude potential obtained by subtracting the acceleration potential from the ionospheric potential. Between the panels the resulting potential pattern is indicated for the different altitudes regions. The three different cases correspond to northward, southward, and west-north-westward background electric fields, respectively. After Marklund (1984).

but are associated with a parallel electric field in the auroral acceleration region. As discussed in Section 4.1.1 (see also Block and Fälthammar, 1990), there are several ways to determine the potential drop associated with this parallel field; for example by measuring the downward electron beam energies. Figure 6.6 shows an example of such a measurement by the AMICIST sounding rocket payload (Hallinan et al., 2001). By differentiating the measured acceleration energy, the transverse electric field associated with the acceleration potential was estimated. (Some smoothing was performed due to the inherent generation of noise in numerical differentiation). The shaded regions indicate the extent of three discrete arcs, as determined from optical data. Note how the arcs are associated with negative space charge, or equivalently with converging electric fields, whereas the regions between the arcs are associated with positive charge, i.e., diverging electric fields.

In order to correctly determine the high-altitude potential distribution, one should subtract the high-altitude acceleration potential from the ionospheric potential (obtained by integrating the ionospheric electric field across the arc.) The ionospheric potential is typically smaller than the high-altitude potential by a factor of the order of ten, and thus contributes only to a light modification of the shape of the high-altitude potential. Figure 6.7 shows an example of such a determination for three arcs associated with mainly northward, southward and westward ambient electric fields, respectively, assuming a regular U-shaped acceleration potential. Note how in the first two cases the U-shaped potential is distorted asymmetrically, and how the lower portion of the potential distribution is turned into an S-shaped potential.

6.1.3 A Synoptic Case Study of an Evening Arc

Even though the preceding conclusions about arc electrodynamics allow categorization of many independent arc observations, they should be compared with simultaneous remote sensing and in situ observations well above the ionosphere. Until quite recently no complete and simultaneous observations existed of both ionospheric and magnetospheric particle fluxes and associated current flow. This is not surprising if one considers the actual probability of the event that an auroral arc drifts over an incoherent scatter radar, which is operating in the correct mode, at the same time when a suitable low-altitude satellite traverses the field lines connected to the same arc. Nevertheless, such a rare constellation of events occurred on February 19, 1998, between 18:40 and 19:00 UT, when the FAST satellite passed over a slowly southward drifting auroral arc. The arc was located only a few tens of km to the north of the tri-static EISCAT UHF radar in Tromsø (see McCrea and Lockwood, 1997, for more details about the EISCAT facility). During this evening, the EISCAT UHF radar was operating in a field-aligned, high-resolution data-collection mode in support of an EISCAT-FAST satellite observing campaign. The EISCAT data were taken at a resolution of 10 sec, but later post-integrated to 2 minutes to achieve analyzable data, in particular in the region of low ionospheric ionization adjacent to the auroral arc. The two remote antennas in Kiruna and Sodankylä intersected the field-aligned Tromsø beam at a lower F-region altitude of 275 km, to derive the three-dimensional electric field vector in the purely $\mathbf{E} \times \mathbf{B}$ drifting plasma regime, without collisional neutral wind contamination.

As the conjunction between FAST and EISCAT on this evening was predicted, the radar experiment could be specially configured as just described. However, the timely appearance of an auroral arc just at the right location was a mere coincidence. The auroral arc was overflown by FAST at 18:42-45 UT when it was located only slightly to the north of EISCAT. The arc continued its southward drift and passed through the field-aligned radar beam only 5-8 minutes later. So we can compare two cuts through the three-dimensional patterns of currents, precipitation and electric fields, one in the ionosphere and one close to the acceleration region at 4000 km altitude. While FAST traversed this system from north to south between 18:42 and 18:47 UT, EISCAT scans it from south to north between 18:40 to 19:05 UT. Only at about 18:45 UT, just after FAST leaves the visible auroral feature, are EISCAT and FAST probing the same field line.

6.1.3.1 Ground-Based Observations

All-Sky Images. Figure 6.8 shows a sequence of three selected auroral images, at 18:43, 19:02 and 19:13 UT. These images were taken by an all-sky camera at Kilpisjärvi and have been projected onto a plane at 100 km altitude above the map (Syrjäsuo et al., 1998). The intersection of the field-aligned EISCAT antenna beam with the ionosphere at 100 km is indicated. At 18:43 UT a double auroral arc is located just to the north of EISCAT, so that the radar beam probes the region just adjacent to the visible feature. During several minutes before, the arc system had been observed to drift slowly into this region from the north. The top panel contains the ground track of the FAST satellite, which according to this auroral image crossed the double arc structure close to this time, and left it around 18:45 UT. Between 18:43 and 18:55 the arc is nearly stationary. The southward motion then continues, so that the equatormost arc element crosses the EISCAT beam at about 19:00-19:05 UT (central image shown at 19:02 UT). After this passage through the EISCAT beam the arc remains southward of EISCAT until an auroral activation at 19:10 UT results in a poleward expansion and a westward drifting auroral loop, which brings a different type of visible aurora, emanating from the initial arc, into the vicinity of the EISCAT beam (bottom panel at 19:13 UT). At this time the FAST satellite is already located on latitudes too far to the south to monitor any of these temporal changes.

From this sequence of images it becomes obvious that, even though the arc evolves with time, the satellite passes over the same auroral feature just to the north of EISCAT at 18:45-18:46 UT, which finally drifts through the EISCAT beam at 19:02-03 UT. The region to the south of the visible aurora, i.e., the region most interesting for this study, is probed by FAST and EISCAT at the same time, however.

Incoherent Scatter Radar. The EISCAT observations during this period are shown in Figure 6.9. Standard EISCAT data consisting of field-aligned altitude profiles of electron density, electron temperature and ion temperature are presented in the top three panels. Measurements of the horizontal electric field, as derived from remote UHF receiver stations at a tri-static [1] intersection altitude of 275 km, are displayed in the bottom panel. In this case the horizontal electric field is almost purely poleward directed, perpendicular to the approaching auroral arc. The initial electron density is indicative of a normal background ionosphere with moderate ionization, leading to average height-integrated Hall and Pedersen conductivities of 6 and 4 S, respectively. Shortly after 18:40 UT and definitively at about 18:44 UT the electron density decreases drastically at all altitudes (the height integrated conductivities decrease to $\Sigma_H = 0.2$ S and $\Sigma_P = 0.5$ S), and at the same time the ion temperature (third panel) is seen to increase from 1200 K to more than 2500

[1] Tri-static means that three radar receiving stations point to the same region in space.

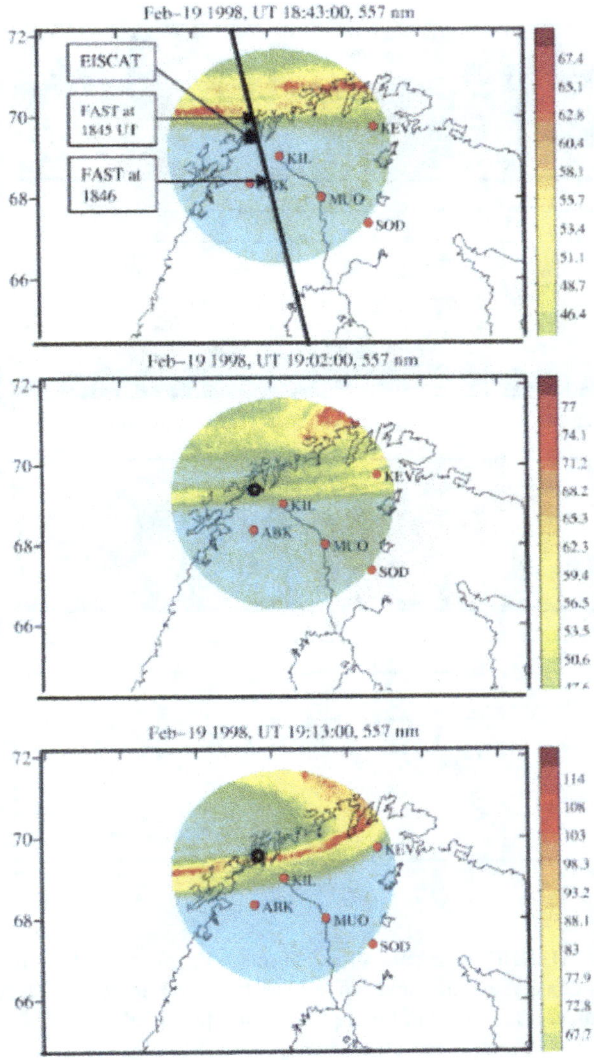

Figure 6.8. Rectified auroral all-sky camera images from Kilpisjärvi, for three selected times during and after a passage of the FAST satellite over Scandinavia. The location of EISCAT, a few selected cameras and magnetometers are given in all panels, and the locations of FAST during portions of its orbit are given in the top panel, the auroral image closest to the situation encountered by FAST. Figure provided by H. Opgenoorth.

K. This increase in ion temperature is usually attributed to the existence of a strong electric field, leading to strong convection and consequent ion-neutral frictional heating in the collision-dominated E-region. Such an electric field enhancement from about 50 $\mathrm{mV\,m^{-1}}$ at 18:40 UT to 150 $\mathrm{mV\,m^{-1}}$ at 18:55 UT is seen in the bottom panel.

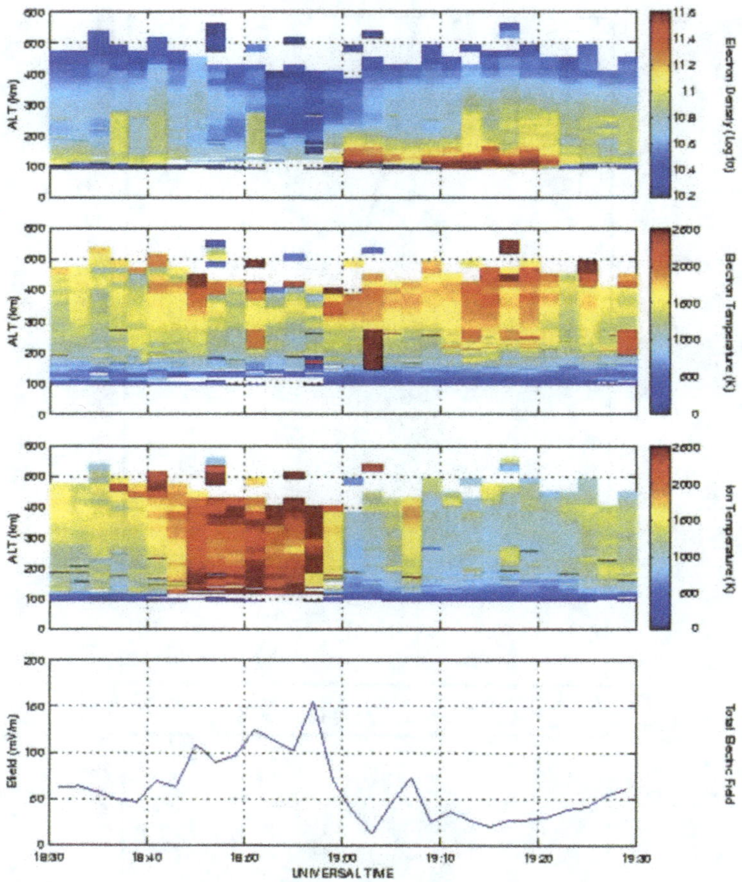

Figure 6.9. EISCAT UHF radar data above Tromsø. From top to bottom: field-aligned altitude profiles of electron density, electron and ion temperature, and horizontal electric field at an altitude of 275 km, perpendicular to the arc in Figure 6.8. Figure provided by H. Opgenoorth.

Another effect of the strong electric field is visible in the electron temperature at 18:40 UT (second panel). As described in more detail by Opgenoorth et al. (1990), and Williams et al. (1990), very strong ionospheric convection can heat the electron gas via E-region instabilities. However, this effect is seen to disappear very quickly in the EISCAT data as the extremely low electron density after 18:45 UT makes the interpretation of the incoherent radar spectra close to impossible. The ion heating and depletion of ionospheric electron density, accompanying the intense electric field, increase slowly with time until 19:00 UT when the visible auroral arc enters the radar beam. The intense precipitation immediately produces strong ionospheric ionization and conductivity (Σ_H and Σ_P increase to 25 S and 11 S, respectively), and suppresses the electric field to values lower than the background convection field observed prior to the event, e.g., at 18:30 UT. The ion heating also ends abruptly, and the ion gas is now even colder than the background ion temperature. The high-altitude increase in electron temperature may be attributed to low-energy (secondary) electron precipitation that commonly accompanies the primary, keV electron precipitation responsible for auroral arcs.

After 19:06 UT, the visible auroral arc leaves the radar beam towards the south, and, for a brief interval, the electron density decreases again, the electric field increases and the ion temperature recovers to moderate values. At 19:10 UT, a large-scale auroral activation transforms the arc into a westward drifting loop that again fills the radar field of view with effects resulting from energetic precipitation, including enhancements in the Hall and Pedersen conductivities to 65 S and 16 S. However, the resulting features are qualitatively very different from the main focus of this section — the quiet drifting arc.

If one assumes that the arc drifts at a constant speed, then the length of time that the effects of strong electric fields remain in the radar field of view (from 18:40 to 19:00), compared with the duration of the arc itself (from 19:00 to 19:06 UT), would give the incorrect impression that the equatorward region of return current is three times wider in latitude than the arc precipitation region. Rocket data have shown, in fact, that the in situ width of the downward FAC channel adjacent to auroral arcs is typically comparable in width or narrower than that of the upward current channel of the arc. Careful analysis of the entire auroral film for this event revealed that the arc drifted equatorward from about 18:30 UT onwards, until it reached a latitude to the north of EISCAT. Then its motion seems to have stalled during the FAST overpass, until it finally crossed the EISCAT beam with its southern arc element (not both arcs) just after 19:00 UT. Thus the apparent width of the region of strong electric field stems from the immobility of the structure from about 18:45-19:55 UT. The FAST data described below confirms this interpretation.

Magnetometer Array. So far we have only discussed the ionospheric closure current, i.e., the northward directed Pedersen current between the two regimes of upward and downward directed FACs within and adjacent to the arc. However, the northward directed electric field—weak in a region of high conductivity, and strong in the region of low conductivity—also gives rise to a wide region of eastward Hall current, under the entire three-dimensional current system.

The temporal and latitudinal development of this eastward Hall current as observed by a chain of magnetometers in Scandinavia is shown in Figure 6.10. The data corresponds to the magnetic field disturbance right below the ionospheric current itself, utilizing the 'upward continuation' formalism described by Amm et al. (1999). Clearly a region of eastward current, which is about 2.5° in latitude wide, is seen to drift equatorward together with the arc in Figure 6.8. The equatorward drift starts at about 18:35 UT, enhances at 19:00 UT and finally evolves into a poleward expansion with the formation of an auroral loop at 19:13 UT. The latitude of

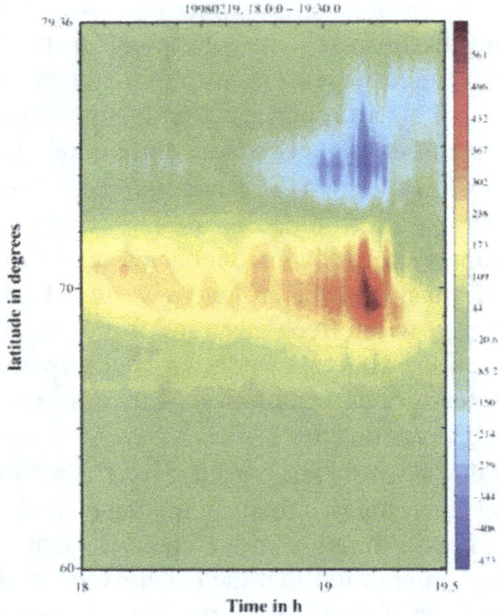

Figure 6.10. Upward continued magnetic disturbances at ionospheric levels, as derived from magnetic disturbances on the ground, along the MIRACLE network of magnetometers in Scandinavia. Figure provided by H. Opgenoorth.

EISCAT (about 69°) is engulfed in the main eastward current regime after about 18:40 UT, when the first effects of the downward FAC region are observed.

6.1.3.2 In Situ Observations

The relative widths of the adjacent up- and downward field-aligned current channels is clearly revealed by the FAST satellite traversal through the entire arc structure. Figure 6.11 shows selected FAST data from 18:40 to 18:48 UT. In correspondence with the auroral images and the EISCAT data discussed above we see that FAST crossed a double auroral feature (two adjacent 'inverted V' events) between 18:43 and 18:45 UT. The arc visible in the top panel of Figure 6.8 corresponds to the FAST data obtained between 18:44 and 18:45 UT, i.e., only the second half of the second inverted V shows up. Here the energy of the quasi-mono-energetic peak in the precipitating electron distribution (at 0° pitch angle) reaches values of 10 keV (see Figure 6.8a) within the auroral arcs. The direction of the field-aligned current is upwards, corresponding to a positive slope in the mainly east-west magnetic field component, i.e., tangential to the arc. Judging from the linearity of the slope in B, the current density must have been fairly homogeneous, except near the gap between the two arcs just after 18:43 UT (see also Figure 6.11b). In the region of precipitation, the electric field projected onto the FAST trajectory is nearly zero, while it was moderately southward in the polar cap north of the arc. After 1845 UT, which is just above EISCAT, FAST enters a region of different plasma character. The field-aligned current direction rotates downward, the energetic electrons disappear, and the electric field is strongly enhanced and points northward at about 50 $\mathrm{mVm^{-1}}$. As there is a mapping factor of 2 between FAST and the ionosphere, this value is in good agreement with $\approx$ 100 $\mathrm{mVm^{-1}}$ inferred by EISCAT in the ionosphere. As noted in the discussion of the EISCAT data, the electric field is strongest directly adjacent to the auroral arcs; its magnitude then progressively decreases during the subsequent two minutes of the satellite's equatorward traversal. This duration and, therefore, the inferred width, is comparable to that of the double auroral arc.

Within this wide region of downward field-aligned current FAST reveals two spatial regions in which distinctive structures are present in all parameters. The most dramatic effect is the occurrence of upward directed (180° pitch angle), suprathermal electrons with energies up to a few 100 eV. Within both events, from 18:45:40 to 18:46:10 UT and 18:46:30 to 18:46:40 UT (but clearest in the first one), the slope of the magnetic field measurement increases. From Figure 6.11 we see that, in fact, the field-aligned current density in both cases approaches or exceeds 1 $\mu\mathrm{A\ m^{-2}}$. Within both regions FAST sees clearly upward accelerated field-aligned electrons, consistent with a field-aligned potential drop below the spacecraft. Furthermore the suppression of the enhanced electric field during both suprathermal streams, as compared with the more continuously decreasing EISCAT measurements, indicates a decoupling of the ionosphere and magnetosphere somewhere along the field lines below the satellite.

Figure 6.11 also shows that the FAC inferred from the magnetic field is generally more intense (2-3 times larger) than the FAC obtained from the first moment of the electron distribution function at energies exceeding 8 eV. This result indicates that electrons with energies below 8 eV carry a substantial portion of the field-aligned current.

6.1.3.3 Synoptic Interpretation

Figure 6.12 summarizes the general features of the observations for this event, which are also generally valid for other evening sector auroral arcs. The relationships are also valid for morning sector arcs but with the structure mirrored north to south as a consequence of the southward directed convection electric field prevailing in the morning sector. The current system associated with an auroral arc consists primarily of the upward field-aligned current associated with energetic electron precipitation and often an adjacent downward current.

The EISCAT data presented have added another facet, the reaction of the winter ionosphere to downward field aligned currents. At the interface of the downward current with the poleward Pedersen current, ionospheric electrons are pulled upward and a corresponding number of ions is transported poleward by the transverse

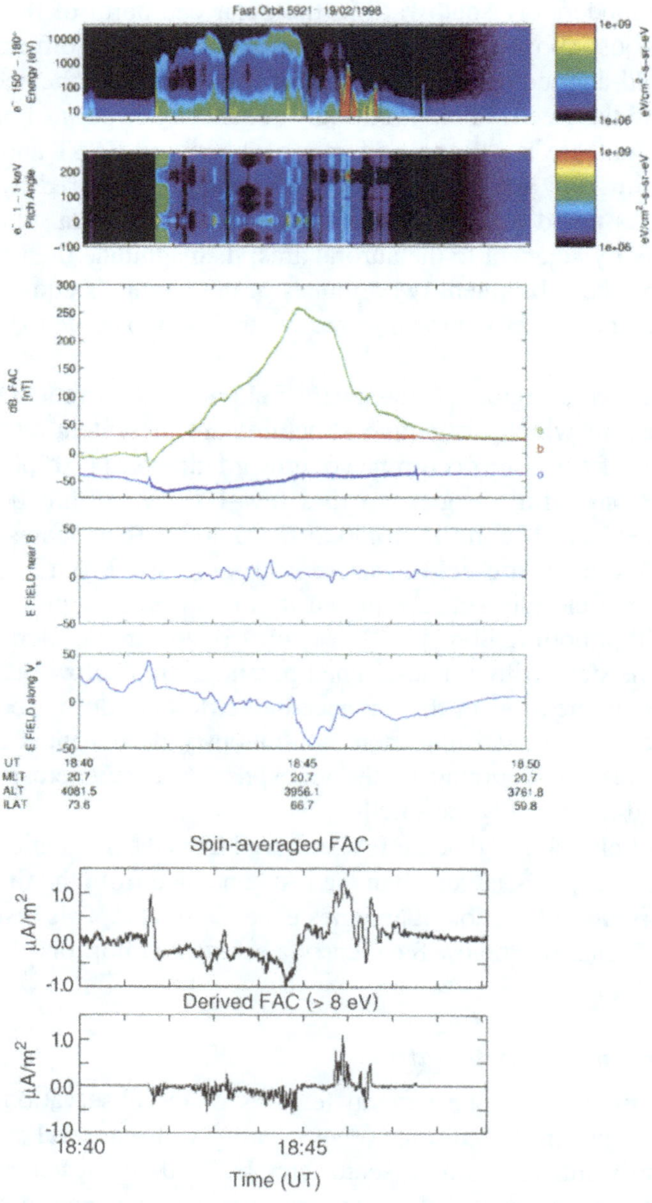

Figure 6.11. Top five panels: FAST measurements of electron energy spectra, pitch-angle distributions, three components of the magnetic field, and electric field parallel to the ambient magnetic field and in the direction of the satellite velocity vector (essentially north-south). Lower two panels: field-aligned current densities at the FAST satellite calculated from the gradient in B and from the integrated electron flux at energies above 8 eV, respectively. Figure provided by H. Opgenoorth.

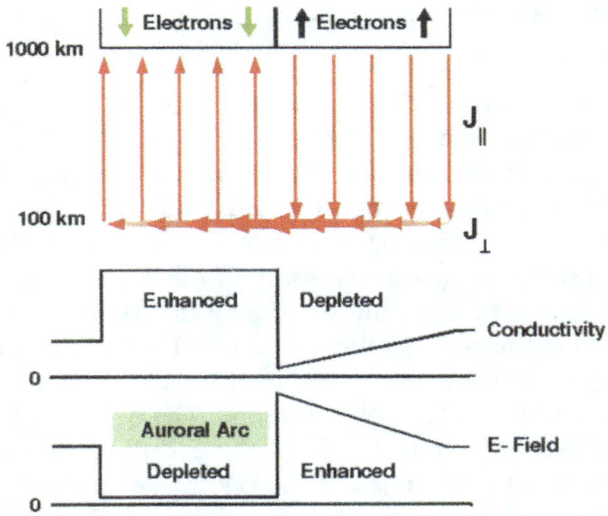

Figure 6.12. Sketch of the geometry and relative intensity of magnetospheric field-aligned currents, ionospheric closure current, height-integrated conductivity, and corresponding horizontal electric field (see text for details).

electric field, resulting in a plasma depletion. This is a non-linear effect. While the Pedersen current is kept constant, $E_\perp$ must grow inversely to the density or conductivity depletion in the Pedersen current layer (~100-150 km). The higher the magnitude of $E_\perp$ the faster the depletion proceeds, while the plasma is moving poleward. Therefore, $E_\perp$ must grow continuously from the equatorward edge of the current system ($j_{||}$ downward), until it drops drastically where the upward $j_{||}$ sets in, with its associated electron precipitation and sudden conductivity increase. The electric field profile shown in the fifth panel of Figure 6.11 reveals this effect very clearly. Of course, if the ionosphere was illuminated and photo-ionization took place, then the effect would be masked.

The combination of ionospheric and magnetospheric data within and around an auroral arc has shown that there is always a more extended current system associated with the visible auroral features (cf. Marklund et al., 1982). The optical features are strongly dependent on the electron flux at the upper end of the spectrum and therefore more concentrated than the upward current region. As the nonlinear modification of Σ_p and $E_\perp$ proceeds, a feedback on the magnetospheric source plasma must arise. The auroral arc is thus a fundamental element of the interaction between the ionosphere and magnetosphere. Its electrodynamic properties clearly show that the ionosphere is an active participant in the larger distributed current system.

6.2 Westward Traveling Surge

The westward traveling surge (WTS) is one of the most spectacular and dynamic auroral forms. It represents the westernmost part of the westward protruding substorm auroral bulge after its formation by an auroral breakup, typically around magnetic midnight. The surge moves westwards with velocities of 1-10 $\mathrm{km\,s^{-1}}$, with the largest velocities occurring immediately after the breakup. During the later phase of a substorm, the expansion speed of the WTS tends to decrease, and eventually it can become almost stationary during the late recovery phase, before it decays in the evening sector. The lifetime of a WTS varies as much as the lifetime of substorms, ranging from several tens of minutes to a few hours. Often a poleward movement of the WTS is observed, too. The typical WTS velocities are significantly larger than the surrounding local $\mathbf{E} \times \mathbf{B}$ drift velocities. A typical display of the WTS as seen by the Freja satellite UV imager is shown in Figure 6.13. Westward of the surge which corresponds to position 8 to 9 in the figure, a stable arc is visible which is associated with the auroral oval prior to the substorm disturbance. Close to the WTS, this arc can be intensified due to the enhanced convection caused by the expanding substorm front, and occasionally develop into an 'auroral horn' or 'surge horn'. In the wake of the surge located eastward of the WTS itself (which is also called 'surge head'), a diffuse auroral background is overlaid by a rich structure of discrete auroral features. The poleward boundary of the wake is typically marked by a bright discrete arc which connects to the poleward edge of the WTS. The WTS itself appears as a spiral form with clockwise winding spiral arms, and areas of less intense emission between these arms.

The WTS is the western terminator of the substorm electrojet flowing westwards inside the substorm auroral bulge, and is located at the westward edge of the substorm current wedge. Strong upward FACs of more than 2 $\mu\mathrm{A\,m^{-2}}$ diverting this electrojet back into the magnetosphere have been found to be associated with the surge (see detailed analysis examples below), as well as hard electron precipitation of energies up to 100 keV. The WTS may occur quasiperiodically in time, related to multiple substorm onsets, but it does not show a regular spatial periodicity such as, e.g., omega bands discussed in Section 6.3.

The typical signature seen in ground magnetometers during the passage of a WTS is shown in Figure 6.14, where the center of the surge passes overhead between stations MIK and RDS of the north-south chain of the Scandinavian Magnetometer Array (SMA) shown. The X component of the magnetic field (pointing geomagnetically north) shows a steep negative excursion at the surge's passage which persists in its wake, indicative of the substorm electrojet region. Equatorward of the surge, smaller positive X components are seen which reflect an eastward di-

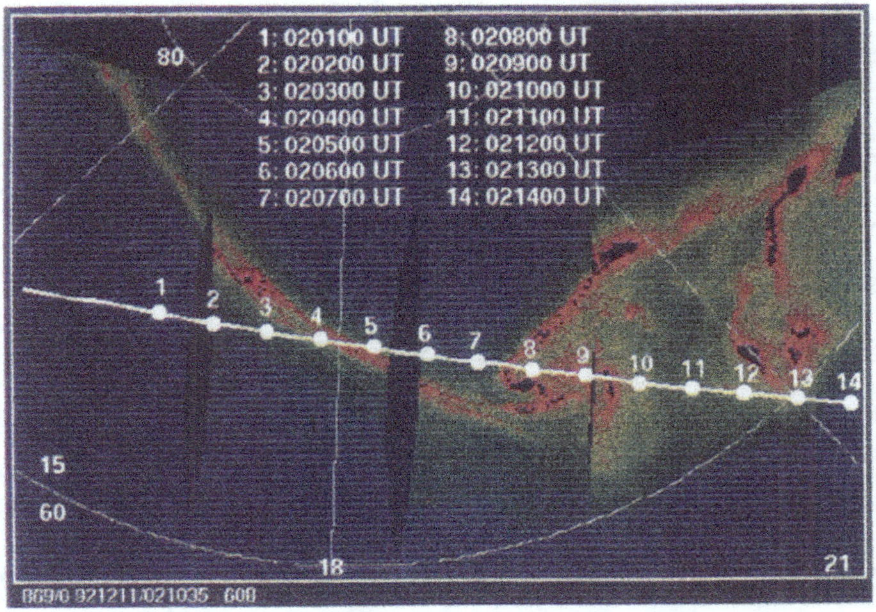

Figure 6.13. Auroral display of a WTS, as recorded by the Freja UV imager on December 11, 1992 (labels 7-10), with superposed ionospheric footpoints of the Freja satellite at 14 consecutive times. On the right-hand side of the figure, an auroral streamer (see Section 6.4) is seen. After Marklund et al. (1998).

rected equivalent current[2] flow. The Y component (pointing geomagnetically east), as observed immediately under the surge, shows first a positive and then a negative deflection, such that together with the X component observations an anticlockwise equivalent current pattern around the head of the surge emerges. This feature and the purely positive B component deflection equatorward of the surge indicate a net upward FAC associated with the WTS. In the Z (vertical downwards) component, immediately under the surge (near RDS) first a small positive and then a negative deflection is measured, while poleward or equatorward of the WTS, the Z excursions are purely positive or negative, respectively.

6.2.1 Three-Dimensional Ionospheric Current System

While qualitative sketches of the current system associated with the WTS have been proposed since the early times of substorm research (e.g., Akasofu et al., 1969; Kamide and Akasofu, 1975), the first detailed, quantitative models of the spatial electrodynamics of the WTS were developed during and after the International

[2]Equivalent currents are deduced from ground magnetic perturbations and represent currents flowing in a 2D layer of the ionosphere (see e.g., Lühr and Schlegel, 1994, for a more comprehensive description).

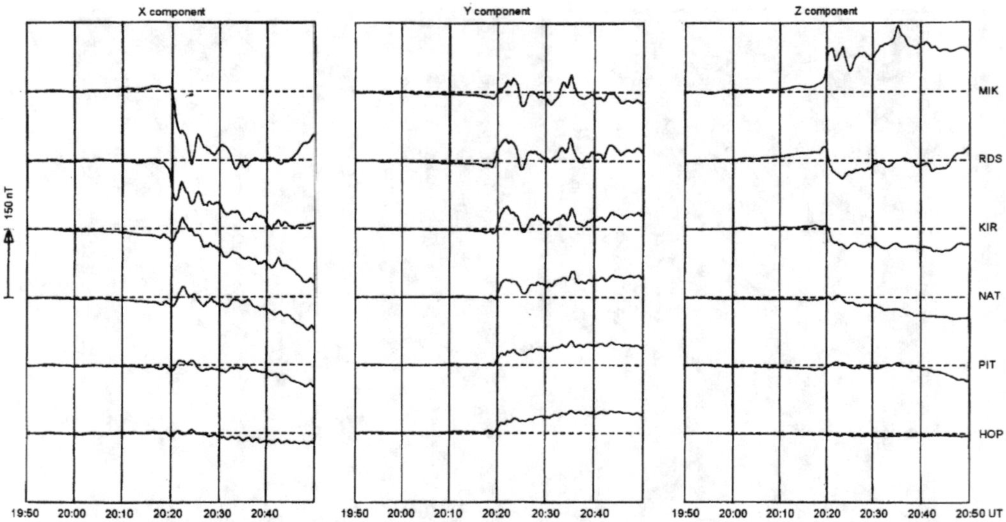

Figure 6.14. Ground magnetic signature of a WTS, recorded on September 24,1979. At about 20:20 UT, the surge passes over a meridional chain of the Scandinavian Magnetometer Array. Left panel: X (northward) component; Middle panel: Y (eastward) component; Right panel: Z (vertical downward) component. After Opgenoorth et al. (1983a).

Magnetospheric Study (IMS, 1977-1979). For this purpose, data of the SMA was combined with STARE coherent scatter radar and all-sky camera measurements (Inhester et al., 1981; Opgenoorth et al., 1983a). Additionally, during some events electric field measurements on the Barium-GEOS rocket passing equatorward of the surge and in its wake were available. These additional data were crucial, since the STARE radar can only measure electric fields above a threshold value of about 17 mV m^{-1} (e.g., Cahill et al., 1978), which is larger than the typical electric field in substantial parts of the surge and the wake, due to the electric field suppression by high conductances. Generally, northward directed electric fields equatorward of the auroral bulge and the stable arc west of it, and southward directed fields poleward of this demarcation line, were observed. In the vicinity of the surge, the electric field is directed radially towards it. While the magnitudes of these electric fields amount to 20-40 mV m^{-1}, the rocket data add the information that inside the wake, the field turns southwestwards and becomes considerably smaller, with magnitudes around 5-10 mV m^{-1}.

6.2.1.1 Model of WTS

A model of a WTS is shown in Figure 6.15 (Opgenoorth et al., 1983a). The electric field measurements were modeled as being associated with negative column charges placed at the arc westward of the WTS, in the surge itself, and along the northern and southern boundary of the auroral bulge. Additionally, weaker positive and negative column charges at the northern and southern boundary of their modeling area, respectively, represent the ambient electric field (panel A). The resulting model electric field (panel B) contains all of the features observed as described above. The authors completed their model by finding distributions of the ionospheric conductances such that the three-dimensional current system resulting from them and the given electric field produces a ground magnetic effect that agrees with the magnetometer observations as closely as possible. The resulting conductance model (panel C) shows Hall conductances of 25 S in the center of the surge and its wake, which decrease steeply at the poleward and equatorward border of the bulge. Westward of the bulge, following a similarly steep decrease, a background Hall conductance value of 1 S was assumed. The Hall to Pedersen conductance ratio modeled ranges from 1.25 in the areas of highest conductances towards 1 in the region westward of the bulge. The horizontal ionospheric current distribution (panel D) following from this model consists of purely westward directed currents with magnitudes up to 600 μA m^{-1} in the wake of the surge, that are diverted into the magnetosphere in the WTS by concentrated upward FACs with maximum current densities of 7 μA m^{-2}, carrying a total of 10^6 A out of the ionosphere (panel E). Smaller downward FACs are also present along the poleward and equatorward border of the auroral bulge, and along the stable arc in the west. Around the WTS, the horizontal currents form an anticlockwise spiral, and are dominated by Hall currents circulating around the surge due to the convergent electric field structure there. This structure determines the typical ground equivalent current pattern observed during a WTS passage as described above. The resulting current system of the WTS was interpreted in terms of a Cowling channel (see Figure 3.3) caused by the slab of enhanced conductances in the surge and wake, through which the substorm electrojet currents flow westward until they terminate and diverge back to the magnetosphere in the WTS. This essential feature is also referred to as the 'slab model'.

More direct measurements of conductances, electric fields and currents associated with the WTS became available, albeit at a single point, with the EISCAT incoherent scatter radar. The EISCAT results show that inside the surge and its wake, the conductances were underestimated and the electric field was overestimated by the earlier models presented above. An example of such measurements is shown in Figure 6.16 (Kirkwood et al., 1988). During the passage of the WTS, Σ_H shows a peak of about 60 S, while the electric fields observed are very small (< 5 mV m^{-1}). Regions in the wake of the WTS covered by discrete arcs show considerable fine structure both in the conductances and electric fields: here the largest Σ_H values of about 90 S are observed. While E is very small inside the discrete arcs, it can locally reach larger values up to about 20 mV m^{-1} in adjacent regimes. In the diffuse auroral region of the wake, which is not covered by discrete arcs, the Hall conductance is almost uniform with values around 30 S. Both in the surge and its wake, the measured Hall to Pedersen conductance ratio values reach 2-3,

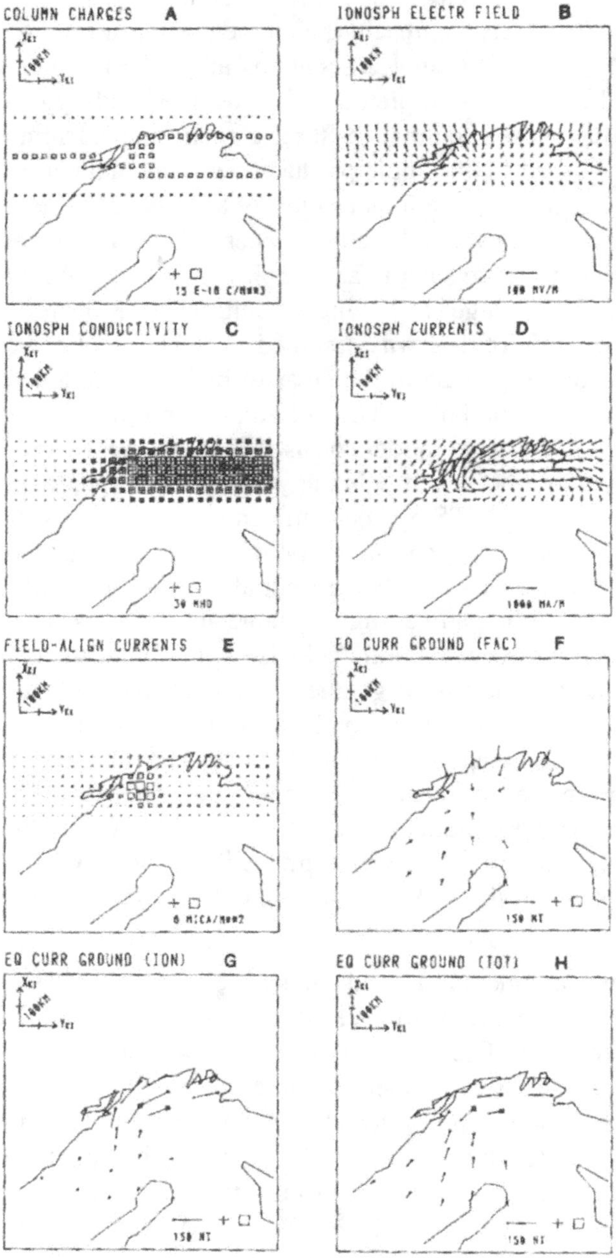

Figure 6.15. WTS model of Opgenoorth et al. (1983a) illustrating model distributions of ionospheric parameters in the vicinity of a westward travelling surge. Panel A: Distribution of field-aligned columns of positive (cross) and negative (square) charges. Panel B: Ionospheric electric field, modeled by the charge distribution shown in panel A. Panels C-E: Modeled conductances, horizontal currents, and field-aligned currents. Panels F-H: Ground equivalent currents, as caused by field-aligned currents only (F), as caused by horizontal currents only (G), and total (H).

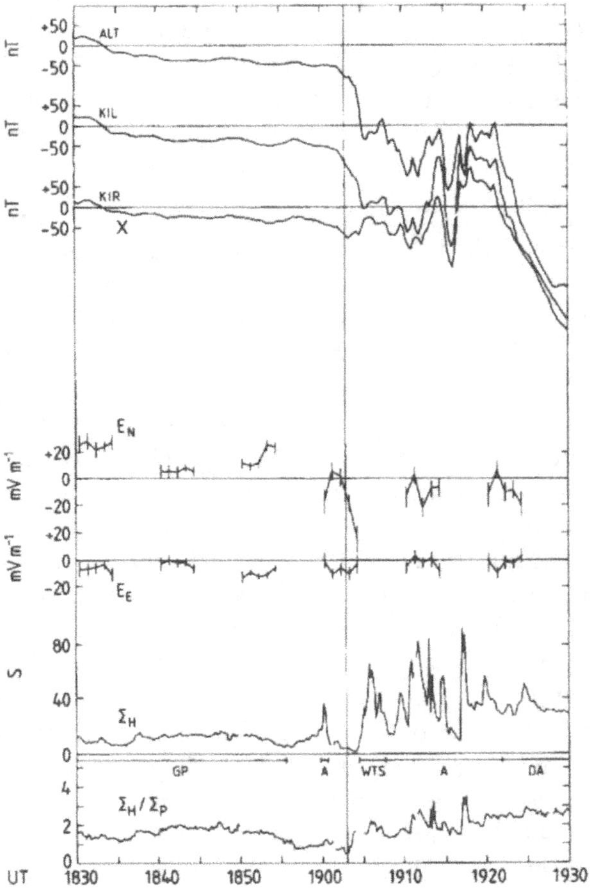

Figure 6.16. Ground magnetic records of three auroral magnetometer stations, and EISCAT measurements during a pass of a WTS at about 19:05 UT on March 25, 1986, over northern Scandinavia. From Kirkwood et al. (1988).

thus being larger than in the earlier models. In another case study using EISCAT, a sharp Hall conductance increase to about 160 S and a simultaneous increase of the Hall to Pedersen conductance ratio to a value close to 6 during the passage of a WTS, indicating very hard particle precipitation, was observed (Aikio and Kaila, 1996). Also in this study, the largest Σ_H values near 200 S were not found in the surge itself, but in the area covered by discrete arcs in its wake.

Based on these results, the WTS models have been modified by leaving their geometry essentially untouched, but changing the electric fields and conductances according to the results of the EISCAT studies (Amm, 1995). The spatial resolution of the revised model, like that of the earlier ones, is 50 km, mostly due to the limited resolution of the ground magnetometer data. Therefore, they can represent the features seen by EISCAT on this scale, but cannot resolve the details on a kilometer scale. Results of a multi-instrument study including EISCAT for a WTS passage (Lühr et al., 1998) agree with the updated model in all respects, except that the Hall to Pedersen conductance ratio in the surge was only about 1.5 in their case.

6.2.1.2 WTS Electrodynamics

With their increasing availability, space missions largely contributed to the understanding of the electrodynamics of the WTS. Satellite data from the DE 1 and DE 2 missions were used to analyze the current system of the substorm auroral bulge (Fujii et al., 1994; Hoffman et al., 1994). While the magnetometer data from the low orbiting DE 2 was used to infer the FAC, DE 1 was taking auroral images from a higher orbit. To construct a two-dimensional model from the different one-dimensional satellite paths of different events, the authors divided the bulge and the area immediately west and east of it into 6 longitudinal sectors based on the auroral images, and placed each orbit into one of these sectors. Therefore, the latudinal resolution of their model, i.e., along the trajectories of the satellites, is very high, while the longitudinal resolution is essentially given by the width of the different sectors. Figure 6.17 shows results for the FAC properties of the whole substorm bulge. Both the WTS and most of the wake area are contained in sector 3 in Figure 6.17. For the surge sector, as well as for the neighboring sectors, poleward of the bulge they find a sheet of downward FACs. These currents are connected via southward flowing Pedersen currents to a narrow sheet of upward FACs at the poleward edge of the bulge. While the total current of these two sheets almost exactly compensate each other, the average upward FAC densities, the authors inferred for the upward FAC sheet are due to its narrowness with the upward current densities larger (13.7 μA m^{-2}) than the downward current densities (8.3 μA m^{-2}), since the upward region is narrower. The main net upward FACs in the surge result from two other upward FAC sheets located a little more than 100 km equatorward of the bulge's poleward boundary: a narrow one of about 20 km width carrying intense FACs of about 12 μA m^{-2}, and a broader one of about 170 km width with moderate FAC densities of about 2.4 μA m^{-2}. In the domain of these three upward FAC sheets located at the poleward side of the surge and its wake (cf. Figure 6.17), the largest eastward plasma flows are observed.

Equatorward of those sheets, a region characterized by smaller and mixed FACs follows, in the poleward part of which the plasma flow speeds often are very small. This region is sometimes called the 'stagnation region'. Larger westward flows are again observed in the equatorward part of this mixed FAC region close to the equatorward boundary of the bulge, and in the region of downward FACs equatorward of the bulge. This latter region also contributes to the feeding of upward currents in the bulge via northward flowing Pedersen currents. The total net upward current of 118 kA that was calculated for the surge sector using the DE 1/ DE 2 results is somewhat smaller than the estimated values of the earlier models. However, the satellite-based results include net upward FACs not only in the WTS and its immediate vicinity like the earlier ones do, but also in sectors 2-4 of Figure 6.17. Generally, their results stress the importance of local, meridional ionospheric current closure in addition to the divergence of the westward substorm electrojet in the WTS and its wake. However, they do not exclude the possibility that an (incomplete) Cowling channel (see Section 3.3) is present in the poleward part of the bulge. Their analysis of single passes indicates that the current systems in the bulge are more filamentary than the continuous current sheets as assumed in the earlier models. Hall and Pedersen conductances from the electron flux spectra measured by the DE 2 satellite have been estimated by Gjerloev and Hoffman (2000), using the same data set as discussed above. The results of their surge passes generally agree with those inferred by EISCAT: in the surge's head, Hall conductance peak values above 100 S are inferred, and the Hall to Pedersen conductance ratio ranges between 2 and 3. However, satellite data indicate a more complex relation between conductances and electric fields such that areas of high conductance are not always colocated with weak electric fields.

Using their conductance model together with a model of the convection electric field derived from the same satellite data set, Gjerloev and Hoffman (2002) composed a self-consistent model of the horizontal currents and FAC (calculated from the divergence of the horizontal currents) over the whole substorm auroral bulge. For the WTS region, they conclude that the large upward FAC resulting from their model are fed by three sources: by the drainage of the westward electrojet, the drainage of the eastward electrojet via poleward flowing currents, and by local currents flowing equatorward from a sheet of downward FAC at the poleward flank of the westward electrojet. However, in contrast to the classical WTS models, the electrojet is not flowing into a single, well defined region of upward FAC, but its western termination is rather due to a continous drainage of the electrojet over an extended area in longitude, caused by an imbalance in the adjacent upward and downward FAC sheets. Although the model of Gjerloev and Hoffman (2002) has a much smaller longitudinal resolution than the classical models, and some uncertainty is introduced due to superposition of different events and smearing out caused by the statistics applied, their results strongly suggest that the WTS plays an important, but not exclusive role in diverting the substorm currents back to the magnetosphere.

While the space-borne studies discussed so far are mainly statistical in nature, Weimer et al. (1994) analyzed a single pass through the head of the surge, using the same combination of satellites as the previous authors. In contrast to them, Weimer et al. (1994) used the moments of the electron velocity distributions measured by the particle instrument of DE 2 to calculate the FACs. By a comparison of the FAC obtained along the satellite path with the auroral UV emission as seen by DE 1, Weimer et al. (1994) converted this emission to a spatial FAC density. The result-

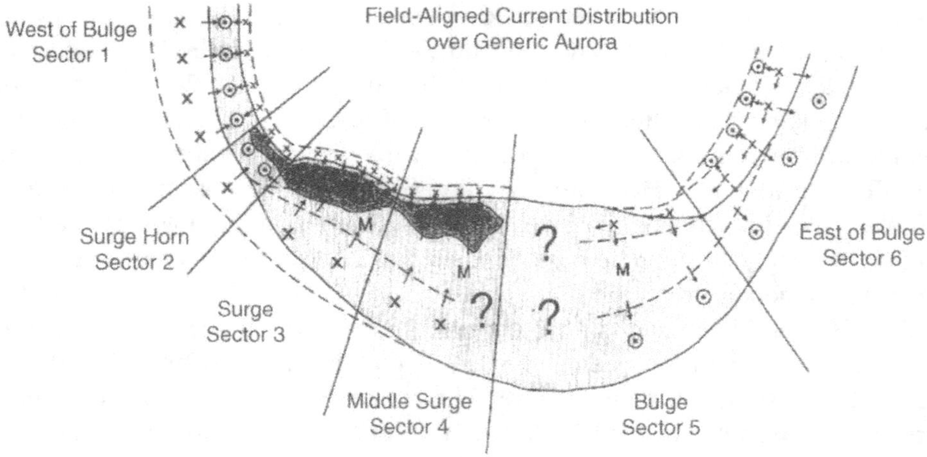

Figure 6.17. Model of the field-aligned current distribution for the entire substorm auroral bulge by Hoffman et al. (1994). The WTS is contained in sector 3. After Hoffman et al. (1994).

ing of FACs and plasma flow vectors are shown in Figure 6.18. Upward directed FACs are shown as shaded areas, where darker shading corresponds to a larger FAC density. In the northern part of the WTS, a region of highly concentrated upward FACs of up to 20 μA m^{-2} is observed. The upward FACs in the surrounding areas have smaller magnitudes of 1-5 μA m^{-2}. No downward currents were detected by the particle instrument, perhaps a result of incomplete energy coverage. In terms of the plasma flow velocities, the results of the instantaneous DE 2 satellite overpass mostly agree with the statistical results. The most poleward part of the bulge, the plasma flow is westward, whereas in the polewardmost part of the bulge, the flows turn to a southeast direction. The authors interpret this as an indication of the presence of a Cowling channel in the poleward portion of the WTS and its wake, since the corresponding southwest electric field is exactly what would be predicted in such a situation. Equatorward of this region, the flows rapidly turn to a northeast direction, followed by a stagnation region of very small flows and electric fields. Significant northwest plasma flows are seen in the eastward electrojet domain equatorward of the bulge, in agreement with the previous studies.

The analysis of the Freja satellite path through the WTS shown in Figure 6.13, reveals both the fine structure of the current systems inside the surge, and the importance of local current closure. Figure 6.19 shows the satellite's measurements between points 7 and 10 as marked in Figure 6.13. Close to the northwestern edge of the WTS between positions 7 and 8, a narrow and intense upward FAC region with a peak current density of 30 μA m^{-2} is observed (calculated from the satellite's magnetometer data under the assumption of an infinite east-west aligned current sheet). This upward FAC region is found to be fed locally by a broader and thus somewhat weaker downward FAC area immediately westward and ahead of

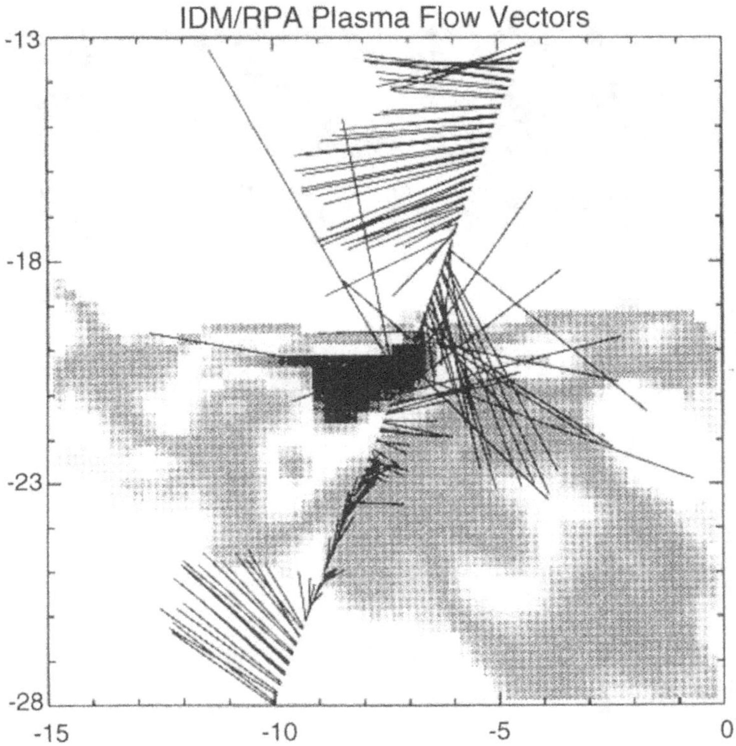

Figure 6.18. Plasma drift velocity vectors and field-aligned currents (shaded areas correspond to upward currents, darker shading means larger magnitude) obtained for a DE 2 pass overhead a WTS on September 25, 1981. After Weimer et al. (1994).

the surge which exhibits FAC densities up to 20 μA m^{-2}. In this area, the electric field points predominantly southward, with a peak value of 300 mV m^{-1} close to the bulge's boundary as measured at about 1700 km altitude. Following the satellite path southeastward from the upward FAC spike, a region of mixed but mostly upward FACs is present in which the electric field shows strong fluctuations. During the remainder of Freja's path through the WTS and its wake, the electric field is small, increasing along the path from almost zero values in the eastern part of the surge's head to values between 5 and 12 mV m^{-1} in its wake. Such a depressed electric field is a signature of large ionospheric conductances, in accordance with the EISCAT studies. In the wake, the field direction is predominantly westward. The FACs in these parts of the surge are small, except for occasional passages through discrete arcs in the wake (not seen in Figure 6.19), where downward FACs between the arcs compensate narrower upward FACs associated with them, again suggesting local closure of current filaments. In the vicinity of these arcs, convergent electric fields with peak magnitudes of 100 mV m^{-1} are measured.

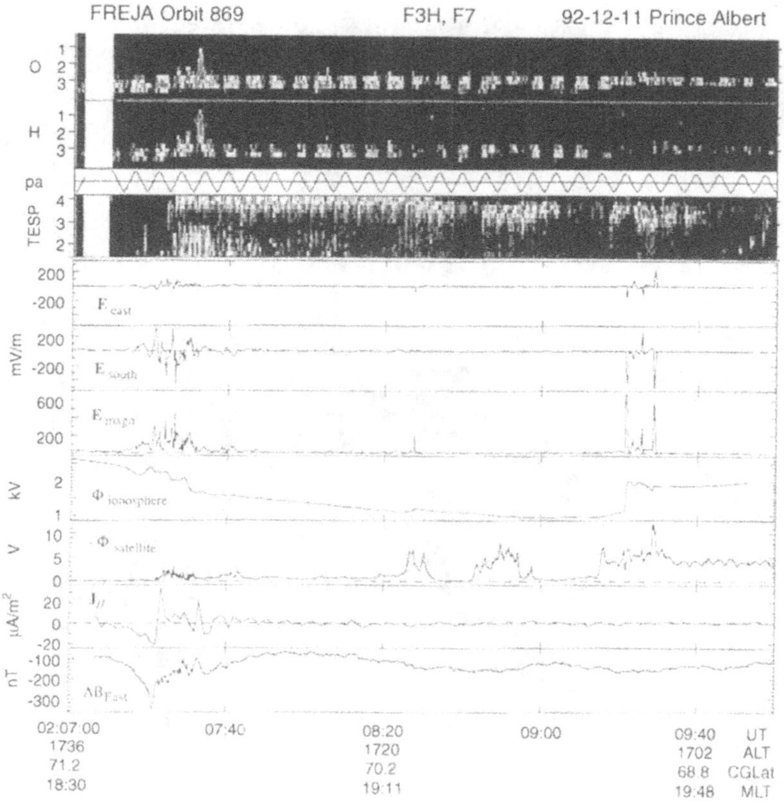

Figure 6.19. Freja energy spectra, electric field, electric potential, field-aligned current, and magnetic field data for the pass overhead a WTS shown in Figure 6.13. Around 02:07:20 UT, immediately before the satellite reaches the WTS, strong downward field-aligned currents are observed. From Marklund et al. (1998).

These results are in some contradiction to the models in which esentially the whole WTS and wake region is governed by a Cowling channel which carries the westward substorm electrojet to the head of the surge where it then diverges into the magnetosphere. However, the Freja data do not exclude the possibility that such a channel exists along the arc that extends in northeast direction from position 8 along the poleward portion of the bulge, in accordance with the DE 1/DE 2 satellite results. Generally, while the picture drawn by the classical models might still be correct if regarded as integrated over a scale size of about 100 km, the more recent studies clearly show that the smaller-scale electrodynamics of the WTS are much more dynamic and complicated, and require further research to be fully understood.

6.2.2 Associated Particle Precipitation

As can be expected from the concentrated upward FACs that were found in the head of the surge by all studies presented above, intense electron precipitation is present inside the WTS. An early study using the DMSP-32 satellite revealed that the head of the surge is characterized by a flat energy spectrum, harder than $dJ/dE \propto E^{-1}$, where J is the electron flux in $(\mathrm{cm}^2\,\mathrm{s\,sr})^{-1}$, and E the energy in keV (Meng et al., 1978). This characteristic extended up to the detector energy limit of 20 keV. Obviously, a substantial portion of the electron precipitation inside the WTS occurred above this limit. Similarly, also in the WTS studies using the Freja and the DE 2 spacecraft, it is clear that many of the precipitating electrons have larger energies than the detector limits at 25 keV (compare Figure 6.19, fourth panel). An electron flux peak at 20.3 keV was found in the head of the surge by DE 2. Ion heating is observed at the equatorward boundary of the intense upward FAC region.

In a different approach, electron spectra can be obtained by inversion of EISCAT electron density measurements (Olsson et al., 1996b). The resulting spectra (Figure 6.20, curves for 18:47:20-18:47:55 UT) confirm the main particle features inferred by earlier studies, and show the flat spectra to continue with considerable fluxes at least up to an energy of 100 keV. A shoulder is observed in the spectrum near 10 keV, i.e., in the energy range that causes strong E-layer ionization and therefore an increased Hall conductance.

The WTS as the westernmost border of the evolving substorm auroral bulge is a region in which strong upward flowing FAC are diverted into the magnetosphere. These FACs are carried by energetic precipitating electrons which cause large values of the Hall and Pedersen conductances. The FAC are fed both by horizontal currents of the westward substorm electrojet that are closed remotely, and by local current loops fed by downward FAC in the immediate vicinity of the WTS. Some important yet unresolved issues related to the WTS are the relative importance of these two feeding mechanisms, the exact role of the Cowling channel mechanism for the current transport in the region eastward of the surge, and the fine structure of the arc systems inside the surge.

6.3 Omega Bands

Auroral omega bands are periodic, wave-like undulations of the poleward boundary of the morning side diffuse aurora occurring in the recovery phase of substorms (see Figure 6.21 for a typical auroral display as seen from ground-based all sky cameras, and Figure 6.22 for a 'street' of omega bands along the post midnight sector as viewed by the DMSP F6 satellite). The name omega bands was originally chosen by Akasofu (1964) with respect to the dark areas between the auroral waves or tongues, which resemble the form of the inverted capital Greek letter Ω. The luminous tongues tend to be narrower and have larger latitudinal extent for

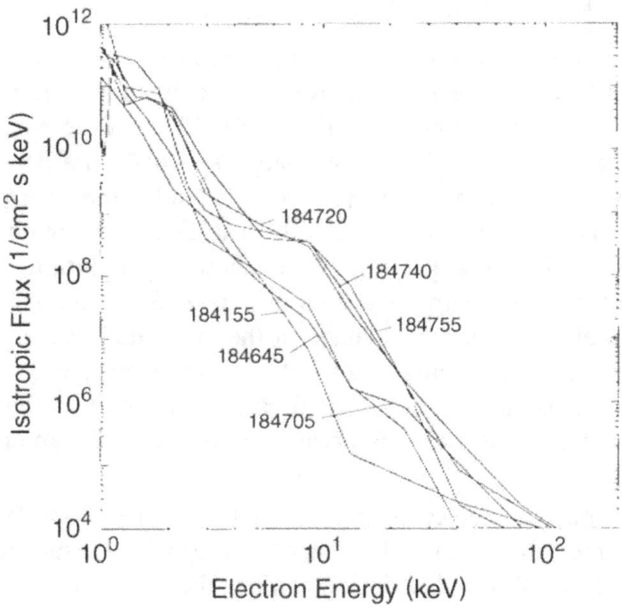

Figure 6.20. Spectra of electrons precipitating into a WTS, obtained by inversion of EISCAT electron density measurements. The resulting spectra show a shoulder at about 10 keV. From Olsson et al. (1996b).

more intense substorms, and may occasionally develop into auroral torches. Inside the tongues, which are primarily deformations of the diffuse auroral boundary, discrete aurora can be found as well, especially at the tongues' poleward boundary. Pulsating aurora has been observed inside torches. The typical longitudinal as well as latitudinal extent of the tongues amounts to about 400-500 km. Omega bands have been found to grow simultaneously over a longitudinal range of several MLT hours, and have also been observed simultaneously in both hemispheres (Mravlag et al., 1991).

The auroral forms drift eastward with velocities ranging between 400 and 2000 $\mathrm{m\,s^{-1}}$, with a tendency for increasing drift speed with time. These velocities have been found to agree closely with the local $\mathbf{E} \times \mathbf{B}$ drift velocity. During their eastward drift, the shape of the auroral forms is essentially preserved, thus indicating a stationary structure in the frame of reference comoving with the forms. A ground magnetometer positioned below the auroral omega bands which periodically pass over the station measures Ps6 pulsations, i.e., magnetic disturbances with periods of 5-40 minutes, and amplitudes that may vary from 10 to over 1000 nT. The causal relation between the auroral and the magnetic phenomena was first pointed out by Saito (1978). A typical ground magnetic signature of omega bands is shown in Figure 6.23: the pulsations are most prominent in the magnetic Y (eastward) com-

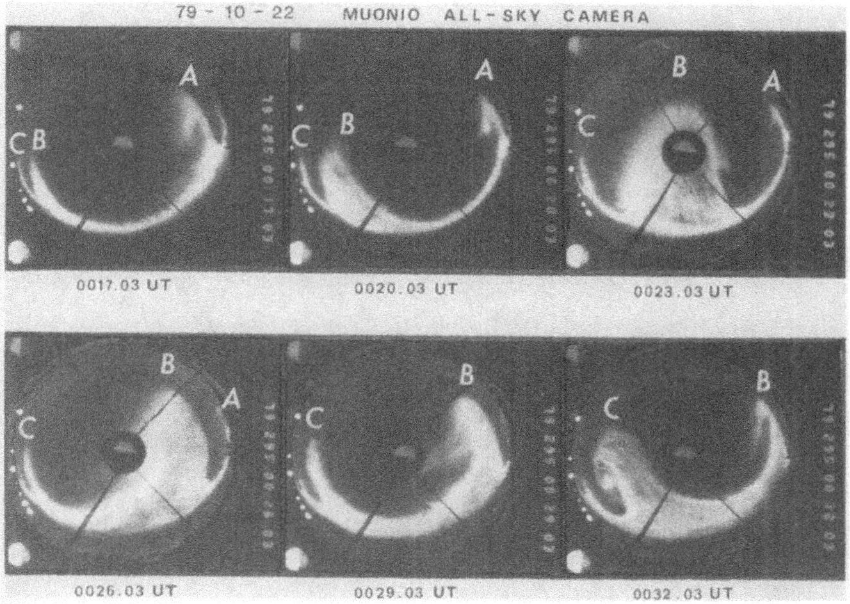

Figure 6.21. Typical auroral display of an omega band, as seen from an all-sky camera in northern Scandinavia on October 22, 1979. From Opgenoorth et al. (1983a).

ponent, in which the maxima are more pronounced than the minima. As the vertical dashed lines indicate, the maxima in Y occur between respective minima and maxima in the Z component related to the same omega band, where the Z (vertical downward) minima are leading the Y maxima by a phase shift of approximately 90 degrees. In the X (northward) component, the pulsations are also often visible, but they are typically quite disturbed due to the temporal variation of the substorm electrojet intensity during the recovery phase.

6.3.1 Three-Dimensional Ionospheric Current System

To explain the ground magnetic field disturbance associated with the omega bands, two basic three-dimensional current systems have been proposed: the first, here called 'system 1', is a north-south oriented Cowling channel (cf. Section 3.1.2) in which a concentrated horizontal currents is flowing southward (Kawasaki and Rostoker, 1979). It is fed by downward flowing FAC on the poleward side and diverted by upward flowing FAC on the equatorward side of the channel. In this model, the electric field is assumed to be uniform, but the conductances are enhanced inside the Cowling channel. In an alternate model, here called 'system 2', the conductances are assumed to be uniform, but the electric field has a negative divergence in the center of the auroral tongues and a positive one in the dark areas between them (Gustafsson et al., 1981). Consequently, the ionospheric Pedersen currents are

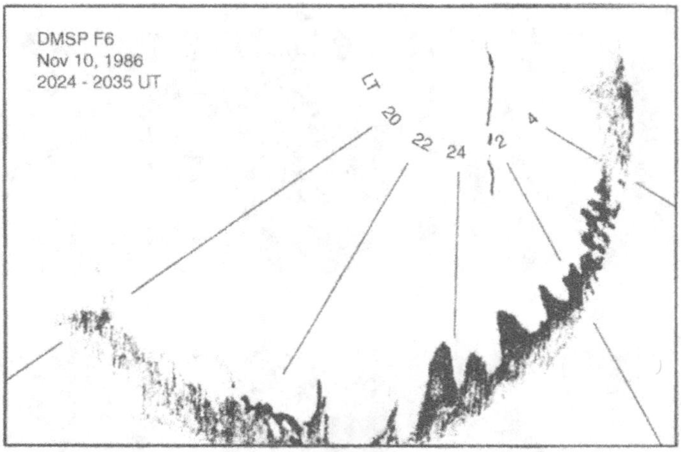

Figure 6.22. Typical display of a chain of omega bands along the midnight to morning side auroral oval, as seen from the DMSP F6 auroral imager on November 10, 1986. From Yamamoto et al. (1997).

converging or diverging in the tongues or in the dark areas, respectively, leading to upward FACs inside the tongues and downward FACs in the regions in between. Since for uniform ionospheric conductances the combined magnetic effect of Pedersen currents and FACs is invisible below the ionosphere (e.g., Fukushima, 1976), the ground magnetic effect is completely caused by the Hall currents which are circulating around the centers of the electric field divergences, counterclockwise inside the auroral tongues and clockwise in the dark auroral holes. Both current systems described are assumed to be comoving with the auroral form and have to be regarded as being superposed on the background westward electrojet.

Although the ground magnetic fields associated with the two current systems have subtle differences, it is difficult to distinguish them with a single latitudinal chain of magnetometers. A clear distinction was achieved using simultaneous two-dimensional data from the Scandinavian Magnetometer Array and the STARE coherent backscatter radar (Opgenoorth et al., 1983b). Most important, the radar electric field data showed a convergence towards the auroral tongues and corresponding divergences in the dark areas. Since the radar did not receive backscatter from the center of the tongues, Opgenoorth et al. (1983b) applied an electric field model which comprises all features of the observations. They then modeled the ionospheric conductances such that the ground magnetic field calculated from their model agreed as closely as possible with the magnetometer observations. In this process, a Hall conductance distribution was obtained that shows a maximum of about 16-20 S in the center of the tongues, radially decreasing outwards towards a background value of 4 S in the dark areas. A constant Hall to Pedersen conductance ratio of 2 was assumed. The final model composed from the SMA data is shown in

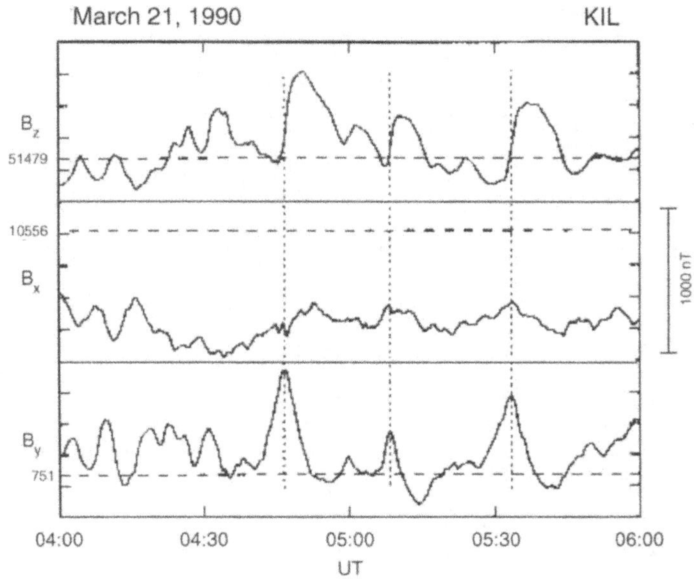

Figure 6.23. Ground magnetic signatures of three consecutive omega bands, measured by the EISCAT magnetometer cross on March 21, 1990. The typical phase shift between maxima in the Y (eastward) and X (northward) components is indicated by vertical dashed lines. From Lühr and Schlegel (1994).

Figure 6.24. The horizontal (upper panel) and field-aligned (middle panel) current distributions mainly resemble system 2. However, due to the non-uniform conductance distribution, the combined ground magnetic effect of the Pedersen currents and the FACs associated with them does not completely vanish, but results in southward directed ground equivalent currents (lower right panel) which add to the dominant equivalent current pattern due to the Hall currents and their FACs (lower left panel). While the latter is symmetrical with respect to the flanks of the tongues, the former part leads to the characteristic asymmetry that the positive magnetic Y component peak at the western flank of the tongue is more pronounced than the negative bay at its eastern flank. This asymmetry is even more apparent in the true (not equivalent) ionospheric currents. While at the eastern flank of the tongues, the northward component of the Hall currents almost equals the southward component of the Pedersen currents, at the western flank the southward components of both currents add up.

More direct conductance measurements, though pointwise, became available with the EISCAT incoherent scatter radar. With this instrument, Hall conductance enhancements of up to 50 S and Pedersen conductance peaks up to 20 S were observed in the center of the omega bands, see Figure 6.25. While these maxima occur in the center of the tongues, the strongest southward current is observed by

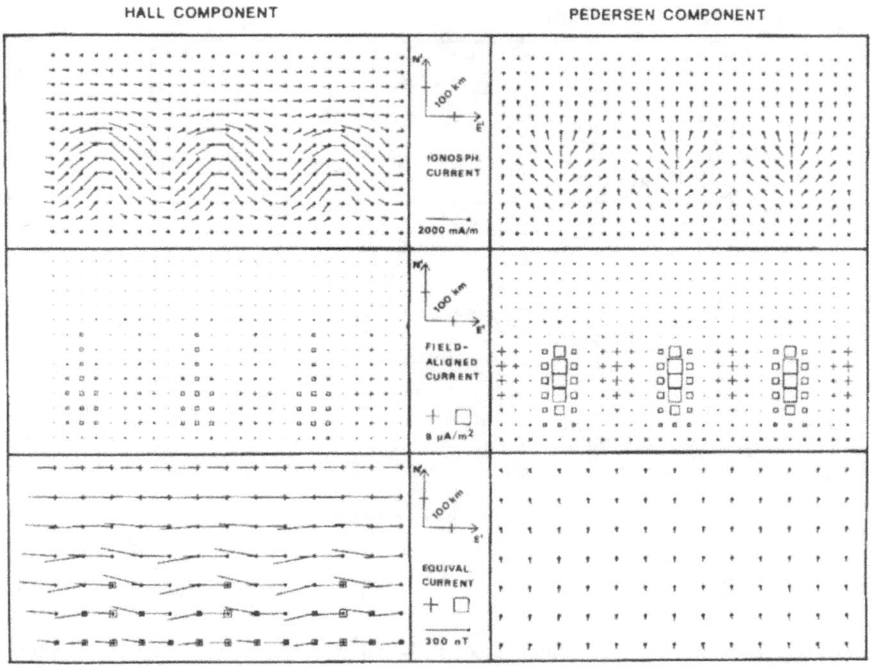

Figure 6.24. Model of three consecutive omega bands (Opgenoorth et al., 1983b). Upper panel: Horizontal ionospheric currents. Middle panel: Field-aligned currents. Lower panel: Ground equivalent currents. For each panel, the Hall component is shown on the left side, and the Pedersen component on the right side.

EISCAT a few minutes later, i.e., at the western flank of the tongues. Although the authors do not present a quantitative two-dimensional current model from the measurements, their qualitative picture consists of FACs with opposite polarity in the tongues and the dark regions with Hall currents circulating around them, consistent with system 2.

Using a multi-epoch analysis approach, pointwise conductance and electric field measurements from EISCAT and data from the EISCAT magnetometer cross were combined into a spatial model of the omega band's ionospheric electrodynamics (Buchert et al., 1990). The two-dimensional FAC distribution was modeled using empirical relations between the conductances measured by the radar and enhanced upward FACs. Finally, Ohm's law, combined with the condition of current continuity, was solved for the electric field. The model was composed such that it reproduces both the EISCAT measurements of E and $\Sigma_{H,P}$ and the ground magnetometer data satisfactorily. A uniform Hall to Pedersen conductance ratio of 3.2, which is the measured value in the center of the tongues, was chosen. The resulting model, which comprises only the auroral tongues but not the dark areas, is shown

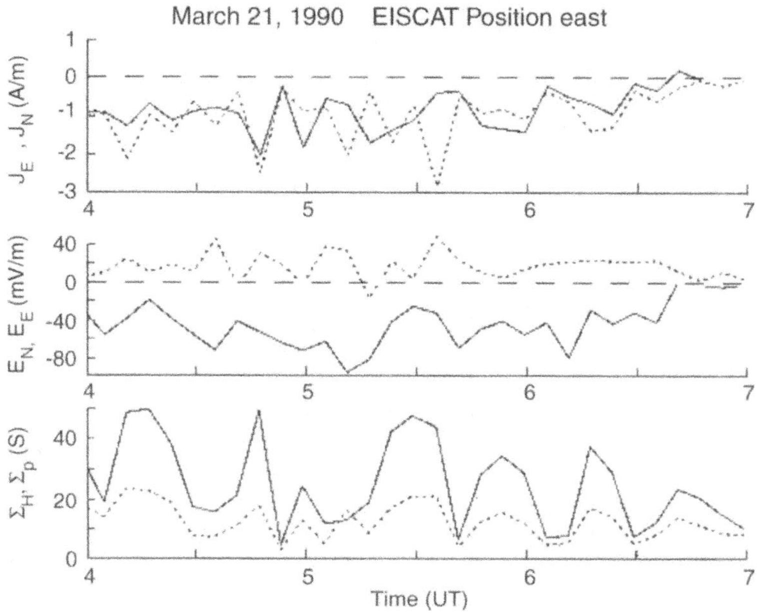

Figure 6.25. EISCAT measurements of currents, electric fields, and conductances, during the pass of several omega bands in March 21, 1990. Solid and dashed lines refer to the first and second parameters respectively. From Lühr and Schlegel (1994).

in Figure 6.26. Also in these results, the most intense upward flowing FACs are located in the center of the tongues, consistent with system 2, having peak magnitudes of about 1.8 μA m^{-2}. Equatorward of the tongue, upward FACs of up to 1.2 μA m^{-2} associated with the undisturbed band of diffuse aurora are seen, and weak downward FACs exist near the poleward boundary of the model area. The conductances peak in the center of the auroral omega bands, with Σ_P values of about 20 S and corresponding Σ_H values of about 60 S, in good agreement with the results of Lühr and Schlegel (1994). In the diffuse auroral band equatorward of the omega band, the peak conductances are only little smaller. The characteristic patterns of the Hall, Pedersen, and total ionospheric currents mainly agree with the results that were obtained in the SMA period. Note that the center of convergence of the electric field in the EISCAT-based model is not exactly in the center of the auroral tongue, but displaced eastwards, so that eastward electric fields and Pedersen currents are present in the largest part of the tongue.

A more realistic, spatially varying distribution of the Hall to Pedersen conductance ratio was added to this model by Amm (1996). They analyzed the resulting current system in terms of curl-free and divergence-free current parts. The separation of the total ionospheric current system into the curl-free (J_{cf}), divergence-free (J_{df}), and Laplacian parts (J_{lap}) has two main objectives: 1) Dividing up cur-

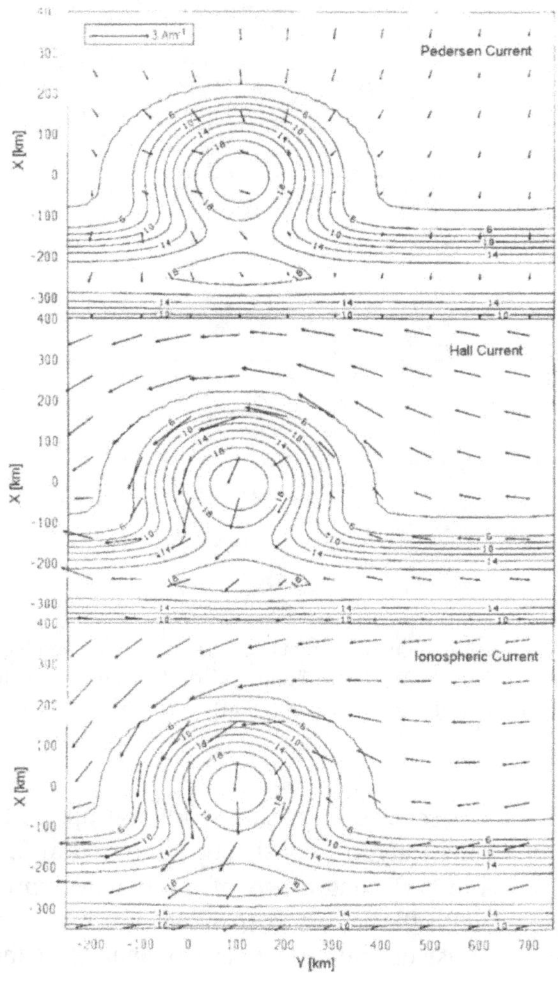

Figure 6.26. Omega band model of Buchert et al. (1990). The model is composed for an auroral tongue only, not for the dark parts between the tongues. Upper panel: Pedersen currents. Middle panel: Hall currents. Lower panel: Total horizontal currents. Isocontour lines in all panels: Pedersen conductance.

rents caused by distant sources (Laplacian part) from those which are due to local sources inside the analysis region (curl-free and divergence-free parts), and 2) Dividing up the horizontal currents that are connected to field-aligned currents and together with them do not cause any magnetic effect on the ground (curl free part) from those which close within the ionosphere and cause, together with the mostly almost uniform contribution from the Laplacian part, the ground magnetic disturbance (divergence-free part).

Figure 6.27 shows the Pedersen (left side) and Hall (right side) parts of those currents that do not have either curls or divergences in the model area (J_{lap}, first panel), which have finite curls, but vanishing divergence (J_{df}, second panel), which have finite divergences, but zero curls (J_{cf}, third panel), and of the FACs (fourth panel). Note the different vector scalings and the fact that the J_{cf} currents are magnetically invisible on the ground since their effect cancels out with that of the FACs, independently of the conductance distribution (Amm, 1997). The background westward electrojet with its westward flowing Hall and southward flowing Pedersen currents are represented in the J_{lap} part. The most intense current part is $J_{df,Hall}$ which again has the geometry as expected for the Hall currents associated with system 2. However, in the region around x=0 km, y=50 km at the western flank of the tongue, the J_{cf} parts of the Hall and Pedersen currents are pointing in opposite directions canceling each other out almost exactly, while the Hall and Pedersen parts of J_{df} are flowing nearly parallel, and both perpendicular to J_{cf}. This is exactly the situation to be expected for an northwest-southeast aligned Cowling channel at the western flank of the auroral tongue, which is, however, not fed and diverged by FACs at its northwestern and southeastern edges, but caused by a spatial concentration of the J_{df} currents in the channel. The FACs at the sides of the channel can be seen in the fourth panel: While the oppositely directed FACs associated with Hall and Pedersen currents at the southwestern part of the tongue cancel each other out, at the other flank of the channel the FACs associated with it are colocated with the FACs due to the converging E structure in the center of the tongues, thus leaving a net upward FAC there. In summary, this detailed analysis indicates that the main current system associated with the omega bands is of system 2 type, but at its western flanks, locally a system 1 type current configuration exists.

A recent case study using the SuperDARN and STARE radar data, together with ground magnetometer data, confirms the previous results in that the largest southward flowing horizontal currents and upward flowing FACs are located at the western flank of the omega band (Wild et al., 2000). In addition, it was found that the downward flowing FACs between the tongues are not symmetrically organized, but are most intense just outside of the tongue's eastern flank, being aligned with the boundary of the auroral luminosity.

The essential FAC structure found in the studies mentioned above, as well as the auroral morphology of the omega bands, was reproduced by a two-dimensional particle simulation of Yamamoto et al. (1997), in which the interchange instability acts as a primary generator of the omega band formation.

6.3.2 Particle Precipitation

Figure 6.28 shows a typical result of an analysis of DMSP-F6 satellite data passes over the morningside auroral oval, and a comparison of the precipitating electron fluxes and spectra for situations with and without omega bands (Lyons and Fennell, 1986). The diffuse aurora equatorward of the omega band shows hard spectra, i.e.,

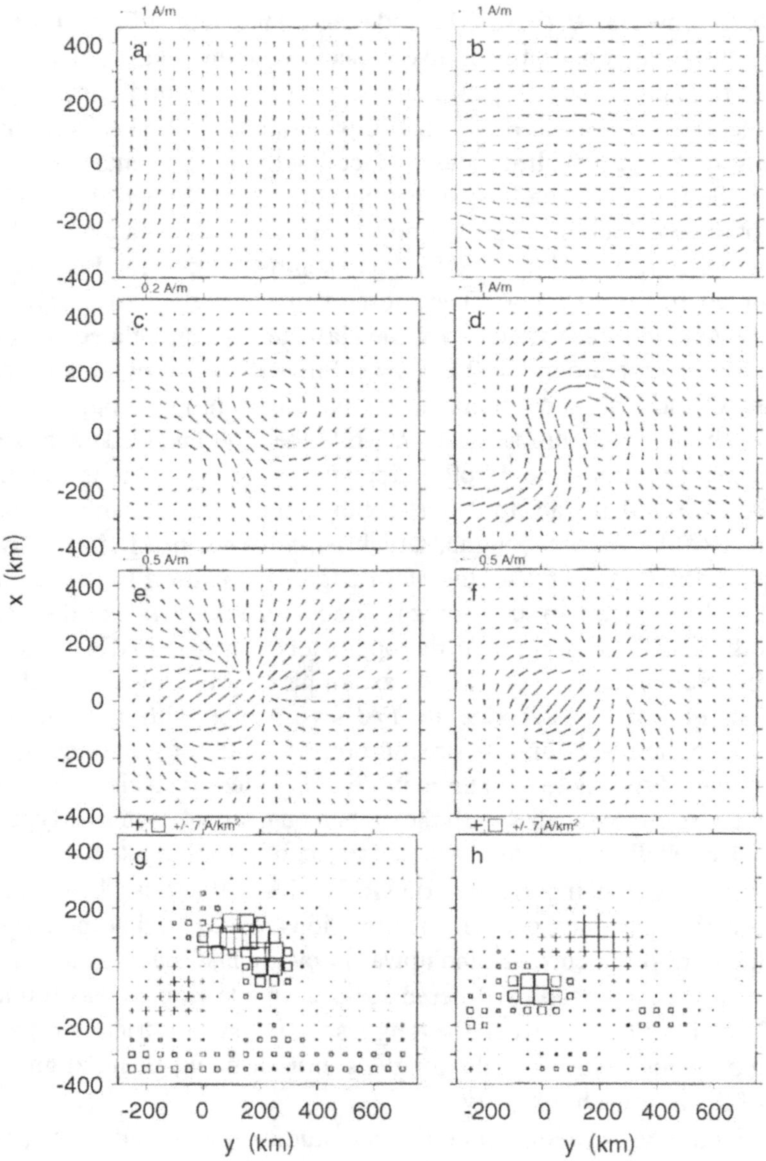

Figure 6.27. Decomposition of the currents of the omega band model by Amm (1996) into the part without divergences or curls in the modeling area (Laplace field, corresponding to the background electrojet; panels a and b), the part caused by curls in the modeling area only (panels c and d), the part caused by divergences in the modeling area only (panels e and f), and the field-aligned currents (panels g and h). Left column: Pedersen part of each current system, right column: Hall part. For more details see text.

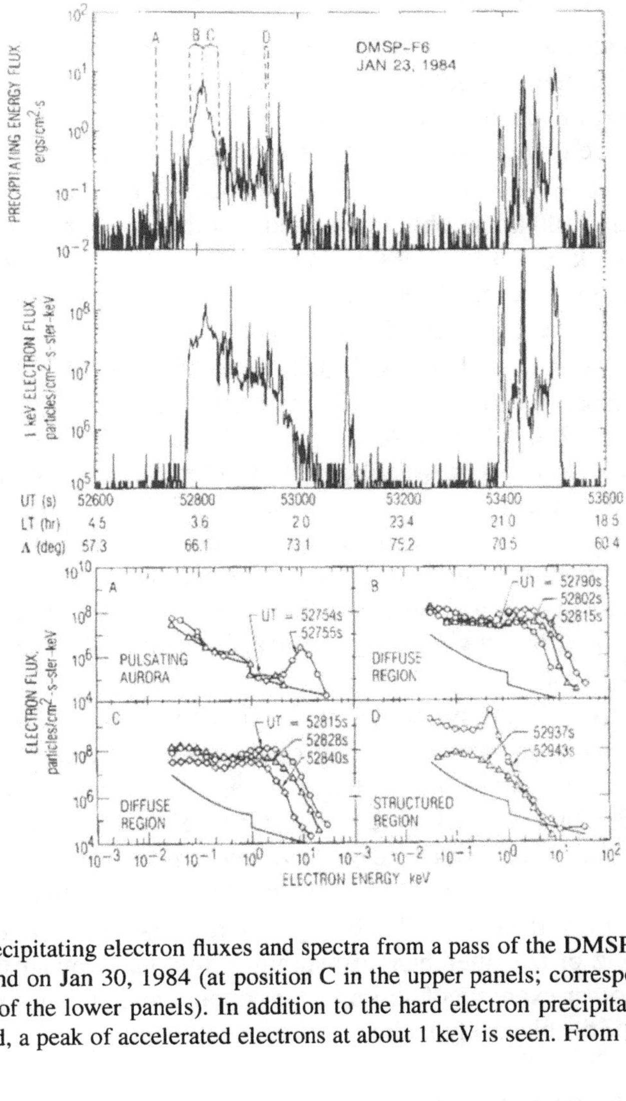

Figure 6.28. Precipitating electron fluxes and spectra from a pass of the DMSP-F6 satellite overhead an omega band on Jan 30, 1984 (at position C in the upper panels; corresponding spectra are shown in panel C of the lower panels). In addition to the hard electron precipitation of the diffuse auroral background, a peak of accelerated electrons at about 1 keV is seen. From Lyons and Fennell (1986).

a remarkable decrease in the spectral components is only observed above about 10 keV. Large total electron fluxes of about 5-10 mW m^{-2} are measured, but no signs of electron acceleration are apparent. On the other hand, the structured aurora poleward of the diffuse auroral belt is associated with acceleration peaks at about 1 keV. Inside the omega bands, the properties of both regions are combined such that on top of the hard diffuse auroral precipitation background, acceleration-related spectral peaks at about 1 keV are measured, consistent with the upward flowing FACs in the center of the auroral tongues.

6.3.3 Conjugate Magnetospheric Observations

Only a few simultaneous conjugate ionospheric and magnetospheric observations of omega band-related phenomena are available so far. Mapping of ionospheric observations into the magnetotail indicates magnetospheric source regions of the omega bands close to the geostationary orbit (e.g., Pulkkinen et al., 1991).

The Geotail satellite observed quasiperiodic oscillations of the plasma flow with amplitudes around 100 km s^{-1} both in v_Y and v_X close to the mapped location of an omega band (Pulkkinen et al., 1998). A possible explanation of these observations is that an expanding plasma sheet causes large-scale oscillations seen within both the plasma sheet and the morningside ionosphere.

Recently, an event where both the GOES 7 and SCATHA satellites were located close to the mapped location of an omega band at about 6 MLT was studied by Jørgensen et al. (1999). While SCATHA, located at about 6 R_E distance, was measuring magnetic field pulsations of about 10 min period, in which the field configuration switched between dipolar and tail-like, GOES7 at 6.6 R_E did not measure any pulsations and remained in a tail-like magnetic field configuration throughout. The authors interpret their measurements in terms of a corrugated current sheet near the geostationary orbit. While SCATHA periodically crosses the inner edge of this current sheet, the slightly more tailward located GOES 7 remains always inside the current sheet.

The ionospheric current systems associated with auroral omega bands can be described as meandering westward electrojets flowing at the poleward border of the undulating diffuse aurora. At the western flank of the auroral tongues, a north-south aligned Cowling channel current is superposed on this configuration, which leads to an asymmetric pattern of the electric fields and currents inside the tongues, as well as of the ground magnetic disturbance they cause. While theoretical treatments of the growth mechanism for omega bands exist which mainly invoke the Kelvin-Helmholtz or the interchange instability, this mechanism is still not fully understood. Another example of an important open question is the possible role of neutral winds for the electrodynamics of the omega bands.

6.4 Auroral Streamers

Auroral streamers are finger-like, mainly north-south-aligned auroral forms that evolve at the poleward boundary of the auroral oval from where they elongate equatorward. When the streamers reach the oval's equatorward boundary, they often develop a hook-like shape at their equatorward edge, before they finally decay into a bright patch of diffuse aurora (see Figure 6.29 for a typical sequence of evolution).

Streamers have been found to be a feature of a wide (about 6-10°) auroral oval, occurring both during substorm and non-substorm periods. In both cases, they have been reported as ionospheric signatures of bursty bulk flows (BBFs) in the central plasma sheet of the magnetospheric tail (e.g., Henderson et al., 1998; Lui et al.,

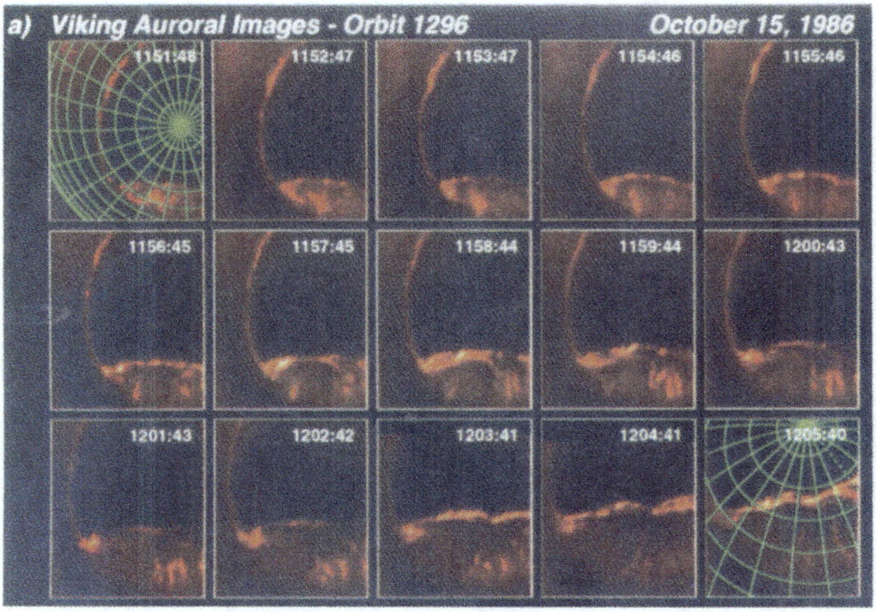

Figure 6.29. Evolution of several auroral streamers inside a substorm auroral bulge, as seen from the Viking auroral camera on October 15, 1986. From Henderson et al. (1998).

1998; Sergeev et al., 1999; Lyons et al., 1999b, see also Section 8.3.3). Although the north-south auroral forms now denominated as streamers have been known since the 1960's (e.g., Akasofu, 1968), only the discovery of their connection to BBFs led to a number of more detailed studies on streamers. Therefore, the results this section tries to summarize are based on a substantially smaller basis of studies than are available for the WTS, or omega bands (Sections 6.2 and 6.3). Consequently, there are more yet unresolved issues related to streamers, as compared to the other auroral forms mentioned. As an important example, it is not yet clear whether the streamers' ionospheric electrodynamics are similar during substorm and non-substorm background conditions. There exist phenomenological differences between the streamers in both situations. Therefore we treat 'substorm streamers' and 'non-substorm streamers' separately. Substorm streamers have also been called 'auroral fingers' and 'N-S aurora', while the phenomenon called 'polar boundary intensifications' (PBI) can be regarded as similar to non-substorm streamers.

Substorm streamers occur during the expansion phase of substorms inside the substorm auroral bulge, where in many cases several streamers evolve simultaneously with an azimuthal separation of 150-500 km, as is the case in the example in Figure 6.29 (for another substorm-streamer example, see Figure 6.13, close to the right border). The first streamers are typically observed about 10 minutes after the

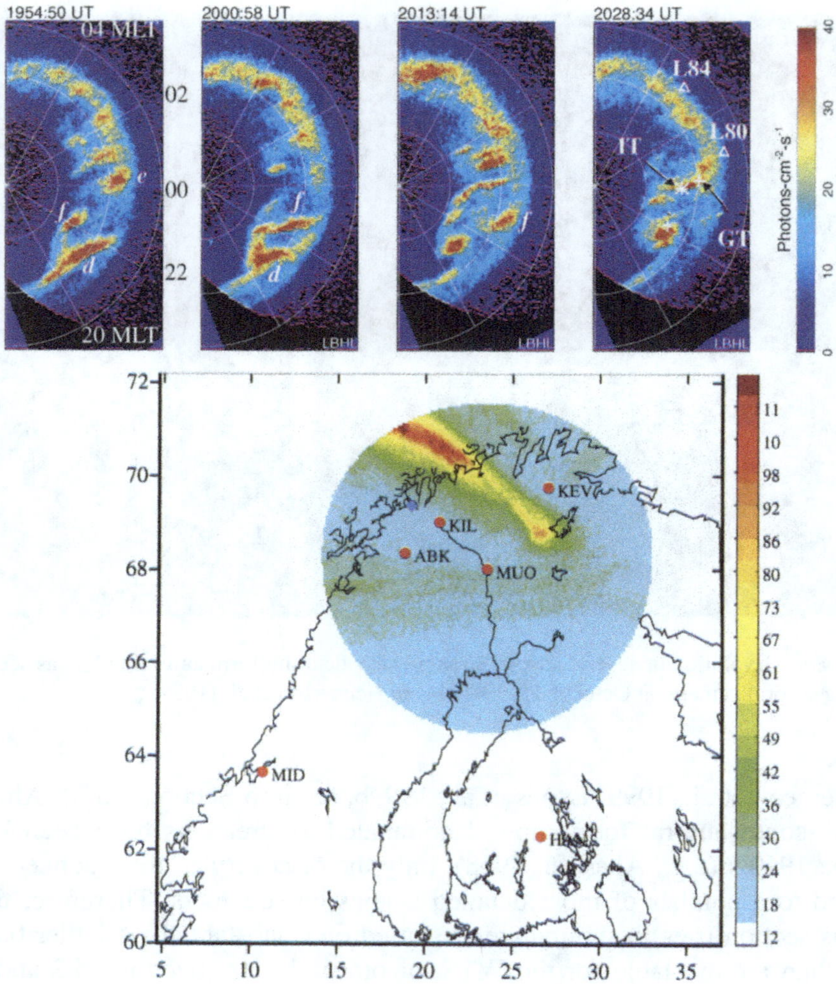

Figure 6.30. Polar UVI image of several auroral streamers during a period of steady magnetospheric convection on December 10, 1996, without substorm activity (upper panel). Streamer d of the upper panel, as seen from an all-sky camera in northern Scandinavia (lower panel).

breakup, and their appearance follows a recurrence interval of 5-15 min. Substorm streamers are most commonly observed in the 21-23 MLT sector, and typically show a westward motion of about 1.5 km s^{-1}.

In contrast, the occurrence of non-substorm streamers appears to be evenly distributed in the 20-03 MLT sector, and no systematic preferred direction of motion has been inferred by the event studies hitherto carried out. An example of non-substorm streamers distributed over the whole auroral bulge during an steady magnetospheric convection period is shown in Figure 6.30 (top). They may sometimes deviate from the meridional orientation by more than 45 degrees (see, e.g., streamer d in Figure 6.30), so its distinction from an auroral arc is not straightforward, especially not from all-sky camera images alone (compare 6.30 (bottom) which shows an ASC image of a part of streamer d mentioned above, taken in Northern Finland). As Figure 6.30 shows, during the convection bay that lasted 1 hour, a few dozen of streamers formed the dominant dynamical auroral pattern. For one of the streamers, the association with a BBF was directly confirmed by observations of three magnetospheric spacecraft (Interball, Geotail, and the geostationary LANL 1991-080).

Both streamer types have in common that their latitudinal extent of more than 500 km is much larger than their longitudinal extent, which is typically of the order of 100-150 km, but can be as narrow as 20 km. The dominating component of their motion is equatorward with velocities of 1-5 km s^{-1}.

6.4.1 Ground Magnetic Signatures

Figure 6.31 shows two examples of the two-dimensional ground magnetic signature of non-substorm streamers (from events on December 21-22, 1995), during which the Wind satellite was on magnetic field lines conjugate to the IMAGE magnetometer chain, and measured plasma parameters in the tail (Kauristie et al., 2000). The equivalent current vectors shown are calculated with respect to a baseline just before the streamer events, so that the background magnetic activity does not show up in the vectors. The Wind magnetic footpoint in the ionosphere is marked by an asterisk. During the event at 00:16 UT (Figure 6.31, right panel), Wind monitored a strong Earthward plasma flow burst in the tail, indicative of a satellite position close to the center of a bursty bulk flow. On the ground, an counterclockwise equivalent current vortex is seen, with a negative z (vertical downward) magnetic component. The satellite footpoint is located right inside this vortex. In contrast, during the event at 23:59 UT (left panel), Wind measured a tailward flow burst, as expected for a satellite position in the return flow area just outside an Earthward bursty bulk flow. For this event, the IMAGE magnetometers showed a clockwise equivalent current vortex with a positive magnetic z component in its center, with the Wind footpoint located at the eastern edge of this vortex. From these and other similar events it can be concluded that the ground magnetic signature of non-substorm streamers is a double vortex in the equivalent currents, with the clockwise vortex on the eastern and the counterclockwise one on the western side (see Figure 6.35, panel c). This double vortex has been interpreted as an image of a similar double vortex convection structure in the equatorial plane of the magnetospheric tail associated with the plasma flow bursts there (Figure 6.35, panel a). Non-substorm streamers have also been observed during geomagnetically almost quiet periods (Lyons et al., 1999b).

The ground magnetic signature of substorm streamers can easily be covered by the dynamic evolution of the substorm and its current wedge. However, a number of studies agree in that a negative bay in the magnetic X component (positive north-

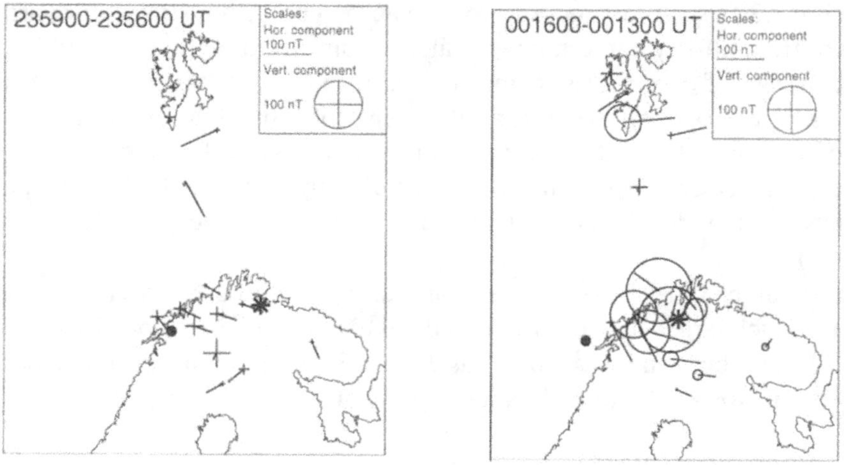

Figure 6.31. Differential equivalent currents (with respect to a baseline shortly before the events shown) of two non-substorm streamer events on December 21-22, 1995, recorded with the IMAGE magnetometer network. The vertical downward magnetic component is marked by crosses (positive) or circles (negative), respectively. The magnetic footpoint of the Wind satellite is marked by an asterisk. From Kauristie et al. (2000).

ward) is associated with the passage of an auroral streamer, i.e., an intensification of the westward substorm electrojet (Lühr et al., 1998; Amm et al., 1999; Solovyev et al., 1999). In all these studies, the most obvious ground magnetic effect has been observed in the Y component (positive eastward). Figure 6.32 shows the magnetic signature of a street of streamers passing westwards over the IMAGE magnetometer network as composed into a set of double vortices for each streamer. Positive Y components (or southward equivalent currents) are associated with the auroral forms, while negative Y components (northward equivalent currents) are observed in the area between two successive auroral streamers (see the J_{SF} vectors for the equivalent currents). A minimum in the Z magnetic component is observed at the eastern edge of the auroral form.

Pulsations in the Pi2 and Pi3 range have been found to be associated with both non-substorm and substorm streamers (Kepko and Kivelson, 2000; Lyons et al., 1999c; Solovyev et al., 1999).

6.4.2 Current Systems

For non-substorm streamers, no detailed studies of the ionospheric electrodynamics including distributions of conductances, true ionospheric currents, and field-aligned currents, have been conducted yet. For a substorm streamer, such an analysis has been carried out using combined data of the STARE radar, the Scandinavian Magnetometer Array, and all-sky cameras for an event on December 2, 1977, at 2200 UT above northern Scandinavia (Amm et al., 1999). This day was charac-

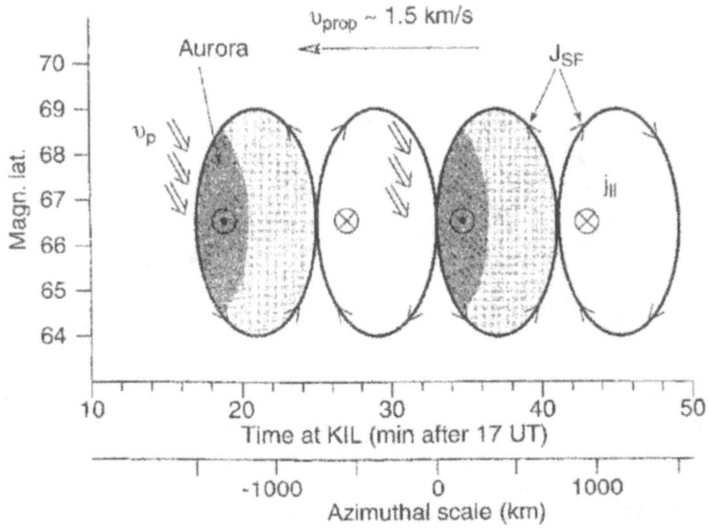

Figure 6.32. Model of the electrodynamics of auroral streamers, composed from an event study by Lühr et al. (1998). The region of the auroral forms is shaded dark, the ionospheric plasma flow is shown by open arrows, the equivalent currents (denoted as J_{sf}) by dark elliptical arrows, and field-aligned currents by encircled crosses (downward currents) and circles (upward currents). From Lühr et al. (1998).

terized by very strong geomagnetic activity, with a Kp value of 7 from 15-24 UT. Several substorm onsets with ground magnetic disturbances of more than 1000 nT took place, during which a number of auroral streamers with strong ground magnetic signatures of the order of 500 nT appeared. The results of the analysis are shown in Figure 6.33, where the shaded area close to the western boundary of the analysis area marks the position of the auroral streamer. Colocated with the auroral form is a strip of very large Hall conductance values of about 140 S, with a peak of over 200 S in the center of the streamer. Eastward of it, the conductance decreases gradually until it reaches values around 30 S at in the eastern part of the analysis area, which are typical for the background westward substorm electrojet. Since the electric field measured by the STARE radar is pointing almost uniformly southwestward throughout the analysis region, and is only slightly suppressed in magnitude inside the streamer, the horizontal ionospheric currents are strongly enhanced in the strip of enhanced conductances, with a maximum sheet current density of 3500 mA m^{-1} at the conductance extremum. These enhanced horizontal currents are fed by a sheet of downward flowing FACs eastward of the streamer, and diverted back to the magnetosphere by an upward FAC sheet at the streamer's western boundary. While the downward FACs are more distributed, reaching maximum current densities of 15 μA m^{-2}, the upward FACs are highly concentrated, with a very large maximum current density of 25 μA m^{-2}. The cross-structure size of the

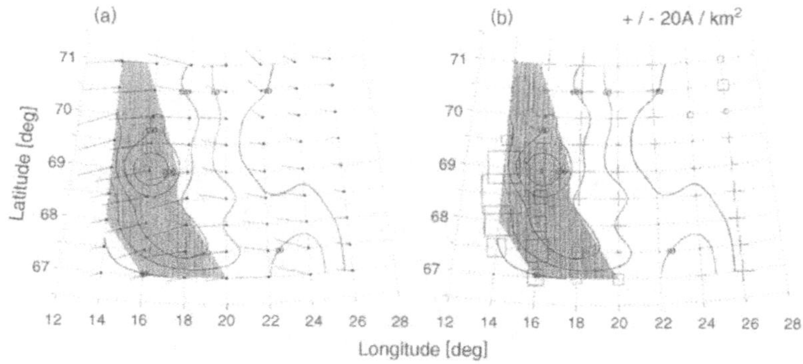

Figure 6.33. Horizontal ionospheric currents, field-aligned currents, and Hall conductance for a substorm streamer event on December 2, 1977, at 2200 UT. In both panels, the position of the auroral streamer is marked by the shaded area, and the Hall conductance (in S) is shown by iso contour lines. Left panel: Horizontal ionospheric currents. Right panel: Field-aligned currents (crosses denote downward currents, squares upward currents). From Amm et al. (1999).

FAC system is about 250 km. The two sheets of FACs are not exactly north-south aligned, but appear to be slightly tilted into a northwest-southeastern direction.

In order to examine how the typical vortical ground magnetic signature of the streamer is produced, the resulting total horizontal current system (Figure 6.33) has been split up into the parts which are caused by divergences and curls, respectively, inside the analysis region (curl-free part and divergence-free part), and the part of the total current that has neither curls nor divergences inside that area (Laplace part), similar to what was done on page 287. The result shows that the Laplace part represents the mainly uniform background westward substorm electrojet (Figure 6.34a). The curl-free part is dominated by the strong southwestward current flow between the two FAC regions mentioned above, constituting the strip of enhanced current densities inside the streamer (Figure 6.34b). This current part, together with the field aligned connecting current, does not cause any magnetic effect below the ionosphere and is therefore invisible to magnetometers alone. In addition to the background electrojet, the ground magnetic effect is produced by the divergence-free currents (Figure 6.34c). They clearly show a vortical structure in the southern part of the streamer, as consistent with the observations, and also reproduce the typical northward equivalent current component (equal to a negative ground magnetic Y component) eastward of the auroral form, in the area between successive streamers. Thus, in the presence of strong ionospheric conductance gradients, equivalent current vortices on the ground can be produced even in case of an almost uniform electric field.

The results of this detailed analysis qualitatively agree with the findings of an earlier study using the IMAGE magnetometer chain, the EISCAT incoherent scatter radar, and all-sky cameras for their analysis of two substorm streamer events on

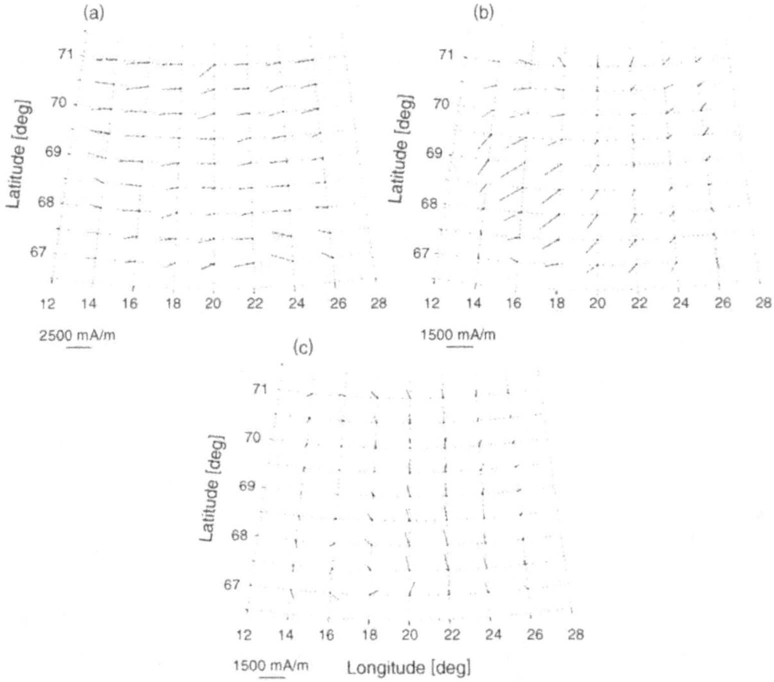

Figure 6.34. Decomposition of the ionospheric currents of the auroral streamer shown in 6.33 into the part without divergences or curls in the modeling area (Laplace field, corresponding to the background electrojet; upper left panel), the part caused by curls in the modeling area only (lower panel), and the part caused by divergences in the modeling area only (upper right panel). Since the ground magnetic effect of the currents shown in the upper right panel is canceled out by field-aligned currents, most of the streamer's current system is invisible by ground magnetometers alone. From Amm et al. (1999).

October 11, 1993, between 17 and 18 UT (Lühr et al., 1998). Similar to the STARE radar in the study discussed above, EISCAT also observed almost uniformly southwestward directed electric fields, equivalent to southeastward directed flows, during both streamer events, with only some suppression of the electric field magnitude during the passage of the auroral forms. Simultaneously, after subtraction of the background electrojet, IMAGE showed alternating clockwise and counterclockwise equivalent current vortices on the ground. These results are comprised into the sketch shown in Figure 6.32. Upward FACs are attributed to the auroral form which is located on the western edge of the counterclockwise equivalent current vortices (J_{sf} in their sketch). Downward FACs are located in the region of northward equivalent current at the western flank of the clockwise equivalent current vortices. The estimated cross-structure size of the FAC system, about 750 km, is considerably larger than the one obtained by the SMA/STARE analysis.

6.4.3 Role of Bursty Bulk Flows

Bursty bulk flows (BBFs) themself as the magnetospheric side of the streamer phenomenon are treated in Section 8.3.4. In this section the above findings are briefly compared with the predictions of a model for the ionosphere-magnetosphere coupling of the BBFs. The most frequently used model is the 'bubble model' originally proposed by Chen and Wolf (1993), which has found observational support by a number of later studies (e.g., Sergeev et al., 1999, 2000a; Kauristie et al., 2000). These authors have illustrated the ionosphere-magnetosphere coupling using the bubble model as shown in Figure 6.35. The BBF is a narrow channel with depleted plasma density and pressure ('bubble'), but increased magnetic pressure. As the gradient and curvature currents are smaller inside the bubble, the cross-tail current is inhibited. Thus space charges of opposite sign will accumulate at the dawnward and duskward sides of the bubble. This leads both to a double vortex convection pattern with Earthward flow inside the bubble and tailward return flow at its flanks, and to a pair of FACs of opposite sign in the centers of the convection vortices at the bubble's boundaries (Figure 6.35a; for more details see Section 8.3.4). Mapping this FAC pair to the ionosphere and assuming uniform conductances, an azimuthally separated double vortex of ionospheric convection (Figure 6.35b) is expected, which leads to an oppositely oriented pair of Hall current vortices. In this model, these vortices are responsible for the observed ground magnetic effect (Figure 6.35c).

However, this ionosphere-magnetosphere coupling model for auroral streamers differs from the observations in that the ionospheric convection does not show a double vortex structure, but is rather uniformly southeast directed, as the studies presented above for which ionospheric electric field data was available show. Consequently, the Hall currents also do not follow a double vortex pattern. This observed convection signature is not inconsistent with the ground magnetic observations of equivalent current vortices, since the vortical structure is caused by gradients of the conductance.

The ionospheric current system associated with auroral streamers can be essentially described as a small current wedge with westward flowing horizontal currents, similar to the substorm current wedge, but with a scale size of only a few hundred kilometers. The upward FACs of this wedge are located at the north-south auroral form, and are particularly concentrated at its western border, while the downward FACs are more distributed and located eastward to northeastward of the auroral form. Since the study of auroral streamers is a young subject, the subdivision into substorm and non-substorm streamers is still tentative. The data available indicates that for substorm streamers, the ionospheric convection does not change its configuration toward a double-vortex structure, as one would expect from a direct mapping of the well-established convection pattern in the magnetosphere. It is unclear yet why the convection patterns in the magnetosphere and in the ionosphere are decisively different. Among the many other open questions

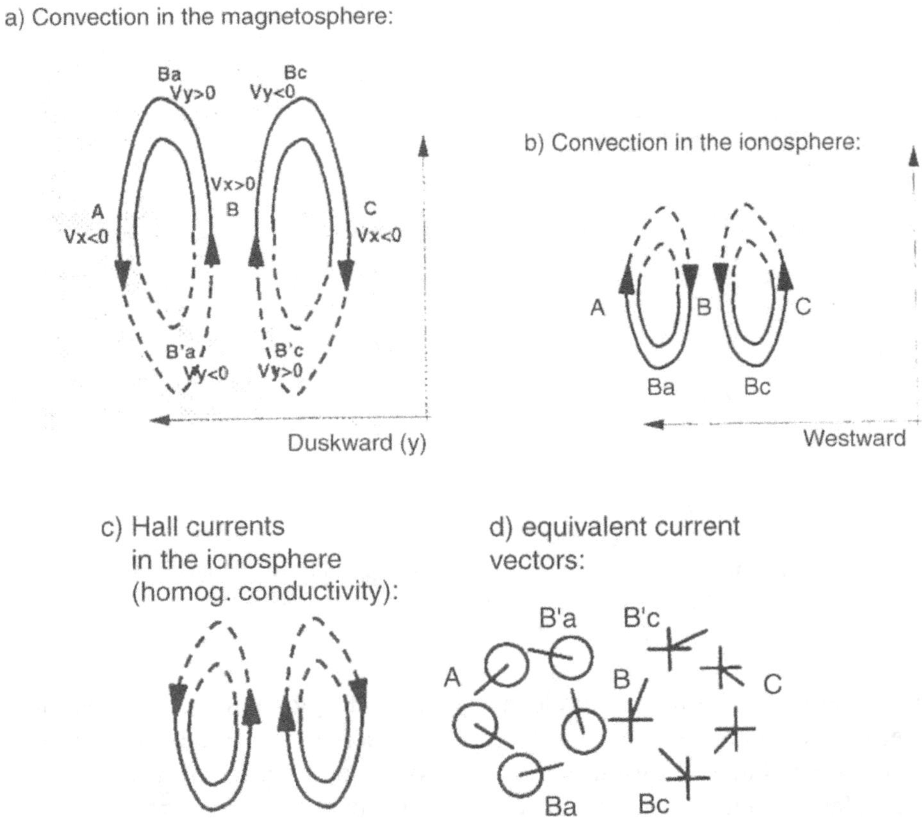

Figure 6.35. Model of magnetospheric convection (a), ionospheric convection (b), ionospheric Hall currents (c), and ground equivalent currents (d) of a streamer by Kauristie et al. (2000).

related to auroral streamers is, e.g., how the streamers dissolve when they intrude into the dense plasma domain of the inner magnetosphere.

6.5 Polar Cap Aurora

The preceding sections have concentrated on the electrodynamics of the auroral forms occurring within the oval proper. The region poleward of the auroral oval, conventionally called the polar cap, also plays hosts to several different kinds of auroral arcs, known variously (and often imprecisely) as the polar cap aurora, polar cap arcs, Sun-aligned arcs, transpolar arcs, or the theta aurora. A recent review by Zhu et al. (1997) covers the observational and theoretical literature on the polar cap aurora in great detail. Here we will provide a summary of the observed properties of the different sorts of polar aurora, discuss the local electrodynamics of the polar cap aurora, as reflected by in situ particle and field data, and describe the current

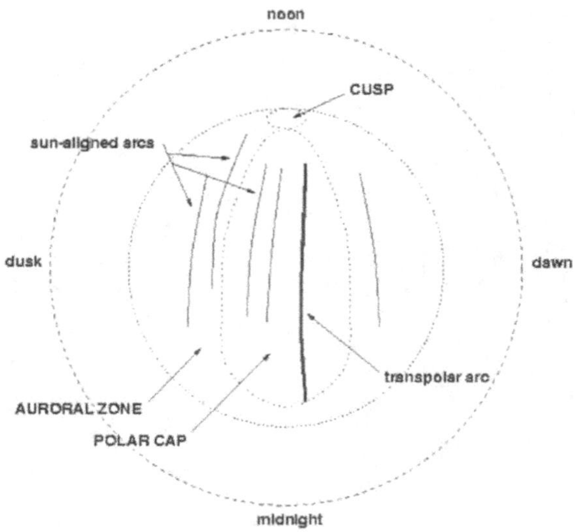

Figure 6.36. Schematic diagram of the various types of optical polar cap aurora and their relationship to the auroral oval, cusp, and polar cap. Figure provided by J. Bonnell.

state of attempts to tie this local electrodynamics to global-scale ionospheric and magnetospheric phenomena. The interested reader is encouraged to consult the above referenced review for further observational and interpretative details.

The polar cap aurora has been recognized as a distinct auroral phenomenon for well over a century. Some of the first systematic studies of the high-latitude aurora that demonstrated their distinct behavior and morphology can be found in Mawson and Chree (1925) and Davis (1960). Figure 6.36 is a schematic diagram showing the various sorts of polar cap aurora. The polar cap aurora occurs at high magnetic latitudes, typically greater than 80-85 degrees, usually during quiet geomagnetic times corresponding to periods of northward interplanetary magnetic field (IMF B_z positive). More active geomagnetic periods correspond to a polar cap devoid of arcs, except for the occasional transpolar arc. The polar aurora arcs are typically called 'Sun-aligned arcs' because they line up more parallel to the noon-midnight line than to the auroral oval. This behavior is shown in Figure 6.37, which shows the traces of 59 sun-aligned arcs observed from Vostok Station, Antarctica. Rather than following the curve of the oval, these arcs strike out from the cusp and dayside nearly parallel to the noon-midnight line. In addition to arcs appearing at high-latitudes across the polar cap, bright sun-aligned arcs appear in the auroral oval, which expands poleward on the dawn and dusk sides during northward IMF conditions. A bright auroral arc often forms near the constricted poleward edge of the auroral oval on the dayside, creating the 'horse-collar' auroral configuration (Hones et al., 1989). Sub-visual sun-aligned arcs, also called 'F-region arcs' (We-

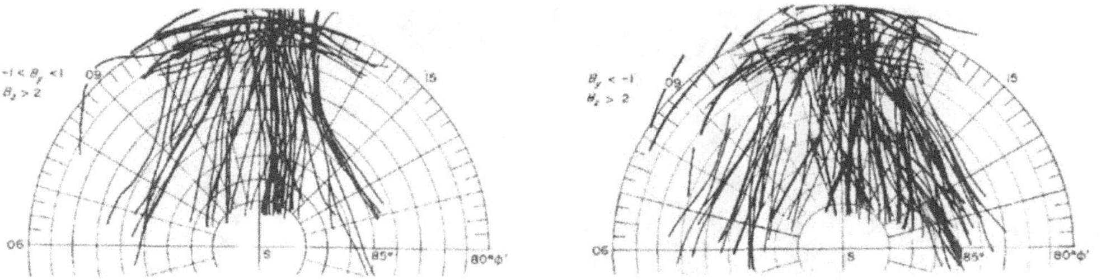

Figure 6.37. Traces of 59 sun-aligned arcs observed from the ground at Vostok Station, Antarctica for IMF B_z positive and B_y near zero (left) and negative (right). From Gusev and Troshichev (1986).

ber and Buchau, 1981), can appear in both the auroral zone (i.e., on nominally closed field lines), and on open field lines in the polar cap, and there are often multiple sub-visual arcs present in the polar cap at any one time (Valladares et al., 1994). A particularly spectacular, but rare auroral configuration is the 'transpolar arc' or 'theta aurora'. An ultraviolet satellite image of this configuration is shown in Figure 6.38, showing how this bright, large-scale auroral feature can extend all the way from the nightside to the dayside auroral oval. Often the point of insertion of the transpolar arc into the nightside auroral zone is a bright spot (Murphree et al., 1987).

6.5.1 Polar Auroral Arc

The bright polar cap arcs exhibit very dynamic behavior. Many of the transpolar arcs split from the dawn or dusk side ovals and move poleward (Huang et al., 1989; Cumnock et al., 1997). Other examples of transpolar arcs have been observed to grow out of the nightside oval towards the dayside (Craven et al., 1986; Murphree et al., 1989). These arcs brightened and faded at speeds of up to 5 km s^{-1} along their length, and remained connected to the dayside for only a few minutes. New transpolar arcs can appear at places widely separated along the oval from previous arcs. Craven et al. (1991) implied that the arcs can reverse direction after moving across the polar cap, as well as move in opposite directions in the northern and southern hemispheres. Spann et al. (1998) and Newell et al. (1999) have shown a case in which the two independent bars of a rare 'double theta aurora' merged into a single bar, crossed the polar cap, split again, and returned across the polar cap. Simultaneous satellite observations of the transpolar arc in both hemispheres have been made (Obara et al., 1988), as well as simultaneous optical observations in one hemisphere and particle precipitation data in the other (Huang et al., 1989). Such observations suggest that the transpolar arcs can appear in conjugate pairs, supporting the theory that the transpolar arc is a closed-field line phenomenon. (see Section 6.5.3 for a discussion of the magnetospheric configurations believed to be responsible for the transpolar arc).

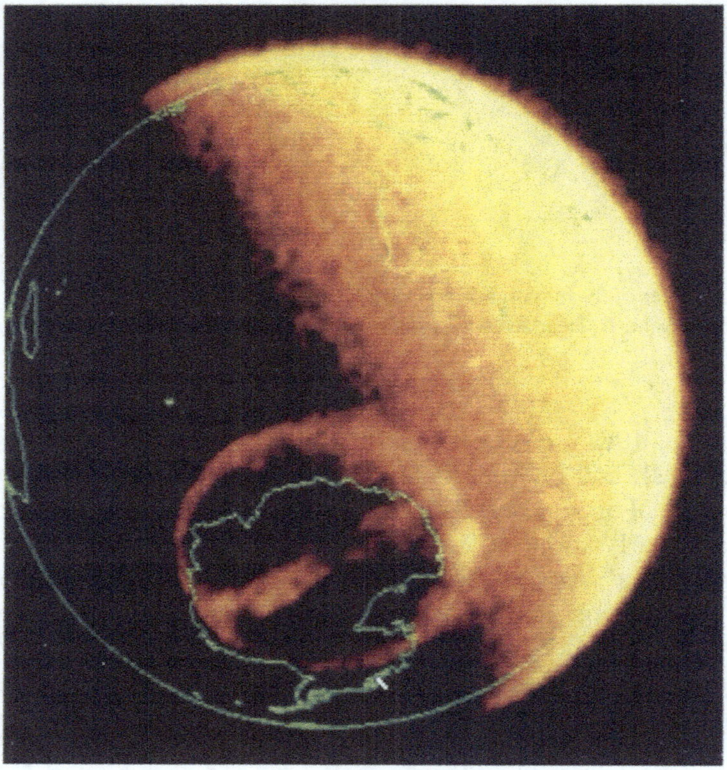

Figure 6.38. Ultraviolet image of the theta aurora over Antarctica taken by the DE-1 SAI at 00:22 UT on 11 May 1983. Figure courtesy of L. A. Frank and 'DE-1 SAI Photo Gallery'.

Sub-visual sun-aligned arcs can display similarly dynamic behavior, as well as long lifetimes. A study of all-sky imaging photometer (ASIP) data taken by a network of stations within the polar cap found that sub-visual arcs occur roughly half the time (i.e., for all periods of IMF northward), and are caused by weak fluxes of hundreds of eV electrons, based upon the dominant 630.0 nm emission (Carlson, 1990). The lengths of the sub-visual arcs were at least 1000 km, with widths of tens of kilometers. An individual sub-visual arc may persist for hours, and the decay of the arcs tends to proceed from the dayside to the nightside in response to a southward turning of the IMF (Valladares et al., 1994; Rodriguez et al., 1997), with a delay of 15-30 minutes for open field line arcs, and of 60-65 minutes for closed field line arcs. No studies have established the conjugacy or lack thereof of the sub-visual arcs.

In addition to the regularities in the morphology and behavior of individual polar cap arcs noted above, one can also find statistical correlations between the occurrence, location, and motion of all the polar cap aurora and the direction of the IMF, although the nature of that dependence is still a subject of debate. The statistical studies of IMF influence on polar cap arc behavior have been carried out using satellite or ground-based imaging systems, and as was pointed out by Zhu et al. (1997), the rather different results of these statistical studies may stem from each imaging system selecting a particular class of arcs (brighter arcs in the satellite studies, fainter in the ground studies), or from the actual existence of two different classes of polar arcs with different behaviors and origins (bright vs. faint). The probable existence of two sorts of the fainter arcs on the basis of in situ electron and ion precipitation properties (Obara et al., 1996b,c; Bonnell et al., 1999) could muddle the statistical influence of the IMF as well.

6.5.2 IMF Dependence

Nonetheless, there are some clear trends in the influence of the IMF on polar cap arcs. The presence of auroral arcs poleward of the oval clearly requires a northward IMF orientation. This was first demonstrated by Lassen and Danielsen (1978), and confirmed by Carlson (1990) in a study of sub-visual sun-aligned arcs. One possibly confusing part of this dependence on northward IMF is the approximately one hour delay in the advent of sun-aligned arcs after a northward turning of the IMF (Troshichev et al., 1988), as well as the persistence of those arcs for up to 30 minutes after a southward turning of the IMF. These delays in the IMF response of the sun-aligned arcs suggest that the Sun-aligned arcs are tied to the global-scale response of the magnetosphere to the IMF, rather than being a more localized, direct response, such as that observed in the cusp aurora (Sandholt et al., 1998). The correlation between the occurrence of polar cap aurora and subdued magnetic and auroral activity (Davis, 1963; Ismail et al., 1977; Lassen and Danielsen, 1989, for example) is probably simply a manifestation the northward IMF requirement (northward IMF lies at the root of all three, rather than one being the cause of the others). Recent studies of the occurrence of transpolar arcs have challenged the requirement for northward IMF, claiming that instead a B_z north-to-south or B_y sign-flip transition of the IMF orientation is required (Cumnock et al., 1997; Newell and Meng, 1995) (see the discussion below on theories of the origin of transpolar arcs).

Some studies have found that the initial location of polar cap arcs depends upon the sign of IMF B_y, while others have found no such dependence. DMSP optical and Viking UV images showed that polar cap arcs preferred the morning or dawn sectors for B_y negative, and the evening or dusk sectors for B_y positive (Gussenhoven, 1982; Elphinstone et al., 1990; Makita et al., 1991). Ground-based studies of subvisual arcs have found no such preference for location, however (Lassen and Danielsen, 1978; Rairden and Mende, 1989; Valladares et al., 1994). The effect of B_y on the dawn-dusk motion of the polar cap arcs is more obvious, with dawnward motion for B_y positive and duskward motion for B_y negative (Rairden and Mende, 1989; Valladares et al., 1994).

Some dependence of the occurrence of polar cap arcs on the IMF B_x orientation was expected because of how polar rain precipitation in a given hemisphere depends upon the IMF B_z and B_x polarities. Satellite UV images have shown that in the northern hemisphere, B_x towards the Sun (polar rain preferred) resulted in weak arcs without a dawn-dusk preference, while for B_x away from the Sun (polar rain unlikely) a single morning sector arc was preferred (Elphinstone et al., 1990). Ground-based studies of sub-visual arcs have found no such dependence, however (Valladares et al., 1994).

On the whole, the cleaner IMF dependence of the brighter polar cap and transpolar arcs suggests that they are more closely tied to the global response of the magnetosphere and ionosphere to varying orientations of the IMF during northward conditions, while the smaller-scale, fainter arcs may represent flow irregularities endemic to all regions of the magnetosphere for northward conditions. To proceed beyond such a qualitative statement about the origins of polar cap arcs, one needs to consider the in situ particle and field observations within and around the polar cap arcs. Such data provided by many satellite studies has proven invaluable in discriminating between the various possible source regions of the arcs, and for illuminating the local electrodynamics of the arcs.

6.5.3 Plasma Sources

The precipitating particle distributions causing the bright polar cap aurora generally resemble in temperature and ion composition those found in the auroral oval and are thus associated with a plasma sheet source population (Meng, 1981). The precipitation in the transpolar arcs also has a plasma sheet source, leading to the conclusion that the field lines threading the arc are closed (Peterson and Shelley, 1984; Menietti and Burch, 1987; Frank and Craven, 1988), in contrast with open field lines threading the remainder of the polar cap (Gussenhoven and Mullen, 1989). 'Confused' convective patterns are associated with the theta aurora, although in some cases, antisunward convection appears on either side of the transpolar arc while sunward convection is found within the arc (Frank et al., 1986). An initial theory of the origin of the bright arcs near the auroral oval was formulated by Meng (1981). The presence of a bright arc at high-latitudes was tied therein to the asymmetric ballooning of the dusk or dawn sides of the plasma sheet for non-zero IMF B_y during B_z northward conditions, as is shown in Figure 6.39B.

The apparent extension (in transpolar arcs) of closed field lines threading the plasma sheet into the open field lines of the polar cap led to the development of models of the theta aurora where a finger of the nightside plasma sheet extends into the lobe, leading to a bifurcation of the open lobe flux and an extension of an aurora feature from the nightside auroral zone towards the dayside as shown in 6.39A (Bythrow et al., 1985; Kan and Burke, 1985; Frank et al., 1982; Frank et al., 1986). Such a model was supported by the observation of nightside plasma sheet protrusions into the lobe (Huang et al., 1987, 1989). However, none of these

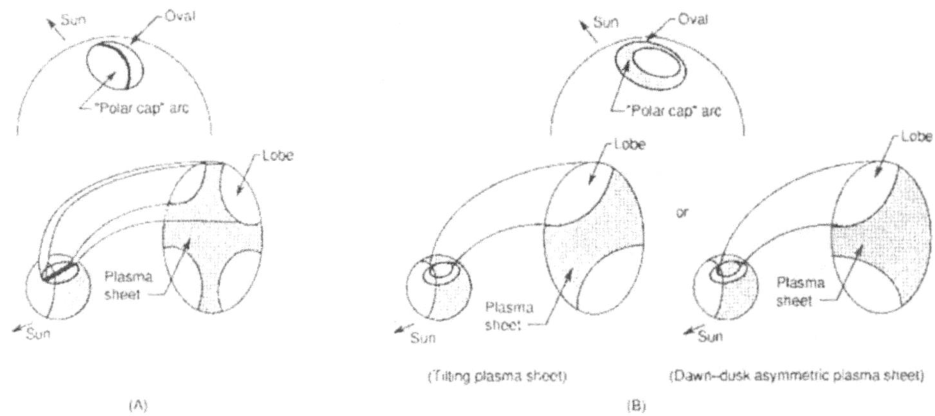

Figure 6.39. Schematic representations of the various transpolar arc models. A: Lobe bifurcated by plasma sheet and/or closed field lines. B: Asymmetric plasma sheet. From Makita et al. (1991).

models went on to develop a theory for why the bifurcation of the lobe occurred, nor what would drive the field-aligned current systems associated with the observed convection and auroral luminosity.

More recent observations showing that the theta aurora apparently occurs during IMF B_z north-to-south or B_y sign-flip transitions (Newell and Meng, 1995; Cumnock et al., 1997) have led to a new model of theta aurora formation based on the isolation of the poleward edge of the auroral oval in the polar cap (Newell and Meng, 1995; Chang et al., 1998; Newell et al., 1999). Under these models, a slab of closed field lines at the poleward edge of the auroral oval is separated on the dayside by new open flux generated by a discontinuous shift of reconnection site to lower magnetic latitudes or different magnetic local times. The plasma sheet does not respond by protruding a finger of plasma into the lobe in these models, however. Such a separation of closed flux from the oval has been observed in both simple models of magnetotail twisting in response to time-varying IMF B_y, and more comprehensive global MHD simulations of the response of the magnetotail, plasma sheet and ionospheric convection patterns to a B_y sign-flip (Kullen, 1999). These models where the transpolar arc is created by the response of the magnetosphere topology and convection to a sudden change in the IMF's orientation will naturally produce conjugate responses in the two hemispheres, and are thus consistent with the apparent conjugacy of the transpolar arcs themselves.

6.5.4 Global Electrodynamics

Attempts have been made to integrate the presence of a transpolar arc into the global electrodynamics of ionospheric convection for northward IMF. In contrast to earlier models that usually tried to relate the transpolar arc to a particular gradient or boundary in the ionospheric convection pattern (Lyons, 1985), more recent mod-

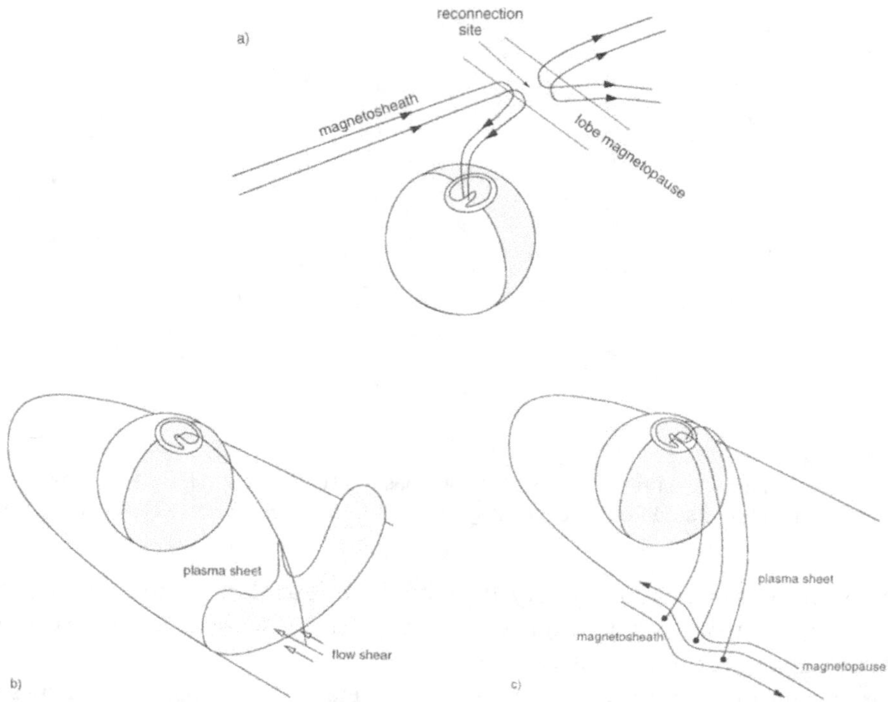

Figure 6.40. Schematic representation of sub-visual arc source/generator region models. a: Lobe reconnection. b: Plasma sheet velocity shears. c: LLBL surface waves. Figure provided by J. Bonnell.

els have found a more subtle relationship between the arc and the convection. For example, the global electrodynamics model of Marklund et al. (1991) demonstrated that sunward or anti-sunward convection could obtain at points along the length of the transpolar arc, with the sense of convection dependent upon the strength of the local perturbation relative to main flow set up by the global Region 11, Region 2 and NBZ current systems. At each point, the transpolar arc was a region of shear in the flow, as can be expected from the observation that $\nabla \cdot \mathbf{E} < 0$ in nearly all polar cap arcs (see Obara et al. (1993) for some counter-examples). This sort of context-dependent electrodynamic response may be contrasted with the fairly clear convection jets observed in the sub-visual arcs discussed below. Such a response is also consistent with the variety of convection signatures reported for the transpolar arc – confused, sunward, or anti-sunward – as well as a non-local closure of the upward current in the arc via an imbalance in the global current pattern. The arc itself is always a locus for shear in the convection (barring any significant modification of the ionospheric conductivity), but it is only when the field-aligned current closes locally that the arc becomes a locus for an obvious, spatially confined flow feature (Marklund and Blomberg, 1991).

It is unclear where the source and generator regions for the sub-visual arcs lie. In situ measurements of electron precipitation and convection electric fields associated with sub-visual arcs have found a good correlation the apparent source temperature for the electron precipitation and the direction of convection observed within the arc. Electron precipitation with a plasma sheet source ($T_{\|}$ of hundreds of eV with a keV of field-aligned acceleration) corresponded to sunward convection, while electron precipitation with a magnetosheath source ($T_{\|}$ of tens of eV with a few hundred eV of field-aligned acceleration) corresponded to anti-sunward convection (Obara et al., 1996a,b). In each case, the region of electron precipitation corresponded to the region of $\nabla \cdot \mathbf{E} < 0$, as had been found in earlier electrodynamical studies of sun-aligned arcs (Carlson, 1990).

The convection in these arcs often appears to be a channel (i.e., a spike in the electric field), rather than a shear in a larger region of flow, indicating that most of the upward field aligned current associated with the arc is closing locally, either to dawnward or duskward of the arc, rather than transiting the entire polar cap and closing through the oval current systems (Obara et al., 1996b; Bonnell et al., 1999). The confused convection systems often reported for northward IMF, especially on the nightside may be a manifestation of this more local mode of current closure. Furthermore, the local pairing of up- and downward currents in the small-scale arcs serves as a constraint upon any model of their generator region.

Presumably the sub-visual arcs with plasma sheet sources and sunward convection correspond to flow shears in the plasma sheet, much like their brighter cousins. Bonnell et al. (1999) examined several models of the source/generator region for the anti-sunward-convecting sub-visual arcs in light of in situ particle and field data from the FAST satellite. These models are shown in Figure 6.40, where the lobe reconnection model was proposed for the anti-sunward-convecting arcs in particular, while the plasma sheet velocity shears (Kan and Burke, 1985) and LLBL surface waves (Rodriguez et al., 1997; Elphinstone et al., 1990) had been proposed for polar cap arcs in general. It was found that the most likely model was one where a reconnection site on the lobe magnetopause acted as the source/generator region for the arcs (shown in Figure 6.40 a). Other models proposed for sun-aligned arcs based on plasma sheet velocity shears and LLBL surface waves (Figure 6.40b,c) disagree with particle precipitation, the maximum ion transit time from source to satellite based on the spatial offsets between the electron and ion precipitation, and the shape of convection (jets vs. bipolar shears) observed in the arcs.

A quantitative time-dependent magnetosphere-ionosphere (M-I) coupling model has also been developed to explain the commonly occurring multiple sub-visual polar cap arcs (Zhu et al., 1996, and references therein). Under this model, the non-linear interaction between ionospheric conductivity modification due to electron precipitation, reflection of Alfvén waves, and generation of Alfvén waves via conductivity modification lead to periodic precipitation regions without requiring analogous periodicity in the magnetospheric generator region.

Chapter 7

Theoretical Models

In the previous Chapters we have presented many aspects of the morphology and microphysics of the aurora. The purpose of the present chapter is to describe theoretical models that try to put these into wider context. As there is yet no model that encompasses all aspects, the chapter will necessarily be rather heterogeneous.

Recognizing the crucial importance of large field-aligned electrostatic potential drops for the generation of the aurora, we start out with a description of a static model of $E_\parallel$ and the resulting density cavity (Section 7.1). This model includes a numerical simulation of the variation of the field-aligned electric field with altitude along an auroral flux tube, thereby reproducing many observational features, including properties of the phase space distribution of the auroral electrons.

The field-aligned electric fields are closely related to the field-aligned auroral current system and its generation by magnetospheric sources. In Section 7.2 a model is described where these currents are related to the vorticity of a boundary layer flow that includes the coupling of this flow to the resistive ionosphere. Such boundary layers are the magnetopause boundary layer, the tail current sheet, or the plasma sheet boundary layer. Depending on the properties of the layer as a current or voltage generator, the coupling to the magnetosphere varies.

The distinction between current and voltage generators is most apparent when considering time-dependent models of the auroral current system. In this case, the generator is a mechanism for launching Alfvén waves into the system, and the response of this generator to Alfvén waves that are reflected from the ionosphere and acceleration region determines the evolution of the current system. Section 7.3 describes a number of models that have attempted to simplify the notion of a generator for the auroral field-aligned currents. In addition to pure current and voltage generators, these models have parameterized the response of the generator in various ways. Of course, it should be realized that these models are oversimplifications, and that a fully dynamical model of the generation of field-aligned currents and Alfvén waves is necessary.

The transport of energy and information from the generators to the auroral region to power the field-aligned potentials is discussed in Section 7.4. It seems certain that the transport is provided by large-scale plasma or magnetohydrodynamic waves, among which Alfvén waves play a dominant role. Resonant mode coupling in the inhomogeneous magnetospheric plasma, field-line resonances and propagation in the ionospheric resonator all contribute to this energy and information transfer by Alfvén waves. Their dynamics causes time and space variations in the aurora.

Section 7.5 investigates the dynamics of an arc when it moves transversely across the magnetic field in the auroral zone. The motivation for this approach comes from the energetics of auroral arcs implying a depletion of the free magnetic energy content of the involved flux tubes within a few Alfvén transit times. Analysis of this situation combines elements of the Alfvén wave and static field pictures into a single entity known as the stationary Alfvén wave. Such a wave can generate spatial structure on small and mesoscales, and provides a possible mechanism for the generation of quasi-static $E_{\|}$ and density depletions. While these models of Alfvén wave dynamics can give the evolution of auroral currents and parallel electric fields, they do not directly address the question of the acceleration of auroral particles.

Section 7.6 describes models of how waves with short perpendicular wavelengths can lead to bursty electron acceleration, resulting in electron distributions that have a broad energy spectrum but have a narrowly field-aligned pitch-angle distribution.

Section 7.7 describes three-dimensional magnetohydrodynamic (MHD) simulations of the dynamic evolution of the magnetotail related to substorms. These simulations illuminate a large number of global aspects of the coupling between the magnetosphere and the ionosphere, the formation of the auroral bulge, the creation of the substorm current wedge and the spatial and temporal distribution of the current systems, generator regions, and stress transmission. This section also addresses the possible role of small-scale structures in generating auroral arcs, from a large-scale view.

The chapter ends with a discussion of the essential elements of an integrated auroral theory in Section 7.8. Energy flow, current filamentation and the feed-back of the auroral energy and stress release onthe dynamics of the generator are being analyzed. Finally, the question is pursued which role auroral arcs play within the global magnetospheric plasma transport, in particular during substorms.

7.1 Static $\mathbf{E}_{\|}$ Models

We begin this chapter on theoretical models of auroral acceleration by considering a simple case: a one-dimensional, static model of the parallel electric field in the upward current region. The term 'simple' is relative, because the auroral acceleration zone is tremendously complex. Though limited in many respects, static solutions have one advantage: they can consider the fully kinetic evolution of particle distributions, a consideration that is beyond most time-dependent models of large-scale processes.

As discussed in Section 3.3, it has been established that a parallel electric field can drive a current against the magnetic mirror force (Knight, 1973). This relationship has proven to be sound, but it does not provide a theoretical description of how parallel electric fields are supported self-consistently in a collisionless plasma or how they are distributed (Borovsky, 1993a). In this section we discuss the distribution of the parallel electric fields along the auroral field lines following several studies which found quasi-neutral solutions (Chiu and Schulz, 1978; Stern, 1981; Ergun et al., 2002). In situ observations regarding parallel electric field were reported in Section 4.1.3, and their altitude distribution was discussed in Section 5.1.3.

There are several observational aspects of the upward current region that were introduced in earlier chapters that we rely on. The auroral cavity clearly stands out in the observations as a distinct region (see Figure 4.7 in Section 4.1.2.2). In this region one sees an ion beam of *ionospheric* origin and precipitating electron fluxes of *magnetospheric* origin. The plasma density, as measured by Langmuir probes, is extremely low, often less than 1 cm^{-3}. The auroral cavity is distinguished by having a parallel electric field above (at higher altitudes) and below.

A matter of debate has been the make-up of the plasma in the auroral cavity. The instruments on the earliest missions did not have sufficient accuracy or resolution to determine the absolute density, so it was conjectured that a 'cold' ambient plasma population could strongly influence the electrodynamics of the cavity. As discussed in Section 4.1.2, the FAST observations have tipped the scales in favor of no 'cold' plasma. This somewhat esoteric but very important finding is crucial for quasi-neutral solutions. The auroral cavity is clearly a buffer zone between the ionosphere and the magnetosphere not just highlighted by, but often dominated by *ionospheric* ions and *magnetospheric* electrons.

7.1.1 Numerical Simulations

We want to find a solution to the one-dimensional, static Vlasov-Poisson equations (3.39) - (3.41) that represent the coupling between the magnetosphere and ionosphere with an applied current and a static field-aligned potential drop. Since the mirror force is ultimately responsible for the parallel electric field, the equations need to consider at least two dimensions in velocity space. Thus, static, 1-D spatial, 2-D velocity, Vlasov codes are well-suited to search for large-scale, self-consistent solutions of the parallel electric field (Chiu and Schulz, 1978; Stern, 1981; Ergun et al., 2002). A description of similar, non-static Vlasov codes can be found in Cheng and Knorr (1976).

The basic idea is as follows. Electron and ion distribution functions are prescribed at the boundaries of a large spatial region which has a mirroring magnetic

Species	$n\ cm^{-3}$	Temp.	Type	Boundary
Ionospheric O^+	2×10^5	0.5 eV	Fluid	Left
Ionospheric H^+	100	0.5 eV	Fluid	Left
Scattered (e^-)	1	1 keV	Fit to Data	Left
Secondaries (e^-)	30	100 eV	Power Law	Left
Ionospheric (e^-)	$\sim 2\times10^5$	0.5 eV	Boltzmann	Left
Magnetospheric H^+	0.5	5 keV	Maxwellian	Right
Magnetospheric e^-	0.5	1 keV	Maxwellian	Right
Trapped (e^-)			Filled	

Table 7.1. Boundary conditions for the simulations described in this section.

field. The ionosphere is on one side, the magnetosphere on the other. The potentials at the boundaries are also prescribed and the region is initiated with an estimated potential distribution, $\Phi(z)$. The distribution functions are broken into $N_v \times N_v$ velocity-space elements ($N_v = 70$ in these simulations), each of which is treated as a fluid. Trapped electrons, an important part of the distribution (see later) are filled in with the same phase space density as the magnetospheric electrons as a function of energy, up to a value of αf_{max}, where f_{max} is the maximum phase space density of the magnetospheric electrons. The parameter $\alpha \simeq 0.02$ is adjusted to best match the observed (FAST) distributions. Using the prescribed distributions at the boundaries and $\Phi(z)$, the velocity space distributions are calculated at $N_z = 99$ spatial locations between the boundaries. An error is defined as:

$$\xi(z) = \nabla^2\Phi + \frac{e}{\varepsilon_0}(n_i - n_e) \tag{7.1}$$

where n_i and n_e are the ion and electron densities derived from the distributions. $\Phi(z)$ is then iteratively adjusted to minimize $\xi(z)$ which yields a static solution. If the grid spacing, δz, is much larger than λ_D, then it can be shown that the $\nabla^2\Phi$ term is negligible, which essentially results in a quasi-neutral solution. The static code enforces adiabatic evolution (no velocity-space diffusion) and therefore yields the Knight current-voltage relation (Eq. 3.37). Gravitational potential is included which confines the dense, cold ionospheric ions to low altitude.

Figure 7.1 displays the results of the simulation. The horizontal axis represents geocentric distance in Earth radii and the vertical axis represents density or potential. The left hand side is the ionospheric boundary at which cold O^+, cold H^+, cold electrons, electron secondaries, and scattered primaries are prescribed (Table 7.1).

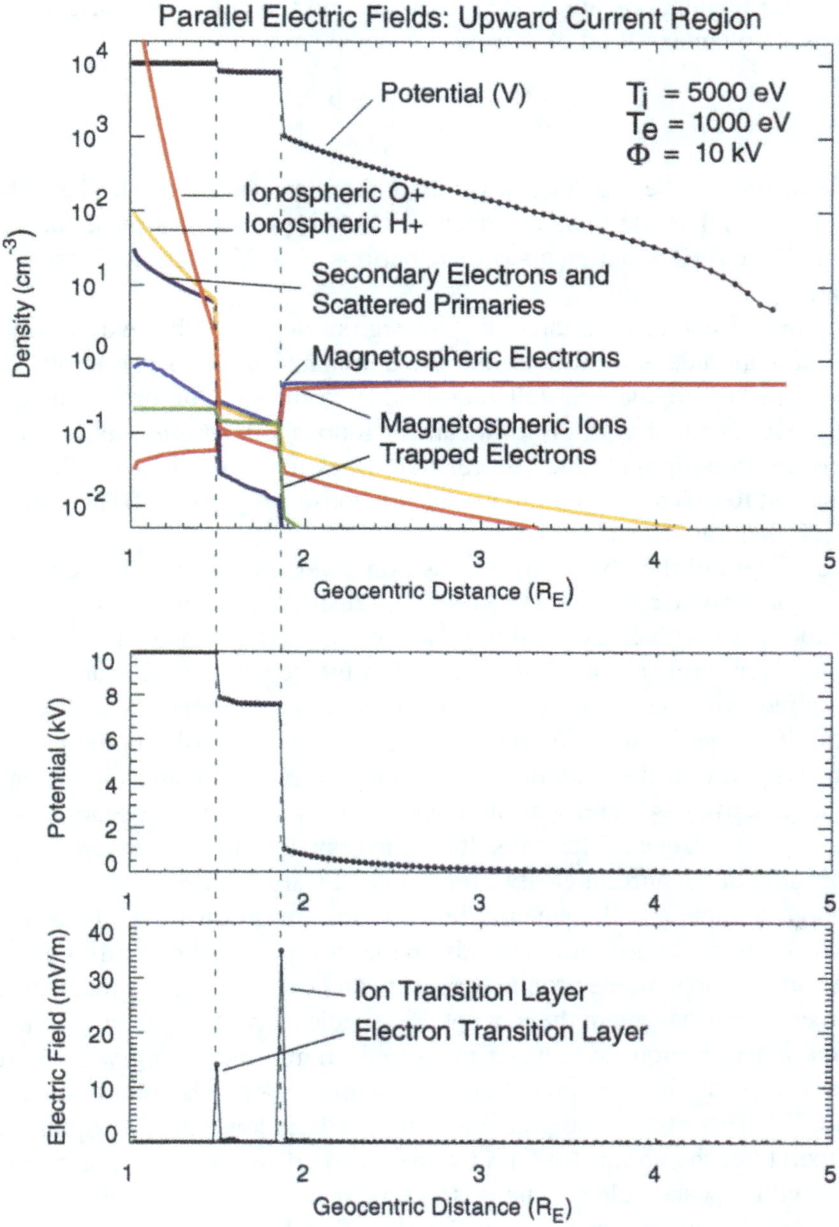

Figure 7.1. The results of a 1-D spatial, 2-D velocity, large-scale Vlasov simulation. Top: Ionospheric O^+ (orange), ionospheric H^+ (yellow), electron secondaries and scattered primaries (blue), and cold electron distributions are specified at the left boundary (ionosphere). The ionospheric electrons (dark blue) and ions (red) are specified at the right boundary (magnetosphere). The trapped electrons (green) are filled in with the same phase space density as the magnetospheric electrons as a function of energy, up to a value of αf_{max}, where $\alpha = 0.02$. The circles represent the potential, $\Phi(z)$. Center: The potential $\Phi(z)$ on a linear scale. Bottom: The one-pixel averaged electric field $E_{||}$ on a linear scale. Figure provided by R. E. Ergun.

The cold ion populations are treated as fluids and the cold electron population is treated as a Boltzmann fluid of density :

$$n_e = n_0 \exp\left(\frac{e\Phi}{k_B T_e}\right) \tag{7.2}$$

which conforms to the ion density via an ambipolar electric field. The right-hand side of Figure 7.1 is the magnetospheric boundary which has prescribed plasma sheet ion (5 keV H^+) and electron distributions (1 keV) that are consistent with observations.

In Figure 7.1 one can see three distinct regions separated by two transition layers. The low-altitude auroral field line are dominated by ionospheric plasma. The cold O^+ and H^+ populations fall rapidly due to the gravitational potential. The auroral cavity is filled with an ion beam of ionospheric origin and plasma sheet electrons and the high-altitude auroral field lines are controlled by plasma sheet electrons and ions. The simulation results are consistent with FAST, Polar, and DE results (cf. Section 4.1.3).

The thickness of the low-altitude transition layer, called the 'electron transition layer', is the grid size ($\delta z \approx 250$ km in the displayed simulations $>> \lambda_D \approx 1$ km). Simulations with δz as small as ≈ 125 km have the same result. The electron transition layer forms at the altitude at which the density of electron secondaries and scattered primaries equals the cold ion density, i.e., between $\simeq 2000$ and $\simeq 5000$ km. The ion density altitude profile plays a strong role. If the ionospheric ion density is lower, the altitude of the electron transition layer is lower. If the ionospheric density is increased, the electron transition layer appears at a higher altitude and is weakened. These results are in consort with the seasonal variations of the altitude of the auroral cavity (Temerin and Carlson, 1998a).

The high-altitude parallel electric field, or 'ion transition layer', is located near $\approx 1\ R_E$ in altitude. The location depends on the total potential drop, the current density, the magnetospheric electron temperature and density, and the magnetospheric ion temperature. Changes in the ionospheric species have less effect. The potential above the ion transition layer has a monotonic ramp, increasing with decreasing altitude. The abrupt jump in potential, once again, is one grid point in the example in Figure 7.1. However, if the potential jump, $\Delta\Phi$, is lower than the plasma sheet ion temperature, the abrupt (one grid size) potential jump is no longer seen. It is replaced with a peaked electric field structure spread over a distance of order 0.1 R_E. The low density region between the transition layers is the auroral cavity; it has a low-level parallel electric field. The electric field maintains a quasi-neutral environment between two species whose densities evolve in altitude differently.

7.1.2 Theory and Data

Figure 7.2 compares the measured electron distributions (cf. Section 4.1.1 and Figure 4.4) in the auroral cavity with those in the figure on identical scales. The bound-

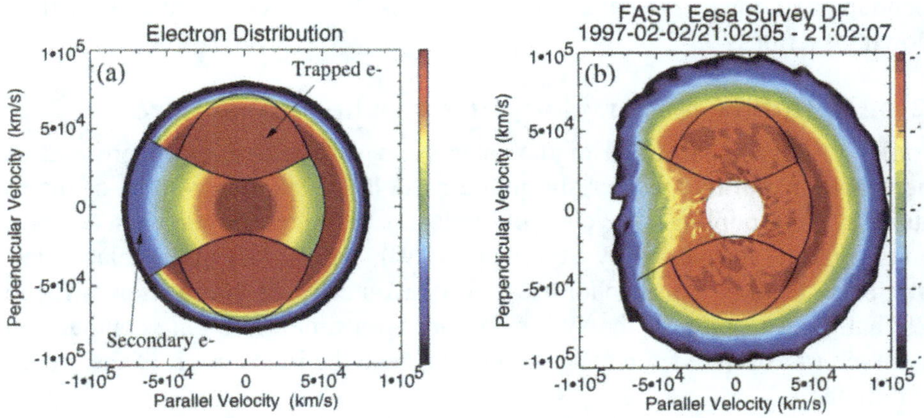

Figure 7.2. Left: The electron distribution from the adiabatic simulation. The boundaries between the magnetospheric, secondary and scattered primary, and trapped electrons are marked. Positive (negative) parallel velocities correspond to downward (upward) motion. The partial ring at 7×10^4 km s^{-1} represents magnetospheric electrons that have been accelerated by the ≈ 7.5 kV potential drop above the cavity. Subsequent motion and mirroring in the converging magnetic field explains the formation of this ring. The downward moving electrons with a cut-off near 5×10^4 km s^{-1} are primaries that were first backscattered below the cavity, retarded by the potential drop below the cavity and finally turned around by the potential drop above the cavity. Right: An electron distribution in the auroral cavity as measured by FAST. The two distributions are quantitatively similar except in the region of secondary and scattered primary electrons. Velocity space diffusion may account for the differences between the distributions. See also Figure 7.1. Figure provided R. E. Ergun.

aries between the accelerated, trapped, and secondary/scattered primary electrons are marked on the simulation. The same boundaries, scaled to accommodate slightly lower energies (96%), are marked on the measured distributions for reference. The distributions are quantitatively similar. The main difference is the phase space density of the secondary electron region, which in the FAST data has been increased by velocity space diffusion.

The large-scale representation of the parallel electric fields associated with discrete arcs based on quasi-neutral solutions to the Vlasov-Poisson equations gives us some insight in interpreting the observations. The 1-D spatial, 2-D velocity static Vlasov simulation above is applicable to discrete arcs with large acceleration potentials (≥ 5 keV). The results are strongly supported by observations (cf. Section 4.1.3.4).

The static, one-dimensional Vlasov-Poisson models such as discussed here have met with great success and are very useful in providing a framework for further studies, but are severely limited. The simulation results may not be unique, the trapped electron population was imposed by an empirical formula, velocity-space diffusion has not been accounted for, and the resulting potential structures may not be stable. The large-scale results, however, provide a basis to investigate the time-dependent, self-consistent parallel electric fields. Such models are presented in the following sections.

7.2 Boundary Layer Coupling to the Ionosphere

In order to develop a model of the interaction of the magnetosphere and ionosphere, it is useful to consider simplified models that illustrate some of the basic features of this coupling. One class of models is based on the early viscous boundary layer model developed by Sonnerup (1980), which was further elaborated by Lotko et al. (1987). This model uses the quasi-steady approximation, which implies that the evolution of the system occurs on a time scale long compared with the Alfvén transit time from the boundary layer to the ionosphere, of the order of a few minutes.

As mentioned in Section 3.2.2.2, this model treats the simplified case where $\mathbf{B}$, p and ρ are constants, so that the current divergence relation (Eq. 3.23) reduces to

$$B\frac{\partial}{\partial s}\left(\frac{j_{\|}}{B}\right) = \frac{\rho \mathbf{B}}{B^2} \cdot \nabla \times \left(\frac{d\mathbf{v}}{dt}\right) = \frac{\rho}{B}\frac{d\Omega_{\|}}{dt}, \tag{7.3}$$

where $\Omega_{\|} = \hat{\mathbf{b}} \cdot (\nabla \times \mathbf{v})$ is the vorticity of the flow parallel to the background magnetic field and s is the coordinate along the magnetic field direction.

For a convective flow given by the $\mathbf{E} \times \mathbf{B}$ drift, the electric field can be written in terms of the gradient of a potential. The vorticity is then given by

$$\Omega_0 = \hat{\mathbf{b}} \cdot \left[\nabla_\perp \times \left(\frac{\mathbf{E} \times \mathbf{B}}{B^2}\right)\right] = -\frac{1}{B}\nabla_\perp \cdot \mathbf{E}_\perp = \frac{1}{B_0}\nabla^2_{\perp 0}\Phi_0, \tag{7.4}$$

where we have neglected the gradient in the magnitude of B. The subscript 0 has been introduced in the last form of this expression to emphasize the fact that these quantities are evaluated in the source region. This potential Φ_0 represents the source potential for the ionospheric coupling equation:

$$\left(1 - \lambda_I^2 \nabla^2_{\perp I}\right)\Phi_I = \Phi_0 \tag{7.5}$$

Recall that $\lambda_I^2 = \Sigma_P/K$ is the magnetosphere-ionosphere coupling scale length, evaluated at the ionosphere. This quantity was already introduced in conjunction with Eq. (3.48), but termed λ_{ES} to emphasize the electrostatic context. Using current continuity in the ionosphere, this can be written in terms of the field-aligned current at the ionosphere:

$$\left(1 - \lambda_I^2 \nabla^2_{\perp I}\right) j_{\|I} = -\Sigma_P \nabla^2_{\perp I}\Phi_0. \tag{7.6}$$

The subscript I in this equation indicates that the relevant quantities are evaluated at the ionosphere; however, in Eq. (7.4), the gradient is evaluated in the boundary layer. In order to connect these two, a metric scale factor must be introduced into the formulation. The scale factor is determined by the ratio of the distance between two field lines in the two different regions. Note that this scale length depends in general on the direction over which the gradient is taken.

Consider a flux tube whose cross section in the ionosphere is a square. For an orthogonal mapping of filed lines as in a dipolar field, if the field lines going through the four corners of the square are traced up to higher altitudes, the cross section of this flux tube at higher altitudes will be rectangular. Thus, if x and y are the two directions perpendicular to the field line, the scale factors h_x and h_y in the two directions will not be the same. Even for a non-orthogonal mapping, a generalized scale factor can be defined in general. The magnetic flux going through the flux tube is then given by $\Phi_B = h_x h_y B$. Note that the conservation of magnetic flux on the flux tube implies that $h_x h_y \propto 1/B$. If we then take the derivative of some quantity at two different altitudes, we can write $h_{xI}(\partial/\partial x_I) = h_{x0}(\partial/\partial x_0)$ and similarly for y. If the Laplacians in Eq. (7.6) are taken at an angle φ with respect to the x axis, we can write $h_I^2 \nabla_{\perp I}^2 = h_0^2 \nabla_{\perp 0}^2$, where the general scale factor could be written as $h^2 = h_x^2 \cos^2\varphi + h_y^2 \sin^2\varphi$. Then Eq. (7.6) can be written as

$$\left(1 - \lambda_I^2 \nabla_{\perp I}^2\right) j_{\parallel I} = -\Sigma_P \frac{h_0^2}{h_I^2} \nabla_{\perp 0}^2 \Phi_0 = -\Sigma_P B_0 \frac{h_0^2}{h_I^2} \Omega_0. \tag{7.7}$$

The effect of the ionosphere on the boundary layer flow can be considered by combining Eq. (7.7) with equation (7.3), as was considered by Lotko et al. (1987). We can write Eq. (7.3) as an evolution equation for the vorticity in the form,

$$\frac{d\Omega_0}{dt} = \frac{B_0^2}{\rho_0} \frac{\partial}{\partial s} \left(\frac{j_{\parallel 0}}{B_0} \right) = \frac{B_0^2}{\rho_0 H} \frac{j_{\parallel 0}}{B_0}, \tag{7.8}$$

where in the last step, we have replaced the field-aligned derivative with a scale height H. Note that the current scales as $j_{\parallel 0}/B_0 = j_{\parallel I}/B_I$. If we now operate on Eq. (7.8) with the operator $1 - \lambda_I^2 \nabla_{\perp I}^2 = 1 - \lambda_0^2 \nabla_{\perp 0}^2$, where $\lambda_0^2 = (h_0^2/h_I^2)\lambda_I^2$ is the M-I coupling length mapped to the generator, we obtain

$$\begin{aligned} \left(1 - \lambda_0^2 \nabla_{\perp 0}^2\right) \frac{d\Omega_0}{dt} &= \frac{B_0^2}{\rho_0 H B_I} \left(1 - \lambda_I^2 \nabla_{\perp I}^2\right) j_{\parallel I} \\ &= -\frac{\Sigma_P B_0^2}{\rho_0 H} \frac{h_0^2 B_0}{h_I^2 B_I} \Omega_0 = -\nu_0 \Omega_0, \end{aligned} \tag{7.9}$$

where the damping constant ν_0 is given by

$$\nu_0 = \frac{\Sigma_P B_0^2}{\rho_0 H} A = \mu_0 v_{A0} \Sigma_P \frac{v_{A0} A}{H} = \frac{\Sigma_P}{\Sigma_{A0}} \frac{v_{A0} A}{H}. \tag{7.10}$$

An anisotropy factor has been defined as $A \equiv h_0^2 B_0 / h_I^2 B_I$. Note that for an isotropic flux tube, $h_x = h_y = 1/B^{1/2}$, and so $A = 1$. In this last form, we have introduced the Alfvén conductivity in the generator, $\Sigma_{A0} = 1/\mu_0 v_{A0}$ (Eq.3.54). Note that Eq. (7.9) shows that the ionospheric conductivity acts as a frictional term on the magnetospheric convection, and so the vorticity will decay exponentially at a rate given by $\nu = \nu_0/\left(1+k^2\lambda^2\right)$, where either subscript, I or 0 could be applied to k and to λ, and $k^2 = -\nabla_\perp^2$. This expression implies that, at short spatial scales, the presence of a parallel electric field (finite λ) decreases the damping from the ionospheric conductivity. Thus, parallel electric fields decouple the convection in the outer magnetosphere from the ionosphere.

The effects of viscosity and convection can also be added to the total time derivative of Eq. (7.9). The resulting equation for the vorticity becomes

$$\left(1-\lambda_0^2\nabla_{\perp 0}^2\right)\left(\frac{\partial}{\partial t}+\mathbf{v}_0\cdot\nabla_{\perp 0}-\eta\nabla_{\perp 0}^2\right)\Omega_0 = -\nu_0\Omega_0, \tag{7.11}$$

where η represents the viscosity. In this case, the total damping of the system can be written as

$$\nu = \eta k_0^2 + \frac{\nu_0}{1+k_0^2\lambda_0^2} \tag{7.12}$$

This expression indicates that, at a certain length scale, the damping of these vortices is a minimum, with viscous damping dominating at small scale lengths and ionospheric damping dominating at larger scales. For parameters discussed by Lotko et al. (1987), this minimum damping scale occurs at $k\lambda \sim 10$, corresponding to a wavelength somewhat shorter than the M-I coupling scale itself. This mechanism implies that convection is strongest at length scales of a few tens of kilometers for typical parameters. The ionospheric damping rate was determined by Lotko et al. (1987) as $\nu_0 = 3.8\times10^{-3}$ s^{-1}, corresponding to a damping time of $\approx$ 260 s. The minimum damping rate is about a factor of 10 smaller, indicating that such vortices may survive nearly 1 hour at the minimum damping scale.

This simple model illustrates many of the essential features of quasi-static magnetosphere-ionosphere coupling on an auroral flux tube, including the effects of parallel electric fields. The model is primarily applicable to incompressible flow in boundary layer regions, where the plasma pressure is not important. In other circumstances, particularly in the midnight region during substorms, plasma compressibility and, therefore, pressure gradient forces become important. Such models will be discussed in later sections of this chapter.

7.3 Current and Voltage Generators

Many auroral models implicitly assume a generator for the field-aligned current systems that exist in the outer magnetosphere. Such a generator must tap the kinetic energy of the magnetospheric plasma and convert it to the electromagnetic energy of the auroral current system (see Section 3.2.2). This electromagnetic energy is transmitted as a Poynting flux to the acceleration region and ionosphere, where it is either dissipated or returned to the outer magnetosphere as a reflected

Alfvén wave. On a phenomenological level, one often speaks of current or voltage generators. While these concepts are useful in visualizing the generator process or for boundary conditions on models that focus on the lower altitude auroral processes, it should be recognized that these concepts are oversimplifications of the actual current generation process.

The distinction between generators that deliver a fixed current and those in which the voltage is fixed is most useful in considering time-dependent models of the auroral current system. These limits represent two extreme cases. In one case, the perpendicular current, determined by the curl of the magnetic field, is fixed, and the magnetospheric convection pattern changes in order to accommodate this amount of current. On the other hand, one could consider the convection pattern to be fixed. Since the convection pattern is given by the $\mathbf{E} \times \mathbf{B}$ drift, this is equivalent to assuming a potential pattern in the outer magnetosphere. In this case, the perpendicular current adjusts to maintain the convection. These two extremes may be visualized by a current analogy, in which the generator is a battery with an internal resistance. If the internal resistance is small compared to the load on the circuit, the load will receive a fixed voltage; whereas if the load is small compared to the internal resistance, a fixed current, given by the emf of the battery divided by the internal resistance, will be delivered to the load. An important distinction between current and voltage generators is that the ionospheric dissipation will be proportional to the Pedersen conductivity in the voltage generator case, while it is inversely proportional to the conductivity if the current is held fixed.

7.3.1 Pure Current or Voltage Generators

The differences between voltage and current generators becomes most clear when one considers the reflection of Alfvén waves that propagate from the ionosphere back to the generator. In the case of the voltage generator, the reflection must be in such a sense that the electric field remains constant at the generator, i.e., the reflected wave must have an electric field opposite in sign to the incident wave. Similarly, a wave reflected from a current generator must have a current, and thus also a magnetic perturbation, with the opposite sign as the incident wave. Since the wave Poynting flux, $\mathbf{S} = \mathbf{E} \times \mathbf{B}/\mu_0$ is opposite for the incident and reflected waves, either the electric or magnetic field but not both must reverse sign upon reflection. This implies that a wave reflected from a voltage (current) generator has the same sign of current (electric field) as the incident wave.

The evolution of current and electric field for both current and voltage generator when a wave is launched from the generator and is incident on a conducting ionosphere has been considered by Lysak and Dum (1983), assuming a highly conducting ionosphere. This has already been discussed in Section 3.4.2. For a voltage generator, the imposed electric field propagates toward the ionosphere, where it is reflected in accordance with Eq. (3.64). An observational example of this is shown in Figure 4.47. The reflected wave has the opposite electric field as the incident wave, and so the net result is a reduced electric field. This reduced electric field propagates back toward the generator, where is it again reflected so as to maintain the imposed electric field. After a number of reflections, the imposed electric field is maintained along the entire flux tube. The field-aligned current is increased on each bounce until it reaches the value given by the ionospheric Pedersen conductivity (see Eq. 3.61) times the divergence of the imposed electric field. (Note that in this discussion we assume that the fields are mapped along magnetic field lines as described in the previous section).

The current-driven run evolves much differently. Since the Alfvén conductivity is lower than the ionospheric Pedersen conductivity, the initial electric field is much higher than the steady-state value. Upon reflection from the ionosphere, the electric field is reduced as before, and the current is increased above the imposed value. The reflection from the generator reduces the current back to the imposed value, and the electric field is reversed in sign. Again after a number of bounces, the transient fields are damped out by the ionospheric conductivity and a steady state is achieved. Note that in the current-driven case, the electric field and current oscillate with a period of $4\tau_A$, where τ_A is the travel time of the Alfvén wave from the generator to the ionosphere, while for the voltage generator, the electric field oscillates at $2\tau_A$. If one considers the generator to be on the equatorial edge of a closed field line, the current generator corresponds to the fundamental oscillation of the field line, while the voltage generator corresponds to the first harmonic (see Section 7.4.2 for more discussion of resonant waves on closed field lines).

7.3.2 Generator Conductivity Models

A way of bridging the gap between the two opposite limits of voltage and current generators is to introduce a generator conductivity, Σ_G, which plays the same role as the internal resistance of a battery in the circuit analogy given above. Such a conductivity was introduced by Sonnerup (1980) in order to represent the influence of ionospheric coupling on boundary layer flow. This conductivity may also be considered to result from Alfvén wave propagation in the generator region, resulting in the deceleration of the convection in the generator (Lysak, 1985). If one considers the reflection of Alfvén waves from such a generator, an expression similar to that for the ionosphere (cf. Eq. 3.64)

$$R_G = \frac{\Sigma_A - \Sigma_G}{\Sigma_A + \Sigma_G} \tag{7.13}$$

is found, where Σ_A is the Alfvén conductance (cf. Section 3.4.1). Thus, voltage generators have a reflection coefficient of -1, while current generators have a reflection coefficient of $+1$. (Recall that this is the reflection coefficient for the electric field). Use of this conductivity is particularly useful in numerical models, where by setting the conductivity equal to the Alfvén conductivity, an absorbing boundary condition ($R = 0$) can be implemented. Again invoking the circuit analogy, it can be seen that high generator conductivity (low internal resistance) corresponds to a voltage generator while low conductivity (high internal resistance) corresponds to a current generator.

The generator conductivity model leads to a modification of the magnetosphere-ionosphere coupling length scale, $\lambda_{ES} = \sqrt{\Sigma_P/K}$, deduced from Eq. (3.48). That calculation was based on a voltage-driven model where the magnetospheric potential was fixed. A more general model can be considered in which the total current emanating from the generator can be written as in the form:

$$j_{\parallel} = j_{\parallel 0} + \Sigma_G \nabla^2 \Phi_0 \tag{7.14}$$

If this equation is combined with the ionospheric current continuity Eq. (3.42) and the linear current-voltage relation (3.37), the current can be written as:

$$\left[1 - \frac{\Sigma_G \Sigma_P}{K\left(\Sigma_G + \Sigma_P\right)} \nabla^2\right] j_{\parallel} = \frac{\Sigma_P}{\Sigma_P + \Sigma_G} j_{\parallel 0} \tag{7.15}$$

Thus, this equation implies a modified magnetosphere-ionosphere coupling scale length given by

$$L = \sqrt{\frac{\Sigma_G \Sigma_P}{K\left(\Sigma_G + \Sigma_P\right)}} \tag{7.16}$$

Note that this length scale reduces to λ_{ES} for the voltage generator case (i.e., $\Sigma_G \gg \Sigma_P$). For a current generator ($\Sigma_G \ll \Sigma_P$), the coupling scale length becomes $L = \sqrt{\Sigma_G/K}$, which is smaller than λ_{ES}. Thus, small-scale auroral structures are more likely to result from current generators than voltage generators.

It is important to note that the 'generator' region can be considered to be the magnetospheric region where the field-aligned currents are closed. If interfering Alfvén waves or nonlinear effects cause the current system to be closed in a region of high Alfvén speed, the effective generator conductivity would be the Alfvén wave conductivity in the generator region. This conductivity is quite low due to the high Alfvén speed in this region, and so currents closing in this region are likely to be associated with small-scale auroral arcs. On the other hand, open field lines that are connected with the solar wind would be more consistent with a voltage generator, since the Alfvén speed is low and thus the Alfvén conductivity is high in these regions.

7.3.3 Tau Generator Model

A different way of parameterizing the generator was introduced by Vogt et al. (1999). In their model, they considered the mechanical driving force for the current (e.g., the pressure gradient) to be fixed in the generator process, as opposed to pure current or voltage generators that keep the field-aligned current or electric field, respectively, fixed. This fixed force is balanced with a given cross-field current (e.g., the diamagnetic current). Denoting the height-integrated current due to this force as $\mathbf{J}_f$, and including waves incident and reflected from the generator region, they found an equation for the curl-free generator electric field in the form

$$\frac{\partial \mathbf{E}}{\partial t} = \frac{1}{\tau}\left(\frac{\mathbf{J_f}}{\Sigma_A} + 2\mathbf{E}^+ - \mathbf{E}\right) \tag{7.17}$$

where τ is the Alfvén travel time across the generator and $\mathbf{E}^+$ is the electric field of the wave incident on the generator. Note that this expression shows that in the absence of forcing or any incident waves the electric field will decay away on the time scale τ due to the emission of Alfvén waves that carry away the plasma momentum.

Note that an isotropic gradient pressure force cannot produce a curl-free electric field. Pressure gradients can cause plasma motion, converting internal energy to kinetic energy. This flow energy can then directly be converted into magnetic energy. An increase of the plasma internal energy, which is related to an increase of the diamagnetic current, is often caused by slowing down the plasma flow. Overall, the energy source for the generation of field-aligned current in the generator region is not the plasma internal energy, but rather the relative kinetic energy of the plasma (e.g., Song, 1998; Song and Lysak, 2000).

The reflection of Alfvén waves from this type of generator can be shown to give a reflection coefficient (Vogt et al., 1999):

$$R_G = \frac{1 - i\omega\tau}{1 + i\omega\tau} \tag{7.18}$$

where ω is the complex frequency of the incident wave. Note that this reflection coefficient is a function of frequency, in contrast with a fixed conductivity model, Eq. (7.13), in which the reflection coefficient was independent of any of the wave properties. The τ generator model thus shows that for low frequency waves ($\omega\tau \ll 1$), the generator acts like a current generator, while for higher frequency waves ($\omega\tau \gg 1$), a voltage generator. Since $\tau = h/v_A$, where h is the height of the generator region, this is consistent with the generator conductivity model since low Alfvén speed is consistent with a voltage generator in all cases. It might be noted that one drawback of the τ generator model is that the reflection coefficient is always non-zero, and so a perfectly absorbing boundary condition cannot be modeled in this way.

7.3.4 Dynamic Generator Models

The physical processes occurring in the generator region, such as the generation of field-aligned currents and the transfer of the energy and momentum into the generator region cannot be explained by these oversimplified models of current and voltage generators. In such models, the generator is treated as a confined region, where the variation of the electric field along the background magnetic field is neglected. In reality, the generator region cannot be well defined as a confined region around the equatorial plane or in the tail. There is no rigid interface between the generator and the rest of the auroral current system, and the electric field and magnetic field in the generator region are not constants. In order to determine the properties of the generator, a more detailed study of the physical phenomena occurring in the generator region and its effect on the whole auroral current system is necessary.

For a quasi-steady model, the generation of field-aligned currents is often associated with large-scale convection. The generation of field-aligned current is determined not only by the driving force in the generator region, but also by the frictional drag of the ionosphere (e.g., Haerendel, 2001b). In a collisionless plasma, the internal resistance of the generator caused by resistivity is negligible. Therefore a large-scale, steady state generator is approximately a voltage generator. In a dynamical case, the perturbed electric and magnetic fields are carried by means of a shear Alfvén wave. The internal resistance of the generator plasma corresponds to a wave impedance, and the generator conductivity is related to Alfvén conductance; thus, such dynamical generators are often approximately current generators (e.g., Lysak, 1985).

The physical process of the generation of field-aligned current filaments carried by shear Alfvén waves has been described by Song and Lysak (1989a). They pointed out that not all mechanisms of field-aligned current generation have been included in the traditional descriptions of field-aligned current generation given by e.g., Hasegawa and Sato. (1979); Vasyliunas (1984). For example, the compression-related perpendicular current satisfies $\nabla_\perp \cdot \mathbf{J}_\perp = 0$, and so does not enter into the traditional formalism as a source for field-aligned current generation. Nevertheless, field-aligned current filaments can be generated from a divergence-free perpendicular current if a localized breaking of the frozen-in condition occurs. In this process, a field-aligned current is generated that comes out of one ionosphere and goes into the conjugate ionosphere, is generated.

This process can be caused by nonlinear wave mode conversion (e.g., Song and Lysak, 2000). When fast mode wave packets impinge on a current sheet, the kinetic energy carried by the wave packets is converted into electromagnetic energy, the current sheet becomes thin and intense, and localized and time-dependent breakdown of the frozen-in condition can occur if the threshold conditions are satisfied (Song and Lysak, 1994). This nonlinear mode conversion from the compressional to the shear mode is a current dynamo process. The interaction of compressional fast mode wave packets with a current sheet can be described as the interaction of two magnetic flux tubes surrounded by current loops. This interaction results in the formation of twisted flux ropes after reconnection. This process can be described in terms of the magnetic helicity, $K = \int \mathbf{A} \cdot \mathbf{B} dV$, which describes the magnetic field self-linkage (e.g., Moffatt, 1969). Two flux tubes have relative mutual helicity $K = \pm\Phi^2$ before the interaction, where Φ is the magnetic flux carried by the tube and the sign is determined by the right hand rule (e.g., Berger and Field, 1984). By magnetic helicity conservation, each twisted flux tube formed after reconnection carries a half turn twist (e.g., Berger and Field, 1984; Song and Lysak, 1989b). The twisted magnetic structure radiates away as Alfvén waves carrying field-aligned current. This kind of generator is basically a current generator, which irreversibly generates field-aligned current filaments. In addition to directly generating field-aligned currents, this process provides a source for viscous and magnetic drag, which causes an enhanced large-scale convection producing the large-scale discharge field-aligned current. The irreversibly generated field-aligned current filaments overlap with the large-scale steady field-aligned current to form the total field-aligned current system (Song, 1998).

Since information about the state of the ionosphere and the auroral acceleration region is carried back to the generator region by Alfvén waves on the Alfvén transient time scale, generator dynamics should be discussed in the context of the whole auroral current system. Auroral processes, including both MHD wave-related and convection-related processes, operate at widely varying spatial and temporal scales. The physics of these processes has still not been completely determined, and future work must consider this important unsolved problem in order to achieve a more complete understanding of the auroral current system.

7.4 Resonant Alfvén Waves

Alfvén wave propagation in the magnetosphere is strongly influenced by the spatial structure of the magnetospheric plasma and magnetic field and by natural boundaries such as the ionosphere, plasmapause and magnetopause. This structure supports an Alfvén continuum, which can be resonantly excited, for example, by solar wind buffeting of the magnetosphere and magnetotail activations. At least three types of resonant Alfvén oscillations are important in auroral dynamics.

The first occurs in regions where the transverse gradient in Alfvén speed is steep, e.g., in the low-latitude and plasma sheet boundary layers. Resonant MHD mode coupling is relatively efficient in these regions and gives rise to large, earthward-directed Alfvén Poynting fluxes that deposit energy at low altitude. These Alfvén wave emissions tend to be broadband in frequency.

The second type of resonant oscillation has a discrete frequency spectrum and occurs on magnetic field lines with both ends anchored in the ionosphere. This configuration gives rise to standing rather than travelling Alfvén waves. It is a special case of resonant mode coupling known as 'field line resonance'. Field line resonances achieve large amplitudes even in regions where the Alfvén speed gradient is weak because the reactive energy in a driven resonance is confined in the transverse direction by magnetic guidance and in the parallel direction by the conducting ionosphere. Typical periods of field line resonances at auroral latitudes range from one to ten minutes, corresponding to frequencies of 1–20 mHz.

The third type of resonant oscillation occurs in the topside ionosphere and low-altitude magnetosphere where the plasma density decreases exponentially with altitude. The resulting steep increase in Alfvén speed traps Alfvén waves in the region below the Alfvén speed peak and above the conducting ionosphere, as seen in Section 4.3. (The formation of an auroral acceleration region near 1 R_E altitude also facilitates Alfvén wave confinement.) Typical periods of oscillation for this resonator range from one to ten seconds. The discrete spectra of both auroral resonator oscillations and field line resonances are found in auroral observations.

7.4.1 Resonant Mode Coupling

The absorption of energy by Alfvén waves via resonant mode coupling occurs at points x_0 where the frequency ω of an externally excited fast compressional wave or incompressible surface wave matches the local Alfvén frequency, $\omega_A = k_\parallel v_A(x_0)$. The bandwidth and spatial extent of resonantly excited Alfvén waves depends on the frequency and wavenumber spectrum of the external driver.

In outlining the basic theory of resonant mode coupling, we avoid the complexities of the magnetic geometry of the magnetosphere and assuming a slab geometry, where the field lines are straight and point in the z direction, with the ionosphere lying in the xy plane. The y direction is taken to be the azimuthal direction in the magnetosphere or the dawn-dusk direction in the magnetotail, with the x coordinate projecting to magnetic latitude in the ionosphere. MHD mode coupling occurs in both finite- and zero-β plasmas. For simplicity, we proceed from the cold plasma approximation with the electric and magnetic fields described by

$$\frac{\partial \mathbf{b}}{\partial t} = -\nabla \times \mathbf{E}, \tag{7.19}$$

$$\frac{\partial \mathbf{E}_\perp}{\partial t} = v_A^2 \nabla \times \mathbf{b} \tag{7.20}$$

Eq. (7.20) is derived from Ampere's law and one-fluid force balance, and identifies the perpendicular current as a polarization current. At this point, we assume ideal MHD with $E_\parallel = 0$.

If we assume field quantities vary as $\exp i(k_\perp y + k_\parallel z - \omega t)$, then the components of these equations separate into two groups, one containing E_y and b_x, and the other containing E_x, b_y, and v_z. This first group has the dispersion relation $\omega = k_\parallel v_A$, and thus corresponds to the shear Alfvén wave, while the second has $\omega = k v_A$ (because the sound speed is zero in a cold plasma) and corresponds to the fast mode. The slow mode has been eliminated by the cold plasma approximation.

If we now keep the x derivatives and allow for a possible variation of the Alfvén speed in the x direction, Eqs. (7.19) and (7.20) can be combined into a single equation for the b_z component:

$$\frac{d^2b_z}{dx^2} - \frac{dK^2/dx}{K^2 - k_\parallel^2}\frac{db_z}{dx} + \left(K^2 - k_\parallel^2 - k_\perp^2\right) b_z = 0, \tag{7.21}$$

with the definition, $K(x) = \omega/v_A(x)$.

First consider the case of no spatial variation of the Alfvén speed, $dK^2/dx = 0$. Then it can easily be seen that there are two classes of solutions. When $K^2 > k_\parallel^2 + k_\perp^2$, the solutions for b_z are oscillatory in x, with a wave number $k_x^2 = K^2 - k_\parallel^2 + k_\perp^2$, or rewriting this in the form of a dispersion relation, $\omega^2 = (k_x^2 + k_\perp^2 + k_\parallel^2)v_A^2$, which is recognized as the dispersion relation for the fast (compressional) mode. This should not be surprising since b_z is the compressional component of the magnetic field. When K^2 is very large compared to the other wave numbers, we can include the effect of weak inhomogeneity by means of the WKB theory. It can be seen that the position where $K^2(x) = k_\perp^2 + k_\parallel^2$ is a turning point of the solution. Clearly when $K^2 < k_\perp^2 + k_\parallel^2$, we will have exponential solutions to the wave equation, provided we are still far from the singular point where $K(x) = k_\parallel$. The evanescent wave structure in this region has the form of a surface wave.

At the singularity, $\omega = k_\parallel v_A(x)$, corresponding to the magnetic surface where the Alfvén dispersion relation is satisfied. The wave equations (7.19) and (7.20) can be used to find

$$E_x = -\frac{\omega k_\perp}{K^2 - k_\parallel^2} b_z \tag{7.22}$$

This shows that if b_z is finite at the resonant location, E_x will go to infinity, as will the corresponding magnetic component b_y. Thus, at this point, in the absence of any dissipation, it appears that the transverse wave solution becomes infinite. This is the phenomenon of resonant mode coupling, first discussed in the MHD context by Southwood (1974) and Chen and Hasegawa (1974).

Resonant mode coupling channels a portion of the energy contained in isotropically propagating compressional waves (or transversely propagating surface waves) into parallel propagating Alfvén waves. Poynting's theorem in the simple slab geometry considered here relates power absorption to the x component of the energy flux transmitted to the resonance layer by fast compressional and/or surface waves. If we assume an absorption layer of thickness $x_2 - x_1$, and let $\mathbf{J}$ and $\mathbf{E}^*$ represent the current density and complex conjugate of the electric field, then the power W absorbed per unit area in the yz plane is

$$\frac{dW}{dt} = \frac{1}{2}\mathrm{Re}\int_{x_1}^{x_2} \mathbf{J}\cdot\mathbf{E}^*\,dx = \frac{1}{2}\frac{\omega_0}{\mu_0}\mathrm{Im}[E_y^* b_z]_{x_1}^{x_2} \tag{7.23}$$

The last expression may be evaluated in terms of background parameters and the boundary conditions at x_1 and/or x_2 using the modified Bessel function solutions (Chen and Hasegawa, 1974) derived from Eqs. (7.21) and (7.22). A graphical form

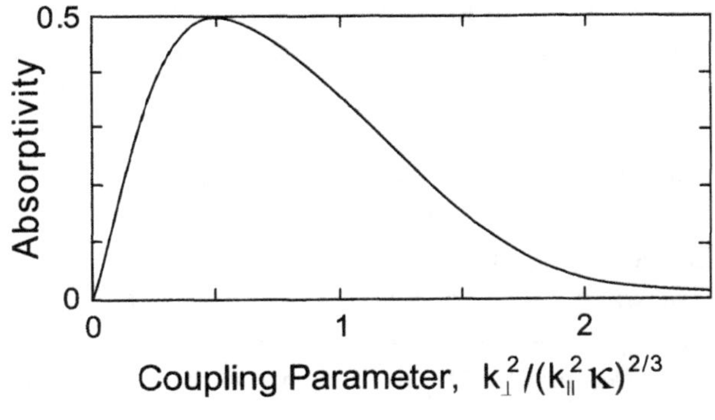

Figure 7.3. Absorptivity as function of the coupling parameter $k_\perp^2(L_n/k_\parallel^2)^{2/3}$. Adapted from Kivelson and Southwood (1986).

of the result is shown in Figure 7.3. Absorptivity is the fractional power absorbed per wave cycle. The coupling parameter $k_\perp^2(L_n/k_\parallel^2)^{2/3}$ depends on the density gradient length scale, $L_n = n_0(\mathrm{d}_{n0}/\mathrm{d}x)^{-1}$, assumed to be constant in calculating the absorptivity shown in the figure.

The peak in the absorptivity at $k_\perp^2(L_n/k_\parallel^2)^{2/3} \approx 0.5$ is a consequence of two competing effects. At large $k_\perp$, the amplitude of the evanescent fast/surface wave at resonance is exponentially small, rendering energy transfer inefficient. In the opposite limit of small $k_\perp$, mode coupling between the Alfvén and fast/surface waves is weak, thereby reducing the absorption. For $k_\perp = 0$, the modes are uncoupled because the plasma motion in the fast mode, purely in the x direction, creates no magnetic pressure forces in the y direction and the transverse mode motion, purely in the y direction, creates no magnetic compression (Kivelson and Southwood, 1986).

To give some idea of characteristic values, consider mode conversion in the PSBL where $L_n \approx 1000$ km. Then a magnetotail disturbance spanning say 5 $R_E$across the tail and 6.5 $R_E$in the magnetic field-aligned direction (GSM x direction) yields a coupling parameter of 0.5. A compressional driver smaller than 5 $R_E$in the cross tail direction or larger than 6.5 $R_E$in the tailward direction would not reduce the absorption as strongly as a longer cross tail or shorter tailwise wavelength.

The resonantly excited Alfvén waves develop increasingly fine scale structure with time due to a phenomenon called 'phase mixing'. Consider an Alfvén wave with a long perpendicular wavelength which propagates along field lines in a nonuniform plasma. Assume that at some time $t = 0$, the phase front of the wave is perpendicular to the magnetic field. Since the Alfvén speed has a gradient across the field lines, the phase front will propagate down the field line more rapidly in the region of fast phase velocity than in the slower regions. Thus, the phase front will become inclined to the magnetic field, and will become more inclined as time goes by. It can be easily seen that, after a time t, this wavefront will develop perpendicular structure with a characteristic length scale on the order of $L_\perp \sim (\omega'_A t)^{-1}$, where $\omega'_A = k_\parallel dv_A/dx$ is the gradient in the resonant Alfvén frequency (Allan and Wright, 2000).

The evolution of fine-scale structure in the resonant Alfvén wave is ultimately limited by effects not contained in the MHD approximation. The most important of these in the outer magnetosphere is thermal dispersion, which is activated when the transverse wavelength approaches the length scale $\ell = (\rho^2 L_n)^{1/3}$ where $\rho = \rho_i(3/4 + T_e/T_i)^{1/2}$ is an effective ion gyroradius. T_e and T_i are the electron and ion temperature and $\rho_i = (2T_i/m_i)^{1/2}/\omega_{ci}$ is the ion gyroradius. The resulting 'kinetic' Alfvén wave (Hasegawa, 1976) acquires a parallel electric field (cf. Eq. 3.77) and a finite perpendicular group velocity $v_{g\perp}$. However, the wave is still magnetically guided to a high degree because, typically, $v_{g\perp}/v_{g\parallel} \ll 1$. In effect, the perpendicular group velocity induced by plasma thermal motion is the medium's way of resolving the resonance – it enables 'radiation' of energy away from the resonant surface and eliminates the logarithmic singularity at the resonant surface. The resulting waveform in x involves a combination of Airy functions.

Evidence for kinetic Alfvén waves in earthward-directed, Alfvén Poynting fluxes, including field-aligned electron acceleration by the wave parallel electric field, has been found in recent satellite measurements in the mid-altitude, plasma sheet boundary layer (Wygant et al., 2002), as discussed in Section 4.4.2. Satellite measurements in the low-altitude magnetosphere adjacent to and equatorward of the polar cap boundary, corresponding to the low-altitude projection of the LLBL in the dayside auroral zone and the PSBL on the nightside, also provide evidence that dispersive Alfvén waves propagate vast distances from their source regions without encountering overwhelming losses. As discussed in Section 3.4.3, kinetic dispersion is supplanted at low altitude by inertial dispersion.

Conservation of energy for an Alfvén wave propagating into a region of increasing magnetic field and variable plasma density requires $(\mathbf{E} \times \mathbf{b}) \cdot \mathbf{B_0}/B_0^2$ to be constant along the flux tube. If dispersive effects and the parallel inhomogeneity are weak, WKB-theory (see Section 3.4.3) can be applied. Then the relation between the wave electric and magnetic fields is approximately $E_\perp = v_A b_\perp$; energy conservation then implies that $E_\perp \propto (v_A B_0)^{1/2}$. The wave electric field is therefore largest in the low-altitude magnetosphere near the peak in v_A. Eq. (3.77) indicates that the parallel electric field of inertial Alfvén waves should also be large in the same region. This effect explains the development of electron inertial-scale fine structure in the suprathermal electron beams observed at low altitude near the polar cap boundary (Chaston et al., 2002b), as was discussed in Section 4.4.2.

7.4.2 Field Line Resonances

The characteristic length of a (closed) field line in the magnetosphere varies between about 10 and 30 R_E, and a typical Alfvén speed in the equatorial magnetosphere is about 1000 km s^{-1}. Consequently, waves with a period exceeding 1 minute must have a wavelength comparable to the length of the field line. Observations of field line resonances were presented in Section 4.4.3.2. Dungey (1955) first noted that standing Alfvén waves with periods on the order of minutes could be established on geomagnetic field lines. In a simple model where the field-aligned inhomogeneity is ignored and the ionosphere can be considered as a perfect conductor, upon which the tangential electric field must vanish, the parallel wavelength $\lambda_{\parallel}$ of such waves is quantized: $n\lambda = 2\ell_F$. ℓ_F is the length of the field line between ionospheric endpoints, and n is a positive integer. The fundamental mode with $n = 1$ has one-half wavelength between ionospheric boundaries, the second harmonic, $n = 2$, has a full wavelength, and so on. The corresponding eigenfrequencies are

$$f = v_A/\lambda_{\parallel} = nv_A/2\ell_F \tag{7.24}$$

Since the Alfvén speed in the equatorial magnetosphere beyond the plasmapause decreases with increasing radial distance, while the field line length increases, the fundamental frequency of these standing waves should decrease with increasing magnetic shell parameter L. This behavior is observed in the ultra-low-frequency band on low-inclination, elliptical orbit satellites that traverse the magnetosphere more or less radially. Historical reviews of this phenomenon have been presented by Hughes (1994) and Glassmeier (1995).

Field line eigenmodes can be driven by various sources in the solar wind, magnetosheath and magnetotail. One mechanism that has been studied extensively involves the Kelvin-Helmholtz instability at the magnetopause. This instability takes the form of an incompressible surface wave with an amplitude that decays exponentially with increasing distance from the magnetopause boundary. Southwood (1974) and Chen and Hasegawa (1974) independently realized that the frequency of this decaying wave should match the eigenfrequency of a nearby field line eigenmode. The mathematical description for the harmonically driven eigenmode problem in the cold plasma approximation using a 'box model' for the magnetosphere is the same as that given in Section 7.4.1 except that ω and $k_{\parallel}$ take on the integer values noted above in order to satisfy the boundary conditions at the ionosphere.

The singular solutions that occur on field lines with eigenfrequency matching the frequency of the driver are called 'field line resonances'. The field line resonance model explains a number of peculiar properties of discrete-frequency micropulsations observed by ground-based magnetometers (Samson and Rostoker, 1972). In particular, the polarization of the oscillations switches sense at noon, consistent with a Kelvin-Helmholtz source that propagates westward in the morning and eastward in the afternoon. And the polarization reversal observed at the latitude of maximum amplitude can be traced to the singularity that distinguishes the field line resonance.

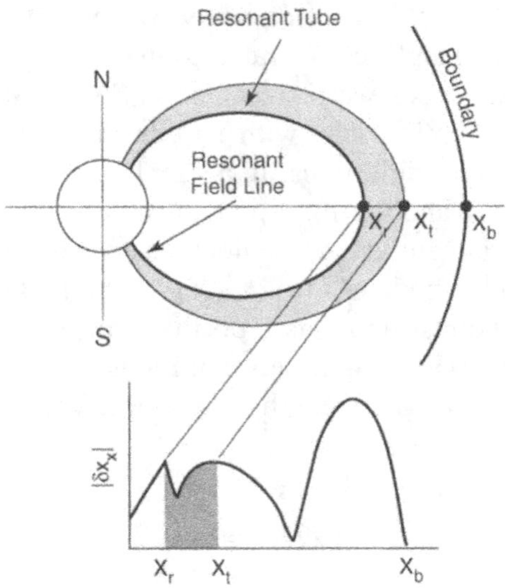

Figure 7.4. Schematic representation of a dawn-dusk cross-section through the magnetospheric cavity viewed, for example, from the tail (upper portion). The sketch below is a characteristic radial profile of the MHD wave amplitude for B_x, as predicted by the theory of Zhu and Kivelson (1988). Points of interest include the position of the magnetospheric boundary (x_b), the turning point of the compressional wave (x_t), and the resonance position (x_r). The evanescent fast mode tunnels energy to the resonance in the shaded region.

Although the basic field line resonance model is successful in predicting the wave polarization and amplitude, it does not explain why only specific frequencies are strongly excited as field line resonances. Kivelson and Southwood (1985) elaborated the earlier theory by noting that, in a compressible plasma, the Kelvin-Helmholtz instability could directly excite a propagating fast mode wave rather than the incompressible surface wave. These fast mode waves propagate isotropically throughout the plasma and can form global eigenmodes of the entire magnetosphere. In a non-uniform plasma, this compressional mode couples to the shear Alfvén mode, giving rise to the field line resonance. The radial profile for a shear Alfvén resonance coupled to a second harmonic, global compressional mode is illustrated in Figure 7.4. Since the Kelvin-Helmholtz instability excites a range of perpendicular wave numbers, the frequency of this global mode has to be greater than the cutoff frequency given by $(k_\parallel^2 + k_\perp^2)^{1/2} v_A(\mathbf{x})$ in order to propagate to the position x_t. The perpendicular wavelength is set by the Kelvin-Helmholtz instability and the parallel wavelength by the length of the field line. This frequency is greater than the Alfvén resonance frequency, $\omega_A = k_\parallel v_A(\mathbf{x})$. Consequently, fast mode cutoff must occur at greater radial distances in the magnetosphere than that of the resonant field line. Thus, the compressional wave becomes evanescent before reaching the field line resonance, and must tunnel energy to the resonance.

This basic model is supported by ground-magnetometer observations of waves in the expected frequency range and by satellite observations (e.g., Takahashi and McPherron, 1984; Anderson et al., 1988, 1990) showing that the frequency of the toroidal (i.e., azimuthal) component of the magnetic field varies with L shell, while the frequency of the compressional (parallel to B_0) component is independent of position.

Numerical models describing the coupling between compressional modes and field line resonances in dipole magnetic geometry, with ambient density varying in both the parallel and perpendicular directions, have been developed by Lee and Lysak (1991, 1994) and Proehl et al. (2002). These more realistic numerical calculations produce spectrograms of the frequency of each component of the wave field as a function of the radial and azimuthal coordinates, presented in a form similar to the spectrograms obtained from low-inclination orbit satellites. Diagnostics for energy flow in these simulations show that MHD wave power flows radially inward from a magnetopause surface oscillation as a compressional wave to the resonant flux surface, is diverted azimuthally at the resonant surface, and becomes field-aligned in the resonant surface by coupling to the magnetically guided Alfvén wave at locations where the shear mode is in azimuthal phase quadrature with the compressional driver. Such studies give some indication of the circuitous paths energy flow can take before reaching the auroral ionosphere; and they vividly illustrate the ambiguities one can encounter in trying to identify the magnetospheric source region of auroral processes based on single point measurements. One-dimensional eigenvalue calculations have also been implemented in more complex magnetic geometries representative of the magnetosphere (Singer et al., 1981; Rankin et al., 2000). These calculations provide useful insights into the origins of anomalously low eigenfrequencies observed by ground-based instruments (Ruohoniemi et al., 1991; Ziesolleck and McDiarmid, 1995; Mathie and Mann, 2000).

Studies of field line resonance phenomena in an auroral dynamics context focus on the physical processes that limit the secular behavior of resonantly excited Alfvén waves. For fundamental torsional waves standing along field lines between northern and southern ionosphere, it has been found observationally that the north-south width of the resonance is often broad, spanning about $1-2°$ in latitude, when ionospheric Joule heating is the primary dissipation mechanism (Walker et al., 1979; Glassmeier et al., 1984). The field-aligned current carried by the resonant Alfvén wave must close in the ionosphere via a Pedersen current and the resulting Joule dissipation damps the standing wave. The damping decrement is given by $\nu = \omega_A \ln(1/|R|)$, where R is the ionospheric reflection coefficient given by Eq. (3.64) and $\omega_A = k_\parallel v_A$ is the resonant Alfvén wave frequency (Ellis and Southwood, 1983). Note that in the usual case $\Sigma_P \gg \Sigma_A$, so the damping rate becomes $\nu = (2\Sigma_A/\Sigma_P)\omega_A$. This damping can be heuristically included in the model by replacing the wave frequency ω by $\omega - i\nu$. Thus the Alfvén wave number $K^2 \to K^2 - 2i\omega\nu/v_A^2$. When the damping is small compared to the wave frequency, it may be neglected for the most part, except near the resonance. The addition of a negative imaginary part to K^2 resolves the MHD singularity and provides a damping mechanism for the resonance.

The width of the resulting field line resonance is determined by a balance between the competing effects of ionospheric Joule dissipation, which tends to broaden the resonance, and phase mixing which tends to form narrow structure as described in the previous section. The saturated width of the resonance can be estimated by equating the ionospheric damping rate ν given above to the inverse time scale for phase mixing, $L_{\perp} d\omega_A/dx$. This provides the estimate, $L_{\perp} \sim \nu/(d\omega_A/dx)$. A more detailed analysis of the widths of field line resonances (Mann et al., 1995) confirms this expectation.

There are also indications from ground-based observations that some field line resonances contract to much smaller widths than can be explained by saturation due to ionospheric Joule heating. Some resonances exhibit structure at length scales smaller than the 40-km scale resolution of HF radars (Walker et al., 1992); and 5-minute time-scale oscillations in 10-km scale auroral arcs observed from the ground have been interpreted as evidence for dispersive-scale resonances (Samson et al., 1996). Figure 4.49 shows an example of a 50-km wide resonance measured by a polar orbiting satellite with embedded structure in the electric field and field-aligned current at length scales on the order of 5 km. The downgoing electron energy flux measured by the satellite is the in situ signature of a 10-km scale auroral arc recorded by an all-sky camera sited below the satellite at Gillam, Manitoba (Lotko et al., 1998).

A variety of collisionless processes become operative when the resonance contracts to small transverse scales. These include dispersion due to electron inertia (Wei et al., 1994; Streltsov and Lotko, 1995), electron pressure and finite ion Larmor radius (Streltsov et al., 1998; Rankin et al., 1999a), anomalous resistivity (Streltsov and Lotko, 1999) associated with microturbulence induced by the wave field-aligned current, a reactive 'nonlocal impedance' (Rankin et al., 1999b) sustained by trapped electrons, and Landau (Lysak and Lotko, 1996) and transit-time damping (Koechlin and Samain, 1971). All are potentially relevant on auroral field lines. As discussed in Section 3.4.3, finite electron inertia and pressure give rise to wave parallel electric fields (Eq. 3.77) and auroral particle acceleration. For the event study shown in Figure 4.49, both types of dispersion together with anomalous resistivity were found to influence the structure of the resonance (Streltsov and Lotko, 1999). The 50-km scale width of the resonance is regulated by anomalous resistivity, while the 5-km scale 'electrostatic shock' structure evident at 04:26:01 UT in the figure is essentially an inertial Alfvén wave.

In general, competition between the various nonideal effects noted above determines how the resonance will saturate and how it impacts the auroral ionosphere and low-altitude magnetosphere. Ionospheric dissipation is proportional to the root mean square (rms) electric field amplitude whereas dispersion sets in when the resonance contracts to dispersive length scales. Saturation via dispersion is tantamount to radiation limitation because, as noted in Section 7.4.1, dispersion enables a perpendicular group velocity of the shear Alfvén wave. Narrow resonances are expected in regions where the transverse gradient in the Alfvén speed is large. Thus relatively broad resonances, essentially one-fluid MHD structures without appreciable auroral electron acceleration, are indicative of saturation by ionospheric Joule heating. Such resonances are expected in regions where the perpendicular gradient in v_A is relatively gradual. In contrast, thin resonances, accelerated electron precipitation resulting from dispersively sustained parallel electric fields, and auroral arcs are expected in regions where the transverse gradient in v_A is relatively steep. This picture is complicated, however, by the effects of anomalous and inertial resistivity which depend on the intensity of the field-aligned current density and, therefore, on both the amplitude and the width (via $\nabla \times \mathbf{B}$) of the resonance. Further study is needed to determine the ambient conditions and driving mechanisms that favor one saturation mechanism over another.

A curious feature of dispersive field line resonances arises from the change in sign of the perpendicular group velocity in the inertial and kinetic limits. Since the low-altitude section of the dispersive resonance is dominated by electron inertia while the higher-altitude region is dominated by ion and electron thermal effects, the resonant Alfvén wave propagates (in the perpendicular direction) toward the low Alfvén speed side in the kinetic regime and toward the high Alfvén speed side in the inertial regime. When these two counteracting effects are exactly balanced in a round trip transit between northern and southern ionosphere, the resonant Alfvén wave folds back on itself in a figure-8 pattern (Streltsov and Lotko, 1995, 1996). In this case, narrow field line resonances can achieve large-amplitudes. This effect has been verified in a large-plasma laboratory device (Mitchell et al., 2002).

When the amplitude of a field line resonance becomes sufficiently large, nonlinear effects influence its evolution and saturation. One important such nonlinear and nonresonant effect results from the action of the ponderomotive force excerted by a large amplitude localized wave on the plasma. For instance, a wave of large amplitude $\mathbf{b}$ has nonvanishing wave pressure tensor $\mathsf{P}_b = \mathbf{bb}/\mu_0$. Its gradient $\nabla \cdot \mathsf{P}_b = \mathbf{F}_{b,\mathrm{pmf}}$ is a force that acts on the plasma. This force is called ponderomotive force and contains contribution from the pressure gradient and from the wave stresses which correspond to forces and torsions acting on the plasma-magnetic field configuration.

Ponderomotive effects resulting from the nonlinear $\mathbf{J} \times \mathbf{B}$ term are capable of producing field-aligned ion outflows and density depletions in the topside ionosphere with accompanying density enhancements at higher altitudes where the resonant magnetic-field amplitude is small (Li and Temerin, 1992; Allan, 1993). Ponderomotive effects also mediate nonlinear coupling between the slow and shear Alfvén modes (Rankin et al., 1995), which, in turn, modifies the frequency of the field line resonance. Nonlinear electron inertia is activated when the electron bulk flow becomes comparable to the local Alfvén speed. This effect can produce large parallel electric fields at perpendicular length scales much greater than the electron inertial length (Rönnmark and Hamrin, 2000; Streltsov et al., 2002). Field line resonances can also sustain parasitic or secondary instabilities. The large velocity shear across the resonance promotes a Kelvin-Helmholtz instability which can evolve into auroral curls and spirals (Rankin et al., 1993). When large Alfvén wave electric fields are impressed on the E-region ionosphere, the resulting ionospheric activity promotes feedback instabilities (Sato, 1978; Pokhotelov et al., 2002) that pump the ionospheric Alfvén resonator modes described in the next section.

7.4.3 Ionospheric Alfvén Resonator

As noted in Chapter 3, a strong inhomogeneity in the Alfvén speed implies that we cannot use the WKB approximation and that we need to solve the complete wave equation for the wave fields. In the topside ionosphere, there is such a strong inhomogeneity due to the sharp exponential decrease in the plasma density with increasing altitude above a few hundred kilometers. This decrease takes place with a typical scale height of less than 1000 km. Since the magnetic field varies only weakly on such spatial scales, the Alfvén speed increases exponentially with a comparable scale height above the ionosphere. Above about 6000 km altitude, the density decrease becomes more gradual, and the decrease in the magnetic field leads to a slow decrease in the Alfvén speed. Observational characteristics of these regions were discussed in Section 4.4.3.

In order to analyze the eigenmodes of this system, it is useful to formulate the Alfvén wave equations in terms of scalar and vector potentials. The fundamental starting point of the analysis is the Maxwell equations in a dielectric medium, as given by Eqs. (7.19) and (7.20). Vector and scalar potentials are introduced in the usual way by $\mathbf{B} = \nabla \times \mathbf{A}$ and $\mathbf{E} = -\nabla\Phi - \partial\mathbf{A}/\partial t$. If we restrict the analysis to transverse waves with $b_z = 0$, where b_z is the perturbation in the magnetic field parallel to the background field, we can describe the magnetic perturbation totally by the z component of the vector potential. This implies that the perpendicular electric field is described totally by the perpendicular gradient in the scalar potential $\mathbf{E}_\perp = -\nabla\Phi$. If the perpendicular wave length is large, so that the parallel electric field goes to zero (cf. Eq. 3.77), we can write:

$$\frac{\partial A_z}{\partial t} = -\frac{\partial \Phi}{\partial z} \tag{7.25}$$

The system must be closed by a gauge condition. The appropriate gauge condition for this case is analogous to the Lorentz gauge:

$$\frac{\partial \Phi}{\partial t} + v_A^2 \nabla \cdot \mathbf{A} = 0 \tag{7.26}$$

Noting that only the parallel component of the vector potential is non-zero, Eq. (7.26) becomes:

$$\frac{\partial \Phi}{\partial t} = -v_A^2 \frac{\partial A_z}{\partial z} \tag{7.27}$$

It is easily verified that the perpendicular gradient of Eq. (7.27) yields the electric field Eq. (7.20). It should also be noted that the Alfvén speed on auroral field lines can become quite large, indeed, close to the speed of light. In this case, the Alfvén speed in these equations should be set to

$$v_A^2 \rightarrow \frac{v_A^2}{1 + v_A^2/c^2} \tag{7.28}$$

Eqs. (7.25) and (7.27) then represent a complete set of equations describing the shear mode Alfvén wave. The inhomogeneity of the flux tube enters through the Alfvén speed in Eq. (7.27). In order to simplify the problem to the point where we can find analytic solutions, we shall adopt the simple profile (Greifinger and Greifinger, 1968; Trakhtengertz and Feldstein, 1984):

$$v_A^2(z) = \frac{v_{AI}^2}{\epsilon^2 + e^{-z/h}} \tag{7.29}$$

where v_{AI} is the Alfvén speed at the ionospheric altitude, h is the ionospheric scale height, and ϵ is a small parameter. Here it should be recognized that the coordinate z increases in the upward direction. Note that at high altitudes, $z \gg h$, the Alfvén speed simply becomes $v_{AM} = v_{AI}/\epsilon$. Thus the smallness of the parameter ϵ represents the fact that the Alfvén speed in the magnetosphere is much larger than that in the ionosphere. Note that while the density profile given by Eq. (7.29) does not take into account the decrease of the Alfvén speed at high altitudes, it does give a reasonable representation of the Alfvén speed profiles in the region above the ionosphere.

The system given by Eqs. (7.25) and (7.27) with the Alfvén speed profile (7.29) can be analyzed for its eigenmodes by Fourier transforming in time and in the perpendicular dimensions, leading to the wave equation:

$$\frac{\partial^2 \Phi}{\partial z^2} + \frac{\omega^2}{v_{AI}^2}(\epsilon^2 + e^{-z/h})\Phi = 0 \tag{7.30}$$

This equation can be transformed by switching to the independent variable $x = x_0 e^{-z/2h}$, where $x_0 = 2h\omega/v_{AI}$. This transformation leads to the equation:

$$x^2 \frac{d^2\Phi}{dx^2} + x\frac{d\Phi}{dx} + (x^2 + x_0^2\epsilon^2)\Phi = 0 \tag{7.31}$$

This equation can be recognized as being Bessel's equation; however, in the ordinary form of Bessel's equation, the square of the order enters with a minus sign.

Thus, in this case, the Bessel function solutions have an imaginary order, and the solutions to the wave equation take on the form:

$$\Phi = A^{+} J_{ix_0\epsilon}(x) + A^{-} J_{-ix_0\epsilon}(x) \tag{7.32}$$

It can be shown (Lysak, 1991) that these two solutions correspond to the downgoing and upgoing waves, respectively. This follows from considering the large z limit of the Bessel functions, which corresponds to $x \ll 1$. In this limit:

$$\Phi^{\pm} = J_{\pm ix_0\epsilon}(x) \approx \frac{(x/2)^{\pm ix_0\epsilon}}{\Gamma(1 \pm ix_0\epsilon)} \tag{7.33}$$

where Γ is the gamma function. Writing this in terms of the original variable z we find that

$$\Phi^{\pm} \propto e^{\mp i(x_0\epsilon/2h)z - i\omega t} = e^{-i\omega(t \pm z/v_{am})} \tag{7.34}$$

where we have inserted the time dependent phase factor $e^{-i\omega t}$ implied by the Fourier transformation. Eq. (7.34) shows that the two solutions are downgoing and upgoing plane waves at large altitudes.

Determination of the eigenfrequencies for these waves requires the application of the boundary condition at the ionosphere. Noting that the field-aligned current can be written as:

$$j_z = \frac{1}{\mu_0}(\nabla \times \mathbf{B})_z = -\frac{1}{\mu_0}\nabla_\perp^2 A_z \tag{7.35}$$

If we now write the ionospheric boundary condition discussed in Chapter 3 in terms of these potentials, we find

$$\nabla_\perp^2(A_z + \mu_0 \Sigma_P \Phi) = 0 \tag{7.36}$$

The Fourier transform in time of Eq. (7.25) implies that:

$$A_z = -\frac{i}{\omega}\frac{d\Phi}{dz} = \frac{i}{v_{AI}}\frac{x}{x_0}\Phi' \tag{7.37}$$

where $\Phi' = d\Phi/dx$, and evaluating the condition (7.36) at the ionosphere, which corresponds to $x = x_0$, we can write:

$$i\Phi' + \alpha\Phi = 0 \tag{7.38}$$

where $\alpha = \Sigma_P/\Sigma_{AI}$ measures the Pedersen conductivity in units of the Alfvén conductivity at the ionosphere. The boundary condition (7.38) can be considered to be an eigenvalue equation for the parameter x_0, which is essentially the mode frequency scaled by the fundamental frequency in the problem, which is $v_{AI}/2h$. The upper boundary condition can be taken to be a radiation boundary condition, which picks out the $J_{ix_0\epsilon}$ solution for inward propagating waves and the $J_{-ix_0\epsilon}$ solution for outward propagating waves. It can be easily seen that the solutions are parameterized by ϵ, which is always much smaller than one, and α, which may be small or large.

The eigenvalue Eq. (7.38) can be easily solved in the limits of high and low conductivity. First, note that for $x_0\epsilon \ll 1$, the Bessel function solutions can be expanded in order, noting that the derivative of $J_\nu(x)$ with respect to ν evaluated for $\nu = 0$ is $(\pi/2)Y_0(x)$ (Abramowitz and Stegun, 1964, equation 9.1.64). Thus, the solutions can be written:

$$J_{\pm ix_0\epsilon}(x) \approx J_0(x) \pm \frac{ix_0\epsilon\pi}{2} Y_0(x) \tag{7.39}$$

When the scaled conductivity $\alpha \to 0$, the second term in (7.38) can be ignored, and the eigenvalue condition just yields $\Phi' = 0$. Thus, to lowest order the eigenvalues are given by $x_0 = j_{1,s}$, where the values $j_{1,s}$ are the zeroes of $J_1(x)$, since J_1 is the negative of the derivative of J_0. In this limit, the incoming solution has a positive growth rate, since energy is entering the system, while the outgoing solution is damped (Lysak, 1991).

In the limit of high conductivity, $\alpha \gg 1$, the derivative term in (7.38) is negligible, and the eigenvalues are now the zeroes of J_0. Again the incoming solution has a positive growth rate while the outgoing solution is damped. For arbitrary values of the conductivity, the eigenfrequencies make a transition from the zeroes of J_1 to the zeroes of J_0 as the conductivity is increased. The transition takes place at $\alpha \sim 1$, which also corresponds to the maximum damping of both modes. This is not surprising since it corresponds to the point where the Alfvén conductivity at the ionosphere matches the ionospheric conductivity, which corresponds to no reflection in the uniform case as we saw earlier.

This effective resonant cavity also influences the reflection coefficient of Alfvén waves on auroral field lines. The reflection coefficient may be defined as the ratio of the amplitudes of the upgoing wave to the downgoing wave, $R = A^-/A^+$. This ratio must be determined by applying the boundary condition (7.38) to the combined wave solution at the ionosphere (i.e., at $x = x_0$). This condition implies:

$$R \equiv \frac{A^-}{A^+} = -\frac{iJ'_{ix_0\epsilon}(x_0) + \alpha J_{ix_0\epsilon}(x_0)}{iJ'_{-ix_0\epsilon}(x_0) + \alpha J_{-ix_0\epsilon}(x_0)} \tag{7.40}$$

This relation has been calculated numerically for a number of cases by Lysak (1991). This relationship reduces to the usual form (3.64) in the limit of low frequencies, $x_0 \ll 1$. In this limit, $J_{\pm ix_0\epsilon}(x_0) \approx 1$ and $J'_{\pm ix_0\epsilon}(x_0) \approx \pm i\epsilon$, so that $R \approx (\epsilon - \alpha)/(\epsilon + \alpha) = (\Sigma_{AM} - \Sigma_P)/(\Sigma_{AM} + \Sigma_A)$, as in the uniform case. At these very low frequencies, the wavelength of the Alfvén wave is so long that it effectively does not see the cavity and so the reflection coefficient is based on the magnetospheric Alfvén conductivity. On the other hand, for very large frequencies, the reflection coefficient reduces to $R = (1-\alpha)/(1+\alpha) = (\Sigma_{AI} - \Sigma_P)/(\Sigma_{AI} + \Sigma_P)$. This high frequency limit corresponds to the case where WKB theory is valid, and thus the reflection coefficient is described by the ionospheric Alfvén speed, rather than that in the magnetosphere.

In general, the reflection coefficient shows features at the resonant frequencies of the cavity. When the cavity is very deep (i.e., ϵ very small) and the conductivity is high, the reflection coefficient shows a sharp decrease at the resonant frequencies. This may be interpreted in terms of trapping of the wave in the cavity, so that, after multiple interactions with the ionosphere, the wave is eventually absorbed. When ϵ is not so low, e.g., $\epsilon \sim 0.1$, these features can still be seen, although they are not so sharp. When the conductivity is lower, the wave does not reflect as strongly; however, in this case the presence of the cavity can lead to an enhancement of the reflection compared with the low frequency limit, which may be interpreted in terms of a direct reflection of the wave from the Alfvén speed gradient above the ionosphere.

Although this model was calculated using the ideal MHD approximation, it is an easy matter to introduce the two-fluid effects (see Section 3.4.3). First of all, it can be noted that the strong gradients in the Alfvén speed occur at low altitudes, where the electron inertial effect is dominant. This effect implies that the parallel electric field is given by the first term in Eq. (3.70):

$$E_z = \frac{m_e}{ne^2}\frac{\partial j_z}{\partial t} = -\lambda_e^2 \frac{\partial}{\partial t}\nabla_\perp^2 A_z \tag{7.41}$$

where the last step introduces the vector potential as in (7.35) and $\lambda_e = c/\omega_{pe}$ as before. Writing this parallel electric field in terms of the scalar and vector potentials in the usual way modifies Eq. (7.25) to read:

$$(1 - \lambda_e^2 \nabla_\perp^2)\frac{\partial A_z}{\partial t} = -\frac{\partial \Phi}{\partial z} \tag{7.42}$$

In Fourier space, this equation can be written as:

$$\frac{\partial \Phi}{\partial z} = i\omega(1 + k_\perp^2 \lambda_e^2) A_z \tag{7.43}$$

Combining this relationship with Eq. (7.27), we can generalize Eq. (7.30) to read:

$$\frac{\partial}{\partial z}\left[\frac{1}{1 + k_\perp^2 \lambda_e^2(z)}\frac{\partial \Phi}{\partial z}\right] + \frac{\omega^2}{v_A^2(z)}\Phi = 0 \tag{7.44}$$

This equation does not have analytic solutions in general, but numerical solutions have been presented (Lysak, 1993). Note that the profiles of the Alfvén speed and of the electron inertial length are important in Eq. (7.44). In fact, even the perpendicular wave number should be considered to be a function of z since it will change due to the changing size of the flux tube, indicating roughly that $k_\perp^2 \propto B_0$. The results of Lysak (1993) were obtained by integrating Eqs. (7.27) and (7.43) up from the ionosphere, starting with the ionospheric boundary condition (7.36). The resulting solution can then be decomposed into incoming and outgoing wave solutions at the assumed upper boundary, and the reflection coefficient can be determined by their ratio. More recently, Lysak (1998) has presented solutions for the wave polarization by constructing the b_y/E_x ratio (or equivalently, A_z/Φ ratios) in order to make comparisons with spacecraft data.

7.5 Propagating Arc Models

As described in Section 2.3, auroral arcs have been observed to move relative to the background plasma in which they are embedded. That is, they undergo proper motion. This section discusses the possible roles of proper motion in theories of arc formation. There are at least three such roles. Firstly, proper motion provides an effective time dependence in the frame of the plasma as it drifts across static structures, thereby allowing time-dependent physics such as the ion polarization drift (Rothwell et al., 1994) or inductive generation of electric fields (Seyler, 1990). Secondly, cross-arc convection provides a continuous source of fresh plasma to arc field lines (Haerendel, 1983). And finally, it provides energy input into the arc, either in the form of plasma drift kinetic energy (Lysak, 1991) or as magnetic energy (Haerendel, 1994; Knudsen, 1996).

7.5.1 Ionospheric Feedback Models

Atkinson (1970) proposed what is probably the first propagating arc model based on simple conservation laws and an ionospheric Ohm's law. Specifically, the ionosphere is modeled as a height-integrated slab with Pedersen and Hall conductivities $\Sigma_P = NP$ and $\Sigma_H = NH$ respectively, where N is the height-integrated ionospheric density and P and H are proportionality constants. Ionospheric plasma density is factored out so that it can be tracked as a dynamical variable that responds to electron precipitation enhancements. The height-integrated horizontal (nominally northward) ionospheric current is

$$I_x = N(PE_x - HE'_y) \tag{7.45}$$

where $E'_y = E_y - U_x B$, E_y is the eastward electric field in the drifting plasma frame, and U_x is the northward drift velocity of the neutral atmosphere. Thus E'_y takes into account that electric fields in the frame of the neutral atmosphere differ from those that would be observed drifting with the plasma.

Horizontal gradients in height-integrated ionospheric current I_x imply field-aligned currents, $\partial I_x/\partial x = j_\parallel$. Upward FAC's in turn imply enhanced electron precipitation and ionospheric density. Each precipitating electron will not only add directly to the ionospheric density, but will generate γ secondary electrons, where γ depends on electron energy. Combining with the electron continuity equation in the ionosphere $dN/dt = S - \alpha(N^2 - N_0^2)$, S being a source term, α the recombination coefficient, and N_0 a constant, the following relation between electron density variations and ionospheric current results in the stationary limit ($\partial/\partial t = 0$):

$$v_d \frac{\partial N}{\partial x} = -\frac{\gamma + 1}{e} \frac{\partial I_x}{\partial x} - \alpha(N^2 - N_0^2) \tag{7.46}$$

Here N_0 is the undisturbed (boundary) value of the height-integrated ionospheric density. In the stationary case $d/dt = v_d \partial/\partial x$, where v_d is the cross-arc plasma drift speed. The proper motion of the arc generates a cross-field polarization drift in the magnetosphere,

$$J_\perp = \frac{\rho_m}{B^2} \frac{dE_\perp}{dt} \tag{7.47}$$

where again $d/dt = v_d \partial/\partial x$. In other words, plasma drift across horizontally-varying perpendicular electric field structures acts to inject current into flux tubes from the side. In the frame of the drifting plasma, the electric fields appear as time-varying structures. Since the resulting current is proportional to d/dt, it is analogous to the displacement current through a capacitor, $C dV/dt$. For this reason, Atkinson and others modeled height-integrated flux tubes as capacitors with lumped capacitance C. This lumped parameter omits important effects such as Alfvén wave propagation, as described later.

Eqs. (7.45) to (7.47) combine to predict spatially periodic oscillations in the height-integrated ionospheric density N, conductivity and field-aligned current, similar to those associated with multiple auroral arcs (see Atkinson, 1970). Physically, these oscillations are the result of an instability in which an ionospheric density enhancement could, in certain parameter regimes, lead to upward FAC, electron precipitation, and further enhancement of the ionospheric density. While the model does not explain electron acceleration or large electric fields near arcs, it does make specific predictions of the width and spacing of multiple parallel arcs; these predictions are of the order of observed dimensions, however the true cause of multiple arcs is considered unresolved.

A series of subsequent 'ionospheric feedback models' built on the basic idea of Atkinson (Ogawa and Sato, 1971; Sato and Holzer, 1973; Sato, 1978), adding several effects including magnetospheric inductance, a conjugate ionosphere, and anomalous resistivity. As models became more sophisticated they were able to generate field-aligned potentials of several hundred volts, and were able to reproduce certain observations relating to the intensity and location of arc-related current systems. The calculated growth times for these systems, occurring within interhemispheric field-line resonances, are of the order of several minutes. Lysak (1991) showed that much faster growth times are possible as a result of the higher resonant frequencies ($\sim$ 1 Hz) of the ionospheric Alfvén resonator (Section 7.4.3).

The strength of ionospheric feedback theories is that they make specific, self-consistent predictions of auroral current structure based on a well-defined set of assumptions. The basic instability results from arc drift, ionospheric conductivity responding actively to upward FAC, and a model of current closure in the magnetosphere. However, the mere existence of an instability is not sufficient, since the system must be able to provide power to the arcs comparable to observed levels over their lifetime.

7.5.2 Propagating Arcs and Magnetic Energy

As mentioned at the beginning of this section, relative arc-plasma motion provides a way to supply auroral arcs with both plasma and energy. Energy flux Γ_E into the side of the arc can come in the form of drift kinetic energy, magnetic energy, or plasma thermal energy as represented by

$$\Gamma_E = \left(\frac{\rho_m \mathbf{E}_\perp^2}{2B_0^2} + \frac{\mathbf{b}^2}{2\mu_0} + p \right) v_d \tag{7.48}$$

where ρ_m is the plasma mass density, $\mathbf{b} = \mathbf{B} - \mathbf{B}_0$ the magnetic perturbation, p the plasma pressure, and v_d the transverse drift speed of the plasma relative to the arc. The power available to an auroral arc through advection is found by integrating Eq. (7.48) over y and z on the upstream side of an arc.

In a propagating Alfvén wave, the average $\mathbf{E} \times \mathbf{B}$ drift kinetic energy density equals the magnetic energy density, thus the first two terms are equal. Above the ionosphere, the magnetic term generally dominates as a result of the high ionospheric conductivity, which tends to enhance $|\mathbf{B}|$ and reduce $|\mathbf{E}|$, at least in the quasi-static limit. At altitudes below a few R_E, the plasma beta is low ($\beta < m_e/m_i$) and the pressure term negligible.

Haerendel (1994) studied a system in which an auroral arc propagates into a region of stored magnetic energy, drawing power from a region bounded by upward and downward current sheets. The reason for their propagation is the exhaustion of energy stored in form of magnetic shear stresses in the flux tubes pervading the arc. As these shear stresses are released within time-scales of the order of the Alfvén transit time, τ_A, between the generator and the acceleration region, the maximum $(\nabla \times \mathbf{B})_\parallel = \mu_0 j_\parallel$ moves into the stressed field region. Hence the propagation speed must be $v_n = w_{arc}/\tau_A$, where w_{arc} is the effective arc width. Some of the consequences of this propagating system are summarized in Figure 7.5. They include:

Oblique Alfvén Waves. Stresses deep in the magnetosphere propagate toward the ionosphere as shear Alfvén waves. A source moving with respect to the ionosphere leaves an oblique wavefront inclined at a slight angle θ with respect to $\mathbf{B}_0$, where θ is of order v_d/v_A (Mallinckrodt and Carlson, 1978).

Magnetic Field Line 'Thawing'. Field-aligned current carried by oblique Alfvén waves becomes more intense near the ionosphere because of flux tube convergence in a dipole field. At some point this current exceeds a threshold for the production of microinstabilities, allowing for a non-zero field-aligned electric field and breakdown of the frozen-in condition. This situation is analogous to the fracture of a solid material experiencing shear stresses exceeding a critical value for the material. The 'magnetic fracture zone' shown in Figure 7.5 is therefore a region where magnetic field lines break from one another and reconnect; this region is dissipative, with much of the liberated energy being channeled into field-aligned electron acceleration.

Plasma Evacuation. Most of the energy dissipated in the arc is deposited as heat in the ionosphere. A consequence of this heating is the expansion and consequent expulsion of plasma from low-altitude field lines. Figure 7.5 shows how ionospheric plasma is excavated by the front of the arc as it moves through the ionosphere.

Wave Interference and Structuring. The fracture zone is highly structured and dynamic, launching disturbances toward the ionosphere in the form of small-scale Alfvén waves. These waves can reflect both from the ionosphere and the fracture zone, the latter possessing a wave refractive index different than that found above and below it by virtue of its finite field-aligned conductivity (Vogt and Haerendel, 1998). Multiple reflections lead to highly structured fields in the interference region, with modulation of auroral electron precipitation possible. This interference mechanism has been proposed as a source of auroral fine structure (Haerendel, 1983).

7.5.3 Stationary Alfvén Waves

Section 7.5.1 described Atkinson's (1970) theory of how auroral structure and spacing results from simple models of ionospheric current closure, particle continuity, and ion polarization current in stationary drifting structures. Another mechanism producing periodic spatial structure results not from ionospheric feedback, but from plasma drift across (initially) large-scale FAC sheets, retaining the full physics of inertial Alfvén waves (Knudsen, 1996). The resulting 'stationary inertial Alfvén' (SIA) waves are manifested as periodic spatial oscillations of field-aligned electron energy, plasma density, and electric and magnetic perturbations (both parallel and perpendicular to $\mathbf{B}_0$). Field-aligned electric fields and electron acceleration result from magnetic induction arising from plasma drift across stationary current structures. In the non-drifting frame, this effect is taken into account by including a uniform convection electric field, $E_y = v_d B_0$. The magnitude of the field-aligned electric field in this frame is $E_{\|} = \mathbf{E} \cdot \hat{\mathbf{b}}$, which to first order in the quantity b_y/B is written

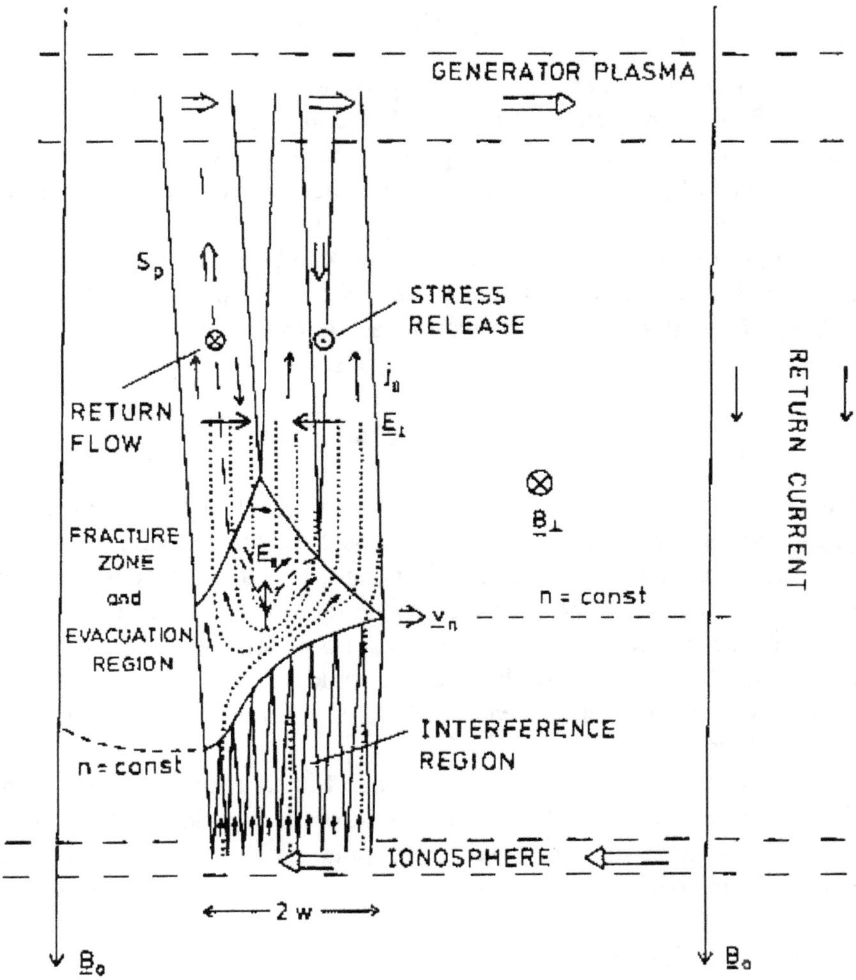

Figure 7.5. Qualitative picture of a magnetic fracture region showing (top) a generator region in motion with respect to the ionosphere, (middle) a 'fracture' region in which concentrated field-aligned currents become unstable, allowing finite potential drops and field-aligned electron acceleration, and (bottom) an interference region characterized by multiple reflections between the fracture region and ionosphere. From Haerendel (1994).

$$E_{\parallel} = E_z + E_y b_y / B_0 = E_z + v_d b_y \tag{7.49}$$

In this geometry the background magnetic field is $\mathbf{B}_0 = B_0 \hat{\mathbf{b}}$; b_y is the magnetic perturbation due to the current sheet.

The parallel electric field described by Eq. (7.49) can accelerate charged particles (and electrons in particular) along magnetic fields according to the fluid momentum equation in the stationary limit:

$$\frac{d}{dt}\mathbf{v_e} = (\mathbf{v_e} \cdot \nabla)\, v_{e\|} = \frac{q_e}{m_e} E_\| \tag{7.50}$$

If electrons were massless, any non-zero $E_\|$ would be neutralized immediately by the resulting field-aligned electron current. Retaining electron inertia allows finite $E_\|$ (hence the term 'inertial' in SIA waves - see Section 3.4.3).

Eqs. (7.49) and (7.50) are coupled by the fact that the latter leads to field-aligned electron acceleration, field-aligned current, $j_\| = q_e n_e v_{e\|}$, and via Ampere's law, $\partial b_y/\partial x = \mu_0 j_z$, results in variations in B_y. This can be seen explicitly by approximating $j_z \approx j_\|$ (accurate to first order in b_y/B), $E_z = 0$, and $\mathbf{v_e} \cdot \nabla \approx v_d \partial/\partial x$ in Eq. (7.50). The latter two approximations are not valid in general, however, the resulting equation is useful for illustrative purposes:

$$\lambda_e^2 \frac{\partial^2}{\partial x^2} b_y = b_y \tag{7.51}$$

The solutions for b_y are exponential on scales of the electron inertial length λ_e. Clearly, exponential solutions do not provide a useful representation of fields across, say, an auroral arc. However, Eq. (7.51) illustrates that the self-consistent interaction between E_y, $E_\|$, $j_\|$, and b_y can lead to stationary electric and magnetic field structures varying on scales comparable to λ_e, consistent with the smallest structures within visible arcs. As described below, solutions of a more complete model are in fact not exponential, but instead oscillate in x, reminiscent of multiple parallel arcs, for example.

Eq. (7.51) results from the assumptions mentioned above, as well as others, such as constant n_e. In reality, ions drifting across static electric field structures will accelerate in the x direction due to the polarization drift. This acceleration will lead to plasma density variations according to the ion continuity equation in x: $\partial(n_i v_i)/\partial x = 0$. A system of equations including (7.49), (7.50), and ion momentum and continuity with the full advective derivative (e.g., $\mathbf{v}_e \cdot \nabla = v_d \partial/\partial x + v_{e\|}\partial/\partial s$) for electrons, where s is the coordinate along the magnetic field) forms a nonlinear set. These equations reduce to one dimension under that assumption that stationary structures are oriented at a small oblique angle θ with respect to the magnetic field, so that spatial variations are related according to $\partial/\partial z = -\theta\, \partial/\partial x$. The fact that stationary drifting field structures are titled slightly with respect to $\mathbf{B}_0$ was described in an early paper on stationary Alfvén waves by Mallinckrodt and Carlson (1978).

Taking the above considerations into account, the resulting equation for SIA waves describes nonlinear oscillations of parallel electron velocity as a function of latitude, x (Knudsen, 1996):

$$\lambda_e^2 \frac{\partial^2}{\partial x^2} u_e^2 = A \frac{U_e}{u_e} + B \tag{7.52}$$

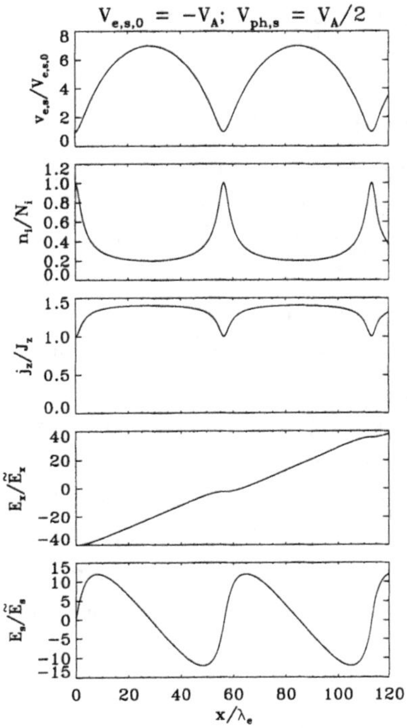

Figure 7.6. Field-aligned electron drift (in units of v_A), normalized plasma density, FAC, and perpendicular and parallel electric field components for an SIA wave propagating anti-parallel to electron drift at $0.5v_A$. From Knudsen (1996).

where $u_e = 1 - av_{e\parallel}$, U_e is an initial value for u_e at x, A, B and a are constants and $v_{e\parallel}$ is the field-aligned electron velocity.

Figure 7.6 plots the field-aligned electron speed, plasma density, field-aligned current, and field-aligned electric field within an SIA wave. In this example, the background current is carried by electrons traveling at the Alfvén speed; SIA waves increase $v_{e\parallel}$ by a factor of seven. Note that the horizontal axis is scaled according to the electron inertial length λ_e. SIA wave periods are of the order of a few λ_e (hundreds of m in the low-altitude auroral zone) in situations where field-aligned wave propagation is parallel to field-aligned electron drift, and of order tens or hundreds of λ_e (tens of km) when these two motions are anti-parallel (the latter being the case shown in Figure 7.6.)

SIA waves suggest a mechanism for self-consistent structure, electron acceleration and density depletion within the aurora. However, it should be noted that the theory in its present form predicts parallel electric fields many times weaker than those observed in the auroral acceleration region. Also, the present theory omits several potentially important effects such as the magnetic mirror force and reflections from the ionosphere. Additional work is required to advance the theory of

SIA waves to the point where careful quantitative comparisons can be made between predictions and theory.

7.6 Transient Acceleration

As described in Section 3.4.3, Alfvén waves can carry a parallel electric field $E_{\parallel}$ if their time variation is sufficiently rapid and their perpendicular scale size $\lambda_{\perp}$ sufficiently small. In particular, the perpendicular wave number $k_{\perp}$ must satisfy $k_{\perp}\rho_s \sim 1$ (kinetic Alfvén wave) or $k_{\perp}\lambda_e \sim 1$ (inertial Alfvén wave). The general term 'dispersive Alfvén wave' is used to refer to either case, the general dispersion relation for which is derived in Section 3.4.1. Observations of such waves are discussed in Sections 4.4.2 and 4.5.1.

A dispersive, transient Alfvén wave is an electromagnetic pulse moving close to the Alfvén speed. If the pulse amplitude is high enough, the wave can accelerate ambient electrons to (1-2)v_A, with corresponding energies of several hundred eV. To determine the pulse amplitude necessary to accelerate electrons, consider the wave front as a moving potential barrier. Initially stationary electrons will be trapped or reflected by a moving wave pulse if the peak wave potential Φ_{max} exceeds the electrons' kinetic energy in the wave frame:

$$q_e\Phi_{max} \geq \frac{1}{2}m_e v_e^2 \tag{7.53}$$

In the wave frame, stationary electrons that are overtaken and trapped by the wave will undergo a net velocity change of $v_{e,f} - v_{e,i} = v_A$, taking negative velocity to signify motion toward the wave. Electrons reflected by the wave pulse will have initial and final velocities of $-v_A$ and $+v_A$, respectively, for a net change - and final velocity in the fixed frame - of $+2v_A$.

An order-of-magnitude estimate of the parallel electric field possible for typical auroral parameters can be made using Eq. (3.77), which, neglecting finite temperature effects, can be written

$$E_z = \mu_0\lambda_e^2\frac{\partial j_z}{\partial t} \tag{7.54}$$

For example, taking λ_e = 1 km, j_z = 10 μAm^{-2}, and using a characteristic time scale of 0.1 s, one obtains $E_{\parallel} \sim 0.1$ mV m^{-1}. The corresponding electric potential drop can reach ~100 V across an Alfvén wavefront extending 1000 km in the field-aligned direction, the latter dimension being consistent with a wave traveling at $v_A = 10^7$ ms^{-1} and ramping from zero to full intensity in 0.1 s. Lower-frequency waves will carry proportionally weaker E_z.

From the above example it is clear that transient Alfvén waves cannot cause the steady, ~10 keV electron acceleration associated with inverted-V's. However, Alfvén waves are almost certainly involved in the acceleration of the soft (100's of eV), highly structured and highly field-aligned electron precipitation known as suprathermal electron bursts or 'STEB'. These bursts are typically found in the dayside cusp/cleft region, in substorm surge fronts, and near the edges of inverted-V's (see page 104).

If the potential barrier presented by the transient wave does not satisfy Eq. (7.53), the wave will pass over the electrons, whose kinetic energy will then change by the net potential drop across the wavefront. Typically, only a small fraction of the ambient electron population satisfies Eq. (7.53) (Hui and Seyler, 1992; Kletzing, 1994). However, observed field-aligned electron fluxes associated with Alfvénic fluctuations can be quite intense (Wahlund et al., 1994), indicating a surprisingly efficient acceleration mechanism. Indeed, Freja and FAST satellite observations suggest the existence of E_z as large as 100 mVm^{-1}, although the possibility of experimental error leaves these results uncertain (Volwerk et al., 1996; Chaston et al., 1999).

Large scale modeling of electron acceleration by Alfvén waves has been considered by test particle models (Kletzing, 1994; Thompson and Lysak, 1996; Chaston et al., 1999, 2000, 2002b; Kletzing and Hu, 2001). This work, in particular that of Chaston et al. (2000, 2002b) and Kletzing and Hu (2001), has shown that the dispersed electron signatures frequently observed in field-aligned electron bursts can be reproduced by the Alfvén wave model. The source of inertial and kinetic-scale structure in the aurora remains one of the foremost unsolved problems in auroral physics. Observations of such effects are discussed in Section 4.4.2.2. One possible mechanism for the production of such structure is phase mixing as described in Section 7.4.1. Another is the ionospheric feedback instability described in Section 7.4.3. In this case, an area of enhanced electron precipitation increases ionospheric conductivity, and thereby the ionospheric reflection coefficient. The parallel electric field E_z from an inertial Alfvén wave adds constructively on reflection from a good conductor; electron precipitation can thereby be enhanced, completing the feedback loop. This process favors smaller scales. However, there is a lower scale size of order 2 km at which reflections become inefficient, due to poor electric field mapping through the ionosphere (Lessard and Knudsen, 2001).

Another proposed source of inertial-scale structure is the stationary inertial Alfvén wave discussed in Section 7.5.3. The spatial period of these oscillations can be of order λ_e or greater. Once inertial-scale Alfvén waves are generated, they can evolve into still smaller-scale structures, for example through steepening and breaking due to advective non-linearities.

7.7 Large-Scale Features and Magnetospheric Sources

Even if auroral features need not be directly related to corresponding features in the outer magnetosphere and magnetotail, they are strongly influenced by the current systems that are built up in the magnetosphere and connect magnetosphere and ionosphere and by changes of the properties of the magnetospheric source region. Specifically the formation of the substorm current wedge (Figure 3.6) has important implications for intense dynamic aurorae during the substorm expansion phase. First, the strong upward field-aligned current at the westward edge of the current wedge is thought to be responsible for intense hard electron precipitation associated with the westward travelling surge (Section 6.2). Second, the heating of plasma enhances direct background precipitation of energetic electrons in the auroral bulge, which is filled by bright luminosity during active poleward expansion. A third effect is the temperature increase in the collapsing inner tail region inside the current wedge, which is combined with a density drop. Both changes work together to increase the field-aligned potential drop associated with a given field-aligned current density through the Knight relation.

Our present view of the formation of the substorm current wedge is largely based on numerical MHD simulations of magnetotail dynamics (e.g., Birn and Hesse, 1996). In these simulations the electric field responsible for the dynamic changes is dominated by the $\mathbf{v}\times\mathbf{B}$ term (Eq. 3.14), that is, with plasma flow. The flow results from a tail reconfiguration associated with magnetic reconnection at a near-Earth neutral line. In the following we will describe this picture, which, at the present time, is not fully accepted. Alternative models, however, are still qualitative, at least on the level of large scales. We note that significant deviations from the MHD picture require sufficiently small scales, as discussed in Section 3.2.

7.7.1 The Substorm Current Wedge

In MHD simulations the features of the substorm current wedge are closely related to the dynamic evolution initiated by the breakup of a thin current sheet formed in the growth phase. This breakup involves magnetic reconnection, which is enabled by some anomalous dissipation modeled as anomalous resistivity (Birn and Hesse, 1996). A section of the plasma sheet is severed and ejected tailward as a plasmoid, while an inner tail portion collapses (dipolarizes). In contrast to early expectations, the diversion of the cross-tail current into the current wedge does not occur at the reconnection site, where the anomalous dissipation acts, but rather as a consequence of the plasma flow earthward from this site. As illustrated in Figure 3.4, the plasma flow carries magnetic field lines earthward. At some point it brakes and is diverted azimuthally away from midnight. The magnetic shear between such distorted field lines and field lines at higher latitude that are not affected by the flow is the manifestation of the presence of field-aligned currents.

Consistent with the discussion in Section 3.2.2, the flow is the crucial element in distorting the magnetic field and building up the field-aligned current system of the current wedge, but it plays only a short, relatively minor role in the diversion of perpendicular into parallel currents, evaluated from the force balance (Eq. 3.23). This diversion hence is largely described by Vasyliunas' Eq. (3.24). As illustrated by Figure 7.7, the dominant effects stem from a distortion of the contours of the flux tube volume V. This distortion is associated with a reduction of the flux tube volume in the inner tail caused by a shortening of magnetic flux tubes by magnetic reconnection. There are two characteristic regions, where the contours cross at a significant angle. In the region between $x \approx -15$ and $x \approx -22$, the pressure

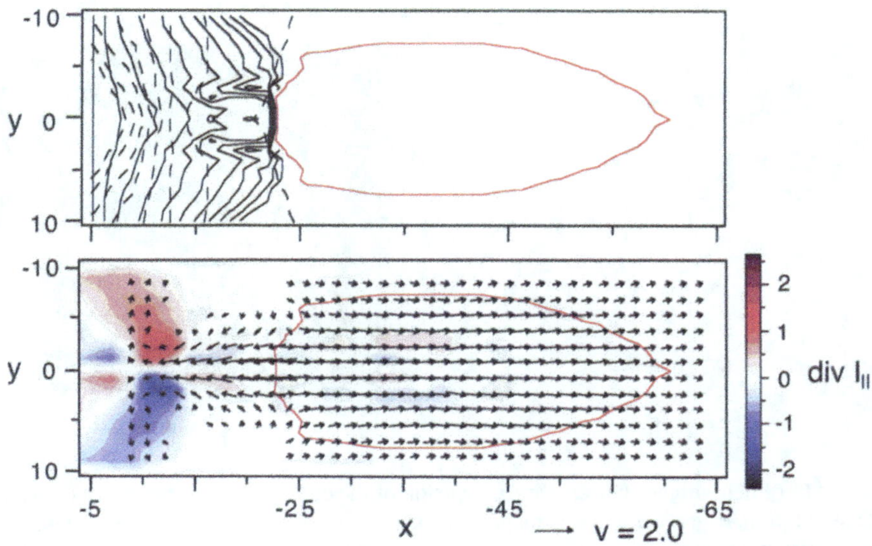

Figure 7.7. Current diversion from perpendicular to parallel currents, as found in a simulation of tail dynamics (top, after Birn and Hesse (1996); bottom, after Birn et al. (1999)) involving plasmoid formation and ejection. The top part shows contours of pressure and flux tube volume in the equatorial plane. The closed red line to the right represents the magnetic neutral line; its left part corresponds to the near-earth X line and the right part to the O line in the center of the departing plasmoid. Solid contours earthward (to the left) of the neutral line are contours of constant flux tube volume, defined by Eq. (3.25), and dashed lines are contours of constant pressure. According to Eq. (3.24), current diversion occurs where these contours intersect at a finite angle. The bottom part shows the color-coded $\nabla \cdot \mathbf{j}_{\|}$ integrated over z and flow velocity vectors in the equatorial plane.

contours go roughly straight across the tail, while the V contours are strongly bent outward. This corresponds to a current diversion of Region 1 sense. In the inner tail, inside of $x \approx -10$, the V contours go nearly straight across the tail, while the pressure contours are strongly curved. This corresponds to a current diversion of Region 2 sense.

Figure 7.7 shows only the integrated effect of the current diversion but does not indicate where along the field lines the major effects occur. A quantitative evaluation of the pressure gradient terms (Birn et al., 1999) shows that the major contributions stem from regions outside the equatorial plane, involving the z gradients, which are located closer to the Earth than the equatorial crossing points of the affected field lines. This is also the reason why the regions where the p and V contours cross each other at a large angle in the top part of Figure 7.7 are located somewhat tailward of the regions of large $\nabla \cdot \mathbf{J}_{\|}$ in the bottom part. The top part represents the current diversion effects mapped along field lines into the equatorial plane, while the bottom panel shows effects integrated along constant x and y.

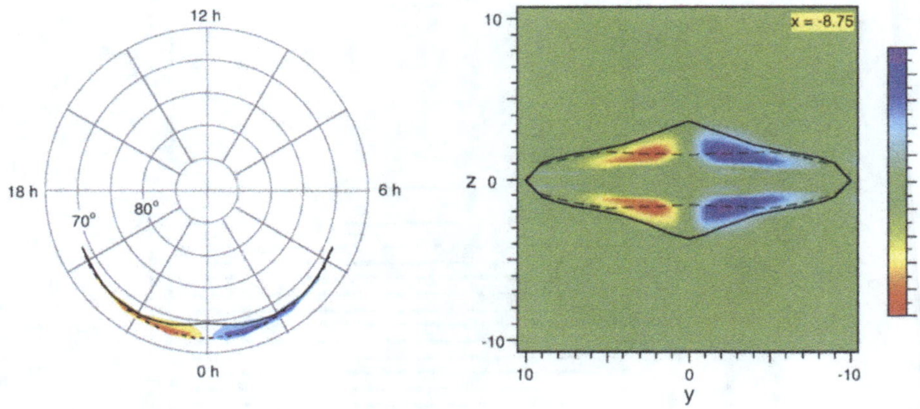

Figure 7.8. Left: field-aligned currents and expansion of the closed field line boundary (separatrix) in a MHD simulation. Right: color-coded magnitude of $j_{\parallel}$ and the separatrix near the earthward simulation boundary and the top part shows these regions mapped to Earth using an empirical magnetic field model for the inner magnetosphere (Tsyganenko, 1987). Figure provided by J. Birn.

7.7.2 Formation of the Auroral Bulge

The formation of the auroral bulge, discussed in Sections 6.2 and 6.4, can also easily be understood on the basis of the reconnection picture. In this picture the poleward edge of the brightly illuminated bulge is associated with the poleward boundary of the closed field line region, which expands poleward when reconnection proceeds to lobe field lines. This is demonstrated in Figure 7.8, which shows the open-closed boundary and the field-aligned current system at the inner tail cross-section of a MHD simulation and its map to the Earth's surface based on an empirical model of the inner magnetosphere. Observations related to the open-closed boundary are presented in Section 4.5.1.

The simulations also demonstrate that the expanding closed field line region is filled with heated plasma with a temperature of about twice the original plasma sheet temperature, while the density gets reduced by a similar factor, which is quite consistent with *in situ* plasma observations. Although one-fluid simulations do not directly include results on electron heating, the fact that the ion heating process in the simulations is adiabatic compression, strongly suggests that electrons should become heated similarly as the ions.

7.7.3 Energy Flow and Conversion

The properties of the generator mechanism in relation to the substorm current wedge and the relative importance of the contributions to $\mathbf{v} \cdot (\mathbf{j} \times \mathbf{B})$ can also be investigated on the basis of the simulation by Birn and Hesse (1996). The results are illustrated in Figures 7.9 and 7.10. Figure 7.9 shows the color-coded magnitude of

$\mathbf{j} \cdot \mathbf{E}$ (top) and the pressure gradient and inertial contributions to $\mathbf{j} \cdot \mathbf{E}$, all integrated over z, together with flow velocity vectors and the magnetic neutral line. Generator regions ($\mathbf{j} \cdot \mathbf{E} < 0$) are indicated by yellow and red and load regions ($\mathbf{j} \cdot \mathbf{E} > 0$) by blue or purple color.[1] The top panel shows that the reconnection site, as well as the region of fast earthward flow, act as a load. Generator regions are present where the flow gets diverted and distorts the magnetic field, leading to the buildup of field-aligned currents. The generator regions hence roughly coincide with the regions where part of the cross-tail current gets diverted to the field-aligned currents of the substorm current wedge. This may not be too surprising, because both the evaluation of generator and current diversion rely on the force balance, Eq. (3.13). We note, however, that the generator evaluation takes only the components of the forces in the direction of the flow into account.

Ohmic dissipation, ηj^2, is highly localized and so small that it would not show in the color representation of Fig. 7.9. Hence

$$\begin{aligned} \mathbf{j} \cdot \mathbf{E} &\approx -\mathbf{j} \cdot (\mathbf{v} \times \mathbf{B}) = \mathbf{v} \cdot (\mathbf{j} \times \mathbf{B}) \\ &= \mathbf{v} \cdot \nabla p + \mathbf{v} \cdot \rho \frac{d\mathbf{v}}{dt} \end{aligned} \tag{7.55}$$

The dominant contribution to $\mathbf{j} \cdot \mathbf{E}$ stems from the pressure gradient term in Eq. (7.55). This is illustrated by the bottom two panels of Figure 7.9, showing the pressure gradient and the inertial contributions as function of x and y, again integrated over z. The inertia term provides a generator mechanism in the region of flow braking, as postulated by Shiokawa et al. (1998). However, this term is overcompensated by a load term from the pressure contribution, so that there is no net generator in this region.

The location of the generator regions in a cross-section of the tail and their relationship to the regions of field-aligned currents is shown in Fig. 7.10. It demonstrates that the generator regions are mainly located above and below the equatorial plane; this is again consistent with the current diversion (Birn and Hesse, 1991). The flow vectors demonstrate that the generator terms (with $\mathbf{j} \cdot \mathbf{E} < 0$) stem primarily from an upward (for $z > 0$) motion against the strong pressure gradient or $\mathbf{j} \times \mathbf{B}$ force at the outer edges of the inner plasma sheet, rather than from motions in the equatorial plane. Although the velocities in the equatorial plane are much larger than the velocities in the z direction, the pressure gradients and $\mathbf{j} \times \mathbf{B}$ forces are typically much smaller.

In a more global view, the generator picture has to be embedded in the overall energy flow picture, which includes the open circuit (Fig. 3.6a). The combined picture consists of a generator in the magnetosheath, which builds up magnetic energy

[1] Note that the value of $\mathbf{j} \cdot \mathbf{E}$, and hence the interpretation of generator and load, are frame dependent. This value generally differs from Ohmic, or more generally, Joule dissipation, $\mathbf{j} \cdot \mathbf{E}'$, which is defined in the plasma rest frame.

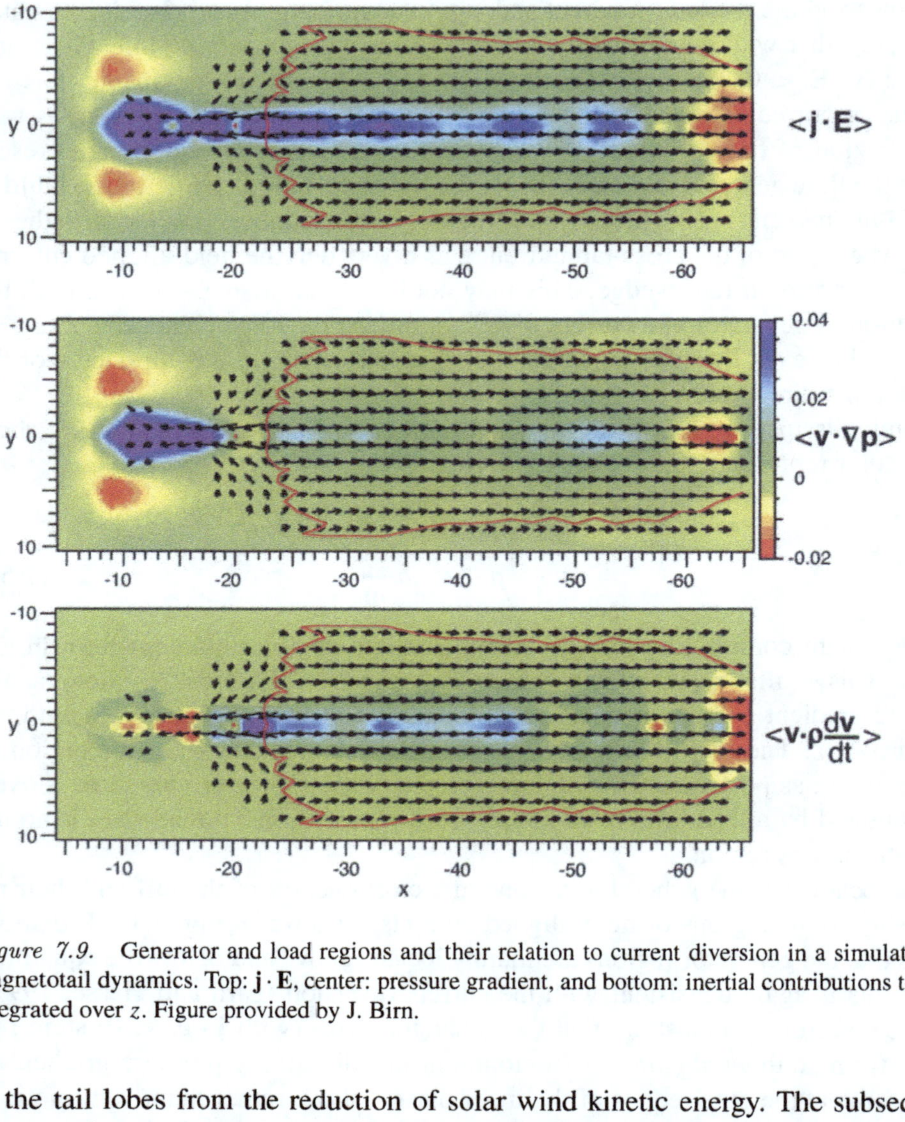

Figure 7.9. Generator and load regions and their relation to current diversion in a simulation of magnetotail dynamics. Top: $\mathbf{j} \cdot \mathbf{E}$, center: pressure gradient, and bottom: inertial contributions to $\mathbf{j} \cdot \mathbf{E}$, integrated over z. Figure provided by J. Birn.

in the tail lobes from the reduction of solar wind kinetic energy. The subsequent energy release converts magnetic energy back to kinetic energy and (to a smaller extent) into heat, in addition to the transport of the magnetic energy via Poynting flux. The slowing of the plasma flow then leads to a compression and further conversion to heat and back to magnetic energy. In the near-Earth generator regions in the tail, locally, thermal energy is converted to electromagnetic energy, which drives the current wedge system. Since there is no direct conversion of thermal into magnetic energy (Eqs. 3.28 and 3.29), this process requires approximate balance of the Lorentz force by the pressure gradient and the flow kinetic energy as a 'catalyst' (Eq. (3.31)). A non-local analysis is necessary to determine the ultimate energy source for the current generation. The energy release and conversion func-

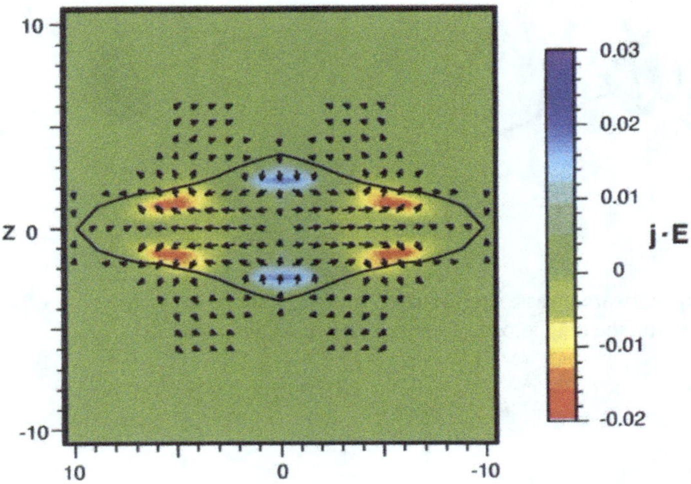

Figure 7.10. Generator and load regions in a simulation of magnetotail dynamics in the $y-z$ plane at $x = -8.75$. The color shows the magnitude of $\mathbf{j} \cdot \mathbf{E}$, positive (blue) representing load, negative (red) generator regions; solid lines indicate the boundary of the closed field line region (separatrix) and the arrows show the velocity vector perpendicular to the magnetic field, projected into the plane at $x = -8.75$. Figure provided by J. Birn.

tions via plasma and field transport, i.e., it involves the flow velocity $\mathbf{v}$, as shown in Eqs. (3.31) and (3.32), but mostly in a quasistatic sense as a mediator rather than a significant energy source or sink. Ultimately, the energy that flows to the Earth is dissipated in the resistive ionosphere.

7.7.4 Stress Transmission

Current generation is but only one aspect of the coupling between the magnetosphere and the ionosphere. In addition the stresses and flows are transmitted between the two regions in contact. Figures 7.11 and 7.12 illustrate aspects of stress transmission incorporating the current circuits of Figures 3.6a and 3.6b. Figure 7.11 incorporates the open current wedge of Figure 3.6a, completed in the simplest possible way by a current loop from dawn to dusk along the tail magnetopause. The other loop in Figure 7.11 also connects the ionosphere with the magnetopause, however, in a more direct way that does not involve the interior magnetotail. Both loops illustrate the role of the (Region 1 type) field-aligned currents in transmitting tailward stresses excerted by the solar wind to the ionosphere, where they provide equatorward drag.

Similarly, Figure 7.12 illustrates the transmission of stresses from the tail to the ionosphere. Current loop (A) is identical to the perturbed system in Figure 3.6b. This loop shows its role in transmitting stresses built up by the earthward flow in the

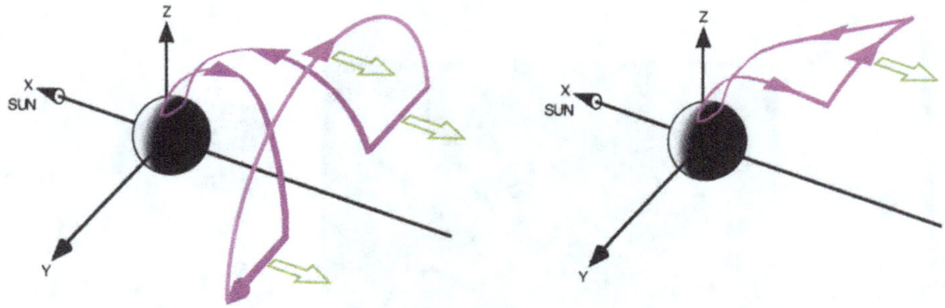

Figure 7.11. Equatorward stresses transmitted to the ionosphere by Region 1 type field-aligned currents connected with the magnetopause. Figure provided by J. Birn.

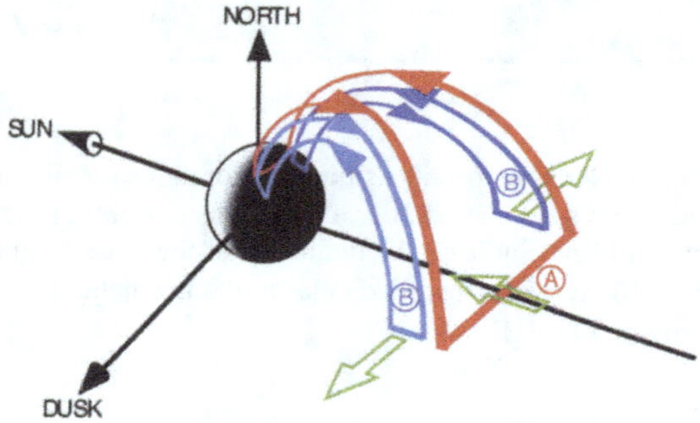

Figure 7.12. Internal tail stresses transmitted to the ionosphere by Region 1 type and Region 2 type field-aligned currents associated with the substorm current wedge. Figure provided by J. Birn.

tail, also providing equatorward drag in the ionosphere. The field-aligned currents are again of Region 1 type. The other loop (B) in Figure 7.12 combines Region 1 and Region 2 type field-aligned currents. Such a loop was postulated by Boström (1964) and Haerendel (1990) and demonstrated in a simulation of tail dynamics (Birn et al., 1999). It serves to transmit stresses built up by azimuthal motion in the tail, leading, in this simple form, to azimuthal stresses in the ionosphere.

We should note that the simple separation between equatorward drag, associated with tailward stresses at the magnetopause or earthward stresses in the magnetotail, and azimuthal stresses in both magnetotail and ionosphere is an oversimplification. In reality, the current loops may be coupled, additional loops may be involved, the currents do not form simple closed loops, and the mapping between the different regions may be distorted, so that the association between stresses is more complicated than suggested here. Figures 7.11 and 7.12 should be taken as a symbolic representation of the coupling rather than a representation of the actual currents.

Figures 7.11 and 7.12 may also serve to illustrate the role of the auroral region in releasing the stresses transmitted to the ionosphere. In the simplest model of magnetosphere-ionosphere coupling, a continuous flow of plasma along the magnetopause or in the magnetotail leads to a continuous increase of stresses in those regions, and subsequently in the ionosphere, combined with an increase of the currents represented by the simple loops in Figures 7.11 and 7.12. For a given ionospheric resistivity, an increase of ionospheric currents is associated with an increase of the electric field. Ultimately, this electric field will match the electric field imposed by the external plasma motion and a steady state could be reached in which the field lines are dragged over the ionosphere in a steady fashion. However, particularly in the regions of upward going field-aligned currents, the precipitation of electrons leads to an increasing conductivity. This conductivity increase causes a more effective field-line tying and thereby a further increase of the field-aligned currents. Consequently, the ionosphere may not be able to permit enough drag in order to release or at least limit the imposed stresses. Under such conditions a different region might take over the role of stress release. The auroral cavity region at $\sim$3000-6000 km altitude is characterized by extrema of several important physical parameters. This makes it plausible that this region is the preferred site of the buildup of a parallel electric field, which enables such a local stress release.

The stress release might occur in a direct way, which, however, involves a buildup of the stresses in the auroral region. In this scheme, stresses would build up in the auroral region rather than the ionosphere until a local breakdown of ideal MHD permits a slippage of field lines and a release of the stresses. The increase or release of magnetic stresses can be achieved by nonlinear interaction of the incident and reflected Alfvén waves in the acceleration region, where the localized breakdown of the frozen-in condition is the result and also a necessary condition for the nonlinear interaction (Song and Lysak, 2001).

A less direct way would involve an intermediate instability mechanism, such as generalized tearing or reconnection (Haerendel, 1989; Hesse et al., 1990; Otto and Birk, 1992; Birn et al., 1993), driven by the enhanced magnetic shear in the auroral region. In this picture the instability would first break the ideal MHD constraint and would transfer the stresses to the auroral region as illustrated in Figure 7.13.

The stress release would be the consequence of the subsequent plasma transport rather than of the non-MHD process. In either case, the breakdown of ideal MHD is closely associated with the rise of a localized parallel electric field (Schindler et al., 1988; Hesse and Schindler, 1988).

7.7.5 Small-Scale Source Regions

In Section 3.2.2, we have mostly concentrated on large-scale mechanisms for the generation or diversion of currents, corresponding to the basic auroral systems, such as Region 1 or 2 (Iijima and Potemra, 1976) or to the substorm current wedge (McPherron et al., 1973) but not to individual arcs. However, we also mentioned briefly that there might be a relation between tail phenomena and aurora on smaller scales, corresponding to arcs of, say, 10 km latitudinal width in the auroral zone.

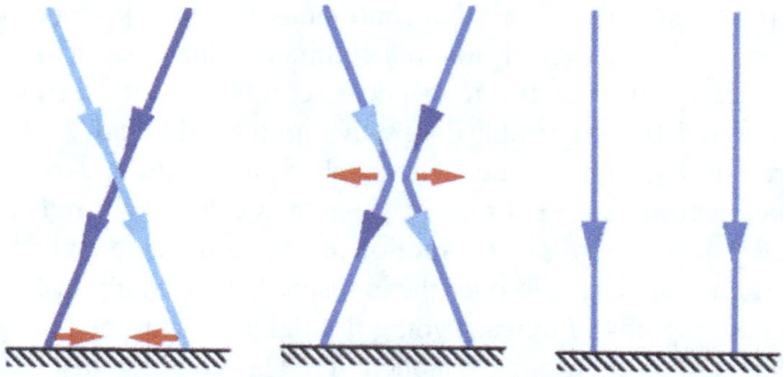

Figure 7.13. Stresses resulting from a generalized tearing mode. Figure provided by J. Birn.

This scale maps roughly to a few hundred km in the north-south direction in the near tail, comparable to an ion inertia length or an ion Larmor radius. As discussed in Section 3.2.1, at this length scale deviations from ideal MHD become important. The dominant term in Eq. (3.14) to affect the electric field is the Hall term, which can be combined with the $\mathbf{v} \times \mathbf{B}$ term to yield Eq. (3.17). When this approximation holds, the arguments in Section 3.2.2 for the buildup or intensification of field-aligned currents remain valid if the plasma flow is replaced by the flow of the electron fluid. Several small-scale regions in the tail have a suitable location for an association with arcs:

The Inner Edge of the Electron Plasma Sheet. In contrast to ions, the electron temperature and pressure typically show a rather steep gradient at the inner edge of the plasma sheet. Although the electron pressure in the plasma sheet is almost an order of magnitude smaller than the ion pressure, a steep gradient could nevertheless render its contribution to current diversion more important in Eq. (3.24).

The Thin Growth Phase Current Sheet. The growth phase of a substorm is characterized by the formation of a thin current sheet in the inner magnetotail (e.g., McPherron et al., 1987; Sergeev et al., 1993; Pulkkinen et al., 1994; Sanny et al., 1994), whose thickness can approach an ion inertia length. Although this current is essentially perpendicular to the magnetic field, one might expect that the three-dimensional structure would also yield a field-aligned component. In fact, if we assume that the duskward current in the thin current sheet is carried primarily by the electrons (Pritchett and Coroniti, 1994; Hesse et al., 1995), the corresponding dawnward electron flow is expected to cause a bending of field lines that is associated with outward field-aligned currents at its inner, low-latitude edge, as illustrated in Figure 7.14 (left). The possible association of this current region with auroral arcs is further supported by the fact that self-consistent Vlasov equilibria (Schindler and Birn, 2002), as well as particle simulations of thin current sheets

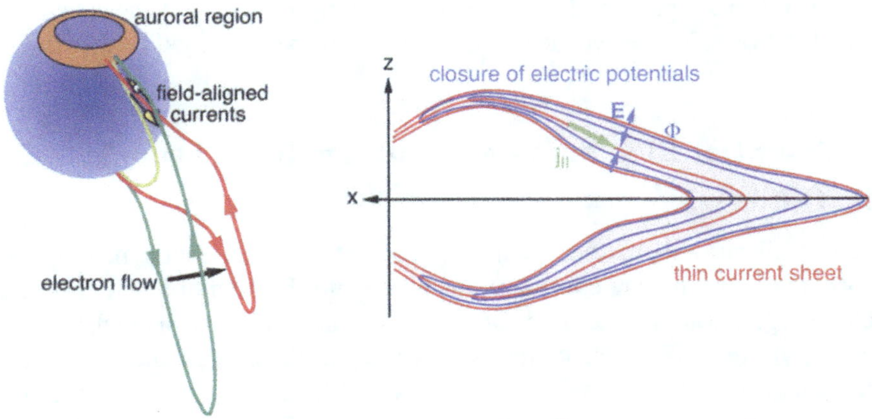

Figure 7.14. Thin current sheet in the near tail; (left) intensified current carried by electrons distorts the magnetic field lines and builds field-aligned currents; (right) the selfconsistent structure requires field-aligned electrostatic potentials, whose closure in the auroral acceleration region causes parallel electric fields and arc brightening. Figure provided by K. Schindler and J. Birn.

(Pritchett and Coroniti, 1995; Hesse et al., 1995), imply the presence or formation of electric potentials that are constant on field lines. The closure of these potentials across the parallel current sheet in the auroral acceleration region would then provide the parallel electric field associated with an arc (Schindler and Birn, 2002), as illustrated to the right in Figure 7.14.

The Near-Earth Reconnection Site. From the discussion in Section 3.2.1, it is clear that the onset of collisionless reconnection requires sufficiently small scales. Simulations of magnetic reconnection that include the Hall term indeed show thin layers of field-aligned currents (Yamagishi et al., 2000), which could be associated with arcs. The mechanism for building up the currents is quite similar to the one associated with the thin current sheet illustrated in Figure 7.14.

The Plasma Sheet Boundary Layer. The boundary region between plasma sheet and lobes is often characterized by the presence of fast field-aligned flows with a separation of ions and electrons (e.g., Takahashi and Hones, 1988; Onsager et al., 1991). The difference between ion and electron flows directly corresponds to field-aligned currents, which are also manifested by magnetic field shears (Elphic et al., 1985).

Bursty Bulk Flows. These have been shown to be closely related to small-scale auroral features such as north-south aligned arcs (Henderson et al., 1998), poleward boundary intensifications (Lyons et al., 1999b), and auroral streamers (Nakamura et al., 2001a). Although the estimated cross-tail extension of these flows of a few

R_E is still macroscopic, smaller scales presumably exist in the shear layers at the edges of the flows. These events and structures will be discussed further in Chapter 8.

7.8 Essential Elements of an Integrated Theory

7.8.1 Outline

Practically all auroral theories treat only part of the total problem. For instance, most often the currents are considered as given and the attention is focussed on the particle energization and the fields and waves which are instrumental in achieving this. There are models which address solely the parallel energy and mass exchange, others focus on the transverse electric fields and heating by shear flows, although parallel and transverse fields are intimately related. The microphysics of the auroral acceleration regions is often treated without reference to the source of energy. In fact, the latter one is one of the most neglected elements in the context of auroral arc formation. Even less work has been done on the impact of arc formation on the source plasma. In short, it is fair to say that we are still far from a selfconsistent theory, even on the macroscopic scale.

What would be the essential elements of such a theory? We approach this question by considering what role auroral arcs play in the context of magnetospheric plasma transport. Here, we must realize that not all aurora is of the same kind. All aurora is due to energetic particle impact on the upper atmosphere, but the reasons for the precipitation and the source of energy can differ. One may distinguish three major classes, diffuse aurora, Alfvén wave driven and convection related aurora. None of these three classes are well understood, and many researchers may not agree with this classification. In continuing our lines of thought with respect to the third class, and in the end contrasting it with the two other classes, the reasons for this classification may become clearer.

The magnetosphere is constantly being stirred by momentum and energy transfer from the solar wind. Forces can be applied to magnetospheric flux tubes either by flow braking of the solar wind passing along a slightly open magnetopause or by some sort of viscous momentum coupling to the magnetopause boundary layers. The other major driving force is pressure gradients arising from the existence of preferred regions of hot plasma inflow and/or energization, like the near-Earth tail around midnight, most dramatically during substorms. The primary process behind this plasma inflow, be it reconnection or current disruption, seems to have little immediate effect on the aurora, but rather an indirect one via pressure build-up and flow initiation. The overall consequence of this is the generation of two-cell or sometimes four-cell and strongly time varying internal convection patterns in the magnetosphere. This convection is driven against the frictional force exerted at the ionospheric ends of the field lines. The friction is overcome by magnetic shear stresses maintained by continuous supply of transverse momentum from the convecting hot plasma at high altitude. This momentum transport is associated with the flow of field-aligned currents, which appear to be the essential element in the formation of elongated, narrow and structured auroral arcs, in contrast to aurora of the two other classes mentioned above. It is certainly not coincidental that such extended arcs tend to be closely aligned with the convective flow lines.

The friction-controlled magnetospheric convection is usually relatively slow, several $100\,\mathrm{m\,s^{-1}}$ to a few $\mathrm{km\,s^{-1}}$ as measured in the ionosphere. All the same, the associated Joule heating by the closure currents is the greatest energy sink of the magnetosphere (Lu et al., 1995; Baker et al., 1997). If the energy and momentum supply from the solar wind were interrupted for a while, this convection would come to a halt, except for some effects of the atmospheric tides. The magnetic shear stresses and field-aligned currents would die out, there would be no elongated, structured auroral arcs. In the opposite case of very high momentum and energy input, shear stresses and currents are greatly enhanced, and rich auroral displays are observed. The energy flux density due to auroral particle precipitation can exceed that of Joule heating per unit cross-section by up to two orders of magnitude. Thus auroral arcs are traces of enhanced energy release from the magnetosphere. By the same argument they must also be regions of enhanced magnetic stress release. So, we can conclude that auroral arcs constitute a second and at least locally very efficient way of magnetospheric stress release. Haerendel (1992) cast the name 'auroral pressure valve' for this role of auroral arcs.

How is this fast stress release achieved? By decoupling bands of magnetic flux tubes from the frictional ionosphere through parallel potential drops which allow very fast convective plasma motions to exist above the ionosphere free from the latter's frictional control. This is the meaning and role of the U-shaped potentials. The generation of electron and ion beams by field aligned potential drops is only one side of the coin. The other side is the high transverse electric fields, the so-called 'electrostatic shocks' (Mozer et al., 1977) found at several 1000 km altitude. They also manifest themselves as auroral curl or ray motions of typically $10\text{-}20\,\mathrm{km\,s^{-1}}$ apparent velocity (Davis, 1960), whereas no equivalent electric fields are measured at the altitudes of the auroral displays (Haerendel et al., 1996). It is these motions which produce the magnetic stress release, while the field-aligned acceleration takes care of the corresponding magnetic energy release. The auroral mechanism breaks the magnetic connectivity of magnetosphere and ionosphere whereby elastic magnetic energy is shed, quite reminiscent of a fracture process ('magnetic fractures' of Haerendel, 1989).

What are the conditions for and consequences of the magnetic fractures or auroral arc formations? Most important is the existence of high field-aligned electric current densities in the presence of low plasma densities. The currents reflect the interaction of magnetospheric and ionospheric (frictional) forces. The maintenance of intense currents at low densities of current carriers, or against some current driven plasma turbulence, can require the application of potential drops along the current path. This is equivalent to voltages set up in an electrical circuit across a fuse or a resistor. Current continuity is only one of the origins for such field-aligned potential drops. Another one is the need for charge neutrality in the presence of accelerated particle populations. The creation of sufficiently intense currents for such effects to occur is often aided by filamentation and cavitation of the cool plasma in the topside ionosphere. The cavitation appears to be a consequence of upward cold ion injections for want of creating a charge neutral plasma above. Thus, auroral arc formation and ionospheric modifications go together. Other types of modifications occur in the lower ionosphere: density and conductivity enhancements through impact ionization by the incoming auroral electrons, and density depletions as well as enhancements caused by divergences of the transverse and parallel currents. Such density and conductivity inhomogeneities may be instrumental in causing current filamentation.

Probably more important are the modifications of the generator plasma high up in the magnetosphere. Magnetic stress release is achieved by propagating the fast motions or high transverse electric fields set up in the auroral acceleration regions upward into the generator plasma. This modifies the local force balance - magnetic versus mechanical forces - initiates plasma redistributions and leads to loss of internal energy. Hence the name 'auroral pressure valve'. Since the energy flux into the ionosphere above a bright auroral arc is strong enough to dump, within two Alfvén transit times (about 100 s), the equivalent of the free magnetic energy stored inside the flux tube, a resupply of energy is needed if the arc is to persist for longer times. This includes the re-establishment of the released magnetic shear stresses, as well as maintenance and possibly filamentation of field-aligned currents. Such resupply of energy is probably the reason why bright arcs often exhibit brightness modulations and morphological changes with a time scale of a few minutes while maintaining their identity. All of these processes constituting the feed-back of the arc formation on the generator plasma are largely unexplored. Also the flow of energy along the magnetic field lines between generator and auroral acceleration region contains many unknowns.

In summary, essential elements of an integrated theory, apart from the problem of sustaining field-aligned potential drops with all its microphysical aspects, would be: field-aligned current concentrations within a magnetospheric convection zone, low magnetospheric plasma densities, U-shaped potential or magnetic fractures, propagation of the stress releases back to the generator plasma. At the bottom end, modifications of the topside as well as of the lower ionosphere occur. At the top end, one must consider the dynamic impact of the stress release on the generator plasma, including local reductions of internal energy, reshaping of the pressure distribution and resupply of energy and magnetic stresses. The next sections will address some of these topics in more detail.

Before we enter these discussions, a word on the other two types of auroral arcs is in order. One way of enhancing the energetic particle influx into the upper atmosphere is by pitch angle scattering inside the hot plasma sheet at high altitudes. It has been thought since long that this is the origin of diffuse aurora, and beautiful theories have been developed in the 1960's and 1970's (e.g., Kennel et al., 1970). However, more recent improved wave measurements have cast doubts on the validity of these early theories rendering the origin of diffuse precipitation into "one of the major unsolved problems in magnetospheric physics" (Lyons et al., 1999a). The other possibility of energizing downward electron fluxes is through direct acceleration by propagating kinetic Alfvén waves thanks to their attending parallel electric fields (see Section 3.4.3). Aurora resulting from small-scale Alfvén waves has been found predominantly at the poleward boundary of the auroral oval. Such waves seem to be excited by not-yet-understood processes mainly at the edge of plasma sheet and cusp. In both cases, diffuse and kinetic Alfvén wave-driven aurora, there is no need for large-scale field-aligned currents, although the latter carry, of course, currents of small transverse scales.

7.8.2 Energy Flow

In the preceding introduction we distinguished between three classes of aurora, diffuse, Alfvén wave driven, and convection-related aurora. In correspondence with this classification, but extending beyond, we can distinguish four ways of energy flow along magnetic field lines from the outer magnetosphere into the ionosphere. They are symbolized in Figure 7.15.

Ways (1), (3), and (2) correspond respectively to the three classes of aurora. Way (4) does not lead to appreciable light emissions. From a different point of view, ways (2), (3), and (4) are related in that the dominant mode of energy flow occurs in form of electromagnetic energy, a Poynting flux. They differ from each other by the plasma component that absorbs most of this energy. Way (2) is the mode of bright auroral arcs and inverted-V events. The Poynting flux is mainly (not totally) absorbed by magnetospheric electrons (and ions) in the lower magnetosphere. Thereby, the mirror points of the primary electrons are lowered and the current is kept continuous. A smaller fraction of the energy is, however, consumed by the auroral cavitation process, i.e. ion extraction from the topside ionosphere, for the sake of establishing charge neutrality (see Section 7.1). In way (3) the Poynting flux appears to be absorbed mainly by cold electrons (and ions) of the topside ionosphere, as manifested by the dominance of transversely cold (a few eV) strongly field-aligned electron beams. The energy conversion mechanisms is not well understood. It seems to favor arc edges. The origin may well be related to the braking of small-scale Alfvén waves (Hui and Seyler, 1992). Way (4) is the classical Ohmic dissipation of the Pedersen current which closes the magnetospheric field-aligned currents.

The latter path dominates whenever the currents can be carried by the magnetospheric plasma population without the need of post-acceleration. However, as in the transmission of electromagnetic energy by wires (wave guides), the Poynting

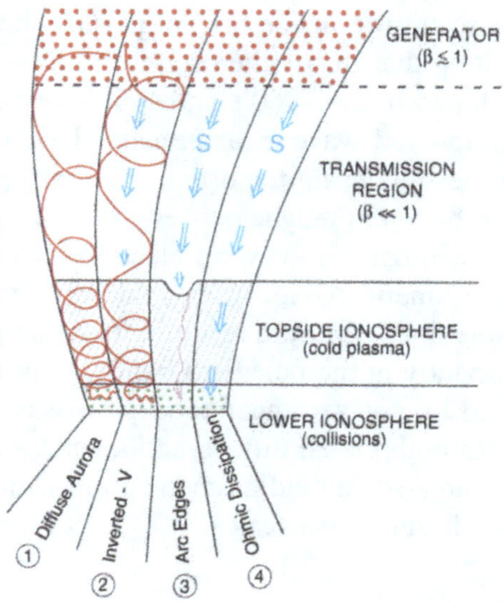

Figure 7.15. Four ways of energy flow from the magnetospheric energy reservoir into the auroral region or directly into the ionosphere. The four ways are distinguishable by their low-altitude manifestations as (1) diffuse aurora, (2) inverted-V events and bright arcs, (3) strongly field aligned fluxes, often at arc edges, and (4) just ohmic dissipation of Pedersen currents. S is the Poynting flux. From Haerendel (2001b).

flux is not restricted to the current paths. This means that also a region separating upward and downward field-aligned currents may be pervaded by a finite, even the major Poynting flux, i.e., $\mathbf{S} = \mathbf{E} \times \mathbf{B}/\mu_0 \neq 0$. The crucial $\nabla \cdot \mathbf{S}$, on the other hand, is non-zero only where a dissipative current flows. In way (4) this is the closing Pedersen current, $j_\perp$ (Figure 7.16). In the two other ways, the Poynting flux is dominantly absorbed by $j_\parallel$. Contrary to these three ways of energy flow from the magnetosphere, the energy is carried mainly by the precipitating energetic particles in case of the diffuse aurora. However, all 4 kinds of energy flow may coexist in various degrees of mixing in any real situation.

Haerendel (2001b) made an attempt to compare the energy conversion rates, $\dot{\varepsilon}_\parallel$, the energy content, W, per unit cross-section of the involved flux tubes, and the respective energy release times, $\tau_E = W/\dot{\varepsilon}_\parallel$. The conversion processes are ionization and excitation of atmospheric particles (way 1 of Figure 7.15), generation of field-aligned beams and/or conical distributions (S), (ways 2, 3 of Figure 7.15), and Joule heating (way 4 of Figure 7.15). Typical flux-tube contents of energy, W, are estimated either in form of internal energy of the hot plasma or as magnetic energy stored in the current systems.

Table 7.3 contains the underlying physical relation as well as the numerical results. In order to be specific, the values given in Table 7.2 have been assumed.

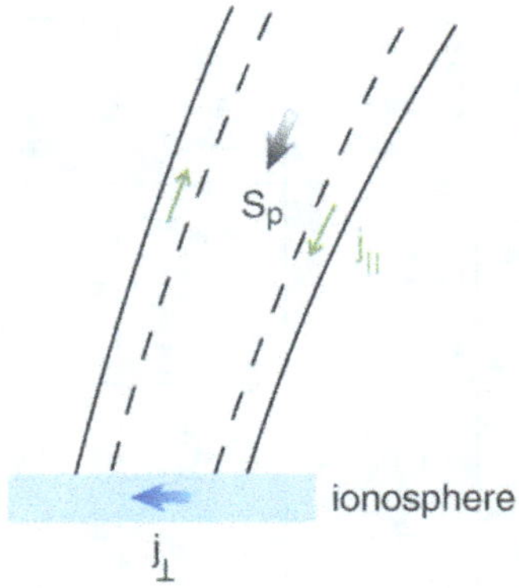

Figure 7.16. Poynting flux, S_p, into the ionosphere between balanced sheet currents.

These values are more or less canonical, but appreciable deviations do occur in both directions.

Furthermore, the following assumptions have been made in evaluating the expressions contained in Table 7.3. The field-parallel extent of the generator plasma in one hemisphere, $\ell_{\|}$, was assumed to be 6 R_E. The effective mirror ratio B_i/B_m, was assumed to be 5×10^2 in (1), where B_i and B_m are the ionospheric and magnetospheric fields respectively, and the width of $j_{\|}$ was taken as 100 km in (2), where a broad inverted-V is envisaged. For the edge effect (3) of a thin arc (5 km), it was assumed that 20% of the magnetic energy content of $0.02\,\mathrm{Jm}^{-2}$ was consumed in the energization and erosion of the cold ionospheric plasma. Of course, some of the above parameters have a broad range of variation (for K and $j_{\|}$ see for instance Table 1 in Olsson et al. (1998)), but the conclusions drawn subsequently will not be qualitatively altered.

The results shown in Table 7.3 are taken from Haerendel (2001b) with a few modifications applied. Here $u = \frac{3}{2}k_B(T_i + T_e)$ is the internal energy density of the hot magnetospheric plasma which is assumed to be isotropic. The magnetic energy

Way	Parameters	
1	$T_i = 1.2\,\mathrm{keV}$	$n_H = 0.4\,\mathrm{cm}^{-3}$
	$T_e = 0.32\,\mathrm{keV}$	
2	$K = 6 \times 10^{-10}\,\mathrm{Sm}^{-2}$	$j_{\|} = 1.5\,\mu\mathrm{Am}^{-2}$
3	$\bar{W}_{el} = 1\,\mathrm{keV}$	$j_{\|} = 1.0\,\mu\mathrm{Am}^{-2}$
4	$\Sigma_P = 5\,\mathrm{S}$	$E_{\perp} = 10\,\mathrm{mVm}^{-1}$

Table 7.2. Input values for the different regions of Figure 7.15. K is the field-aligned conductance of the Knight relation, Eq. (3.38).

	① Diffuse Aurora	② Arcs Inverted V's	③ Arc Edges	④ Ohmic Dissipation
Energy Conversion Rates	$\pi \bar{v} \varepsilon$	$K \cdot \Phi_\parallel^2$	$\frac{1}{e} j_\parallel \cdot \overline{W}_{el}$	$\frac{1}{\Sigma_p} J_\perp^2$
$\dot{\varepsilon}_\parallel$	$\lesssim 10^{-3} \frac{W}{m^2}$	$\gtrsim 10^{-2} \frac{W}{m^2}$	$\sim 10^{-3} \frac{W}{m^2}$	$\lesssim 10^{-3} \frac{W}{m^2}$
Energy Content per Unit Cross-section	$n_H \cdot \varepsilon \cdot \frac{B_i}{B_m} l_\parallel$	$\frac{\mu_0}{2} J_\parallel^2 \cdot l_\parallel$	$\delta \cdot \frac{\mu_0}{2} J_\parallel^2 l_\parallel$	$\frac{\mu_0}{2} J_\perp^2 l_\parallel$
W	$W_i \cong 2 Jm^{-2}$ $W_{el} \cong 0.3 Jm^{-2}$	$1\ Jm^{-2}$	$5 \cdot 10^{-3} Jm^{-2}$	$1 Jm^{-2}$
Energy Release Time $\tau_E = \frac{W}{\dot{\varepsilon}}$	$\tau_{Ei} \triangleq 2$ hours $\tau_{Eel} \cong 10$ min	$\lesssim 100$ sec	5 sec	~ 30 min

Table 7.3. Energy conversion rate, energy content and release times for the four cases illustrated in Figure 7.15. The energy conversion rates range from less than 10^{-3} Wm^{-2} in (1) and (4) to well above 10^{-2} Wm^{-2} in (2). The magnetic energy content, W, in (2)-(4) depends strongly on the assumed width of the upward current sheet, even if measured per unit cross-section, because the magnitude of $B_\perp$ enters quadratically. For this reason, W in (2) and (4) is much in excess of W in (3), where a narrow arc is assumed. The close agreement of the thermal and magnetic energy contents in (1) and (2) implies a beta value of the generator plasma of order unity. This may be a low estimate, in particular for the near-midnight region.

content can be approximated by:

$$W_m \simeq \frac{B_{\perp,i}^2}{\mu_0} R_E L \tag{7.56}$$

Since the energy conversion rate and influx into the ionosphere is

$$\dot{\varepsilon}_\parallel \simeq \frac{B_{\perp,i}^2}{\mu_0} v_{AG}, \tag{7.57}$$

where v_{AG} is the Alfvén speed at the effective lower edge of the generator region ($r_G \approx 5\,R_E$), one gets consistently

$$\tau_E \approx \frac{R_E L}{v_{AG}} \tag{7.58}$$

Table 7.3 shows that in spite of the high energy conversion rates, inverted-Vs could exist for a few minutes without resupply of energy to the generator, whereas

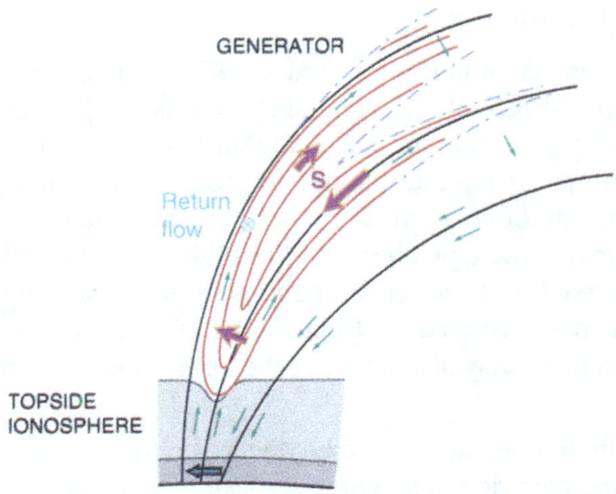

Figure 7.17. Re-investment of part of the downflowing energy is needed to maintain the return flow of the plasma in and above the acceleration region, **green arrows denote currents.**
Figure provided by G. Haerendel.

the erosion process of topside plasma, which we relate to the appearance of broad-band field-aligned electrons, has a rather short time scale. This is consistent with the proposal that arc edges are erosion fronts progressing with a speed of the order of 100 m s^{-1} (Haerendel, 1989, 1994, 1999). Energy release by particle precipitation and ohmic dissipation proceeds comparatively slowly. Most intriguing is the relatively short energy release time for broad inverted-Vs in view of their observed existence of many minutes. It suggests either a $\beta \gg 1$ source plasma or a continuous pumping of energy/hot plasma into the generator region. To realize that there is a problem with the duration of bright auroral arc systems for $> \frac{1}{2}$ hour and to find proper solutions, is still a prime subject in auroral physics.

Relation (7.57) is a simple expression of the downcoming Poynting flux as it emerges from the lower border of the generator region $r_G \approx 5\,\mathrm{R_E}$ and flows along magnetic field lines without reflection, in correspondence with strictly field-aligned currents. In reality, the situation is quite different. Not all of the Poynting flux is consumed by the energy conversion process, $J_\parallel \Phi_\parallel$, but about one quarter is reinvested in driving the return flow of the plasma (opposing $\mathbf{E}_\perp$) which is enforced by the incompressibility of the lower altitude magnetic field. Figure 7.17 is a sketch of the situation. In other words, the return flow of the low-altitude magnetospheric plasma which also manifests itself as apparent motions in TV-imaging of auroral rays or folds (Davis and Hicks, 1964; Davis, 1978; Haerendel et al., 1996) is driven from below by re-investment of part of the total energy influx from the magnetosphere (Haerendel, 1994).

7.8.3 Current Filamentation

Auroral arcs often appear in multiples, and satellite magnetometers often see the signature of multiple sheets of field-aligned currents (e.g., Figure 2.2). Since it seems to be the shortage of current carriers what leads to the occurrence of field-aligned potential drops, it is quite natural to assume that current filamentations and, therefore, concentrations are favorable for the creation of auroral arcs. However, highly structured particle precipitations implying the existence of parallel arcs have also been observed in the absence of any significant current structure (Mozer and Hull, 2001), as demonstrated by Figure 4.12. This implies that depletions of magnetospheric electrons may also underlie the appearance of auroral acceleration regions.

In general, current filamentation must be due to nonlinear interactions of the current and its attending electric fields with the plasma environment. This may occur in the lower ionosphere, where the closure currents flow, in the topside ionosphere, where cold ions can be extracted for the sake of charge neutrality above, and in the generator plasma, where a spatial structure of the pressure distribution may lead to increased divergence of the transverse current. Figure 7.18 shows an example of the first case, an ionospheric modification at the interface of a downward $J_{\|}$ and the continuing Pedersen current. Electrons are extracted upward and ions sideways from this region. This creates a reduction of the Pedersen conductivity and thus an enhancement of $\mathbf{E}_{\perp}$, which in turn leads to a steepening of the front of this perturbation. If projected up to the generator region, it may increase the local divergence of the transverse current and thereby enhance, i.e., concentrate $j_{\|}$.

This effect should be strong in the winter ionosphere where electron supply by photo-ionization is much reduced (see Elphic et al., 2000, and Section 6.1.3.3). Ionospheric modifications also underlie the ionospheric feedback models of auroral arcs (see Section 5.1).

Auroral density cavities in the topside ionosphere (see Section 4.1) are created by ion and electron extractions out of the cool and comparatively dense ionospheric plasma by the high electric field of the auroral acceleration region. The exact nature of this process is not known, nor how these cavities may feed back on the overall electrodynamics of the arc. Haerendel (1999) considered the extraction process in the context of a propagating arc model. It is sketched in Figure 7.19. The density cavity has a sharp, near-vertical boundary at the leading edge of the arc. If the thickness of the boundary is of order of c/ω_{pe}, the electric field can easily penetrate (skin depth) and act on ions and electrons. The ions respond initially mainly to the much stronger transverse field, owing to their high inertia, while the electrons are accelerated downward by the residual potential, $\Phi_{\|}$. The extracted ions are transversely heated to a few 100 eV and subsequently driven upward by the mirror force and $E_{\|}$. The electrons show up as narrow, strongly field-aligned, i.e., transversely cool beams, and are frequently seen at the edges of inverted-V structures.

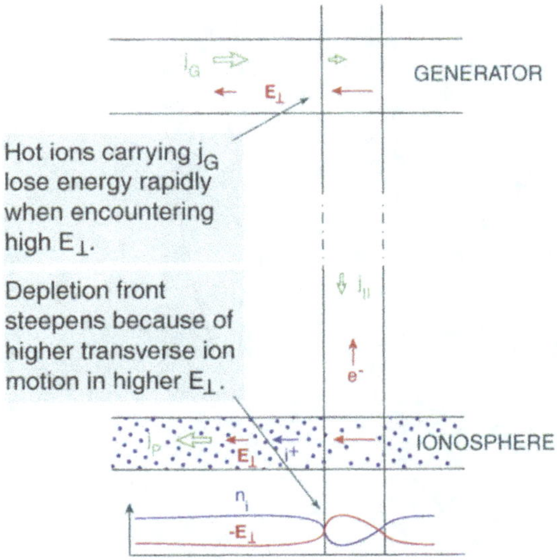

Figure 7.18. Depletion of the night-time E-region at a sink of downward $j_\parallel$ and source of $j_\perp$. The accompanying enhancement of $E_\perp$ acts back on the generator plasma. Figure provided by G. Haerendel.

While the ions move upward, they change the positive charge density in the cavity and above. The need for charge neutrality leads to a modification of the electric field distribution, as discussed in Section 4.1.3.4. In addition, plasma turbulence may be created by the two-stream instability between downgoing magnetospheric electrons and upgoing ions. A resulting anomalous resistivity may add to the local parallel electric field.

This is just one little-explored mechanism. The frequently observed, highly structured transverse electric fields near the bottom of the cavities are witness of the existence of intriguing effects which must impact on the dynamics, i.e., also on the currents, within the whole flux tube above an arc. This subject is certainly a matter of future research.

Another reason for current concentrations may be found in the dynamics of the generator plasma. Since the rate of energy conversion at lower altitudes, whatever the detailed process is, depends on the square of the current density ($\dot{\varepsilon}_\parallel \sim j_\parallel^2$) and since $j_\parallel$ is determined by the divergence of the transverse current in the generator, one wonders whether a linkage between the two can be established. If one takes the view that auroral currents are largely driven by pressure gradients and that pressure gradients in the perpendicular (i.e., in the current) direction determine $\nabla \cdot \mathbf{j}_\perp$ and hence $j_\parallel$, one can easily see why current filamentation can arise from maximization of the lower altitude energy conversion. The higher the latter, the stronger are the associated electric fields. These fields propagate up into the generator plasma, where $\mathbf{E}_\perp \cdot \mathbf{j}_\perp < 0$. The loss of energy of the plasma powering the generator is

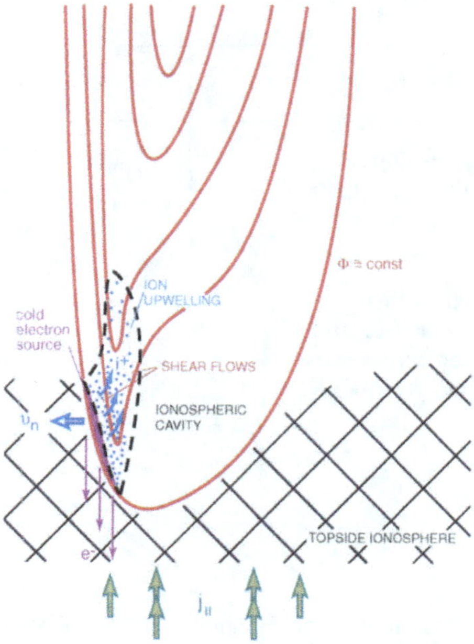

Figure 7.19. Sketch of an auroral cavity. Figure provided by G. Haerendel.

then locally enhanced which would lead to increased structuring of the pressure distribution in the generator. This is, of course, only a general concept which has yet to be explored by theory.

7.8.4 Feedback on Generator

Auroral arcs owe their origin to the acceleration of electrons in field-aligned potential drops. These electric fields are part of a magnetospheric-ionospheric current system. They constitute a mechanism of conversion of magnetic energy stored in the current circuit into kinetic energy of particles. A bright arc consumes the stored energy in an Alfvén transit time ($\tau_A \approx 1$ min). Hence such arcs cannot be regarded as static phenomena with infinite energy reservoirs. They are at best quasi-stationary phenomena, if their lifetime $\tau \gg \tau_A$. The consumption of energy must be communicated with Alfvén speed to the energy reservoir. This is the essence of the auroral arc models involving standing Alfvén waves (Haerendel, 1989, 1994, 2001b; Knudsen, 1996, see also Section 7.5.3). A lifetime $\tau \gg \tau_A$ of a visible arc is attributed to a continuous tapping of the energy reservoir. In this context, the observation of the existence of proper motions of the arc with respect to the plasma frame is of crucial significance (see Section 2.3).

The essence of these considerations is that a front of enhanced $j_{\parallel}$ (standing Alfvén wave) propagates into the interior of a current circuit as shown in Figure 7.5. No physical motion in the arc normal direction is connected with this propagation (v_n). It is solely the consequence of magnetic stress release in a thin layer at the edge of a large current circuit, i.e., a region of strong shear stresses. The detailed structure of the standing Alfvén waves shown in this figure may be ignored. What is important is that, owing to the propagating current front, the force balance in the generator plasma is changed.

Whereas in the interior of the current circuit ($j_\parallel \simeq 0$) this balance can be described by:

$$\frac{\partial p_0}{\partial y} + j_{x0}B_z = 0 \tag{7.59}$$

where y is the direction of slow convection under frictional control of the ionosphere, it changes to:

$$\rho v_n \frac{\partial v_y}{\partial x} = -\frac{\partial \delta p}{\partial y} - \delta j_x B_z \tag{7.60}$$

in the parallel current sheet. Here $\delta p = p_0 - p$, $\delta j_x = j_{x0} - j_x$, and the normal propagation speed $v_n \approx w_{arc}/\tau_A$.

Eq. (7.60) expresses the change of equilibrium, as the transverse current density is being reduced ($\delta j_x < 0$) by diversion along **B**. Accordingly the plasma is being accelerated in the $-\nabla p_0$ direction whereby it loses internal energy. Without a consistent theoretical treatment we can only guess that another time-scale is introduced, the acoustic time-scale

$$\tau_p = \frac{c_s}{\ell_p}, \qquad \ell_p = \frac{p_0}{|\nabla p_0|} \tag{7.61}$$

The ratio τ_p/τ_A will determine the response of the generator, for instance, to which degree the energy reservoir is depleted by the passage of an arc. If $\tau_p \gg \tau_A$, there would be sufficient energy for the generation of parallel arcs.

Another aspect of feedback on the generator has been treated in a simple straight field-line model by Vogt et al. (1999), namely the reflection of Alfvén waves. The main difference with the preceding deliberations is that in that model it is assumed that the driving current in the generator region is not affected by the reflected Alfvén wave and remains constant.

Multiple reflections between generator and ionosphere have been treated by Vogt and Haerendel (1998), using the reflection coefficient at the 'tau generator, (Eq. 7.18) and the combined reflection coefficient of auroral acceleration region and ionosphere (Eq. 3.66). Figure 7.20 gives an example.

This model can also be used to show that certain transverse wavelengths, close to the transient scale-length $\lambda_t = (2R_w K)^{-1/2}$ are preferentially absorbed in the auroral acceleration region ($R_w = \mu_0 v_A$, K = Knight conductance).

What is still completely missing from auroral theory is a consistent nonlinear treatment of the interaction of standing or propagating Alfvén waves with the generator plasma, which includes the propagation of shear waves in a non-stationary medium.

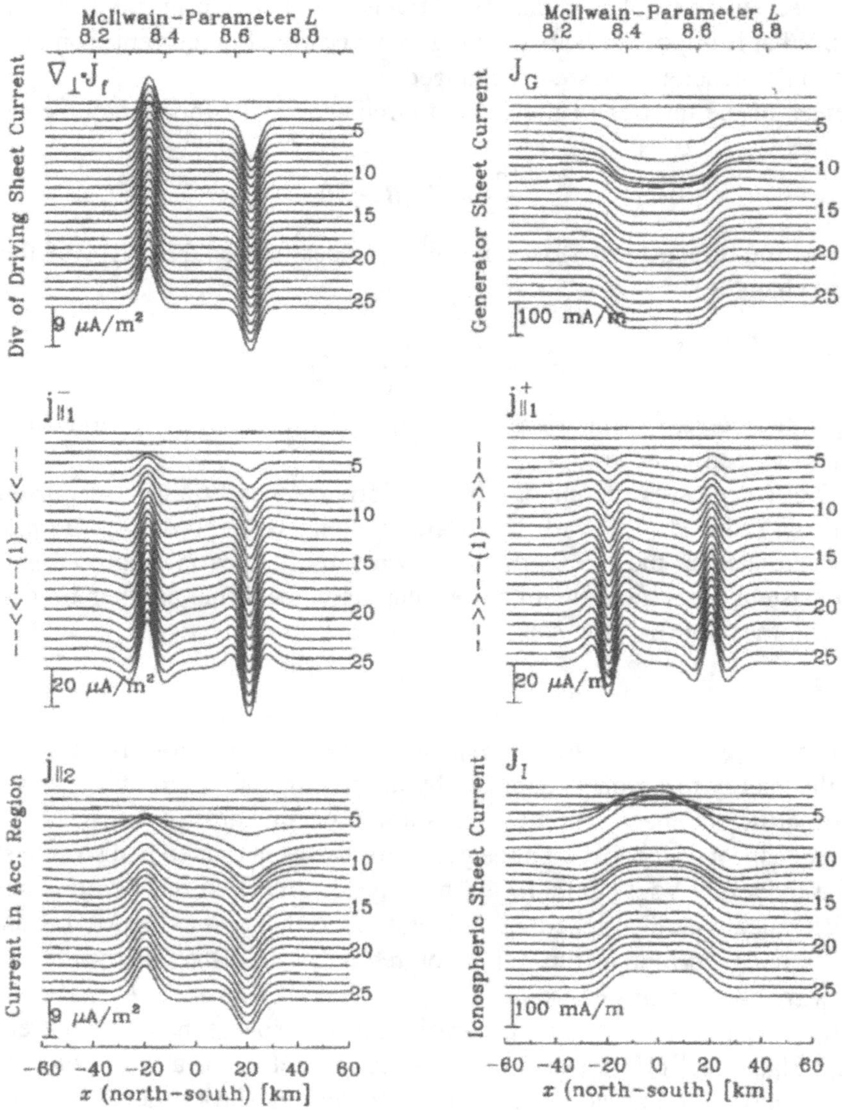

Figure 7.20. Twenty-five current oscillations between generator, auroral acceleration region, and ionosphere. $\mathbf{J}_f$ is the current equivalent to a constant generator force; $\mathbf{J}_G$ is the perpendicular generator current that closes the field aligned currents; $j^-_{||1}$ and $j^+_{||1}$ are the downward and upward field aligned currents above the acceleration region, respectively; $j_{||2}$ is the current within the acceleration region. $\mathbf{J}_I$ is the ionospheric current. The current source is switched on during the initial oscillations. From Vogt (1997).

7.8.5 Global Impact of the Aurora

If we want to understand the role played by auroral arcs in the global dynamics of the magnetosphere, we have to consider first the dynamics in the absence of aurora. The characteristic E-W extension of arcs (in particular on the evening side) is indicative of their connection with the global convection system. We address in particular the substorm situation. As plasma is convected towards the Earth in the midnight sector, it piles up and establishes a force equilibrium with the surrounding magnetic field and plasma. Naturally, a maximum of total pressure is set up with respect to the longitudinal direction, whereas owing to the earthward increasing strength of the dipole field, the pressure gradient still has an inward component. As always, high pressure wants to expand, which is most easily done in the longitudinal direction. The expansive motion has to overcome friction by ion-neutral collisions in the ionosphere, or Ohmic dissipation of the attending current system, which is of the Type II of Boström (1964). This system can be related to the later discovered Region 1 and 2 currents (Iijima and Potemra, 1978, cf. Section 5.1.1.2) by identifying approximately the outer (inner) current sheets of Boström (1964) with the Region 1 (Region 2) currents. This identification is, however, not exact, since an appreciable part of the Region 1 currents is fed from another generator, the solar wind acting on the magnetopause field. Figure 7.21 shows the current system and Figure 7.22 the resulting convection along the auroral oval as part of the global high-latitude plasma circulation system, which is, of course, highly non-stationary. Considerations of energy and momentum balance in the magnetosphere and of the flow of both quantities into the ionosphere, lead to a relation between total pressure drop from midnight to noon, ionospheric friction, measured by the integrated Pedersen conductivity, Σ_p, and the magnetospheric convection speed, U_{con}, at the equator. This relation is given here without derivation:

$$U_{con} \simeq \frac{1}{2\mu_0 \Sigma_p} \frac{\ell_\parallel}{L_{ps}} \ln\left(\frac{P_{tot}^{24}}{P_{tot}^{12}}\right) \tag{7.62}$$

$\ell_\parallel$ is the half length of a field line and L_{ps} the length of the inner plasma sheet from 24 to 12 MLT. With $\Sigma_p = 10\text{S}$, $\ell_\parallel / L_{ps} = 0.25$, and $p_{tot}^{24}/p_{tot}^{12} = 3$, one obtains $U_{con} = 11$ km s^{-1}. This is indeed the order of magnitude of average $\mathbf{E}\times\mathbf{B}$-velocities in the outer magnetosphere.

Evaluation of Eq. (7.62) also implies that it takes several hours to move plasma from midnight to noon, whereas a typical substorm duration is close to 1-2 hours and the initial expansion phase may be as short as 10–20 minutes. This means that the convection pattern and speed change substantially faster than a stationary circulation period would last. The balance of pressure forces and ionospheric friction and associated energy dumping into the upper atmosphere, cannot be the whole story.

Here the aurora comes into play. Aided by filamentation, $j_\parallel$ may become so intense in certain locations that the auroral energy conversion process sets in. In

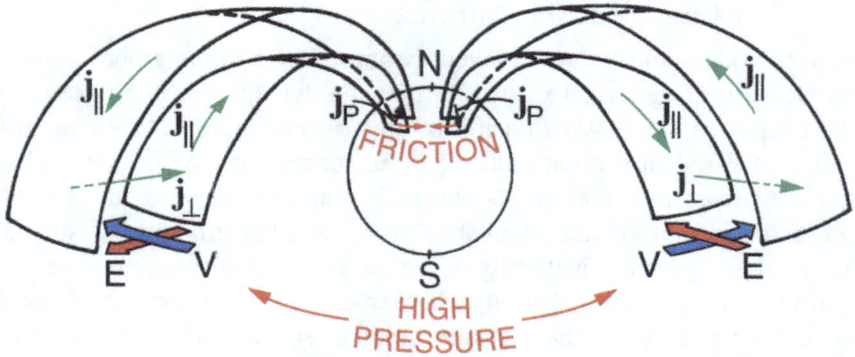

Figure 7.21. Dynamo forces, auroral current system, and resulting convection under frictional control by the ionosphere. After Boström (1964).

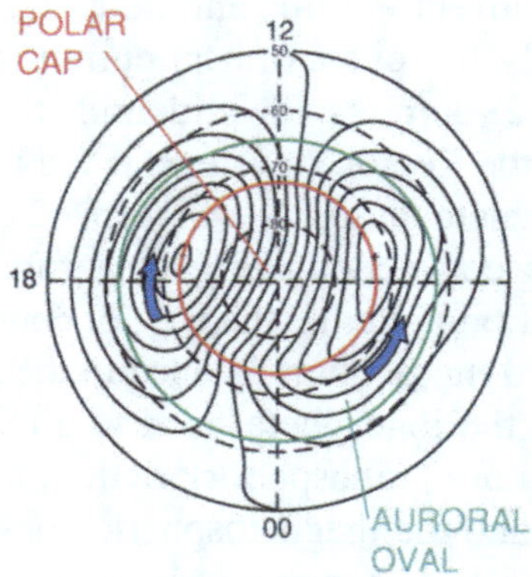

Figure 7.22. High-latitude plasma circulation system at times of an active magnetospheric dynamo (e.g., during substorms).

bands of enhanced $j_\parallel$ (mostly aligned with the convection paths) the energy conversion, i.e., energy release from the hot magnetospheric plasma, is greatly increased. This can be demonstrated by a simple calculation (Haerendel, 2000). Instead of being smoothly distributed over the width w_0 of the source region, the upward current $j_\parallel$ may become concentrated in N discrete bands of the width of auroral arcs, w_{arc}:

$$j_{\parallel,0} w_0 = N j_{\parallel,arc} w_{arc} \tag{7.63}$$

The total energy conversion rate, $\dot{W}_{arc} = w_{arc} N \dot{\varepsilon}_\parallel = w_{arc} N K^{-1} j^2_{\parallel,arc}$, for N auroral arcs, can then be compared with the Ohmic dissipation, $\dot{W}_0 = w_0 \Sigma_P^{-1} J_P^2$, of the ionospheric Pedersen current, $J_p = w_0 j_{\parallel 0}$:

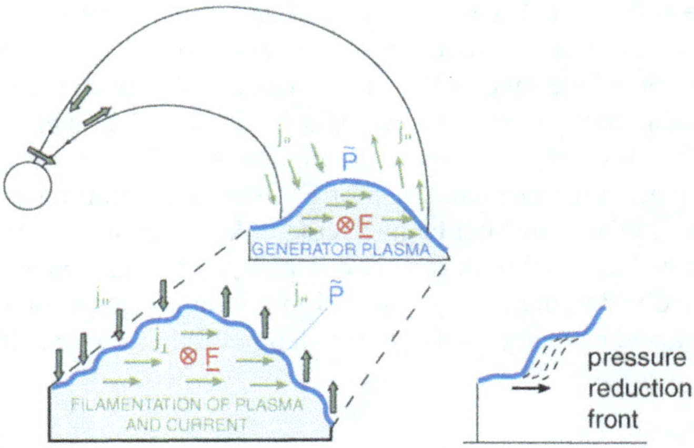

Figure 7.23. Modification of the smooth magnetospheric pressure distribution leading to filamentation of $j_{\|}$ and thereby to enhanced energy conversion in the auroral acceleration region. The progression of the pressure jumps in the plasma frame, as energy is flowing downward, is also indicated. From Haerendel (2000).

$$\frac{\dot{W}_{arc}}{\dot{W}_0} = \frac{L_{MI}}{N w_{arc}} \tag{7.64}$$

Here we have identified w_0 with the M-I coupling length, $L_{MI} = (\Sigma_p/K)^{1/2}$, which is of the order of a few 100 km, whereas w_{arc} is typically 10 km. Eq. (7.64) shows that the energy conversion above a system of parallel arcs can make a substantial or even larger contribution to the magnetospheric energy loss than the ionospheric friction or Ohmic dissipation. For this reason the notion of a magnetospheric pressure valve was introduced (Haerendel, 1992, 2000).

One can thus conclude that the aurora is not just a beautiful and intriguing asset of the dynamic magnetosphere, but a necessary valve that the magnetospheric plasma employs when its pressure and convection speed build up to excessive levels. Of course, there is no distinct level of 'over-heating'. But it is obvious that the energy inflow from the tail during substorms constitutes such a state. Loss of magnetospheric plasma pressure is concomitant with the release of magnetic shear stresses.

Summary

The objective of Section 7.8 was to show which elements need to be included in a successful theory of the auroral phenomena, at least from the macroscopic point of view. The aurora emerges as an essential ingredient in the global redistribution of energy, in particular intensity during substorms. The flow of energy into discrete bright arcs is such that most of the free (magnetic) energy is consumed in one Alfvénic transit time. Thus, to maintain arcs over longer periods the resupply of energy must be addressed. The discreteness of arcs is, as far as we can judge now, related to field-aligned current concentrations whose origin is still poorly understood. The reaction of the magnetospheric plasma to the existence of auroral arcs is a crucial element in the overall magnetospheric dynamics as is the accompanying loss of energy. And, finally, the impact on the ionosphere of auroral precipitation for upward current on the one hand, and of E-layer evacuation for downward currents, on the other hand, must not be neglected either. It is obvious that the overall problem posed by the aurora is extremely complex, quite apart from the host of microprocesses involved in the energy conversion and momentum coupling. There is still a major task and challenge posed to the magnetospheric research community.

Chapter 8

Dynamic Coupling to the Magnetosphere

This chapter deals with dynamical coupling of the auroral ionosphere to the magnetosphere. As discussed in the previous chapters, many important questions in auroral physics can be described within a quasi-static picture. However, even the most static arcs do dissipate energy and this energy dissipation must be supplied by an external energy source, which ultimately is the solar wind-magnetosphere interaction. The coupling, undoubtedly, is a dynamical phenomenon. Furthermore, most auroral phenomena, at least the physically most interesting ones, are far from static. This is true in particular during times of magnetospheric activity, such as storms and substorms. Although a full treatment of the storm or substorm physics is beyond the scope of this book, we discuss in this chapter how the magnetospheric activity is related to the auroral phenomena. The basic ways how the magnetosphere controls auroral dynamics are:

- Formation of auroral plasma sources in the magnetosphere. For example, diffuse aurorae are produced by direct precipitation through processes taking place in the magnetosphere (Lyons et al., 1999a).
- Creation of FAC systems which are associated with parallel electric fields and a host of wave-particle interactions able to energize charged particles.
- Changes in plasma sheet parameters that influence the regime and efficiency of field-aligned acceleration. For example, changes in electron temperature and density affect the current-voltage relationship (see Section 3.3.1)
- Configurational changes which distort the mapping between the magnetosphere and the ionosphere. This causes observable asymmetries in auroral structures and lead to stresses that can cause rapid motions, particle acceleration, etc.

This chapter is organized as follows: Section 8.1 discusses the magnetosphere as a source of auroral precipitation. Section 8.2 deals with dynamics in terms of energy circulation in the solar wind-magnetosphere-ionosphere system. Section 8.3 is a more detailed discussion of magnetospheric control of the auroral phenomena. It is obvious that the ionosphere reacts back, affecting the magnetospheric dynamics, but how this happens and how important it is, has remained rather unclear; these issues are discussed in Section 8.4. This chapter concludes in Section 8.5 with a discussion of some of the important open issues in the relationships between aurorae and magnetospheric dynamics.

8.1 Magnetosphere as an Auroral Precipitation Source

Relationships between the magnetospheric regions and the ionosphere were discussed in Section 5.2, where it was noted that the correct mapping between various regions of the magnetosphere to the ionosphere is still very much an open issue. This makes it difficult to determine definite locations of the source regions of particular auroral forms. The issue is particularly complicated during dynamical development of the magnetosphere. However, it is clear that most of the auroral oval maps to the plasma sheet, and on the other hand, most of the plasma sheet maps to the auroral oval. The plasma sheet is bounded by the plasmapause in the inner magnetosphere, and by the magnetic separatrix between open and closed field lines in the distant magnetotail.

Precipitation from the closed field line regions requires several processes to occur: First, particles must gain access to the right location, they are often accelerated and heated, and finally there must be sufficiently efficient mechanisms, such as pitch-angle scattering and/or acceleration, to remove the particles from the magnetic bottle through the atmospheric loss cone.

Particles may enter the plasma sheet from the mantle, the low-latitude boundary layer, or the ionosphere. While the relative contribution from these different plasma sheet sources has not been adequately determined, particles entering the tail plasma sheet from any of the sources will cross the tail current sheet, where they gain energy, as the cross-tail current and the average electric field are in the same direction ($\mathbf{j} \cdot \mathbf{E} > 0$).

The magnetic field in the lobes of the distant tail, B_{lobe}, is primarily tangential to the current sheet, whereas the current sheet field B_n is small and mainly perpendicular to the current sheet plane. The characteristics of the particle motion within the current sheet depends on the ratio of the two gyro frequencies, ω and ω_c, determined by B_{lobe} and B_n, respectively, or equivalently the ratio κ^2 of the minimum field line curvature radius to the maximum particle gyroradius. The ratio is given by

$$\kappa^2 = \frac{qLB_n^2}{mvB_{lobe}} = \left(\frac{\omega_c}{\omega}\right)^2 \tag{8.1}$$

where q is the magnitude of the charge of the particle, v is particle velocity, and L is the scale length for changes in the tangential component of $\mathbf{B}$ as it reverses in direction across the current sheet. From Eq. (8.1), the value of κ^2 for electrons and protons in the tail can be written as

Region	Electrons	Protons
Distant tail ($L \sim 10^3$ km) $B_n \sim 1$ nT, $B_{lobe} \sim 10$ nT	.014–90 keV	.008–50 eV
Mid-tail ($L \sim 5 \times 10^3$ km) $B_n \sim 5$ nT, $B_{lobe} \sim 20$ nT	.054–350 MeV	.030–190 keV
Near-tail ($L \sim 10^4$ km) $B_n \sim 30$ nT, $B_{lobe} \sim 60$ nT	30–2 × 10⁵ MeV	.017–110 MeV
Geostationary orbit (quiet times)	-	.150–960 MeV

Table 8.1. Range of energies for which strong pitch-angle scattering is expected for different regions of the nightside magnetosphere (based on Lyons et al., 1999a; Sergeev et al., 1983).

$$\kappa^2 = \left(\frac{W_0}{W}\right)^{1/2} \left(\frac{B_n}{1\ \text{nT}}\right)^2 \left(\frac{L}{1000\ \text{km}}\right) \left(\frac{20\ \text{nT}}{B_{lobe}}\right) \tag{8.2}$$

where W is particle energy in eV, W_0 is 0.12 eV for protons and 220 eV for electrons, and B_n, B_{lobe}, and L have been normalized to nominal tail values.

Particle motion for $\kappa^2 \gg 1$ satisfies the guiding center approximation within a current sheet, whereas particles having $\kappa^2 \ll 1$ undergo a serpentine type of motion (called Speiser motion) Pitch-angle scattering does not occur for either of these cases. Particles with intermediate κ^2 values undergo motion that is chaotic, and such particles are significantly pitch-angle scattered during their interaction with the current sheet. Strong pitch-angle scattering which fills the loss cones, has been found to occur for κ^2 values between $\sim$ 0.1 (Lyons and Williams, 1984) and 8 (Sergeev et al., 1983). Since $W \propto \kappa^{-4}$, strong pitch-angle diffusion for each particle species occurs over a range of energies of nearly four orders of magnitude at any given location within a current sheet. The range of particle energies for which strong pitch-angle scattering would be expected for various regions of the nighttime magnetosphere is given in Table 8.1.

Loss of magnetospheric particles by pitch-angle scattering maximizes if the scattering is strong enough to fill the loss cone in less than one particle bounce period, and thus to produce isotropic pitch-angle distributions. As demonstrated in the table above, strong pitch-angle scattering in weak field regions far from Earth readily leads to isotropization of the distribution functions.

Practically all proton precipitation is caused by current sheet scattering of plasma sheet protons, sometimes enhanced by the downward electric field acceleration in return current regions as discussed in Section 4.2. This is confirmed by the fact that the equatorward boundary of proton precipitation in the auroral oval is colocated with the inner boundary of non-adiabatic scattering. Furthermore, the density and temperature of the precipitated protons nicely agree with the corresponding plasma sheet parameters.

However, diffuse precipitation usually covers most of the auroral region, and can extend as far equatorward as the plasmapause (e.g., Gussenhoven et al., 1983; Hardy et al., 1985). The region where most of the structured aurorae are observed is typically embedded within the poleward portion of the region where diffuse auroral electron precipitation is observed. These electrons are isotropic in pitch-angle, have a mean energy that increases from a fraction of keV to a few keV with increasing latitude, and have a relatively uniform distribution within the loss cone. In fact, diffuse electron precipitation throughout the auroral zone is one of the most persistent processes in auroral physics. Most of the electron number flux as well as most of the total energy flux associated with particle precipitation come from the diffuse precipitation (Lyons et al., 1999a).

Apart from the actual mechanism that fills the loss cone, the intensity and spatial distribution of the diffuse precipitation can be understood and qualitatively estimated as a combination of two basic processes. Earthward convection brings plasma sheet particles to the outer boundary of the inner magnetosphere. The convecting plasma is heated and compressed (Borovsky et al., 1998) while it maintains a nearly isotropic Maxwellian distribution with an enhanced high-energy tail (Christon et al., 1991). Roughly at the distance of $X \approx$ 10-12 $R_E$the gradient and curvature drifts start to remove the highest-energy particles from the convecting flux tubes. Because the observations show that both the plasma sheet and precipitating electrons have nearly isotropic distributions, we can use the isotropic approximation to represent the background precipitation.

In order to estimate the auroral electron precipitation energy flux for given magnetospheric parameters, we note that for an isotropic Maxwellian distribution function per unit volume, $f = \pi^{-3/2}\, v_{Te}^{-3}\, \exp(-v^2/v_{Te}^2)$, the precipitated energy flux is given by

$$J_E = \int \frac{mv^2}{2}\, v \cos\alpha\, f\, d^3v = \pi^{-1/2} v_{Te} P_e \tag{8.3}$$

where v_{Te} is the thermal speed and P_e the pressure of the electrons, α is the pitch angle, and the angular integration extends over the downward hemisphere. Thus the integral over the angles yields $\int d\Omega \cos\alpha = \pi$.

The total plasma pressure in the equatorial plasma sheet can be computed by integrating the steady-state force balance equation $\nabla P = \mathbf{j} \times \mathbf{B}$ along the X-axis starting from $X = -30\ R_E$, where the lobe magnetic pressure is taken as a boundary condition. We evaluate the magnetic field and the currents from the empirical T89 model (Tsyganenko, 1989). Furthermore, we assume a proton to electron temperature ratio of 8, thus $P_e = P/8$ (Baumjohann, 1993). Neglecting losses, the adiabatic law $P/P_0 = (V/V_0)^{-5/3}$ yields a scaling relation for the temperature $T/T_0 = (P/P_0)^{2/5}$. After selecting the initial temperature T_0 we then get both v_{Te} and P_e and, consequently, the precipitated energy flux.

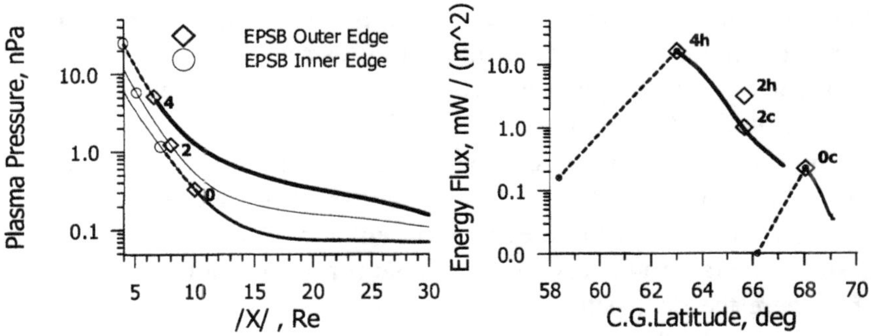

Figure 8.1. Plasma pressure (left) and precipitated electron energy flux (right) at the midnight meridian for Kp = 0, 2, and 4. The diamonds illustrate the maximal precipitation rate and the circles mark the equatorward edge of the precipitation. The calculated precipitation profiles are shown for Kp = 0 and 4 only. The letters *c* and *h* correspond to assumptions of cold (0.1 keV) or hot (1 keV) initial electron temperatures in the plasma sheet at 30 R_E. Both cold and hot cases are shown for the Kp=2 case to illustrate the dependence of the energy flux on the initial temperature. Figure provided by V. Sergeev.

The left panel in Figure 8.1 gives the pressure in the midnight plasma sheet as a function of distance from Earth for Kp = 0, 2 and 4. The pressure balance equation is used to integrate the plasma pressure from the distant magnetotail to the distance where the plasma pressure changes from isotropic to an anisotropic loss cone distribution. This boundary is shown by diamonds in Figure 8.1, and referred to as the electron plasma sheet boundary outer edge. The circles represent the locations of the inner edge of the electron plasma sheet. Two different initial temperatures were used to compute (from Eq. 8.3) the precipitated energy fluxes shown in the right panel: during quiet times the plasma sheet is cool ($T_e \approx 100$ eV) and dense, whereas during disturbed conditions and after substorms it is warmer ($T_e \approx 1$ keV). The right panel of Figure 8.1 shows the precipitated energy flux as a function of geomagnetic latitude. The two solid curves show two extreme cases computed using the Tsyganenko magnetic field model with Kp=0 and low initial temperature (symbol 0*c*), and using the T89 model with Kp=4 and high initial temperature (symbol 4*h*). The diamonds show the locations of the electron plasma sheet boundary outer edge mapped to the ionosphere, using different Kp values and the two temperatures. Note the strong dependence of the energy flux on the plasma sheet temperature, illustrated by the almost a factor of 10 difference shown for the Kp=2 case.

Equation (8.3) can be used for the energy flux estimations to the latitude where the energy of the unstructured electron precipitation maximizes along a given meridian. In an empirical categorization (Newell et al., 1996a), this is called precipitation boundary b2e (see also Section 5.1). Equatorward of this boundary, the energy flux calculation becomes more complicated, because the azimuthal drifts and ionospheric precipitation make the loss-free approximation assumed above invalid. The dashed lines in Figure 8.1 have been added to illustrate the decrease of the flux toward the inner edge of the plasma sheet, defined as the equatorward boundary of electron precipitation (Gussenhoven et al., 1983).

Some important conclusions can be made from Figure 8.1: The precipitated energy flux is always small from the distant and mid-tail regions (less than 1 $\mathrm{mW\,m^{-2}}$). Closer to Earth the precipitation enhances due to plasma heating and compression. The largest energy flux available for diffuse aurorae originates at the boundary where the drift paths of electrons and ions begin to deviate from each other. For quiet times, the computation gives a maximum flux of about 1 $\mathrm{mW\,m^{-2}}$, and for disturbed conditions $\sim 10 - 30$ $\mathrm{mW\,m^{-2}}$. These numbers are in good agreement with the statistics reported by Hardy et al. (1985).

The main factors behind the strong dependence of the auroral energy flux on magnetic activity are the earthward shift of the precipitation boundaries and heating of the plasma sheet. The increase of the tail current and the subsequent stretching of the magnetic field provide additional equatorward motion of the auroral oval. Note that Figure 8.1 is shown for the midnight meridian only, and thus does not illustrate the dawn-dusk asymmetries in the precipitation. The main motivation of this exercise is to illustrate the readily available energy flux in the plasma sheet once there is a strong pitch-angle diffusion mechanism available to fill the loss cone.

Early sounding rocket observations (e.g., Whalen and Diarmid, 1973) showed that precipitation is isotropic up to energies of a few keV, with an increasing anisotropy toward higher energies. However, it has been difficult to explain in physical terms how the loss cone is filled. Fundamentally, the question is how to break the first adiabatic invariant of the electrons, and how to move them into the atmospheric loss cone efficiently enough to account for the precipitation from energies <1 keV to ~ 10 keV over most of the auroral oval. As can be seen in Table 8.1, scattering in the tail current sheet is sufficient in a narrow region at the outer edge of the plasma sheet only, and cannot account for the nearly isotropic precipitation throughout the very much broader auroral oval.

Various wave-particle interactions can also scatter particles into the loss cone. While numerous wave modes have been observed throughout the magnetosphere, only a few wave phenomena are sufficiently intense to account for significant precipitation into the high-latitude ionosphere. As discussed in Lyons et al. (1999a), the diffuse precipitation was thought to occur as a result of wave-particle interactions by a loss cone instability emitting electrostatic cyclotron waves drawing energy from the electron perpendicular energy. This result was based on observations of electron cyclotron harmonic (ECH) waves in the tail current sheet (Kennel et al., 1970). The observed amplitudes were estimated to be large enough to cause the pitch-angle scattering that is required to account for diffuse electron precipitation (Lyons, 1974). However, with the appearance of modern wave instruments, estimates for the wave amplitudes of the ECH waves decreased. Careful analysis of GEOS-1 and -2 data near and within geostationary orbit showed that typical wave amplitudes were actually well below 0.1 mVm^{-1} (Belmont et al., 1983). This was later confirmed with SCATHA and AMPTE/IRM observations (Roeder and Koons, 1989). The present understanding is that the ECH waves cannot account for the observed filling of the loss cone. Because no other mechanism can explain the observed energy fluxes, the equatorial region wave-particle interactions need to be investigated further.

Adiabatic processes, i.e., processes that do not require changing the particle magnetic moment, can also move particles into the loss cone. Weak pitch-angle scattering results from adiabatic gradient and curvature drifts that cross electric equipotentials, but this process is too weak to account for the observed isotropic precipitation.

Whatever the underlying mechanisms for pitch-angle scattering are, the precipitation maximizes if the scattering is strong enough to fill the loss cone in less than one particle bounce period, which then produces isotropic pitch-angle distributions. Current sheet scattering in weak field regions, e.g., in equatorial regions far from Earth, readily gives strong pitch-angle scattering. Pitch-angle scattering by wave-particle interactions can occur at all radial distances from Earth, and can be either strong or weak. Weak pitch-angle scattering moves far fewer particles to the center of the loss cone than to its outer edges. This leads to pitch-angle distributions of the precipitating particles that are peaked near 90° at low altitudes, and hence to significantly lower loss rates as compared to isotropic distributions.

8.2 Magnetospheric Energy Circulation

The magnetospheric dynamics ultimately draws its energy from the solar wind kinetic energy. The most favorable conditions for the energy coupling have been shown to exist during periods when the interplanetary magnetic field has a substantial southward component, i.e., the field is in the opposite direction to the dipole field at the subsolar magnetopause, as noted as early as 1966 (Fairfield and Cahill, Jr., 1966). Thus, reconnection at the dayside magnetopause is considered to be the main coupling mechanism between the solar wind and the magnetosphere. Other processes at the magnetospheric boundaries, jointly called viscous interactions, add to the coupling but their contribution is smaller and less variable over time. This section discusses a variety of dynamic processes that arise during periods of enhanced energy input, and involve the entire magnetosphere and auroral oval. Direct processes at the dayside boundaries and near the cusp are also associated with specific auroral forms, those are treated elsewhere in this book (see Section 4.5).

Magnetospheric dynamic events are important for auroral plasma physics, as they provide most of the energy for the auroral processes. On the other hand, auroral plasma physics is a key element in resolving the magnetospheric energy budget, as a large portion of the energy in dynamic events is dissipated in the high-latitude ionosphere where the aurorae reside.

The coupling between the solar wind and the magnetosphere via dayside reconnection sets up an average convection pattern (as illustrated in Figure 1.8). The plasma enters the nightside magnetosphere from the plasma mantle and through the boundaries at the flank ending up in the plasma sheet. There the plasma drifts earthward under the influence of an average dawn-to-dusk electric field. However, as the plasma β in the plasma sheet is large, the plasma motions are irregular and the electric field is locally strongly fluctuating with amplitudes of up to tens of $\mathrm{mV\,m^{-1}}$. Averaged over the fluctuations, the mean field is small but quite consistently positive with values of about 0.1–0.5 $\mathrm{mV\,m^{-1}}$ inside of about 40 R_E(Nishida and Ogino, 1998; Angelopoulos et al., 1992).

Under quiet conditions during mainly northward interplanetary magnetic field, energy input from the solar wind to the magnetosphere is small, convection is weak, and the magnetosphere is in a quasi-steady state. During such periods, the auroral oval is located at high latitudes and the precipitation is weak. When the solar wind driving is stronger, the energy input is larger than the dissipation in the magnetosphere–ionosphere system, and energy is stored in the magnetotail in the form of enhanced lobe magnetic field energy. This leads to a dynamic situation, where the energy previously stored in the system is later released and dissipated through a variety of processes. During the energy input, the auroral oval brightens and moves to lower latitudes, and the energy dissipation processes are associated with bright auroral forms on a variety of scale sizes.

The flux and plasma transport within the plasma sheet are realised as bursts of localized, high-speed plasma flows. These bursty bulk flows (BBFs) have speeds typically of several hundred km $\mathrm{s^{-1}}$, scale sizes of a few R_E, and time scales of about 10 min (Angelopoulos et al., 1992). Such flow bursts are observed during all geomagnetic conditions ranging from quiet to storm-time, but their occurrence frequency increases substantially during active times. It has been shown that these bursts can account for a major portion of the earthward plasma, energy, and flux transport in the magnetotail (Angelopoulos et al., 1994). These flow bursts are most probably associated with bursty, localized reconnection events. Bursty bulk flows are further discussed in Section 8.3.3.

During periods of enhanced energy input, the response of the magnetosphere depends on the previous dynamic state of the magnetosphere, on the level of the energy input, and on the solar wind and interplanetary field conditions. The configurational changes can be small or large, ranging from small-scale pseudobreakups to global-scale magnetospheric substorms, steady magnetospheric convection events, and magnetic storms (for recent reviews see Baker et al., 1996; Sergeev et al., 1996a,b). It seems that both small and large activations are associated with bursty flow events in the magnetotail, the basic difference being that the smaller events are caused by single flow bursts, whereas the larger events follow as a consequence of a sequence of such bursty activity.

There are no universally accepted definitions for the types of magnetospheric activity: pseudobreakups, magnetospheric substorms, steady magnetospheric convection events, and magnetic storms have distinct characteristics discussed below, but only when they are observed in isolation and when they are in some sense 'typical' events. In reality, there is a continuum of events on different scale sizes, and with the present observational capabilities there are no clear criteria separating e.g., pseudobreakups from small substorms, or large substorms from small storms.

8.2.1 Pseudobreakups

Pseudobreakups were originally defined as small auroral activations where an auroral arc brightened, folded up, and decayed within a few minutes (Elvey, 1957). This process looks much like a substorm onset (see the next subsection), but the activity decays before an expansion phase follows. Later, these breakups were also associated with localized activations in the magnetotail; fast flow bursts, injections near geostationary orbit, and small-scale magnetic field dipolarizations were observed in conjunction with the brightenings (Koskinen et al., 1993; Nakamura et al., 1994). Typically, pseudobreakups occur during slightly disturbed times, such as the substorm growth phase, but they can be observed during all geomagnetic conditions.

Globally, pseudobreakups and larger activations differ in that pseudobreakups do not release large amounts of energy stored in the tail lobes, rather the total dissipation is small and the events are driven by internal plasma sheet energy. Recently, the focus of research on pseudobreakups has been in finding the controlling factors that determine the size of the activation that follows. As both large and small events have similar tail signatures at the first initiation, the following development can be determined either by local plasma sheet parameters in the tail, by global magnetotail configuration, by the state of the conjugate ionosphere, by the interplanetary magnetic field and solar wind conditions, or any combination of these.

Rapid variations in the solar wind or interplanetary field parameters can lead to bursts of activity in the magnetotail (e.g., Zhou and Tsurutani, 1999), and hence part of the pseudobreakups may be associated with an energy state or configuration where the tail is not suitably configured for a larger instability to develop. Another class of pseudobreakup events occurs during the substorm growth phase, again in a situation where the magnetotail is still stable on the large scale (e.g., Koskinen et al., 1993). A limiting case to this group are occasions where the substorm begins with a localized activation and the large-scale magnetotail reconfiguration occurs some time (up to 10–15 min) after the substorm onset (e.g., Pulkkinen et al., 1998). A more detailed discussion of the ionospheric and magnetospheric control of the auroral activity is given in Sections 8.3 and 8.4.

8.2.2 Magnetospheric Substorms

As introduced in Chapter 1, the magnetospheric substorm is a fundamental element in the magnetospheric energy cycle. Typically, substorms can be divided into three phases: The growth phase, the expansion phase and the recovery phase (for recent reviews, see, Baker et al., 1996; Lui, 1996). The process can be described as an energy storage – release process, where energy is first stored in the magnetotail (growth phase), released by a large-scale instability (substorm onset) and dissipated in the magnetosphere – ionosphere system (expansion phase) before a return to quiet state (recovery phase).

Figure 8.2 shows a time sequence during a magnetospheric substorm. The top two panels show the interplanetary magnetic field Z component and the energy coupling parameter ϵ (Perreault and Akasofu, 1978), which is a widely used estimate for the solar wind energy input to the magnetosphere. There was a clear southward turning of the interplanetary magnetic field at about 05:40 UT, which increased the energy input to the magnetosphere. As a result of the enhanced energy input, the polar cap potential started to increase, and continued to increase until a rapid return to the pre-substorm value occurred about one hour later. Similarly, the polar cap area, which is a measure of the energy storage in the tail lobe, shows a continuous increase that starts shortly after the enhanced energy input to the magnetosphere followed by a return to the pre-substorm value. The third panel shows the local electrojet index (analogous to the global AL index) created from magnetograms from the Canadian CANOPUS and the Greenland magnetometer arrays. It is clear that the energy input also led to enhancement of the auroral electrojet pattern, which occurred as a response to the strengthening of the large-scale field-aligned current patterns coupling the magnetosphere and ionosphere. This is evident in the slow, small negative increase of the AL index (see footnote on page 226) after the IMF southward turning. Two substorm activations were clearly seen, one at 06:23 UT and another one at 07:06 UT.

Observations show that the substorm onset is associated with fast earthward and tailward flows in the mid-magnetotail, heating of the plasma sheet, injections of energetic particles at geostationary orbit, the disruption of the thin and intense current sheet in the inner tail, and dipolarization of the inner magnetosphere magnetic field. The fast flows are typically earthward inside of about 25-30 R_E, and tailward further out, and have been associated with localized bursts of reconnection accelerating the particles (Hones, 1979; Baumjohann et al., 1990; Nagai et al., 1998). The cross-tail dimension of the initial flows are rather limited (a few R_E), and they occur close to the field reversal region at the center of the plasma sheet thus being perpendicular to the ambient magnetic field (see also Section 8.3 on magnetospheric control). The particle injections typically have been measured by geostationary satellites, but can at times be observed as far in as 4.5 R_E. The particle energies range from tens of keV to several hundred keV, and the differential flux enhancements are typically more than a factor of 10. The field dipolarization

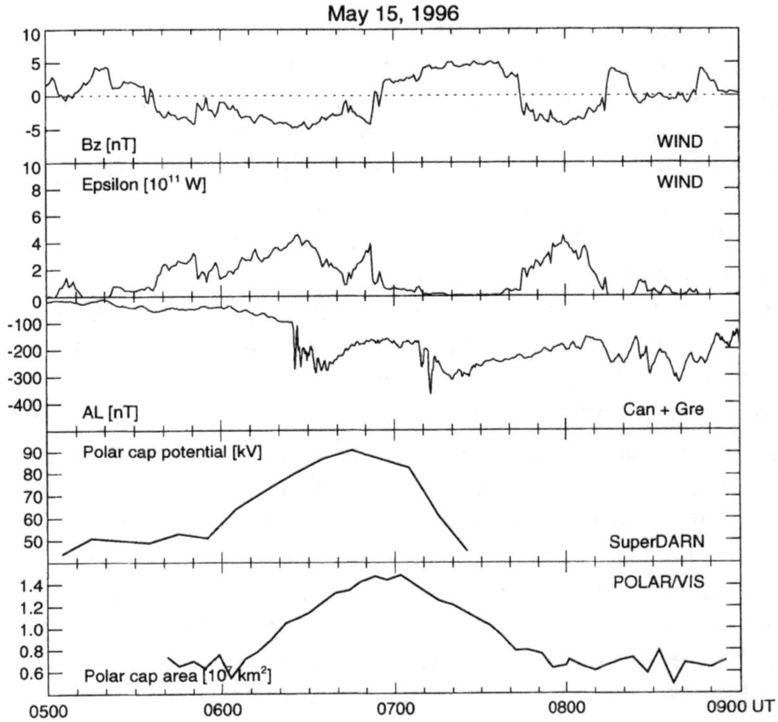

Figure 8.2. Substorm event on May 15, 1996. Panels from top to bottom: interplanetary magnetic field B_Z, ϵ parameter, the local AL index, the polar cap potential, and the polar cap area. From Pulkkinen et al. (1998).

that occurs is stronger than what would follow simply by disrupting the additional cross-tail current within the thin current sheet indicating that part of the effect is due to plasma and flux transport from further down the tail (Reeves, 1998).

During the growth phase, discrete auroral arcs brighten, become more numerous and drift equatorward. At substorm onset, a pre-existing, typically the most equatorward, auroral arc brightens and subsequently active and bright auroral forms appear, forming the auroral bulge (Akasofu, 1964) - see also Figure 5.4. The bulge expands poleward, eastward, and westward by successive formation of new auroral forms ahead of the previous activity and/or by continuous expansion of the bright aurorae. Typically the bulge covers several degrees in magnetic latitude and several hours in local time. Simultaneously with the auroral expansion, the westward electrojet current in the midnight sector increases rapidly, and a strong field-aligned current pattern forms where the net current is downward at the eastern edge of the precipitation region and upward near the western edge of the precipitation region. This current system extending to the magnetosphere is called the substorm current wedge (McPherron et al., 1973) - see also Sections 3.2.2.3 and 7.7.4. Magnetic pulsations in the Pi2 range are observed both at auroral latitudes and at low-latitude stations (see, e.g., Heacock and Hunsucker, 1981).

The top panels of Figure 8.3 show two global auroral images from the Polar spacecraft at the time of the substorm onset. The bottom panels show a magnetic field-aligned mapping (see chapter 5) of the auroral features to the magnetotail. In the ionosphere, the auroral brightening rapidly expands to cover several degrees in latitude as well as several hours in local time. On the other hand, in the magnetotail the auroral brightening maps to the current sheet disruption region near or tailward of the geostationary orbit (see Section 7.7.1). Although we have discussed a single event here, the main features are quite typical for many substorms. From this mapping one can deduce that the auroral activity during substorms is associated with processes in the inner and mid-magnetotail, from geosynchronous orbit outward.

Typically after about 30 min, during the recovery phase, the flows decrease and the tail plasma sheet expands. In the ionosphere, the substorm current wedge decays, the auroral electrojets slowly subside, and the auroral activity moves to the morning sector. Omega bands (see Section 6.3) are often observed near the equatorward edge of the morning sector oval, which relates them with processes in the morning-sector magnetosphere near geosynchronous orbit. A band of bright arcs is often visible several degrees poleward of the main oval; this configuration is called a double oval (Elphinstone et al., 1995). It is interesting to note that the time scales of the recovery phase are very different in different parts of the system: In the ionosphere, the double oval and auroral electrojets decay very slowly, taking up to three hours to return to the quiet-time state. At geostationary orbit, the magnetic field returns to the pre-substorm level within about 1.5 hours (Pulkkinen et al., 1994). Lastly, in the midtail, the plasma sheet expansion after the neutral line retreat occurs only within a few tens of minutes, and the entire recovery is also much faster than in the inner magnetosphere and auroral ionosphere (Hones et al., 1984).

As already mentioned, due to problems in mapping between the magnetosphere and the ionosphere (see Chapter 5), the exact relationship between specific auroral forms and dynamic processes in the tail still remains open. Growth phase auroral arcs map to the inner and mid-magnetotail, but their association with e.g., flow bursts is still an open issue. On the other hand, it has been shown that the auroral bulge during the expansion phase fills the region between the upward and downward currents in the substorm current wedge, and that expansion of the auroral bulge occurs simultaneously with the expansion of the dipolarization region (or current disruption region) in the inner magnetotail (Pulkkinen et al., 1995). An important change occurs when the auroral bulge reaches the poleward boundary of the auroral oval and begins to reduce the size of the polar cap. This has been interpreted as the expansion of progression of reconnection from plasma sheet to open lobe field lines that greatly enhances the reconnection rate, as it is proportional to the Alfvén velocity, which is large in the lobe. At this time, part of the

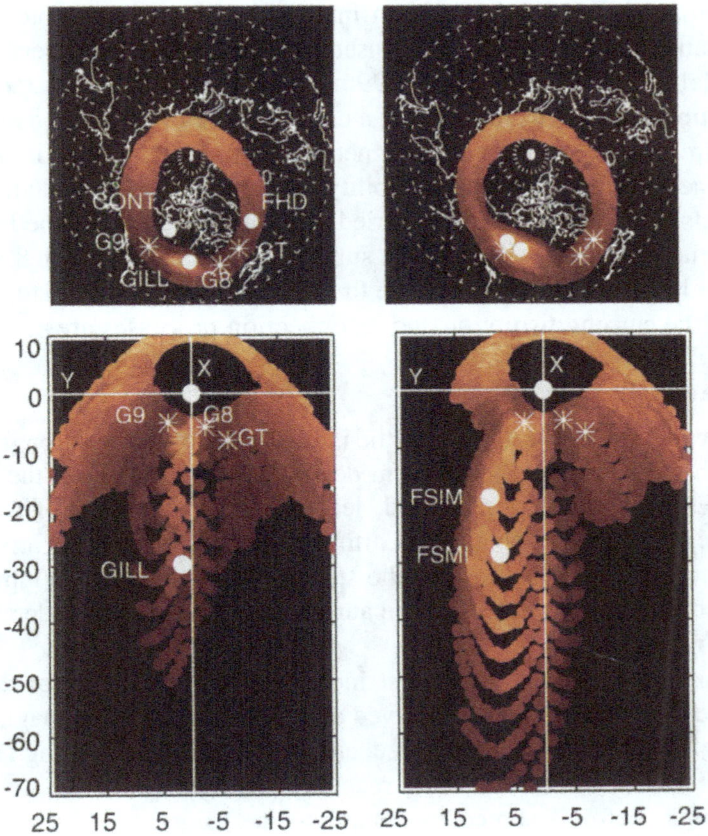

Figure 8.3. Auroral activity during two substorm intesifications at 06:26 and 07:17 UT on May 15, 1996. Top: auroral images from the POLAR/VIS auroral imager. Bottom: mapping of the auroral luminosity to the magnetotail. In these plots, both the northern hemisphere ionosphere and the equatorial plane of the magnetosphere are viewed from above. G8, G9, GT refer to the GOES-8, GOES-9 and Geotail satellites, the other acronyms refer to selected ground-based stations. After Pulkkinen et al. (1998).

plasma tailward of the reconnection region is ejected from the magnetotail in the form of a flux rope or plasmoid. During the recovery phase, the double oval configuration would suggest inner tail (near geosynchronous region) activity as well as a more distant source of precipitation, maybe associated with the tailward retreating plasmoid and reconnection region.

The features outlined above can be observed during almost every substorm event, within a minute of the first onset signatures (which often are the Pi2 pulsations). Theories of the substorm onset process are at present controversial, the main controversy being where and by which process the substorm onset is initiated. The Near-Earth Neutral Line (NENL) model asserts that it is reconnection in the mid-magnetotail initiated by an internal tail instability which creates fast flows and causes the other substorm onset signatures (for the present status of the NENL model see, e.g., Baker et al., 1996; Nagai et al., 1998). On the other hand, current disruption models propose that a current-driven instability develops within the thin current sheet in the inner magnetotail, and reconnection is initiated only after a rarefaction wave from the instability region reaches the midtail (Lui, 1996). Other models (e.g., Kan, 1993) have tried to include the role of the ionosphere in controlling the time and location of the substorm onset (see Section 8.4). However, it is commonly accepted that within the first ten minutes of the substorm expansion phase, both the current disruption and reconnection processes are active.

8.2.3 Magnetic Storms

If the solar wind velocity is large and the interplanetary field is strongly southward for several hours, a geomagnetic storm develops. During storms, the ring current encircling Earth is strongly enhanced, leading to a depression of the equatorial magnetic field at the surface of the Earth by several hundred nT, or by as much as 1% (Burton et al., 1975). Due to the strong solar wind energy input, the polar cap area increases substantially and the auroral oval expands considerably equatorward, often reaching mid-latitudes.

Storms are classified using the Dst index that is an average of deviations of the magnetic north components observed at several low-latitude magnetic stations around the globe. The storm main phase, characterized by decreasing Dst, typically lasts several hours, whereas the storm recovery phase can last from several days up to several weeks before Dst recovers its quiet-time value.

The ring current enhancement during the storm main phase is caused by strong convection, by multiple substorm-like particle injections, and by enhanced outflow of ionospheric ions (Daglis et al., 1999). Whereas it was earlier believed that the ring current was formed by the multiple injections associated with the stormtime substorms, analysis of the correlations between substorm and storm indices suggests that the global convection electric field is the primary mechanism transporting particles inward, whereas the inductive electric fields associated with substorms act to further energize and trap the particles in the closed drift paths (McPherron, 1997).

During storms, the auroral oval brightens everywhere (Figure 8.4) and resides at lower latitudes than during non-storm periods. The precipitating flux of electrons with energies below 1 keV is so much enhanced that the red auroral emissions from oxygen at 630 nm wavelength (see Section 2.4) above 200 km altitude become visually detectable. Furthermore, the precipitating flux is also more energetic, containing significant fluxes of electrons with energies greater than 10 keV. These higher-energy electrons cause red emissions from nitrogen at the lowest border of the visible aurorae at about 100 km altitude. Furthermore, if the ionospheric temperature is sufficiently increased (to above 4000 K), red arcs can also be created

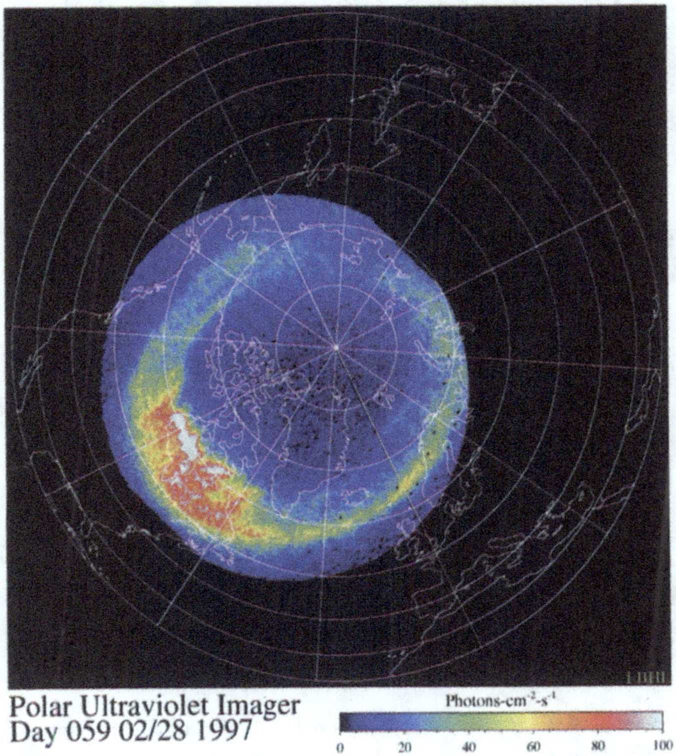

Figure 8.4. Auroral oval during a magnetic storm on 28 February 1997. This is one of the rare storm-time events when the entire oval is in the field-of-view of the UVI imager of the Polar satellite. Figure provided by C. Meng.

without precipitation. The strong interplanetary magnetic field, pressure pulses in the solar wind, and the enhanced magnetosphere–ionosphere coupling lead to increased auroral activity also in the dayside.

8.2.4 Steady Magnetospheric Convection

Steady magnetospheric convection events occur during stable solar wind velocity and steadily southward interplanetary magnetic field (Sergeev et al., 1996a). During such periods, the magnetotail assumes a quasi-steady state where energy is dissipated at the rate it is brought in, so that no explosive energy releases such as substorm onsets occur. The near-Earth magnetotail is typically in a highly stretched configuration, resembling that observed just prior to the substorm onset. The auroral oval is typically very wide and located at relatively low latitudes with a large polar cap area. The aurorae during steady magnetospheric convection are quite similar to substorm recovery phase aurorae, including double oval configuration and isolated auroral activations that do not develop into substorm bulges.

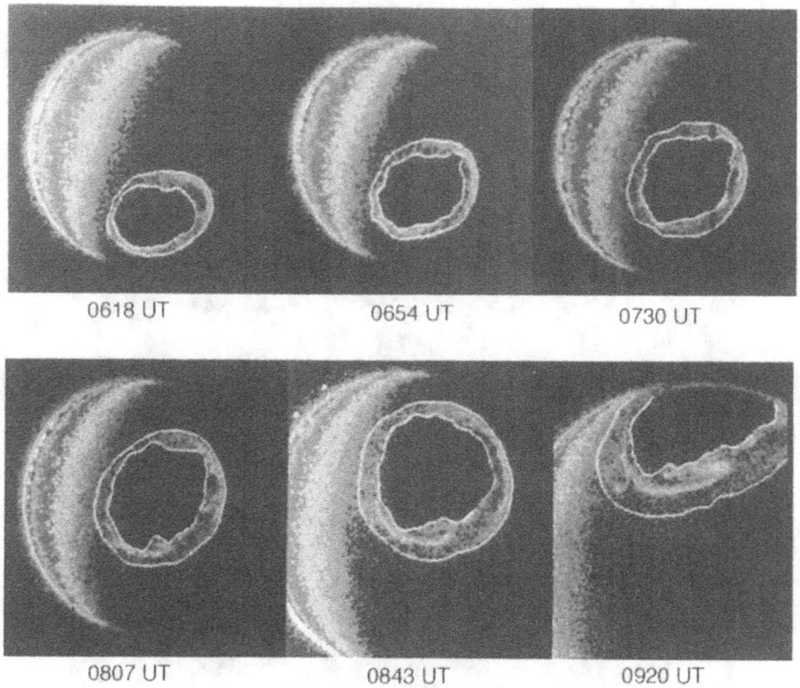

Figure 8.5. Auroral activity during a steady magnetospheric convection period on November 24, 1981. Small-scale activations that did not lead to large-scale expansions were observed throughout the period that lasted more than three hours. From Yahnin et al. (1994).

In spite of the stability of the large-scale system during times of steady convection, the magnetosphere - ionosphere system is by no means inert or quiet during these periods. Bursts of fast flows are observed in the tail, and a variety of auroral activations are seen in the polar ionosphere. As shown in Figure 8.5, the activations typically occur near the poleward edge of the oval and propagate equatorward (Sergeev et al., 1996a; Yahnin et al., 1994). It seems that during such events, the small-scale activations are sufficient to account for the energy dissipation, and the system is in a marginally stable state. Steady magnetospheric convection events end when some variation, either in the solar wind or internal to the magnetotail, triggers a substorm onset, or else the activity gradually subsides without substorm activity.

8.2.5 Magnetospheric Energy Budget

The energy input during the substorm growth phase occurs typically at a level exceeding 100 GW for about one hour, resulting in a typical energy for a substorm of the order of 10^{15} J. During the substorm expansion phase, the energy is dissipated via plasma sheet heating, ring current energetic particle injection, ionospheric Joule heating, and auroral particle precipitation. Part of the energy is returned back to the solar wind in the form of plasmoids ejected tailward subsequent to magnetotail reconnection.

The traditional view of the energy division between the various dissipation avenues was that about 90% of the input energy would be dissipated in the ring current, and 10% in the ionospheric Joule heating, with the remaining contributors being small (Perreault and Akasofu, 1978). Although this result was primarily based on studies of geomagnetic storms, it was assumed to hold for isolated substorms as well. This work and many later ones used the ϵ parameter for evaluating the energy input, and the ϵ parameter normalization was fixed to equal the inner magnetosphere energy dissipation.

Recent work has shown that in addition to the ring current, there are two major energy dissipation avenues. The role of the ionosphere has been found to be much more important than expected earlier, dissipating about half of the available energy both during storms and substorms (e.g., Lu et al., 1998). The energy content of plasmoids has been estimated to be about 2×10^{14} J (Ieda et al., 1998), while the same study concluded that the total energy released tailward during a substorm was about 10^{15} J. Thus, the more recent view indicates that about half of the total energy is dissipated in the inner magnetosphere and ionosphere, whereas the other half returns to the solar wind as plasmoids or plasma flows. The ionosphere is suggested to dissipate at least half of the energy, maybe even more when the auroral precipitation energy is accounted for. Note that this view may require a revision of the scaling of the ϵ parameter, and hence its absolute values cannot be taken as accurate estimates of the total energy input. On the other hand, correlation analyses show that the ϵ parameter is well correlated with magnetospheric activity.

Statistical studies have shown that during isolated substorms, the ionosphere accounts for a larger portion of the energy dissipation than during storm times (Tanskanen et al., 2002). Thus, even though the energy dissipation in the ionosphere is larger during storms than during substorms, its relative role in the energy partitioning is smaller during storms.

The energy budget during steady convection events seems to balance over several hours, with input and output being roughly equal so that no further energy storage occurs. In this case it seems that the energy is mostly dissipated in the ionosphere and as heating of the plasma sheet during the flow bursts. Probably, no large plasmoids occur during steady convection events, as these would necessarily cause reconfiguration of also the inner magnetosphere fields which is not observed (Sergeev et al., 1996a).

Over longer time periods, of course, the energy input and output must be balanced. However, for an individual substorm event - or a pseudobreakup - this needs not be the case. The magnetosphere always has sufficient energy to drive a substorm, thus the energy input is mainly required to initiate the reconfiguration necessary for the substorm onset to occur (Kallio et al., 2000).

8.3 Magnetospheric Control

Traditional cartoons of laminar large-scale plasma and energy circulation in the magnetosphere have little in common with the real convection in the high-beta plasma sheet. Instead of a laminar flow, the plasma sheet displays significant structuring at a number of spatial and temporal scales. The same is true for the large-scale configurational changes in the tail, which occur through complicated structural changes at various scales. This, of course, is reflected in the aurorae and contributes in part to the rich auroral structuring and dynamics. As the magnetotail can only be observed by spacecraft making point measurements, auroral images often provide the only way to visualise and study structural changes and dynamics of precipitation patterns at various levels of detail.

In this section we focus on mesoscale structures in the plasma sheet (scales down to a few minutes and $\sim$ 1 R_E, i.e., much larger than the gyroperiod and gyroradius of the thermal ions). Thus, these structures are large enough to map effectively from the plasma sheet to auroral altitudes and hence contribute to the formation of auroral structures. The choice of particular structures for this section is dictated by the fundamental role they play in the plasma sheet dynamics, transport, reconfiguration and energy dissipation.

8.3.1 Impulsive Plasma Acceleration

One way to modify the auroral intensity or structure is to modify the particle source, that is, to energize the plasma. In the magnetosphere the preferred site of plasma energization is the equatorial region where the magnetic field has its minimum value within a given flux tube. Particularly important is the central part of the tail current sheet (neutral sheet) with very weak B_n and potentially large $\mathbf{E} \times \mathbf{B}/B^2$ velocity. Any process that creates a large amplitude electric field in a relatively large part of this region will accelerate the plasma. In the magnetosphere the largest electric field amplitudes are obtained in the form of short-duration pulses, at time scales of about 1 min.

Pulse-like dissipation combined with a finite propagation speed often produces spatial structures. An important example is related to accelerated ions: Unlike auroral electrons, protons move relatively slowly (about 4 $R_E$per minute at 1 keV energy). Their bounce periods in the outer plasma sheet are often longer than the duration of the electric field pulses. Thus, only a portion of the ions within a flux tube are accelerated. The accelerated particles form ion beams, whose time-of-flight dispersion (if observed) gives estimates of the injection time and the distance to the injection site. Thus, the ion beams can be used as a tool for remote sensing of such impulsive acceleration events. Under the influence of the convective drift across the magnetic field, these beams are also dispersed in space (velocity filtering), forming specific banded structures in the energy spectra (e.g., Boehm et al., 1999; Hirahara et al., 1996).

When observed at mid-altitudes (2-3 R_E), multiple energy-dispersed ion beams are frequent in the poleward part of the auroral oval (Sauvaud et al., 1999) at all local time sectors. Thus impulsive acceleration events must be very frequent in the outer plasma sheet or within the magnetospheric flux tubes connected close to the magnetopause. However, the impulsive acceleration events have been studied most extensively in the nightside magnetotail, where the auroral events have been shown to reflect the impulsive characteristics of one of the fundamental magnetospheric processes, namely magnetic reconnection.

The electron bounce periods are much shorter than the ion time scales. Within the 1-minute time scale the acceleration affects the entire electron population within a given flux tube. However, this does not automatically enhance electron precipitation: during the traversal through the elongated flux tubes in the outer plasma sheet the maintenance of quasineutrality can modify the motion of the electron cloud forcing it to follow the more massive and ($\sim$ 8 times) more energetic ions. During active times, this effect, together with inward convection of ions at the outer boundary of the plasma sheet, is known to displace the poleward boundary of the electron precipitation oval by $1° - 2°$ equatorward from the open-closed magnetic separatrix (see Section 4.5.1).

Large-amplitude impulsive electromagnetic fields can be produced in the magnetosphere by interplanetary shocks hitting the magnetosphere. The auroral precipitation is enhanced by sudden compression of the magnetosphere. The precipitation first increases near noon (cusp) and then quickly propagates along the oval toward dawn and dusk within 3-5 min, consistent with the propagation speed of a shock-induced fast wave in the magnetosphere (Zhou and Tsurutani, 1999). In strong events the electrons can be accelerated to far above auroral energies: In an extreme case, a new inner radiation belt of $\sim$ 15-MeV electrons was formed within 2 minutes (Li et al., 1993).

8.3.2 Pulsed Reconnection

The best observational evidence for reconnection in the near-Earth magnetotail comes from plasma sheet observations of fast tailward flows and plasmoid signatures. A unique observation of strong earthward and tailward flows at the meridian conjugate to an auroral breakup is shown in Figure 8.6.

Within the uncertainty of approximately one Alfvén transit time between the plasma sheet and ionosphere (1-2 min), the observations indicate that plasma sheet reconnection events start simultaneously with auroral breakups and within approximately the same flux tube (Nakamura et al., 2001a). A well-known property of auroral breakups is that the bright aurorae expand poleward while the ionospheric plasma flows toward the auroral bulge, i.e., equatorward. This is also the case for the auroral intensifications at the poleward boundary of a nearly quiet oval (de la Beaujardière et al., 1994). The same effect is well-documented in the magnetosphere where the plasma sheet during a substorm expands outward at a

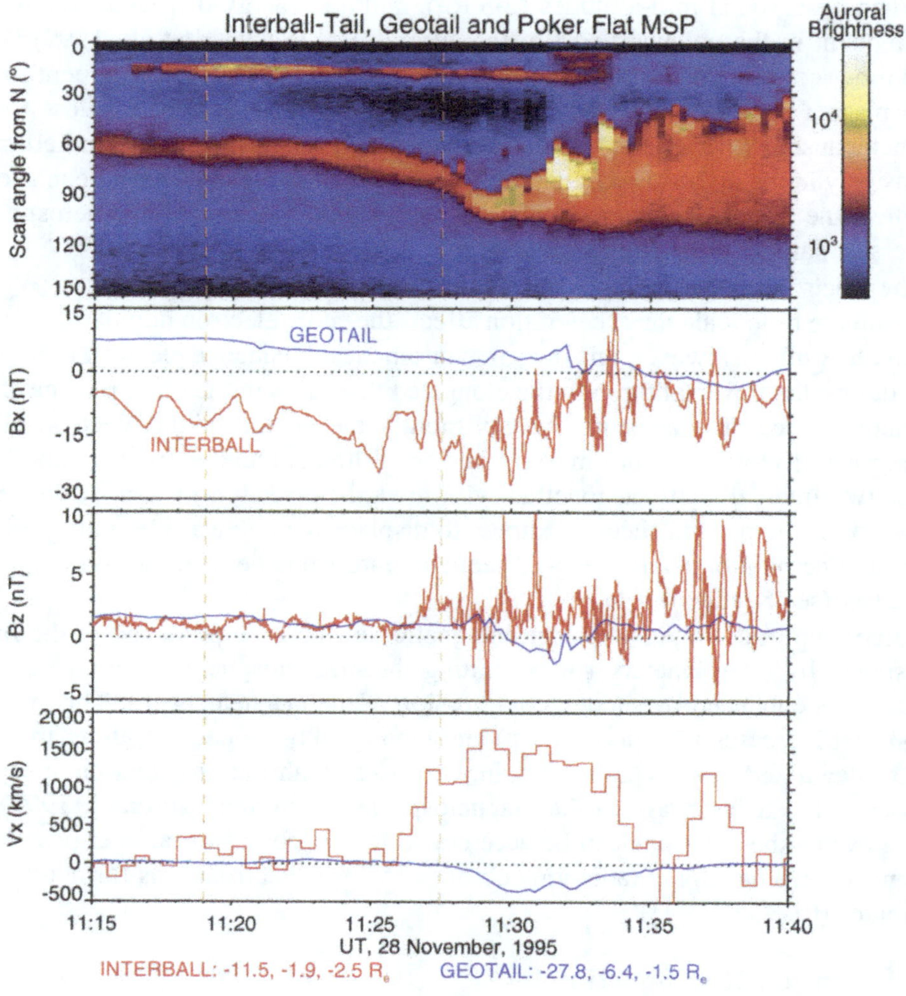

Figure 8.6. Conjugate observations of auroral breakup and of both earthward flow (Interball, at $x \approx -12$ R_E) and tailward flow carrying the plasmoid (Geotail, $x \approx -28$ R_E). From top to bottom: auroral keogram showing the auroral breakup at Poker Flat; magnetic field x and z components and x component of plasma flow velocity. Both spacecraft were close to the neutral sheet and their ionospheric footpoints were mapped to within $\sim$ 0.5 h MLT from Poker Flat zenith. Adapted from Petrukovich et al. (1998).

large speed ($\sim$ 100 km s^{-1}) while the plasma convects inward much more slowly ($\sim$ 10 km s^{-1}). Note that the tailward plasma flow in the magnetosphere is not mapped into the ionosphere, because it is located on field lines that do not connect to Earth. In such events the polewardmost active auroral arcs are directly connected to the reconnection process. The apparent poleward motion of the arcs in the plasma convection frame has been used to estimate the reconnection electric field,

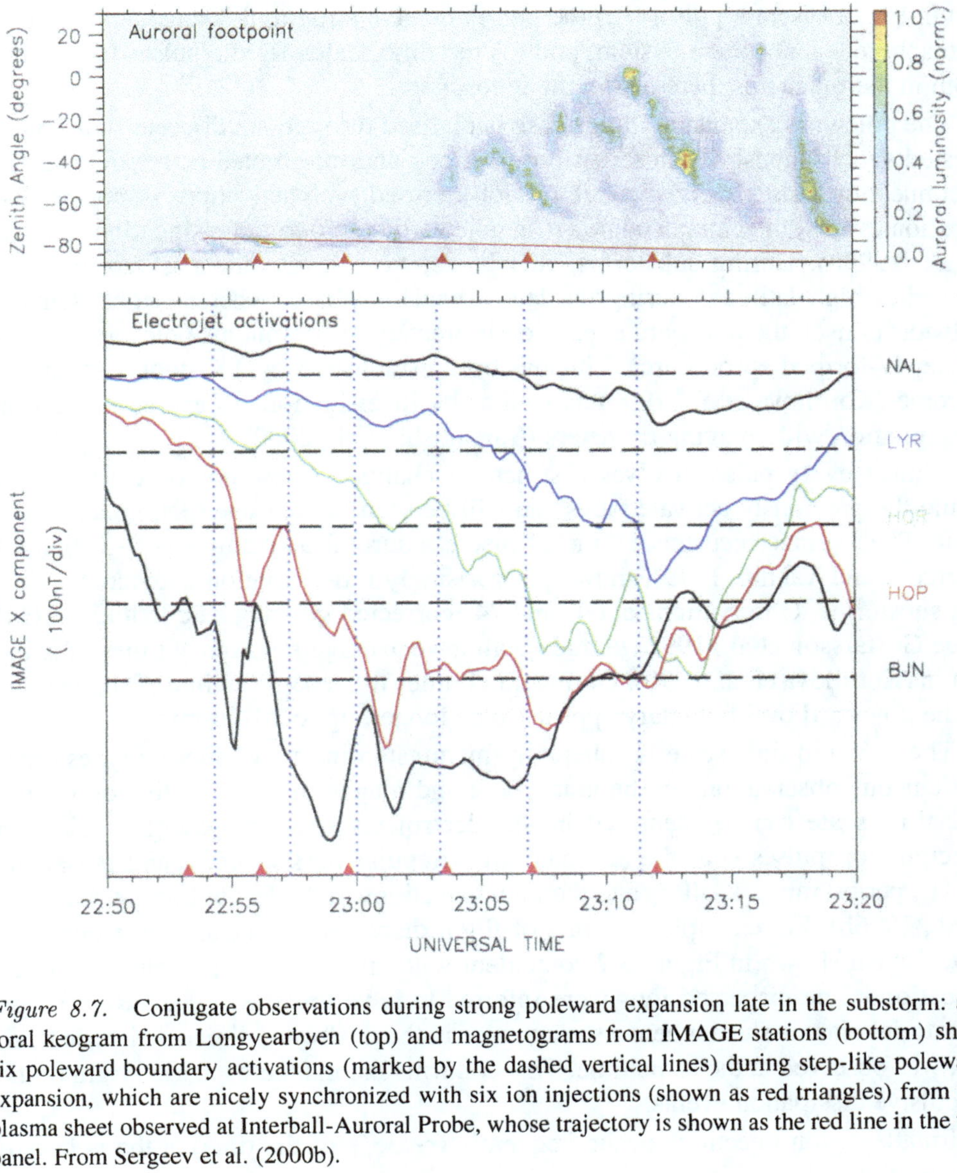

Figure 8.7. Conjugate observations during strong poleward expansion late in the substorm: auroral keogram from Longyearbyen (top) and magnetograms from IMAGE stations (bottom) show six poleward boundary activations (marked by the dashed vertical lines) during step-like poleward expansion, which are nicely synchronized with six ion injections (shown as red triangles) from the plasma sheet observed at Interball-Auroral Probe, whose trajectory is shown as the red line in the top panel. From Sergeev et al. (2000b).

which is a few mV m^{-1} when mapped into the magnetosphere (de la Beaujardière et al., 1994). Such electric fields associated with pulsed flux transport are routinely measured in the plasma sheet during reconnection events such as in Figure 8.6.

The example of poleward motion of the aurora and the electrojet in Figure 8.7 shows three typical features of nightside arc activations (which include, but are not limited to, pseudobreakups, substorm expansion phase onsets, and poleward expansions later in the substorm sequence): (1) the expansion is realized through a number of discrete pulses; (2) the pulses involve structural changes in the magnetosphere – ionosphere system; and (3) the time scales for the pulses is 2–3 min both in the magnetosphere and in the ionosphere.

The poleward expansion in this case is realised through six discrete steps, manifested synchronously in different observations and interpreted as separate strong reconnection pulses. Sergeev et al. (1996b) termed such activations impulsive dissipation events; they are also known as poleward boundary intensifications (Lyons et al., 1999b). During substorms, such a stepwise development is typically observed at high latitudes during the late expansion phase, whereas right after the substorm onset, the poleward expansion is created by continuous poleward motion of newly-formed auroral arcs. This has been established by TV studies of optical aurorae (Kornilova et al., 1990) as well as by investigations of energetic electron precipitation with imaging riometers (Yamagishi et al., 2000).

Almost every pulse involves a structural change: a new poleward arc forms, while the previously activated arcs can still be seen to continue their equatorward drift. The average recurrence time of discrete auroral activations is 2–3 minutes (Sergeev and Yahnin, 1979), showing surprisingly little variation with latitude during substorms. Observations at 64° and 74° corrected geomagnetic latitude (CGM - see Gustafsson et al., 1992) latitude both gave average recurrency times of about 2 min (Kornilova et al., 1990). During quiet times the repetition time of such events at the poleward oval boundary appears to be longer, up to 10–20 min.

The 2–3-min time scale is coherently manifested in many different types of simultaneous observations in the ionosphere and magnetosphere. In the ionosphere this time scale can be identified in the electrojet activations, energetic electron precipitation pulses near the activated arc, irregular bursts, broadband pulsations (Pi1B, period range 1 - 40 s) and the intensifications in the Pi2 pulsations (Sergeev et al., 1996b). For example, six time-of-flight dispersed ion beams were observed in the event shown in Figure 8.7 coincident with six auroral zone poleward intensifications. The dispersed ions were inferred to have travelled 10–20 $R_E$from the equatorial plasma sheet where a strong pulse in the duskward directed electric field, E_y, was observed at the same time. This confirms the magnetospheric origin of the impulsive dissipation events.

Impulsive ion injections occur frequently ($\sim$ 38% of the time) at the poleward part of nightside oval, always in association with stronger or weaker substorm-like auroral events (Sauvaud et al., 1999). Trajectory simulations show that these ions originate from the equatorial magnetosphere (see also Section 4.5). However, the injections occur not only in conjunction with auroral arcs or at the polar cap boundary, but are observed within a wide region even 400–600 km equatorward of the poleward expanding aurorae (see top panel of Figure 8.7). The impulsive acceleration in the magnetosphere accelerates both electrons and ions over a wide range of energies, sometimes up to the MeV range: Belian et al. (1984) reported that 75 % of all observed high-energy (>0.4 MeV) ion injections at 6.6 $R_E$are composed of

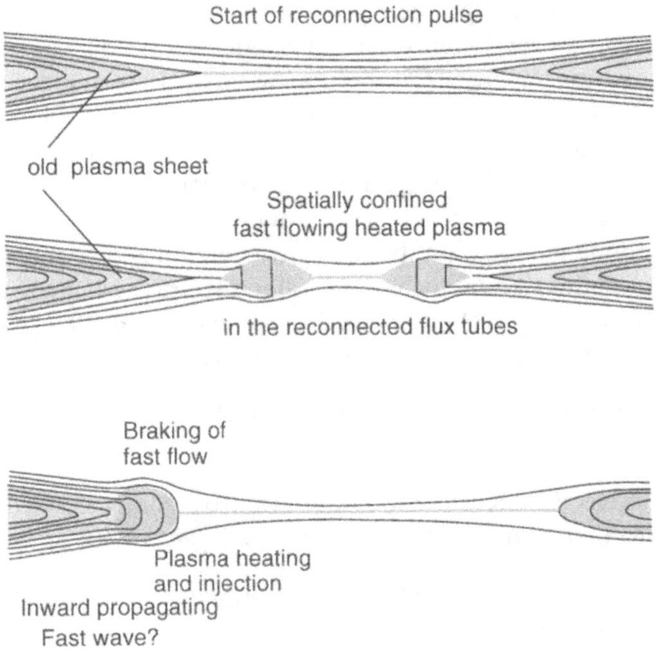

Figure 8.8. Scheme of impulsive reconnection and of interaction of newly reconnected flux tube with closed plasma sheet tubes. From Sergeev et al. (2000b).

superimposed bursts with an average recurrence time of 2.6 min. Such ion injections are coincident with formation of new poleward arcs and Pi1B pulsation bursts (Yahnin et al., 1990) near the onset of intense substorms.

It is possible to interpret the magnetospheric manifestations of the impulsive dissipation events in the framework of an impulsive reconnection model. A specific feature of reconnection pulses is that they generate localized reconnected flux tubes (see the discussion of plasma bubbles in Section 8.3.4), coherent objects, which interact with closed plasma sheet flux tubes as shown in Figure 8.8. If this fast flowing flux tube is suddenly decelerated at the boundary of closed flux tubes, it can send an electromagnetic pulse to the closed field region that can produce a number of effects in the inner magnetosphere in addition to plasma acceleration.

The large peak amplitudes and localized spatial structures (reconnected flux tubes) within the plasma sheet are important specific features of impulsive reconnection. The reconnection pulse can send fast flow jets (see the next subsection) or intense electromagnetic wave pulses into the inner magnetosphere and, therefore, accelerate plasma over a much larger region than covered by the pulse itself. The reconnection pulse can also serve as a broad-band source of compressional wave energy for the inner magnetosphere, possibly exciting Alfvén resonators. These are reasons why the pulsed character of reconnection is so important.

It is as yet unknown what physical process forms the pulsed behavior of reconnection and selects the ~ 1-min time scale, but its occurrence is not limited to magnetotail: Flux transfer events at the magnetopause are evidence of pulsed reconnection, and a 3–8-min recurrence time. Reconnection rates look very spiky in most numerical MHD simulations, but recent large-scale kinetic simulations have also indicated a quasi-periodic reappearance of reconnection at ~ 5-min time scale under continuous solar wind driving (Peroomian et al., 2000).

While it is common to interpret the active arc dynamics in terms of reconnection dynamics and from this to infer the reconnection electric field, there are substantial gaps in our understanding of how the active arcs are formed by the reconnection process. Three possible mechanisms to produce an active arc are:

1. Direct precipitation of accelerated electrons from a newly reconnected flux tube;

2. Field-aligned electron acceleration controlled by an accelerated ion beam injected into the high magnetic field region (Schriver, 1999); or

3. Field-aligned electron acceleration in the Alfvén wave launched from the reconnection region (see Figure 8.9).

None of these models has been developed to the level sufficient for application to the reconnection problem. There is still a large gap in our understanding of the magnetospheric control of auroral phenomena.

Depending on which of these mechanisms is operating, the interpretation of the dynamics can also be different: Propagation effects along the flux tube are important in case 3 (and possibly in case 2). For example, when reconnection proceeds from the center of the plasma sheet toward the plasma sheet boundary, the average plasma density in the reconnected flux tubes gradually decreases. Alfvén waves propagate faster within flux tubes that were reconnected later, and can reach the acceleration region near the ionosphere before the waves launched earlier from reconnection of higher-density flux tubes. In this case the apparent motion in the ionosphere would be equatorward, even though the actual reconnection front in the magnetotail would proceed from the center of the plasma sheet outward (see Figure 8.9).

Elementary activations (reconnection pulses) discussed above need to be sufficiently strong and numerous to provide the global reconfiguration associated with the substorm process. During substorms, intense 1-min pulses are grouped to form substorm intensifications lasting typically 5–10 min. Global auroral observations show that new intensifications may be initiated and develop in a different region than the previous activity, e.g., in a different longitudinal sector. In this way subsequent intensifications expand the region of active aurorae and eventually form the large-scale auroral bulge and the global magnetospheric substorm expansion. Infrequent or weak activations may be found during any other global state of the mag-

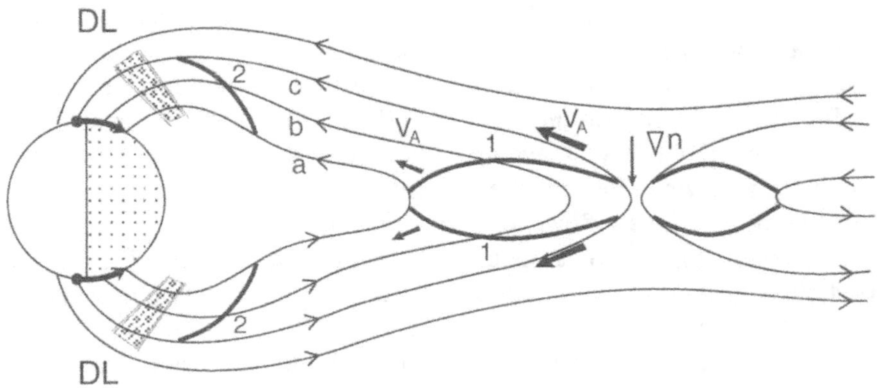

Figure 8.9. Orientation cf an Alfvén wave front launched by impulsive reconnection may change during propagation in the newly-reconnected flux tubes with inward density gradient as illustrated by the lines marked 1 and 2. If the field-aligned electron acceleration occurs mostly near the ionosphere (DL region), an equatorward arc motion will be produced, opposite to the poleward motion of the magnetic field mapping of the magnetic separatrix. After Semenov et al. (1999).

netosphere. An important factor controlling the reconnection electric field, E_R, is the radial position of the activation: as $E_R \sim 0.1\, v_{A0} B_0$ (v_{A0} and B_0 are the Alfvén velocity and magnetic field above the reconnection site), strongest reconnection pulses occur in the inner tail (Sergeev et al., 1996b). This is where the large-scale reconnection region, or near-Earth neutral line is found to form in association with substorm onset.

8.3.3 Sporadic Narrow Plasma Jets

A property of the high-beta plasma sheet is that during active periods it is very turbulent. The fluctuations in observed plasma flows and electric fields much exceed the average values, and the magnetic field fluctuations are comparable to the background field. On the other hand, the plasma pressure and density are less variable. A part of this variability can be attributed to MHD-like turbulence, but its characteristics and significance remain to be studied further. However, there is an important class of variations known as bursty bulk flows, or BBFs, which occur only during a small fraction of the time, but play a fundamental role in the plasma sheet dynamics (Baumjohann, 1993; Angelopoulos et al., 1992).

Bursty bulk flows are episodes of fast plasma flow associated with magnetic field dipolarization that typically last about 10 min. Early studies of BBFs utilized a velocity threshold of $v > 400$ km s^{-1} (Angelopoulos et al., 1992, 1994). Another definition of the same phenomenon (termed rapid flux transfer,RFT, events) has recently been suggested (Schödel et al., 2001), using a threshold of $|V_x B_z| > 2$ mV m^{-1}, corresponding to a localized electric field ten times larger than the average cross-tail electric field. This definition includes also events with lower velocity that occur in the inner magnetotail higher magnetic field regions. Statistical studies based on observations between about 10 and 50 $R_E$in the tail (Baumjohann, 1993; Angelopoulos et al., 1992, 1994; Schödel et al., 2001) imply that the bursty flows are the fundamental transport mechanism in the plasma sheet. Being observed only 5% of time (for RFT), they on average correspond to about 50% of the mass, energy and magnetic flux transport (the percentages vary depending on the thresholds used). Although the bursty flows tend to occur more frequently during the expansion phase of substorms (Angelopoulos, 1996), they are not specifically linked to any particular type of activity, and are observed under conditions ranging from quiescence to strong storms.

At distances of 10–50 R_E, the BBFs (RFTs) exhibit a number of important properties (e.g., Schödel et al., 2001):

- High-speed flows are spiky, with impulses of $\leq$ 1-min duration appearing in groups with average duration of $\sim$ 10 min. Distribution of the durations shows little dependence on distance. The individual BBF lifetime is a few minutes (up to 10 min);

- High speed flows are collimated and the dominating flow component is tail-aligned: $V_x > V_y, V_z$, within a cone of $\sim$ 30°. BBFs are plasma jets with an extension along the tail of about 5–10 $R_E$assuming average values of V_x = 500–1000 km s^{-1} and flow duration of $\sim$ 1 min;

- The cross-tail extent of BBFs has been inferred to be $\sim 2 - 3$ R_E, but this is a difficult parameter to determine from single-spacecraft measurements. Although BBFs are narrow plasma jets, they are not of minor importance in the tail transport: if the electric field is $|\mathbf{V} \times \mathbf{B}| \approx 3$ mV m^{-1}, the potential drop across the jet is about 60 kV, which is comparable to the cross-tail potential drop;

- In the central ($\beta \geq 1$) plasma sheet, the density (and often plasma pressure) drops inside the BBF proper (see Figure 8.10). The total (plasma plus magnetic) pressure shows little change, as the plasma pressure is compensated by the increase in B_z. Therefore, contrary to expectations of compression of plasma convecting earthward, the BBFs are underpopulated flux tubes containing an enhanced magnetic field. Compression effects are also seen at the leading front, and they may dominate at the BBF braking stage;

- The occurrence frequency of intense magnetic flux transfer events (with $E_y >$ 2 mV m^{-1}) shows little change between the distances of 15 and 50 R_E. The occurrence frequency drops quite significantly between 10 and 15 R_E, although some plasma jets penetrate well inside 10 $R_E$and can result in narrow plasma injections detected as close as 6.6 R_E.

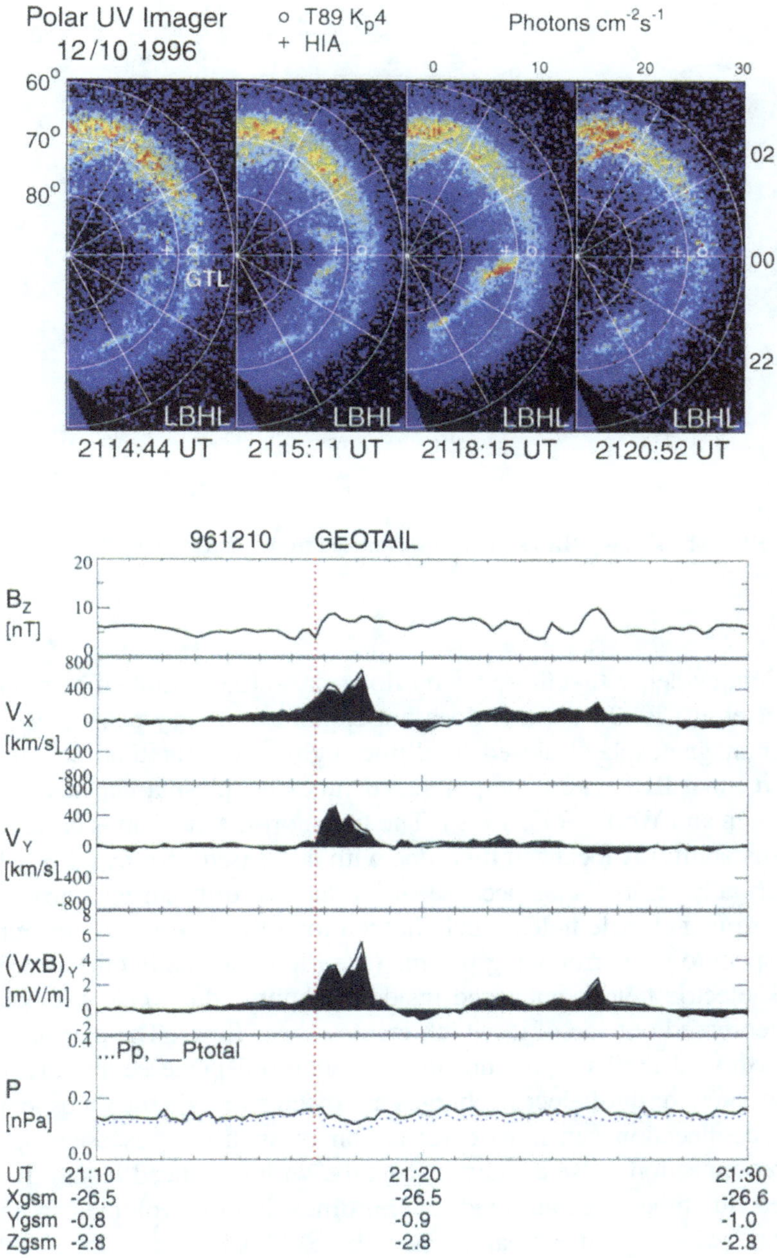

Figure 8.10. Polar UVI observations of localized auroral activation corresponding to isolated flow bursts observed by the Geotail spacecraft in the plasma sheet center. Geotail positions mapped to the ionosphere are shown for two magnetospheric models; observations have been made at the end of a convection bay event. After Nakamura et al. (2001a).

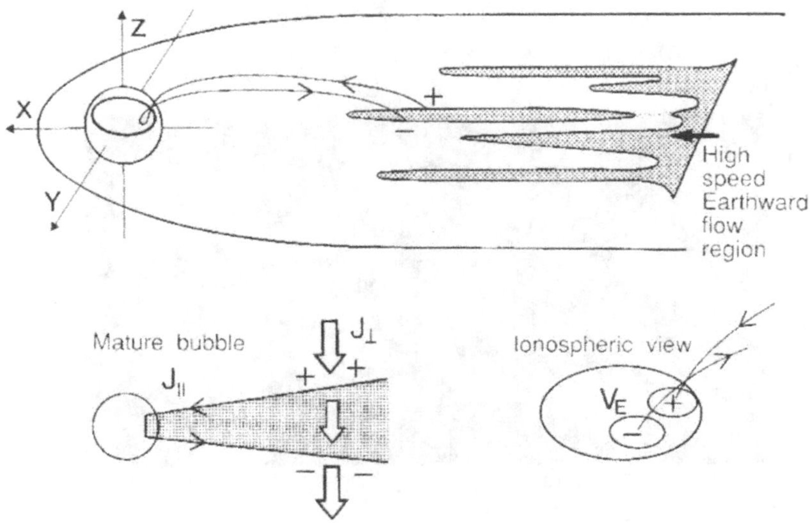

Figure 8.11. Bursty bulk flow (BBF) schematics. Adapted from Sergeev et al. (1996b).

There is clear evidence that these bursty flows are initially formed by reconnection (Øieroset et al., 2000), but this does not explain how they penetrate so deep into the inner magnetosphere closed field line region. A theoretical model based on the property that BBFs are underpopulated tubes or 'plasma bubbles' was developed by Chen and Wolf (1993, 1999). The transport mechanism is basically the interchange instability: A localized flux tube with decreased entropy ($S = pV^\gamma$) is charged electrically on its flanks because of the lack of drift current carriers. The sense of the forming electric field is such that it allows the flux tube to move Earthward with respect to the surrounding plasma (see Figure 8.11). In other words, the dawn-to-dusk electric field is enhanced inside the bubble due to electric polarization, and is decreased outside of it, which creates shear flows. The plasma bubble will move earthward until its pressure in the equatorial plane equals that of the ambient plasma and/or until electric charges are discharged via the ionosphere.

Magnetic reconnection can lead to formation of such decreased-entropy flux tubes: each reconnection pulse creates a flux-tube with a reduced length. If a previously closed flux tube is reconnected several times, both its volume and entropy ($S = pV^\gamma$) are reduced, hence creating the bubble. This is seen in the simulation results shown in Figure 7.7 in Chapter 7, where the flux tube volume is significantly reduced in the Earthward flow channel as compared to the neighboring flux tubes. Plasma heating during the reconnection process does not compensate for the loss of plasma during this process. Reconnection itself can be both impulsive and patchy. The small (2–3 R_E) cross-tail scale size of the bubbles can be controlled either by the reconnection process, or the scale can be selected during the bubble propagation phase.

An essential element of the bubble model is the bubble polarization (flow shear) and propagation of field-aligned currents to the ionosphere. The predicted 3D current system has a Region 1 sense FAC and a pair of transient ionospheric Hall current vortices (Figure 8.11). This forms a basis to search and interpret the ionospheric manifestations of plasma sheet jets. As in situ studies of localized transient structures in the plasma sheet are difficult, the auroral observations can play the major role in interpretation and characterization of these events.

8.3.4 BBF Signatures in the Aurora

Auroral signatures of bursty bulk flows (BBFs) vary according to the plasma sheet state and type of activity. Auroral streamers introduced in Section 6.4 are examples of such signatures. In the present section, the emphasis in on the tail processes that cause the ionospheric phenomena. Starting from observations of auroral breakups, Ieda et al. (2001) generally found tailward fast flows and plasmoids in Geotail data when the spacecraft was in the region of the thin plasma sheet presumably conjugated with the auroral breakup. Nakamura et al. (2001a) studied isolated strong (mostly Earthward) RFTs recorded in the central plasma sheet when $\beta > 1$. The RFTs were found more frequently in the thick active plasma sheet during the substorm recovery phase or steady magnetospheric convection. Under such conditions, pseudobreakups, high-latitude patches, or auroral streamers (see e.g., Figure 8.10) are frequently observed. Activations similar to auroral streamers have also been reported in association with bursty flow events (see Figure 6.30, where three magnetospheric spacecraft observed the passage of a plasma jet from > 40 $R_E$to 6.6 R_E Sergeev et al., 2000a).

In Figure 6.30, three stages of development have been captured for the streamer f: brightening at the poleward boundary (19:54:50 UT), active phase (propagation toward the equatorial oval, 20:00:58 UT), and the last phase with a bright spot remaining at the equatorward oval after the streamer development (20:13:14 UT). The active phase of streamers in the double oval typically last between 3 and 10 minutes which gives a lower estimate for the lifetime of the narrow plasma jets. The equatorward bright spot could be discerned as long as 5–20 min after the intrusion (Sergeev et al., 2001). The latter time may correspond to the lifetime of the local plasma pressure hump in the inner magnetosphere created by the narrow plasma injection.

A remarkable grouping in longitude was found when plotting the spacecraft footpoints as compared to positions of associated pseudobreakups and streamers. In Figure 8.12 the BBF-associated localized auroral activations always occur to the west, within $\sim$ 1 h MLT of the spacecraft footpoint. Furthermore, the Geotail footpoints were further from the auroral structure when the observations were made near the dawn flank of the BBF than when the observations were from its

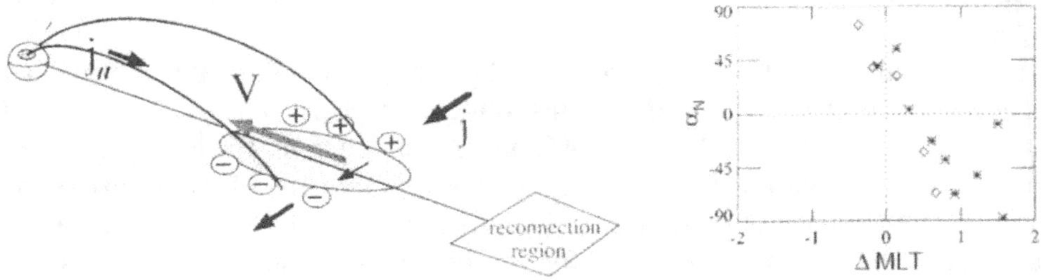

Figure 8.12. Spatial relationship between BBF footpoint and associated auroral activation (Nakamura et al., 2001b). Left: Scheme of a plasma bubble and auropral activation. Right: Local time differences between BBF footpoints and auroral activations for different boundary orientations, α_N, the boundary normal angle in the equatorial plane determined by minimum variance analysis of Geotail magnetic field observations: positive/negative α_n corresponds to the crossing of the westward/eastward flank of the flow burst.

dusk flank. This spatial scale is consistent with the spatial scale of streamer-related current system in the ionosphere (see Section 6.4). When mapped to the tail, this is consistent with the cross-tail size of the BBF of $\leq 3\ R_E$directly obtained from plasma sheet observations.

The ordering of Geotail footpoints with respect to the longitude of auroral activations is a fundamental result for the interpretation of auroral observations. It shows that streamers are not produced by direct precipitation from the BBF proper, but are probably formed by field-aligned acceleration of electrons in the upward field-aligned current region at the westward flank of the BBF. Relative locations and scales of streamer-associated FACs and precipitation near the ionosphere (Section 6.4) nicely fit to this picture. The recent conclusion that the energy flux to power the auroral activation comes in the form of Poynting flux is also consistent with this view. The predictions of the bubble model are thus confirmed, except for the absence of an enhanced westward electric field in the ionosphere as noted in Section 6.4. (Direct mapping of the electric field can be violated in the ionosphere because of inhomogeneous conductivity and/or field-aligned potential drops, and thus this prediction is less certain than the character and continuity of the 3D current system generated by the bubble, which essentially agrees with observations).

This view on the formation of auroral streamers raises a number of questions and, probably, sets limits to the interpretation of auroral observations in terms of plasma sheet processes. The traditional categorization of discrete aurorae into auroral arcs and auroral streamers becomes problematic, especially when using traditional ground-based observations with a limited spatial coverage. Streamer *d* in Figure 6.30 is inclined by about 45° to the meridian, and there are examples of arcs whose orientation is zonal within 30°. Yet these forms start to develop from the poleward oval and intrude into the equatorward oval, which means that they represent clear examples of streamers. Such orientation is not surprising in view of the very different values of radial and azimuthal mapping factors, which deform and rotate the ionospheric image of plasma sheet objects. This mapping issue is critical, because the proper choice of source geometry will drastically affect the formulation and solution of the problem concerning the generation mechanism of auroral arcs.

In case of field-aligned acceleration as the formation mechanism for streamers, not all BBFs, or plasma bubbles, are expected to be visible as streamers. For example, the BBF-related field-aligned current can be roughly estimated from the current continuity and the cross-tail scale size to be of the order of 1 $\mu\mathrm{A\,m^{-2}}$, which is comparable to the background FAC existing in the auroral zone. If the intensity and direction of the background FACs is a discriminating factor, little BBF-related aurorae are expected in the poleward morning oval or in the equatorial evening oval, where the statistical large-scale field-aligned currents are downward, and these activations should be best visible in the evening sector poleward oval and in the morning sector equatorward oval. Therefore, the background conditions may affect the ionospheric manifestations of plasma sheet structures.

8.4 Ionospheric Control

In the previous sections we have discussed the global dynamics and the way magnetospheric activity drives dynamic aurorae. Although ultimately the auroral system is driven by energy coming from the magnetosphere, the ionosphere is more than a passive load in the coupled system.

The ability of the ionosphere to control the magnetospheric processes (tail wagging the dog) is based on physical characteristics of the ionosphere that are different from those in the magnetosphere:

- The ionosphere is collisional which results in horizontal conductivity and drag on the magnetospheric convection;
- The Alfvén speed in the ionosphere is large enough to allow for rapid information transfer across the polar cap in form of magnetosonic waves;
- The neutral wind is able to drive FACs.

8.4.1 Ionospheric Effects on the Magnetosphere

The effects of ionospheric drag have been discussed in Chapters 3 and 7. While the drag can be considered as a passive resistance on magnetospheric flow, it implies that the magnetosphere can supply, for example, FACs only to such regions in the ionosphere where the currents are consistent with the ionospheric electric field and conductivity patterns. Sufficiently strong magnetospheric forcing can modify the ionospheric conditions, but only after a period of information exchange between the ionosphere and the magnetosphere via Alfvén waves. An important consequence of this information exchange is the formation of magnetic field-aligned voltages (parallel potential drops) in regions where more current needs to be drawn than can be carried by the available current carriers without additional acceleration.

Much of the coupled magnetosphere–ionosphere dynamics can be viewed as a result of the differences in ionospheric and magnetospheric flow patterns. These differences lead to build-up of magnetic stresses, whose release is related to the local field-aligned potential differences. Parallel electric fields allow magnetic flux tubes and plasma to slip with respect to each other. In fact, it is important to note that even the ideal-MHD electric field, or plasma flow, mapping between the ionosphere and magnetosphere is not trivial. This applies also to quasi-steady conditions, and much more so to strongly inductive processes, such as the substorm onset, when $\partial \mathbf{B}/\partial t$-induced electric fields cannot be considered small. Since the inductive field does not map along the magnetic field, the resulting plasma flow around the onset region in the tail does not need to map down to the ionosphere in any simple way. While this effect is most obvious during substorm expansion, the inductive electric field is, in fact, comparable to the convective field already during typical growth phase evolution (Toivanen et al., 1998).

The role of the ionosphere as a fast information messenger is epitomised by a southward turning of the IMF as sketched in Figure 8.13 based on observations by several ISTP satellites and ground-based observations on May 15, 1996 (Pulkkinen and Wiltberger, 1999, and references therein). The Polar satellite in the cusp observed electric field and wave effects practically simultaneously with the estimated southward turning of the IMF ($T = 0$). Determination of $T = 0$ was based on observations from three spacecraft (Wind, IMP-8 and Interball) of which one (Interball) was close to the magnetopause. The ions coming from the reconnection region appeared at Polar at $T \approx 3$ min. The effects in ground based magnetometers and in the polar cap potential were recorded at $T \approx 5$ min and the entire polar cap size began to grow at $T \approx 12$ min. Satellites located in the near-Earth tail saw the tail stretching to commence also at $T \approx 12$ min.

This event study is consistent with a statistical study employing empirical AMIE modeling of the polar cap convection pattern (Ridley et al., 1998). While there has been some controversy about the interpretation of the modeling results (e.g., Lockwood and Cowley, 1999; Ridley et al., 1999), the timing figures are important: Statistically, the dayside convection change in the ionosphere was found to begin 8.4($\pm$8.2) min after the estimated IMF turning on the magnetopause. The information of changing electric field was mediated throughout the polar ionosphere perpendicular to the magnetic field in a matter of seconds, or at least faster than the 1-min resolution of the AMIE model. On average, the complete ionospheric convection pattern was reconfigured in about 13 min. The large scatter in the propagation time from the magnetopause to the dayside ionosphere in the statistics of Ridley et al. (1998) was to a large part due to the uncertainty in the determination of the propagation of the upstream effect to the magnetopause and partially due to the relatively small number of events.

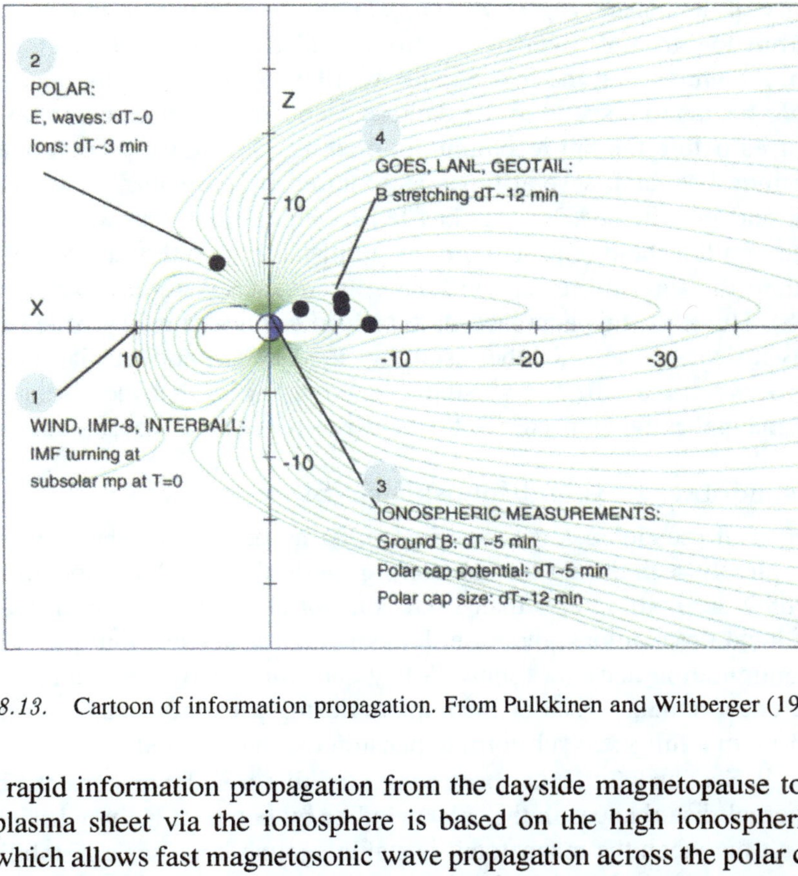

Figure 8.13. Cartoon of information propagation. From Pulkkinen and Wiltberger (1999).

The rapid information propagation from the dayside magnetopause to the midnight plasma sheet via the ionosphere is based on the high ionospheric Alfvén speed which allows fast magnetosonic wave propagation across the polar cap within a few seconds. Under the assumption of ideal MHD, the electric field is mapped to the magnetotail within one Alfvén transit time, the order of one minute. Thus the magnetotail begins to receive information of the change in the dayside as soon as the effect is seen in the dayside ionosphere, only a few minutes after the IMF change at the magnetopause.

At the same time information of changing dayside conditions is transferred tailward along the magnetopause. The solar wind speed is typically 4 R_E/min and thus it takes, for example, 7.5 min to move from $X = +10$ $R_E$to $X = -20$ R_E. Considering the fact that the flow close to the magnetopause is dragged in the subsonic region in the dayside magnetosheath, the information of changing dayside appears to reach the mid-tail plasma sheet roughly in the same time-frame as it happens via the ionosphere. Consequently, the initiation of, e.g., the substorm growth phase involves both the dayside magnetosphere, the ionosphere and the reconfiguration of the near-Earth magnetotail.

As discussed in Section 3.4, the matching of ionospheric and magnetospheric electric fields through Alfvén waves may require several bounces back and forth between the ionosphere and magnetosphere. As the Alfvén speed depends on the mass density of the plasma ($v_A = B/\sqrt{\mu_0 \rho_m}$), the Alfvén transit time (τ_A) varies strongly from low to high L-shells. At the latitudes of Region 1 current, $\tau_A = 1-2$ min. Assuming that the first bounce will begin as soon as the ionospheric electric field begins to respond to the dayside reconnection, it seems to take only a few bounces before the tail reconfiguration starts. During the growth phase the waves continue bouncing several times as required by the enhanced ionospheric convection and growing FACs. This has led to suggestions that when the FACs exceed a certain threshold (e.g., 1 μAm^{-2}), an instability on the auroral field lines would initiate the auroral breakup and the consequent substorm expansion (Kan et al., 1988). However, this approach did not lead to understanding of the auroral breakup, as it does not describe the instability mechanism nor does it address the breakup process in the tail. The present understanding is that the visual auroral breakup is magnetically connected to the near-Earth tail field dipolarization.

8.4.2 Pseudobreakups and Ionospheric State

The existence of pseudobreakups is an intriguing aspect of the substorm process: They are typically seen during the substorm growth phase and it is sometimes argued that they are nothing else than small substorm expansions. From the point of view of local observations, the onset looks similar to a substorm onset, but the global reconfiguration does not follow. A key question is why the disturbance following the onset remains weak or incomplete during pseudobreakups, when after some 10–30 min a full-scale substorm expansion can be initiated.

As noted in Section 8.2.5, it has been suggested that pseudobreakups can be created by external disturbances in the solar wind or by bursts of activity in the magnetotail at a time when the magnetotail is stable against large-scale instability development or at a time when there is not enough free energy in the magnetosphere to drive a substorm (Nakamura et al., 1994). Alternatively, the magnetosphere-ionosphere system may not be ready for a global expansion (e.g., Koskinen et al., 1993) that couples strong field-aligned currents to the ionosphere. This could be caused by inadequate preconditioning of the ionospheric conductivity or of the coupled ionosphere-magnetosphere system.

In a recent study, Amm et al. (2001) made a careful analysis of ionospheric electrodynamics during a growth phase pseudobreakup and showed that the ionospheric conditions around the pseudobreakup could not be distinguished from those of a breakup of a standard expansion phase. However, in this particular case the time difference between the pseudobreakup and full-scale substorm expansion was very small, making its distinction as a pseudobreakup somewhat difficult, although the event occurred to the west of the large-scale substorm current wedge. However, the implication would be that in this case the ionospheric conditions were not the controlling factor in the dynamic evolution of the event.

Comparison of multi-spacecraft observations from the May 15, 1996, substorm with MHD simulations and empirical modeling by Pulkkinen and Wiltberger (1999)

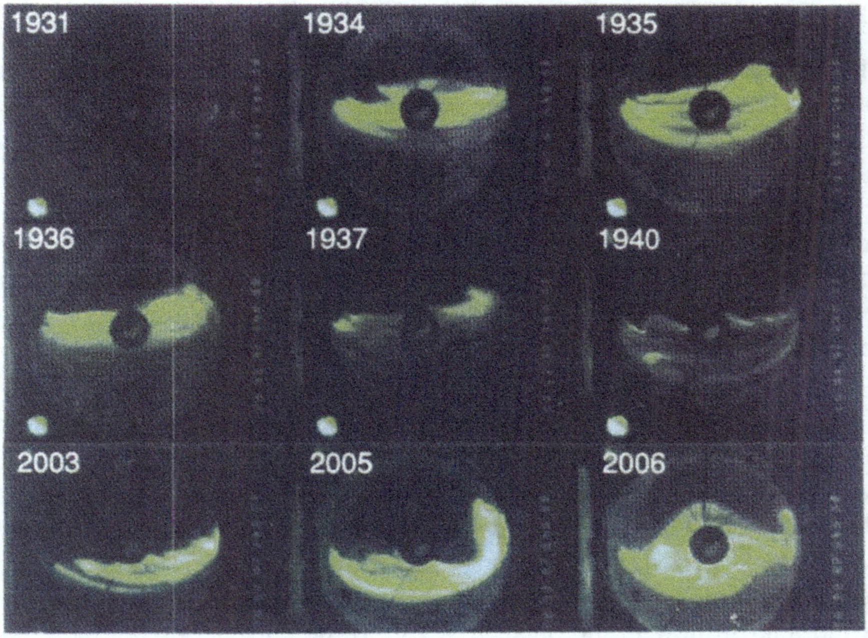

Figure 8.14. All-sky images showing a pseudobreakup on February 14, 1985. The activation at 19:34–19:36 UT fades away after which it takes about 25 minutes until the real expansive phase onset takes place. From Pulkkinen (1996).

suggests that the pseudobreakup was associated with slow reconnection in the mid-tail region before the main current disruption in the near-tail. This event was also one that was closely associated with a substorm onset (see Figure 8.2). In most cases, as also in this case, the pseudobreakups occur on an arc poleward of the most equatorward one, whereas the major substorm onsets take place at the equatorwardmost arcs (Akasofu, 1964). These authors did not address the issue of what controlled the reconnection rate in the tail; why was it slow during the pseudobreakup and faster during the substorm onset, and hence the role of the ionosphere remains open.

So far, no satisfactory threshold energy in the magnetotail for a substorm expansion has been determined. In fact, substantial amount of energy goes into just supporting the tail, as the cross-tail current sheet always consumes magnetic energy. For example, a sheet current of 1 MA/R_E over 40 R_E and through a potential difference of 50 kV yields an energy consumption rate of 200 GW. These figures may vary under different solar wind conditions, but for rather typical conditions the tail current energy sink is in the range 100–1000 GW. Comparing this to a typical ionospheric dissipation rate during the substorm expansion phase of a few hundred GW, the amount of energy does not seem to be the problem, whereas its availability for the process may well be.

8.4.3 Effects of the Neutral Wind

While the high Alfvén speed makes it possible to transfer perturbations rapidly throughout the polar ionosphere, the ionosphere has at the same time significant inertia as the ions are collisionally coupled to the much more massive neutral background. This results in a flywheel effect (Banks, 1972), which can sustain the pre-existing ion motion although the externally imposed electric field already has changed the electron convection. Details of the role of the flywheel are still unknown as the neutral winds are difficult to observe and one has to rely on complicated modeling schemes comprising independent magnetosphere–ionosphere–thermosphere coupling models.

Lu et al. (1995) investigated the effect of neutral winds on the energy transfer and field-aligned currents during the Geospace Environment Modeling (GEM) campaign period March 28–29, 1992, using the AMIE technique (see footnote on page 240) combined with the Thermosphere- Ionosphere General Circulation Model of NCAR (TIGCM, Roble et al., 1988). They found that the wind dynamo had during this period, on the average, 28% negative effect on the ionospheric Joule heating. This was associated with the result that the FACs driven by the neutral wind tend to flow in the opposite sense to the currents driven by the plasma convection. On the average, the neutral wind driven FACs were found to amount to 27% of the total FACs driven by the plasma convection. This is a rather substantial factor in auroral electrodynamics.

In another study Deng et al. (1993) used TIGCM to investigate the flywheel effects after a magnetic storm main phase. During the main phase, when the IMF Z-component was southward, neutral wind-driven ionospheric currents fed FACs in the same sense as in the study by Lu et al. (1995). However, when the magnetospheric driver was shut off in the storm recovery phase, the flywheel FACs became negligible although the flywheel electric field was there for 4–5 hours. Thus the ionospheric feedback to the magnetosphere appears to be most important during the strongest activity. This result is consistent with the observations that the recovery phase of storms and substorms takes a longer time in the auroral ionosphere than in the magnetotail (e.g., Pulkkinen et al., 1994). Note that the storm-time ring current recovery is a separate issue as it is determined by the loss of the ring current carriers not related to auroral electrodynamics.

The dynamic feedback of the neutral wind effects into the magnetosphere remains an open issue. The TIGCM does not take into account the feedback to the overall magnetospheric electric fields. First steps to include this feedback were taken by Peymirat et al. (1998) who developed a model called the magnetosphere – thermosphere – ionosphere electrodynamics general circulation model (MTIE-GCM). The structure of the neutral wind driven FACs was found to be similar to that of Lu et al. (1995) but the contribution was less than 10% of the convection driven FACs. However, the analysis was made for less active conditions (Kp = 2).

Also in this study the true effect on the magnetosphere remained unknown as the model is limited to the dipole field.

8.5 Open Issues

This section briefly discusses open problems in the relationships between aurorae and magnetospheric dynamics. The magnetosphere is less available to direct observations than the auroral ionosphere is, and it is obviously a much more complicated system. Accordingly, in each of the four ways, listed on page 377, of how the magnetosphere affects particle acceleration and precipitation, there are many unresolved issues in understanding the magnetospheric control.

Direct Particle Precipitation. As mentioned in Section 8.1, the main problem is the identification of the microinstability that provides the strong pitch-angle scattering that is causing the equatorial diffuse precipitation of electrons. Acceleration processes range from large-scale convection (which forms radial pressure gradients) to the effects of transient electric fields due to impulsive reconnection, solar wind discontinuities, or other properties (see Section 8.3.1). The latter electric fields are not as well explored by in situ measurements, and the information about them is mostly indirect and comes from the interpretation of velocity-dispersed particle beams.

Creation of FACs Associated with $E_{||}$. Striations, undulations, fragmentation, and filamentation of plasma and current sheets are particular forms of structuring of the plasma as a response to various plasma instabilities. Only a few of the meso-scale plasma instabilities have been studied in detail and identified from observations. The most basic and widely studied are the reconnection instability (associated with flow bursts, see sections 8.2.5 and 8.3) and related interchange instability (which participates in the inward penetration of reconnected flux tubes in the form of plasma bubbles, section 8.3). Still, many basic features, including the origin of the ~2 min time scale of the reconnection pulses and the ~2 $R_E$cross-tail scale of the plasma bubbles remain open issues.

Field-line resonances (cf. Section 7.4.2) in the transition region between dipole-like field and the tail current sheet can also form structured field-aligned currents and electromagnetic fields. They have been proposed to give raise to auroral arcs and pseudobreakups (Samson et al., 1996; Lotko et al., 1998; Voronkov et al., 2000). This effect definitely exists in the magnetosphere, but the amplitudes of the related electric fields and their importance to the auroral processes are still under investigation.

There are many specific auroral structures whose origin is not yet understood. For example, morningside omega-bands (and torch-like structures) that form near the inner edge of the plasma sheet, are not explained by any physical mechanism in the magnetosphere. Azimuthally-spaced auroral forms (Elphinstone et al., 1996), which probably are strings of auroral spirals forming in the nightside, also appear near the inner edge of the plasma sheet. As these are often seen in association with substorm onsets, coupling these forms to magnetospheric processes is important for substorm physics as well. The only suggestion so far is that they are associated with drift-ballooning instability, but the causes or expansion of the instability remain unclear.

Electron Temperature and Density Modifications. The plasma sheet electron density and temperature determine both the regime and proportionality coefficient in the Knight relationship Eq. (3.37). The well-known statistical dependence of field-aligned potential drop and plasma temperature for auroral arcs (see Section 4.1.1.2) can be a manifestation of such a dependence. These plasma sheet parameter changes are sharpest during substorms occurring after quiet periods, when the cold, dense plasma is replaced by hot rarified plasma with changes in both density and temperature by more than a factor of ten each. However, this replacement occurs in a sporadic and patchy way, according to the complicated occurrence scheme of the localized substorm intensification. Therefore, quite often these parameters are very inhomogeneous and show much structure. This grainy structure is difficult to control and leads to a considerable spread of parameters in Knight-type relationships.

Effects on Auroral Forms by Distortions in Mapping. Specific shapes of the auroral forms and high-luminosity regions could be produced by the large localized FACs that deform the magnetic configuration and distort the mapping. Examples of such forms could be the auroral bulge and the westward traveling surge, and in the smaller scale probably also spirals. Whereas the importance of the 3D current systems is obvious, to account for these effects properly, is a big challenge. The main problem is not to model the FACs themselves, but to provide realistic closure of these currents in the magnetosphere, as these currents strongly influence the mapping (Kaufmann and Larson, 1989). Self-consistent modeling of the coupled system is the only way to make progress in this problem.

Role of the Ionosphere. Although the main emphasis of this chapter was on the magnetospheric control of auroral processes, the ionospheric influence on the magnetospheric processes still continues to be an intriguing issue. Besides the aspects discussed in Section 8.4, the role of the ionosphere as a plasma source, and the fast changes in the plasma supply, may be of importance. An intriguing observation was presented (Daglis et al., 1994) that contribution of ionospheric oxygen could increase quickly within a few minutes at the substorm onset or activation. The role of ionospheric ions in the plasma sheet equilibrium and instabilities requires multicomponent plasma studies, and remain to be examined in the future.

Chapter 9

The Aurora as a Universal Phenomenon

The physical processes of the aurora that have been described from both the observational and theoretical side in this book are by no means restricted to terrestrial auroral phenomena, but are of a very general nature. They take place in various different objects and places in the universe when magnetized plasmas of different constitution come into contact. Hot collisionless plasmas in convective motion connected via magnetic fields to cold collisional plasmas can be found in the solar system and in other remote places in the universe. From the auroral effects described in previous chapters we have learned that under such conditions magnetic field-aligned currents, localized space charges, and magnetic field-aligned electric fields and potential drops will be generated causing a wealth of secondary effects. Among those, prompt particle acceleration, plasma heating, and generation of intense electromagnetic radiation are of primary and utmost importance as many of those objects are accessible only by detection of the electromagnetic radiation they emit. As such effects are typical for more remote planetary and astrophysical objects, the present chapter attempts to generalize a few of the auroral processes to a number of these objects which seem particularly suited for the appearance of 'aurora-like' phenomena, or from where information about similar phenomena has been obtained observationally. Identification of such processes may in some cases allow to generalize our knowledge to those objects and to infer about the processes taking part in them.

9.1 Aurora on Other Planets

9.1.1 Solar System Planets

Whereas the aurora on Earth has been part of the human environment as long as people have been living at high latitudes, aurora on other planets have been observed only in the last few decades by means of space-borne instrumentation and the most powerful ground-based telescopes. The first investigations of aurora on planets other than Earth were made by spectrometers and TV cameras on Pioneers 10 and 11 and the two Voyager spacecraft when they passed Jupiter in the 1970s. Figure 9.1 shows a Jovian aurora observed with a TV camera on Voyager 1. The Voyager spacecraft continued to Saturn and recorded aurora there in 1980. Uranus and Neptune also show auroral emissions as discovered by means of Voyager 2 in 1986 and 1989, respectively. The International Ultraviolet Explorer (IUE) and in particular the Hubble Space Telescope (HST) have provided high-resolution observations of aurora on Jupiter and Saturn. In the Solar System all planets with an internal magnetic field and an atmosphere exhibit auroral phenomena. We shall see below that the auroral phenomena differ in many respects on the various planets but they are all similar in that the electromagnetic emissions are generated by the precipitation of energetic charged particles into the upper atmospheres of the planets.

Aurora on Jupiter

Jupiter's aurora has been investigated by numerous spacecraft. Earlier, Jupiter's decametric radio emissions were studied with ground-based (Warwick, 1961) receivers.

Jovian aurora is quite different from aurora on Earth in many important ways. Oxygen emissions dominate in the terrestrial aurora, whereas on Jupiter hydrogen emissions (atomic and molecular) dominate. Spectra in the UV, which match well laboratory spectra of the H_2 continuum in dissociative emission transitions, have been identified on Jupiter for the first time in astronomy (Pryor et al., 1998). These emissions are generated by precipitating electrons. The energetic S^+ and O^+ ions from Io are thought to deposit their energy in the Jovian atmosphere at low heights, where CH_4 prohibits any emissions from escaping into space. Extreme ultraviolet (EUV) emissions are produced over a range of altitudes of 250-1300 km with the bulk of the integrated intensity occurring below 700 km altitude. Primary electron energies appear to be in the several tens of keV range and the total power input to the atmosphere from particle deposition is estimated to be of the order of 10^{14} W (Ajello et al., 1998).

The Jovian auroral belt corotates more or less with the planet instead of being organized in local time as at Earth. That is because the entire magnetosphere of Jupiter out to about 20 R_J is believed to be in complete corotation and the rotation decreases only slowly outwards. However, Cowley and Burke (2001) have provided evidence for the aurora of Jupiter being associated with the departure of the plasma from rigid corotation in the middle magnetosphere.

The morphology of Jovian aurora is different from the terrestrial one also in several other ways. The moon Io plays a major role for the auroral morphology. It generates, by volcanic processes, a time-variable emission of mainly S^+ and O^+ ions and electrons. The magnetic field lines of Jupiter, rotating with the planet, pass Io with a velocity of 57 km s^{-1} and provide a total power of order 10^{12} W available for accelerating charged particles and running an electric current system

Figure 9.1. Aurora on Jupiter observed by the Hubble Space Telescope in January, 1998. Figure courtesy of John Clarke and NASA.

along the magnetic field lines and through the ionosphere of Jupiter. The effects of this appear in UV as a bright spot ahead of the longitude of Io, because the rotation is faster than the orbital motion of Io, and in addition as a narrow faint oval along the magnetic footprint of the Io motion with decreasing emission intensity with increasing distance downstream from Io. The longitudinal distance between Io and the bright spot is variable and mostly larger than expected and does not conform to any existing theory (see e.g., Clarke, 1996; Prangé et al., 1998).

Io has an L value of 6 and represents the low-latitude limit of Jovian aurora. The main auroral oval is at $L \approx 30$. It is quite bright and very narrow; widths as low as 80 km (in the brightness range from 100 kR to $1 - 2$ MR) have been reported (see Prangé et al., 1998).

On the low-latitude side of the main oval there appears to be a dark belt as seen in UV and IR (Satoh et al., 1996; Prangé et al., 1998) but there is also a low latitude spot consisting of series of short segments of concentric arcs. This is generally confined to a narrow range of Jovian longitudes.

The main auroral oval is partly filled with a faint diffuse emission, generally concentrated on the dusk sector. There are also structured emissions observed inside of the main oval and parallel to it. They are fainter than in the main oval and very narrow. The most spectacular feature in the polar cap is a broad bright region extending northward and roughly aligned with a meridian. This transpolar emission, first discussed by Gérard et al. (1994) and later by Prangé et al. (1998) and others, is quite variable and appears to be unrelated to the brightness of the main oval. Some studies have suggested that it corotates with the planet and is related to special features of the magnetic field, but it also seems to change with local time indicating a solar wind control. There is no correspondence to this northern polar cap transpolar emission in the southern hemisphere. This transpolar emission is still far from being understood.

On the whole the Jovian aurora is temporally more permanent and more constant than the terrestrial aurora. However, temporal variations have been reported, even some substorm-like phenomena. They occurred in the dawn part of the main oval and showed up as intensifications of emissions in the UV bands studied. The growth time was of the order of one hour (Ballester et al., 1996).

Although the Jovian auroral region to a large extent corotates with the planet there are many features that indicate magnetic local time dependence and thus solar wind control (such as the above-mentioned dawn intensifications). Obviously corotation and solar wind interaction are both important and differ in importance for different auroral phenomena.

Viewed from remote, one of the strongest signatures of auroral processes that has been identified on Earth is auroral radio emission (cf. Chapter 4). On Earth it is at kilometric wavelengths corresponding to the gyrofrequency which falls into the several 100 kHz range in Earth's auroral zone. At Jupiter the stronger magnetic field shifts this frequency into the MHz range corresponding to hectometric wavelengths. The mechanism of emitting this radiation is based on the electron cyclotron maser process and is closely related to violent processes taking part in the auroral zone, generation of field-aligned electric potential drops and particle acceleration. The emission of hectometric radiation from Jupiter originating at auroral distances of 1-3 R_J has been known for long time (Warwick, 1967; Carr et al., 1983). Part of this radiation is related to the interaction between Io's flux-tube and Jupiter, but the uncorrelated part which is highly variable in time can be attributed to processes similar as those in Earth's auroral zone. The most recent coordinated observations of Cassini on its fly-by at Jupiter to Saturn and the Galileo spacecraft which is in a Jupiter orbit since 1995 (Gurnett et al., 2002) have shown that the uncorrelated hectometric radiation from Jupiter follows the emission from Jupiter in the extreme UV which is a consequence of high energy precipitation into Jupiter's atmosphere during Jupiter's magnetospheric storms or substorms. Indeed, at the time when both emissions were detected on the two spacecraft, strong interplanetary shocks had passed the Jovian magnetosphere which, in analogy to Earth, leads to magnetospheric storm and auroral activity. The detected power flux was very high, a substantial fraction of the energy input by the shock. Such observations suggest that similar processes as on Earth are responsible for the aurora on Jupiter as well.

Aurora on Saturn

Before the encounter of Voyager 1 with Saturn in 1980 there was no unambiguous evidence for the existence of an auroral-type interaction between the magnetosphere and atmosphere on the second largest planet in the solar system. Pioneer 11 demonstrated the existence of a dense atmosphere and a strong magnetic field in 1979 but its UV photometer measurements, showing enhanced emission in its long wavelength channel when the field of view passed the polar edge, could have been due to limb brightening (Judge et al., 1980). Furthermore, the emission intensifications of H Ly-α observed by the IUE spacecraft could not be clearly localized to the atmosphere of Saturn (Clarke et al., 1981). The early attributions of observed radio noise, similar to that from Jupiter, to Saturn were shown to be wrong by the Voyagers (e.g., Schardt and McDonald, 1983). However, the Voyager spacecraft definitely demonstrated that there is an important magnetosphere-atmosphere interaction at Saturn.

Saturn's magnetosphere is intermediate between those of Jupiter and Earth. The surface magnetic field, 21 μT at the equator, is slightly less than that of Earth (31 μT), but much less than Jupiter's (428 μT). It resembles Jupiter's magnetosphere in the extension of the corotating plasmasphere to the magnetopause and in the importance of the moons and rings as sources of plasma and as sinks of energetic particles. In the latter respect it is more extreme than Jupiter's magnetosphere. Because of the much larger distance to the Sun (9.554 AU) the solar wind pressure is so much smaller that the distance of the subsolar stagnation point of Saturn's magnetosphere for typical solar wind conditions is at $20 - 25\,R_S$ as compared to $10\,R_E$at Earth.

The auroral emissions from the Saturnian atmosphere were first investigated by means of the UV spectrometers on Voyager 1 and 2 (Broadfoot et al., 1981; Sandel and Broadfoot, 1981). The auroral emissions were found to extend between latitudes 78° and 81.5°, in reasonable agreement with estimates made by Ness et al. (1981) on the basis of magnetic flux measurements in the Saturnian magnetotail. The emissions observed were found to be from molecular hydrogen excited by electron collisions and they were variable and located so that important solar wind control was inferred. There were, however, also emissions which appeared to peak in intensity when the subsolar meridian was at a certain Saturnian longitude (Sandel et al., 1982), indicating an internal source. However, the Voyager encounters did not provide sufficient data to make possible to discriminate between temporal, longitudinal, and local time variations in the aurora.

Observations of aurora on Saturn by means of the Hubble Space Telescope have recently shown that it is significantly variable on the shortest time scales measured (16 minutes, Trauger et al., 1998). A strong intensification of the aurora was found to have the most prominent features fixed in Saturnian local time between dawn and noon, similar to what has been reported from Jupiter (see above).The peak intensity observed was just below 100 kR. Integrating the northern auroral emissions yielded a total ultraviolet radiated power of about 4×10^{10} W. Comparing the integrated radiated power from the Jovian aurora of a few times 10^{11} to 10^{12} W and auroral intensities from 300 to 3000 kR (Clarke, 1996), the Saturnian figures are quite modest.

Aurora on Uranus

Very little was known about Uranus before the encounter of Voyager 2 in 1986. Not even the rotation speed was well determined. Uranus is unique among the planets in that its rotation axis is only 8° from its orbital plane. At the time of the visit of Voyager 2 the rotation axis pointed nearly along the direction to the Sun and Earth, which happens twice within the orbital period of 84 years.

Up until five days before the closest approach no radio emissions were observed, indicating the absence of a magnetic field and a magnetosphere. On the other hand, IUE had earlier shown unexpectedly bright hydrogen Lyman emissions from the planet (Darius and Fricke, 1981; Clarke, 1982; Durrance and Clarke, 1984) which were attributed to auroral processes and thus required a magnetosphere. It turned out that Uranus has a magnetosphere, but it is quite different from what was predicted before the Voyager 2 arrival, namely one in which the magnetic dipole axis makes an angle of 60° with the rotation axis (Ness et al., 1986) and at the time of the Voyager encounter was displaced towards the nightside so that the maximum surface field is an order of magnitude larger on the nightside than on the dayside (Connerney et al., 1987). The dayside magnetopause radius observed was 18 R_U (Cheng, 1987). However, the H Ly-α radiation was found not to be auroral in nature but a more or less homogeneous luminosity covering the entire sunlit side of Uranus. This 'electroglow' which occurs also on the dayside of Jupiter and Saturn is not well understood but is likely to be associated with coupling between the upper atmosphere and the ionosphere (Clarke et al., 1987; Hudson et al., 1989) rather than between the magnetosphere and the atmosphere.

Curtis (1985) pointed out well before the encounter that the virtual non-existence of radio emissions from the dayside of Uranus would be expected if an efficient convection was set up by the solar wind and he predicted that a strong radio emission associated with nightside aurora would be observed on the dark side of the planet, which was excacly what was observed (Warwick et al., 1986). The nightside aurora was located within a region of $15-20°$ in diameter centered on the axis of the offset tilted dipole on the nightside. Whether it covered the complete cap or was limited to a ring could not be determined (Gulkis and Carr, 1987).

The aurora was found to have similar spectral characteristics as on Jupiter and Saturn and requires electron energies of the order of 10 keV for excitation. The brightness was 1.5 kR for H Ly-α and 9 kR for the H_2 Lyman and Werner bands (Broadfoot et al., 1986). This requires a total power input in the form of electron precipitation on the order of 10^{11} W (Atreya, 1986; Broadfoot et al., 1986; Herbert et al., 1987; Waite et al., 1988; Herbert and Sandel, 1994). The deposition of magnetospheric energy in the upper atmosphere of Uranus in the form of particle precipitation giving rise to aurora is likely to be an important process from an energy budget point of view. It may be a significant contributor to the high exospheric temperature (800 $\pm$ 100 K) found on Uranus (Atreya, 1986; Broadfoot et al., 1986; Herbert et al., 1987). However, Herbert and Sandel (1994) came to the conclusion that the energy contribution from the auroral processes is not sufficient to heat the thermosphere to its observed temperature. Temporal variations of the nightside aurora indicative of substorm-like phenomena were observed (Cheng, 1987; Mauk et al., 1987; Sittler, Jr. et al., 1987). Strong pitch-angle diffusion due to observed strong whistler waves has been estimated to be sufficient to produce the observed auroral emissions (Coroniti et al., 1987). On the dayside no aurora was observed by Voyager 2 nor were any electrons of energies below 6 keV found there (Sittler, Jr. et al., 1987). This may be due to strong convection away from the dayside, as originally suggested by Curtis (1985).

For obvious reasons it has not been possible to carry the investigations of the physical mechanisms very far in Uranus' magnetosphere. The experience at Earth makes it plausible that the ionosphere of Uranus is a strong, if not dominant, source of the magnetospheric plasma from which energetic electrons precipitate into the atmosphere and cause the aurora.

Aurora on Neptune

No H Ly-α radiation from Neptune has been observed from Earth and, as with Uranus, Neptune showed only signs of a weak magnetic field when Voyager 2 approached it in the summer of 1989. At the distance of Neptune (30.11 AU from the Sun) the energy fluxes associated with the solar wind and sunlight are three orders of magnitude less than at Earth. Still the EUV radiation should give rise to an ionosphere.

The particle and field measurements during Voyager's encounter with the Neptune system in August 1989 showed that the planetary magnetic field outside about 4 R_N may be described by an offset (0.55 R_N), tilted (47°) dipole of moment 0.133 G R_N^3 (Ness et al., 1989), where 1G (gauss) = 10^{-4} T. Inside that distance the field is dominated by higher order terms. The plasma densities were generally low (10^{-3} cm^{-3}), except near the magnetic equator where densities up to 1 cm^{-3} were observed. H^+ and N^+ ions with corotating Maxwellian distributions were found to be consistent with the observations (Belcher et al., 1989). Energetic ions and electrons were seen throughout the magnetosphere, while higher energy particles

($>$ 200 keV, corresponding to Van Allen belts at Earth) were confined inside the orbital radius of Triton (Krimigis et al., 1989).

A variety of plasma waves were recorded in the inner magnetosphere, including chorus, hiss, electron cyclotron waves and upper hybrid resonance. Radio wave bursts in the range 100 to 1300 kHz, narrowband and strongly polarized, have enabled determination of the planetary rotation period as 16.11 ± 0.05 hours; smooth emissions in the range 20-865 kHz were also observed (Warwick et al., 1989).

With regard to aurora, spectrally soft electron and ion enhancements during spacecraft passage at high latitudes were consistent with auroral generation and weak auroral-like emissions in H Ly-α were seen on the nightside of the planet (Broadfoot et al., 1989). The measured flux of soft electrons and ions over the polar region and the resulting estimated power input is substantially less than at other planets. The relation of this power input number to the aurora is not yet clear. For more details see Broadfoot et al. (1989).

The Remaining Planets

Under the term 'remaining planets' we collect the Earth-like planets Mercury, Venus, Mars, the small distant planet Pluto, and the various satellites of the big planets, among them Io, Ganymede, Europa, and Titan. Pluto probably has no significant magnetic field nor atmosphere and is thus of no interest. Io is magnetically tied to Jupiter causing well-known disturbances of non-auroral character. Europa has no effect on aurora on Jupiter. Ganymede possesses an own magnetosphere which interacts with the Jovian magnetosphere, but it is unclear whether aurorae can be produced on Ganymede or not.

Mars. At Mars 'auroral' ion beams coming out of the atmosphere have been observed (Lundin et al., 1989) indicating that electrons are accelerated downward, like the primary auroral electrons on Earth. As Mars has no significant internal magnetic field, its field is mainly of solar wind induced nature. The precipitation is thus not concentrated to an auroral zone but may exist over much of the planet. No quantitative estimates of global power input seem to exist but it is likely that the emissions generated by precipitating electrons are too weak to be detectable with existing means.

Venus. At Venus, with no magnetic field and very special atmospheric conditions, what is normally called aurora is not expected to occur. There may be radio emissions, but these are not attributed to anything similar to aurora. Rather it could be radio noise generated in Venus' atmosphere by processes like lightning, cascading of cosmic rays, and from charge exchange processes.

Mercury. Mercury has a relatively weak internal magnetic field and possesses a magnetosphere with accelerated charged particles but no atmosphere to emit the aurora. There will thus be no optical aurora. However, the production of light emission is only one process taking place in an aurora. Auroral processes are mainly due to magnetospheric drivers and their consequences in the vicinity of the planet, i.e., the currents flowing there. On Mercury such currents do not flow in an ionosphere. Rather, they flow in the body of the planet where the conductivity conditions are very different from those in an atmosphere, and are much more stable as they are not exposed to fast atmospheric changes. It is, however, not clear whether or not the currents generated by the presumably violent reconnection processes in the Mercury magnetosphere can close perpendicular to the magnetic field inside the planetary body. Thus it is not clear whether or not processes similar to those in the terrestrial aurora can occur there. The plasma beams released in reconnection in Mercury's magnetosphere probably impinge on the surface of the planet, thereby causing flashes of hard-target bremsstrahlung emission, or X-ray aurora, from the planet. On the other hand, for very strong field-aligned current flow, local instability of the currents may induce anomalous resistivity and parallel electric fields. Whenever these fields cause unstable distribution functions, one may expect radiation emissions from the unstable regions close to the planetary surface. These conclusions are, however, highly speculative, and need confirmation by a future Mercury mission.

9.1.2 Aurora on Exoplanets

At the time of this writing (July 2002), more than 100 planets around other stars than the Sun have been identified from deviations in the star motion. The number is rapidly growing due to modern more sophisticated observations based on micro-eclipsing technique of the central stars by their planets. This technique is hoped to open up the possibility of ultimately detecting even earth-like planets and small planets of the size of Earth, Venus and Mars. All these methods are based on indirect techniques. The planets observed so far have not yet been seen themselves and, of course, direct observations of auroral emissions on such planets are even more unlikely.

The kind of auroral radiation that may give some hope to observe aurora occurrence on exoplanets is the radio emission of the kind of Earth's auroral kilometric radiation (AKR) described in detail in Chapter 4. This radiation is emitted near the electron cyclotron frequency close to the planetary surface. Its frequency thus depends linearly on the exo-planetary magnetic field. Interestingly, its observation would thus be an excellent measure of the magnetic field of the exoplanets. A 1 G source region field should generate emission above 2 MHz. Very strongly magnetized exoplanets of say 1000 G fields should cause radiation in the GHz range.

Contrary to any other frequency range, planetary radio emissions can be as intense as solar radio emissions in their low-frequency range. According to Zarka (1992) the auroral radio emission intensity from the planets is a fraction of 10^{-6} to 10^{-5} of the solar wind energy flux into the cross section of the respective magnetosphere. Large, strongly magnetized planets very close to their parent stars, having a dense or rapid stellar wind, may radiate up to 10^5 times more intense low-frequency radio waves than the Jovian (Zarka et al., 1997) decametric emissions . The possibility of an exoplanet playing the role of Io with the mother star in the role of Jupiter may also give many orders of magnitude higher emission rates than at Jupiter. The total emitted radio power is, however, expected to be significantly lower than that of the auroral emission. This conclusion is supported by the observation that the solar system planetary auroral radio emission is basically controlled by the ram pressure of the solar wind and by reconnection. One thus expects that in the case of exoplanets the control is, in the first place, due to the reconnection imposed by the stellar wind magnetic field (Zarka et al., 2001). Detecting exo-planetary auroral radio emissions above the fluctuation level of the galactic background should be possible up to a distance of about 25 parsecs from the star with radio-telescopes of effective area $10^5\,m^2$, of which several exist. So it appears that we may sometime be able to detect auroral phenomena on planets around other stars in the radio wave range. Future radio telescope arrays such as ALMA, planned in the South American Andes, might even increase this range by up to a factor of ten by increasing the detection area and by using interferometric techniques. Though it will be very hard if not impossible to optically image an aurora outside of the solar system, radio wave observations might provide better possibilities of inferring the existence of aurora in such remote systems, and determining how commonly they occur.

9.2 Auroral Acceleration in Astrophysical Objects

The acceleration of charged particles to very high energies is interesting not only as a physical process. It also is the origin of many astrophysical signals of high information content. Most of these signals are due to electrons interacting with matter, magnetic fields or photons. While these emission processes are mostly well understood, there are still major unknowns with respect to the acceleration of electrons to highly relativistic energies needed to emit photons in the observed energy ranges.

Collisionless shock waves appear to be the most powerful and most often investigated astrophysical acceleration process. Another mechanism is the acceleration of magnetically confined charged particles by stochastic electric fields. A third mechanism is electrostatic acceleration by the electric fields around fast rotating isolated neutron stars (radio pulsars). These objects are, however, rare and the fast rotation short-lived. Highly relativistic plasma jets are generated in rotating magnetic fields of Kerr black holes in the centers of active galaxies. Furthermore, the reconnection process and a host of resonant and nonresonant wave-particle interactions can produce suprathermal particles, although not to extremely high energies in most cases.

In this context it is surprising that the auroral acceleration process has hardly found entry into the astrophysical literature, although its origin, fast release of magnetic shear stresses, should be realized in many astrophysical objects. Since the information available on the magnetic environments of stellar objects is dramatically less than we have in the Earth's environment, simple tools are needed to analyze the potential or even magnitude of auroral acceleration in distant systems. The following (after Haerendel, 2001a) is an attempt to condense the auroral theory into a set of simple relations which can constitute such a tool. Subsequently, a few applications will be treated.

9.2.1 A Condensed Auroral Theory

Basic to auroral theory is the origin of magnetic-field aligned potential drops. In the context of the magnetosphere most theories are based on the Knight-Lemaire-Lyons relation Eq. (3.37). Anomalous resistivity (Section 3.5.1) has been the subject of much experimental and theoretical work in the late 1960s and 1970s, also in the auroral context (e.g., Papadopoulos, 1977), though it has received less attention in the last two decades in modeling of auroral processes.

When addressing other astrophysical objects, one realizes a peculiarity of planetary magnetospheres, i.e., their extremely low plasma density. Stellar atmospheres typically have densities that are many orders of magnitude larger. Is the mirror effect on adiabatically moving electrons still the dominant resistive effect in such high density plasmas, or do other processes take over? The answer to this question can be explored by comparing the inverse conductance or mirror impedance K^{-1} (Eq. 3.38) with an integrated anomalous resistance, $\eta_{an}l$, where l is the effective length over which the underlying wave turbulence is measured. In Eq. (3.83) one must introduce an expression for the anomalous collision frequency, ν_a, which is consistent with the bulk of theoretical, numerical and experimental studies. Based on the work of Papadopoulos (1977) we take

$$\nu_a \simeq (10^{-4} - 10^{-3})\omega_{pe} \tag{9.1}$$

The validity of such a simple approximation rests on the condition that the drift velocity of electrons relative to the ions exceeds a critical threshold dependent on T_e/T_i:

$$v_D = \frac{j_\parallel}{en} \geq v_{crit}\left(\frac{T_e}{T_i}\right) \tag{9.2}$$

We assume that Eq. (9.2) is fulfilled over a length, l. Then we obtain a relation equivalent to the Knight relation (3.36):

$$\Phi_\parallel = \eta_{an} l j_\parallel \tag{9.3}$$

The ratio of the resistive potential (Eq. 9.3) and the potential from the Knight relation (Eq. 3.38) is given by

$$\xi = K\eta_{an}l \tag{9.4}$$

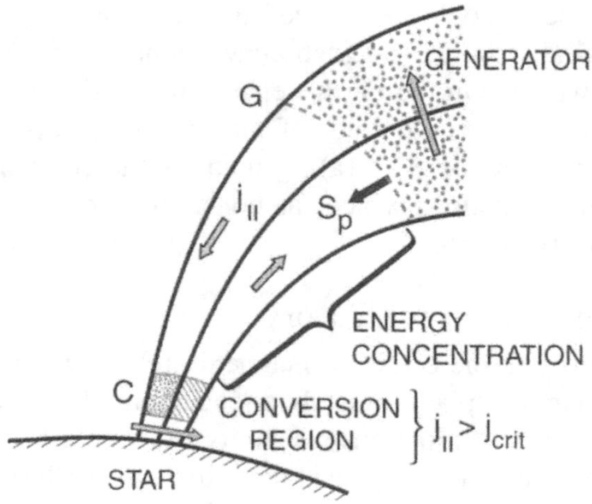

Figure 9.2. Current and energy flow in an auroral circuit. Figure provided by G. Haerendel.

Inserting the above expressions for K and η_{an} (Eqs. 3.38 and 3.83 respectively), with ν_a taken from Eq. (9.1), we find

$$\xi \simeq \nu_{a1} \frac{l}{c_{eH}} \frac{n_H}{n} \sqrt{\frac{n}{1 \text{ cm}^{-3}}} \tag{9.5}$$

where n_H and c_{eH} are the density and thermal velocity of the hot electrons of the generator plasma and ν_{a1} is the anomalous collision frequency given in Eq. (9.1), with ω_{pe} for a density of $n = 1$ cm^{-3}. Typically $\nu_{a1} l / c_{eH}$ is of order 0.1, whereas n_H/n can be of order unity as in the magnetosphere, or the solar corona. In accretion flows, n_H/n decreases as $r^{5/2}$ towards the star (see below). However, even in the latter case the right-hand side of Eq. (9.5) is very large compared with unity thanks to the high density. We can conclude that the mirror force matters only in relatively empty (or tenous) planetary magnetospheres. In the dense plasmas of stellar environments any generated wave turbulence will overshadow the mirror force and render the particle motion nonadiabatic. For a very dense atmosphere, even classical collision processes may become competitive.

The next step is the derivation of energy fluxes, voltages and spatial scales on the basis of the current-voltage relation (Eq. 9.3). Figure 9.2 is a sketch of an auroral current circuit. Energy flows as Poynting flux, S_p, out of the generator (index G) into the auroral acceleration region. If $B_{\perp G}$ is the magnetic perturbation field at the effective exit from the generator region, S_p can be written as

$$S_p = c_{AG} \frac{B_{\perp G}^2}{\mu_0} \tag{9.6}$$

Here c_{AG} is the generalized Alfvén or energy flow speed

$$c_{AG} = v_A \left[1 + \frac{v_A^2}{c^2} \right]^{-1/2} \tag{9.7}$$

which for high magnetic field or low densities approaches the velocity of light, c, and v_A is the Alfvén speed.

If the total electromagnetic energy flux is converted into kinetic energy of beams and heat, the energy conversion rate per unit cross-section, $\dot{\varepsilon}_\|$, is given by

$$\dot{\varepsilon}_\| = j_\| \Phi_\| = \frac{B_C}{B_G} S_p \tag{9.8}$$

Here the convergence of the field lines between generator G and energy conversion region C is taken into account.

Equations (9.6) and (9.8) lead to another current-voltage relation

$$\Phi_\| = j_{\|,C}^{-1} c_{AG} \frac{B_C}{B_G} \frac{B_{\perp G}^2}{\mu_0} \tag{9.9}$$

By virtue of Ampère's law, $B_{\perp G} = \mu_0 j_{\|,G} w_G$, where w_G measures the width of the current filament. Inserting this into Eq. (9.9), one sees that this relation is also linear in $\Phi_\|$ and $j_\|$ like Ohm's law. Combining these equations with proper consideration of the mapping factors for w and $j_\|$ yields an expression for $w_\perp$ in the energy conversion region

$$w_\perp = \left(\frac{\eta_{an} l}{R_{WG}} \right)^{1/2} \tag{9.10}$$

with the Alfvén wave impedance at generator level, $R_{WG} = \mu_0 c_{AG}$.

The desired set of simple relations is constituted by Eqns. (9.6), (9.8), (9.9) and the condition (9.2) for the anomalously resistivity case. From the plasma dynamics in the generator region one may estimate values for $B_{\perp G}$ ($\leq B_G$). The conditions near the central object (density and temperature) would provide a value for $j_\|$. Fundamental to this theory is the conjecture that scales, $w_\perp$, can be generated such that j_{crit} can be reached, albeit in only a small minority of current filaments. Support for this conjecture comes from the observation that current filamentation, caused by MHD turbulence and/or ionosphere-magnetosphere feed-back processes, seems to be unavoidable in low beta plasmas. The generation of smaller and smaller scales would come to an end only when a sudden increase of dissipation is encountered, such as achieved by turning on an anomalous resistivity.

Equation (9.9) written slightly differently as

$$e\Phi_\| = \left(2 \frac{B_C}{B_G} \frac{c_{AG}}{v_{crit}} \frac{n_G}{n_C} \right) \frac{B_{\perp G}^2}{2\mu_0 n_G} \tag{9.11}$$

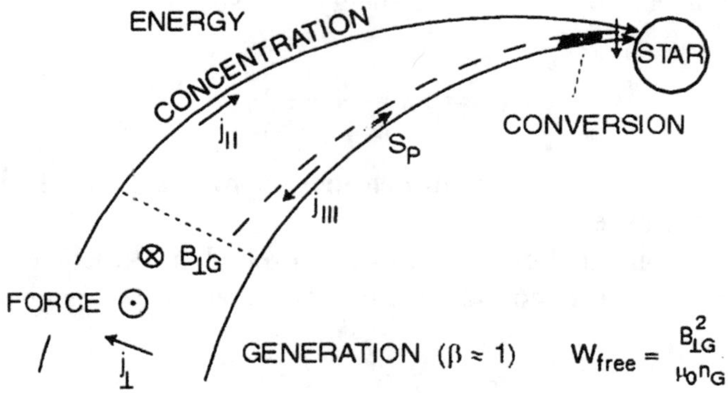

Figure 9.3. Energy generation and concentration before conversion. Figure provided by G. Haerendel.

conveys the reason for the high energies achieved in this process. Here, the term $B^2_{\perp G}/(2\mu_0 n_G)$ is the free energy per particle in the generator region. As demonstrated in Figure 9.3, this energy is channeled as Poynting flux toward the central object whereby it is concentrated by two factors, the field convergence, $B_c/B_G \geq 1$, and $c_{AC}/v_{crit} \geq 1$. The ratio n_G/n_C is either ≈ 1 or may even counteract to some extent the energy concentration, as for instance in accretion columns. In general, however, the value of the bracketed product is rather large.

9.2.2 Application to Solar Flares

Magnetized planets or stars embedded in plasma flows are natural candidates for sites of auroral acceleration. However, internal energy sources like the ones heating a stellar corona may create aurora. This may involve several steps, first driving the dynamics of the generator plasma in the weak outer fields and then, on completely different temporal and spatial scales, investing part of the energy in the build-up of current systems which meet the critical current condition somewhere. Since according to Eq. (9.9), $\Phi_{\parallel}$ depends on B^3 and $n^{-3/2}$, we should not be surprised to find very high energy sources in astrophysical objects.

It can be safely assumed that during solar flares magnetic energy is not only converted into kinetic energy, probably through the magnetic reconnection process somewhere in the corona, but is in part re-invested into magnetic energy by exciting Alfvén waves on various spatial scales. Our conjecture is that in a minority of cases the attending field-aligned currents may reach the critical value, which for $T \simeq 10^6$ K and $n \simeq 10^8$ cm^{-3} would be of order 20 A m^{-2}. This would most likely happen in the concentrated fields just above the chromosphere. Primary magnetic perturbations $B_{\perp G} = 30$ G would imply current widths of order 1 km at the generator and ~100 m at the energy conversion level. Typical potential drops and energy

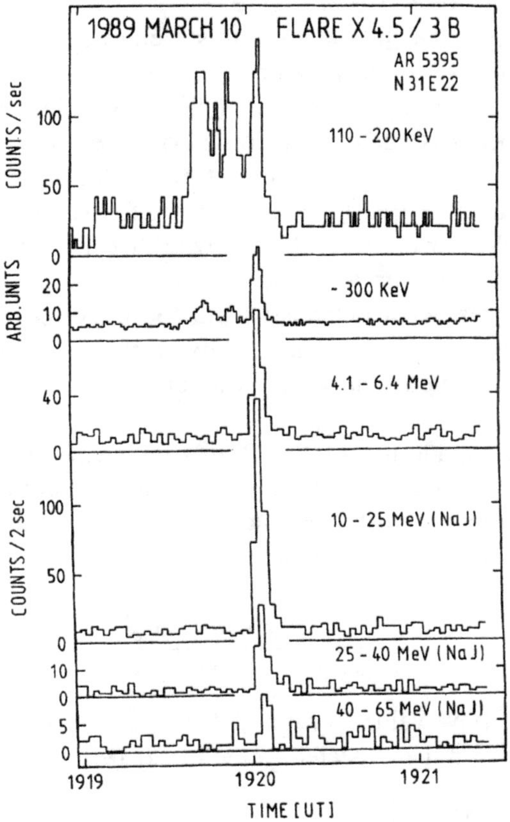

Figure 9.4. X-ray and γ-radiation of the flare on 10 March 1989. From Rieger and Marschhäuser (1990).

conversion rates according to Eqs.(9.10) and (9.9) would then be

$$e\Phi_{\|} = 75\,\mathrm{MeV}\left(\frac{10^8\,\mathrm{cm}^{-3}}{n}\right)^{3/2}\left(\frac{B_c}{10^3\,\mathrm{G}}\right)\left(\frac{B_{\perp G}}{30\,\mathrm{G}}\right)^2 \tag{9.12}$$

$$\dot{\varepsilon}_{\|} = 1.6\times 10^9\,\frac{\mathrm{W}}{\mathrm{m}^2}\left(\frac{10^8\mathrm{cm}^{-3}}{n}\right)^{1/2}\left(\frac{B_c}{10^3\mathrm{G}}\right)\left(\frac{B_{\perp G}}{30\,\mathrm{G}}\right)^2 \tag{9.13}$$

Multiplying $\dot{\varepsilon}_{\|}$ by the typical flare area of $3\times 10^{19}\,\mathrm{cm}^2$ and the small filling factor, $\tilde{f}$, gives a total energy conversion of

$$\dot{W} = 1.6\times 10^{24}\,\tilde{f}\,\mathrm{W} \tag{9.14}$$

These relations show that plausible values for $B_{\perp G}$, B_c and n are capable of explaining the occurrence of gamma-ray emissions with energies up to many tens of MeV and time-scales of ≤ 1 sec. Figure 9.4 shows an example. Not shown is the gamma-ray spectrum which is lacking in structure due to ion lines. According to Rieger and Marschhäuser (1990) the emissions must be electron dominated. But the problem is to generate large fluxes of high energy electrons within a second or so. The above process is capable of doing so since the intrinsic time scale of this process is the Alfvén travel time, thus time scales of approximately 1 s are consistent. A filling factor of $\sim 10^{-5}$ would suffice to explain the total energy flux. Indeed, in one case Wülser et al. (1990) showed that the appearance of the high-energy gamma-ray spike within a normal two-ribbon flare coincided with a new Hα-emission region somewhat detached from the ongoing flare and quite small in extent. Although possible everywhere, only in the flux-tubes intersecting this spot were currents generated that were sufficiently intense to cause auroral-like acceleration.

So far, we have not made use of the Ohm's law (9.3). The width was simply derived from Ampère's law so that $B_{\perp G}$, j_{crit} and $w_\perp$ are consistent. Eq. (9.10) allows us to check whether the postulate that in high density plasmas $\Phi_\parallel$ is sustained by anomalous resistivity leads to plausible values for ν_a. Solving Eq. (9.10) for ν_a from Eq. (9.1) gives

$$\frac{\nu_a}{\omega_{pe}} = \frac{\mu_0 c_{AC} w_{\perp c}^2 \omega_{pe}}{4\pi l} \tag{9.15}$$

With $l = 10^4$ km, the above values for n and T and $w_{\perp c} = 100$ m, we get $\nu_a/\omega_{pe} = 1.4 \times 10^{-3}$. This is within the expected range of Eq. (9.1).

9.2.3 Cataclysmic Variables

Cataclysmic variables (CV) are binary star systems in which one component is a white dwarf and the other a cool main sequence star which fills its Roche lobe[1]. Their periods range from 1.5 to 10 hours which means that the entire system fits into a solar diameter. The matter flowing across the Lagrange point contains considerable angular momentum and would thus form an accretion disk unless impeded by the white dwarf's magnetic field. CVs are quite common, their total number in the galaxy is estimated to be 50 million, of which about 1000 are known.

A subclass of CVs are the AM Herculis systems. In such systems the white dwarf (WD) possesses a strong magnetic field ($\geq 10^7$ G) and the whole system is in bound rotation. The mass flow from the companion star amounts to 10^{13} kg s^{-1} at velocities of 7 km s^{-1}, if the temperature is about 3000 K. This implies a low degree of ionization. The magnetic field encompassing even the companion star is initially unable to stop the accretion flow, but will do so, once the accretion stream has penetrated sufficiently deep into the magnetosphere and its dynamic pressure,

[1] The Roche lobe is defined such that at its surface the counteracting gravitational forces due to both stars effectively cancel each other.

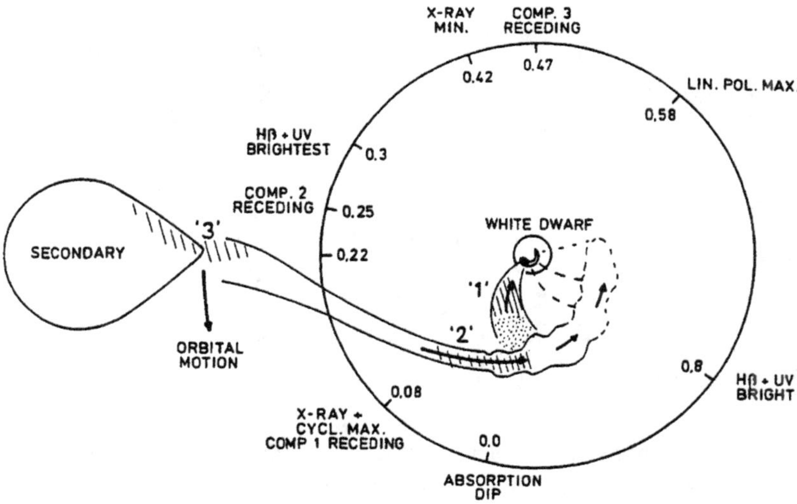

Figure 9.5. AM Herculis systems with accretion flow to a stagnation point and subsequent free fall along field lines to the polar cap. From Beuermann et al. (1987).

increased by gravitational acceleration, is balanced by the faster growing magnetic pressure. Figure 9.5 shows a sketch of the situation. Eventually, the mass flow will follow the field lines onto the polar cap, where it has to pass through an accretion shock.

All of this is known thanks to excellent diagnostic possibilities by eclipse imaging, doppler tomography and broad spectral coverage (Warner, 1995). Here we focus solely on the magnetic tensions and associated currents caused by the slowdown of the accretion flow. A typical distance of the stagnation point, i.e. our generator region (G), is $\sim 7\,\mathrm{R_{WD}}$ with $\mathrm{R_{WD}} \approx 5000$ km. At this point, $B_G \simeq 3 \times 10^4$ G, $n \simeq 10^{14}\,\mathrm{cm}^{-3}$. As the matter falls towards the star, the density must increase as $\sim r^{-3/2}$. Thus, above the accretion shock, $n \simeq 10^{16}\,\mathrm{cm}^{-3}$ and, with $T = 10^5 - 10^7$ K, $j_{crit} = 7\times 10^8$ - 5×10^9 A m^2. Assuming that $B_{\perp G} \simeq B_G = 3 \times 10^4$ G, we find from Eq. (9.11) that $\Phi_\parallel = 3 - 20\,\mathrm{MV}$, $\dot{\varepsilon}_\parallel = 1.3 \times 10^{16}$ W m^{-2} and the total energy flux $\dot{W} = \tilde{f} \times 1.3 \times 10^{28}$ W. The typical widths of such current sheets would be in the range of a few centimeters. The X-ray luminosity is about 10^{26} W. With $\tilde{f} < 10^{-4}$, we expect the gamma ray flux to be at least 2 orders of magnitude lower. This may become detectable with presently planned next generation instrumentation.

It turns out, however, that with the high densities the Coulomb collision rate is approximately one order of magnitude greater than ν_a as derived from Eq. (9.15). So, using the classical resistivity the quoted current density would yield an even higher voltage.

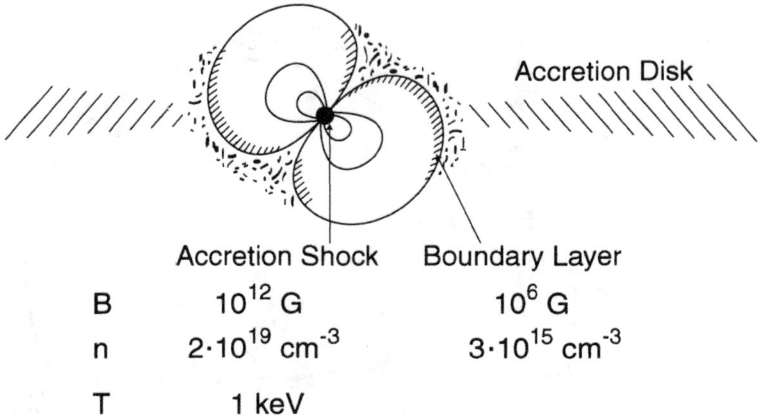

Figure 9.6. Model of the interaction of an accretion disk with a nearly closed neutron star magnetosphere via halo and boundary layer formation. From Haerendel (2001a).

9.2.4 Accreting Neutron Stars

Accreting neutron stars like Her X-1 have an early type, i.e. hot companion star filling the Roche lobe. The matter flowing through the Lagrange point into the gravitational realm of the neutron star is completely ionized. Because of the small radius of the neutron star the magnitude of the magnetic field decreases rapidly and a magnetosphere forms which interacts with a well-developed accretion disk. This interaction occurs where the Keplerian velocity, v_K, of the rotating disk is close to the speed of rigid rotation with the neutron star, ΩR_{MP}. This is typically at a distance of order $100\,R_{NS}$. Many models exist for this interaction (van den Heuvel and Rappaport, 1992). Here we favor a nearly closed magnetosphere with a well defined magnetopause surrounded by a halo in which the Keplerian velocity is largely thermalized and in approximate pressure balance with the magnetospheric field as shown in Fig. 9.6.

In this model, mass enters the magnetosphere by reconnection with the strongly varying magnetic field generated and transported within the accretion disk. Once inside the magnetosphere the plasma fills an internal boundary layer from where it flows along the field lines onto the polar cap. There it reaches near-relativistic flow speed energies the order of $W_p \approx 100$ MeV and forms a shock transition (Fig. 9.7). The column of shocked gas emits X-rays of a characteristic temperature of a few keV. These X-rays cool the infalling electrons.

The accretion disk exerts a torque on the magnetic field which is transmitted to the stellar surface by field-aligned currents. This torque may act in an accelerating or decelerating way, depending on the ratio of $v_K / \Omega R_{MP}$. The main accelerating torque comes from loss of angular momentum of the matter falling along the magnetic field **B** onto the polar cap. Since the torque is transmitted toward the star by

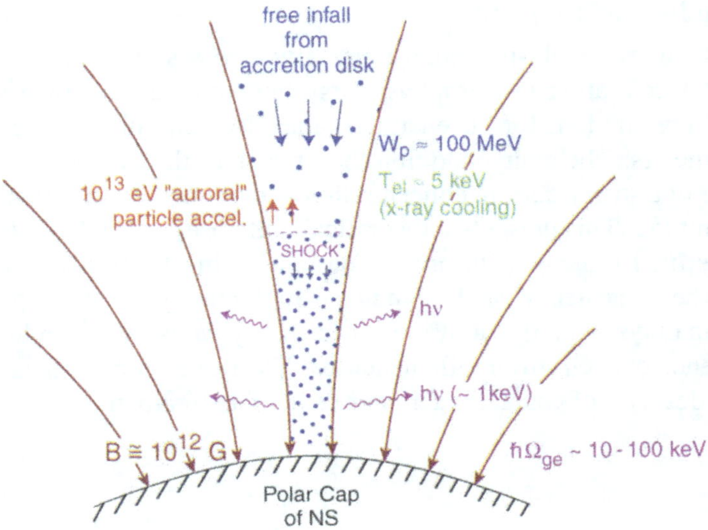

Figure 9.7. Accretion column and shock. The free-fall energy behind the shock is thermalized and radiated with $T_{eff} \simeq 1$ keV. From Haerendel (2001a).

magnetic shear stresses, strong field-aligned currents will arise. We assume that transverse magnetic perturbations of the order of $B \sim 10^5$ nT, close to the field strength in the boundary layer, may exist. Above the accretion shock, the density and temperature would be $n = 2 \times 10^{19}\,\mathrm{cm}^{-3}$ and $k_B T \sim 1$ keV, respectively. The temperature is relatively low due to X-ray cooling. These numbers suggest a critical current density of the order of $2.6 \times 10^{13}\,\mathrm{A\,m}^{-2}$, several orders of magnitude higher than the average current transmitting the shear stresses. If such current concentrations do really exist, Eq. (9.9) yields $\Phi_\parallel \approx 23$ GV, and Eq. (9.8) gives $\dot{\varepsilon}_\parallel = 6 \times 10^{23}\,\mathrm{W\,m}^{-2}$. The total energy converted in this high energy domain would be $\dot{W} = 6 \times 10^{29}\,\tilde{f}$ W. This is to be compared with an X-ray power of 10^{30} W. Even a factor of $\tilde{f} < 10^{-4}$ would thus produce a substantial energy flux that may eventually become detectable.

The width of a typical current sheet turns out to be of order 0.24 cm. If we insert this value into Eq. (9.15) and solve for ν_a/ω_{pe}, we find a value of 1.6×10^{-4}, again quite acceptable. This shows that the resistivity required to sustain $\Phi_\parallel$ in the presence of such a high density of charge carriers can indeed be provided by a current driven plasma turbulence.

The numbers used above imply an energy concentration of 10^3, in the presence of a magnetic field ratio of 10^5 (see Eq. 9.4). The reason is the high mass compression in the accretion flow ($\sim r^{-5/2}$). Any process lowering the mass density in the energy conversion region (bubble formation) would increase $\Phi_\parallel$ dramatically.

Concluding Remarks

The above analysis demonstrates a way to assess the potential existence of auroral acceleration in astrophysical systems outside our own solar system and to obtain predictions for the energies gained by field-aligned acceleration and related parameters. The main condition that must be fulfilled for this to occur is the existence of a strong current concentration and filamentation, often by several orders of magnitude. This presents a theoretical and observational challenge and can hardly be verified directly with present day observing techniques, not even on the sun. But the consistency of the predicted high-energy fluxes, typically manifested as gamma rays, with the above outlined theory can serve as indirect evidence for the presence of such current filamentation. The three astrophysical examples presented here deserve, of course, much more elaborate treatment.

References

Abramowitz, M. and I. A. Stegun: 1964, *Handbook of Mathematical Functions*. Washington, D.C.: National Bureau of Standards.

Aikio, A. and K. Kaila: 1996, 'A substorm observed by EISCAT and other ground-based instruments - Evidence for near Earth substorm initiation'. *J. Atmos. Terr. Phys.* **58**, 5.

Aikio, A., H. J. Opgenoorth, M. A. L. Persson, and K. U. Kaila: 1993, 'Ground-based measurements of an arc-associated electric field'. *J. Atmos. Terr. Phys.* **55**, 797.

Ajello, J., D. Shemansky, W. Pryor, K. Tobiska, C. Hord, S. Stephens, I. Stewart, J. Clarke, K. Simmons, W. McClintock, C. Barth, J. Gebben, D. Miller, and B. Sandel: 1998, 'Galileo orbiter ultraviolet observations of Jupiter aurora'. *J. Geophys. Res.* **103**, 20125.

Akasofu, S.-I.: 1964, 'The development of the auroral substorm'. *Planet. Space Sci.* **12**, 273.

Akasofu, S. I.: 1968, *Polar and Magnetospheric Substorms*. Dordrecht: D. Reidel.

Akasofu, S. I.: 1974, 'A study of auroral displays photographed from the DMSP-2 satellite and from the Alaska meridian chain of stations'. *Space Sci. Rev.* **22**, 1617.

Akasofu, S. I., R. H. Eather, and J. N. Bradbury: 1969, 'The absence of the hydrogen emission H_β in the westward traveling surge'. *Planet. Space Sci.* **17**, 1409.

Alfvén, H.: 1958, 'On the theory of magnetic storms and aurorae'. *Tellus* **10**, 104.

Alfvén, H. and C.-G. Fälthammar: 1963, *Cosmical Electrodynamics*. London: Oxford at the Clarendon Press.

Allan, W.: 1993, 'The ponderomotive force of standing Alfvén waves in a dipolar magnetosphere'. *J. Geophys. Res.* **98**, 1409.

Allan, W. and A. Wright: 2000, 'Magnetotail waveguide: Fast and Alfvén waves in the plasma sheet boundary layer and lobe'. *J. Geophys. Res.* **105**, 317.

Amatucci, W. E.: 1999, 'Inhomogeneous plasma flows: A review of in situ observations and laboratory experiments'. *J. Geophys. Res.* **104**, 14481.

Amm, O.: 1995, 'Direct determination of the local ionospheric Hall conductance distribution from two-dimensional electric and magnetic field data: Application of the method using models of typical ionospheric electrodynamic situations'. *J. Geophys. Res.* **100**, 24173.

Amm, O.: 1996, 'Improved electrodynamic modeling of an omega band and analysis of its current system'. *J. Geophys. Res.* **101**, 2677.

Amm, O.: 1997, 'Ionospheric elementary current systems in spherical coordinates and their application'. *J. Geophys. Res.* **49**, 947.

Amm, O., P. Janhunen, K. Kauristie, H. J. Opgenoorth, T. I. Pulkkinen, and A. Viljanen: 2001, 'Mesoscale ionospheric electrodynamics observed with the MIRACLE network: 1. Analysis of a pseudobreakup spiral'. *J. Geophys. Res.* **106**, 24576.

Amm, O., A. Pajunpää, and U. Brandström: 1999, 'Spatial distribution of conductances and currents associated with a north-south auroral form during a multiplesubstorm period'. *Ann. Geophysicae* **17**, 1385.

Anderson, B., M. J. Engebretson, L. J. Zanetti, and T. A. Potemra: 1988, 'A statistical study of Pc3-5 pulsations observed by the AMPTE/CCE spacecraft magnetic field experiment: Evidence for distinct dayside and flank excitation mechanisms'. In: T. Chang, G. B. Crew, and J. R. Jasperse (eds.): *Physics of Space Plasmas (1987)*. Cambridge, MA, p. 251.

Anderson, B. J., M. J. Engebretson, S. P. Rounds, L. J. Zanetti, and T. A. Potemra: 1990, 'A statistical study of Pc3-5 pulsations observed by the AMPTE/CCE magnetic field experiment, 1, Occurrence distributions'. *J. Geophys. Res.* **95**, 10459.

Anderson, P., W. Hanson, and R. Heelis: 1991, 'The ionospheric signatures of rapid subauroral ion drifts'. *J. Geophys. Res.* **96**, 5785.

Anderson, P., W. Hanson, R. Heelis, J. Craven, D. Baker, and L. Frank: 1993, 'A proposed production model of rapid subauroral ion drifts and their relationship to substorm evolution'. *J. Geophys. Res.* **98**, 6069.

André, M.: 1985, 'Dispersion surfaces and ion wave instabilities in space plasmas'. Technical report.

André, M.: 1986, 'Electrostatic ion waves generated by ion loss-cone distributions in the magnetosphere'. *Ann. Geophysicae* **4**, 241.

André, M., P. Norqvist, L. Andersson, L. Eliasson, A. Eriksson, L. Blomberg, R. Erlandson, and J. Waldemark: 1998, 'Ion energization mechanisms at 1700 km in the auroral region'. *J. Geophys. Res.* **103**, 4199.

Angelopoulos, V.: 1996, 'The role of impulsive particle acceleration in magnetotail circulation'. In: *Third International Conference on Substorms (ICS-3), ESA-SP-389*. p. 17.

Angelopoulos, V., W. Baumjohann, C. F. Kennel, F. V. Coroniti, M. G. Kivelson, R. Pellat, R. J. Walker, H. Lühr, and G.Paschmann: 1992, 'Bursty bulk flows in the inner central plasma sheet'. *J. Geophys. Res.* **97**, 4027.

Angelopoulos, V., C. F. Kennel, F. V. Coroniti, R. Pellat, M. G. Kivelson, R. J. Walker, C. T. Russell, W. Baumjohann, W. C. Feldman, and J. T. Gosling: 1994, 'Statistical characteristics of bursty bulk flow events'. *J. Geophys. Res.* **99**, 21257.

Angelopoulos, V., F. S. Mozer, J. Bonnell, M. Temerin, M. Somoza, W. K. Peterson, H. L. Collin, and B. Giles: 2000, 'Wave power studies of cusp crossings with the POLAR satellite'. *J. Geophys. Res.* **106**, 5987.

Antonova, E. E. and B. A. Tverskoy: 1975a, 'Acceleration of auroral electrons by a stationary double layer'. *Geomagn. and Aeronomy (English translation)* **15**, 563.

Antonova, E. E. and B. A. Tverskoy: 1975b, 'Nature of the electron precipitation band of the inverted V' type and of the Harang discontinuity in the evening sector of the auroral ionosphere'. *Geomagn. and Aeronomy (English translation)* **15**, 85.

Arnoldy, R. L., K. A. Lynch, and J. B. Austin: 1999, 'Energy and pitch angle dispersed auroral electrons suggesting a time-variable, inverted-V potential structure'. *J. Geophys. Res.* **104**, 22613.

Ashour Abdalla, M. and R. M. Thorne: 1978, 'Toward a unified view of diffuse auroral precipitation'. *J. Geophys. Res.* **83**, 4755.

Atkinson, G.: 1970, 'Auroral Arcs: Result of the Interaction of a Dynamic Magnetosphere with the Ionosphere'. *J. Geophys. Res.* **75**, 4746.

Atreya, S. K.: 1986, *Atmospheres and ionospheres of the outer planets and their satellites*. New York and Berlin: Springer Verlag.

Axford, W. I. and C. O. Hines: 1961, 'A unifying theory of high-latitude geophysical phenomena and geomagnetic storms'. *Can. J. Phys.* **39**, 1433.

Bahnsen, A., M. Jespersen, E. Ungstrup, and I. B. Iversen: 1987, 'Auroral hiss and kilometric radiation measured form the Viking satellite'. *Geophys. Res. Lett.* **14**, 471–474.

Baker, D. N., S. J. Bame, W. C. Feldman, J. T. Gosling, R. D. Zwickl, J. A. Slavin, and E. J. Smith: 1986, 'Strong electron bidirectional anisotropies in the distant tail: ISEE 3 observations of polar rain'. *J. Geophys. Res.* **91**, 5637.

Baker, D. N., J. B. Blake, L. B. Callis, J. R. Cummings, D. Hovestadt, S. Kanekal, B. Klecker, R. A. Mewaldt, and R. D. Zwickl: 1994, 'Relativistic electron acceleration and decay time scales in the inner and outer radiation belts: SAMPEX'. *J. Geophys. Res.* **99**, 409.

Baker, D. N., T. I. Pulkkinen, V. Angelopoulos, W. Baumjohann, and R. L. McPherron: 1996, 'Neutral line model of substorms: Past results and present view'. *J. Geophys. Res.* **101**, 12975.

Baker, D. N., T. I. Pulkkinen, M. Hesse, and R. L. McPherron: 1997, 'A quantitative assessment of energy storage and release in the Earth's magnetotail'. *J. Geophys. Res.* **102**, 7159–7168.

Bale, S. D., P. J. Kellogg, D. E. Larson, R. P. Lin, K. Goetz, and R. P. Lepping: 1998, 'Bipolar electrostatic structures in the shock transition region: Evidence of electron phase space holes'. *Geophys. Res. Lett.* **25**, 2929.

Ball, L. and M. André: 1991, 'What parts of broadband spectra are responsible for ion conic production?'. *Geophys. Res. Lett.* **18**, 1683.

Ballester, G. E., J. T. Clarke, J. T. Trauger, W. M. Harris, K. R. Stapelfeldt, D. Crisp, R. W. Evans, E. B. Burgh, C. J. Burrows, S. Casertano, J. S. Gallagher, R. E. Griffiths, J. J. Hester, J. G. Hoessel, J. A. Holtzman, J. E. Krist, V. Meadows, J. R. Mould, R. Sahai, P. A. Scowen, A. M. Watson, and J. A. Westphal: 1996, 'Time-resolved observations of Jupiters far-UV aurora: Comparison of WFPC2 and IUE'. *Science* **274**, 409.

Banks, P.: 1972, 'Magnetospheric processes and the behaviour of the neutral atmosphere'. *Space Res.* **12**, 1051.

Banks, P. and G. Kockarts: 1973, *Aeronomy*. New York: Academic Press.

Banks, P. M., C. R. Chappell, and A. F. Nagy: 1974, 'A new model for the interaction of auroral electrons with the atmosphere: spectral degradation, backscatter, optical emission, and ionization'. *J. Geophys. Res.* **79**, 1459.

Baumjohann, W.: 1993, 'The near-Earth plasma sheet: An AMPTE/IRM perspective'. *Space Sci. Rev.* **64**, 141.

Baumjohann, W., G. Paschmann, and H. Lühr: 1990, 'Characteristics of high-speed ion flows in the plasma sheet'. *J. Geophys. Res.* **95**, 3801–3809.

Baumjohann, W. and R. A. Treumann: 1996, *Basic Space Plasma Physics*. London: Imperial College Press, second edition.

Beach, R., G. R. Cresswell, T. N. Davis, T. J. Hallinan, and L. R. Sweet: 1968, 'Flickering, a 10-cps fluctuation within bright auroras'. *Planet. Space Sci.* **16**, 1525.

Beghin, C., J. L. Rauch, and J. M. Bosqued: 1989, 'Electrostatic plasma waves and HF auroral hiss generated at low altitude'. *J. Geophys. Res.* **94**, 1359.

Belcher, J. W., H. S. Bridge, F. Bagenal, B. Coppi, O. Divers, A. Eviatar, G. S. Gordon, Jr., A. J. Lazarus, R. L. McNutt, Jr., K. Ogilvie, J. D. Richardson, G. L. Siscoe, E. C. Sittler, Jr., J. T. Steinberg, J. D. Sullivan, A. Szabo, L. Villaneuva, V. M. Vasyliunas, and M. Zhang: 1989, 'Plasma observations near Neptune: Initial results from Voyager 2'. *Science* **246**, 1478.

Belian, R. D., D. N. Baker, E. W. Hones, , and P. R. Higbie: 1984, 'High-energy proton drift echoes: Multiple peak structure'. *J. Geophys. Res.* **89**, 9101–.

Belmont, G., D. Fontaine, and P. Canu: 1983, 'Are equatorial electron cyclotron waves responsible for diffuse auroral precipitation?'. *J. Geophys. Res.* **88**, 9163.

Benediktov, E. A., G. G. Getmantsev, Y. A. Sazonov, and A. F. Tarojov: 1965, 'Preliminary results of measurements of the intensity of distributed extra-terrestrial radio frequency emission at 725 and 1525 kHz'. *Kosm. Issled.* **3**, 614.

Bennett, E., M. Temerin, and F. S. Mozer: 1983, 'The distribution of auroral electrostatic shocks below 8000-km altitude'. *J. Geophys. Res.* **88**, 7107.

Benson, R. F.: 1975, 'Terrestrial kilometric radiation'. *Geophys. Res. Lett.* **2**, 52.

Benson, R. F.: 1985, 'AKR: Wave modes, harmonics, and source region electron density structures'. *J. Geophys. Res.* **90**, 2753.

Benson, R. F. and W. Calvert: 1979, 'ISIS 1 observations at the source of auroral kilometric radiation'. *Geophys. Res. Lett.* **6**, 479.

Benson, R. F. and H. K. Wong: 1987, 'Low-altitude ISIS 1 observations of auroral radio emissions and their significance to the cyclotron maser instability'. *J. Geophys. Res.* **92**, 1218.

Berger, M. A. and G. B. Field: 1984, 'The topological properties of magnetic helicity'. *J. Fluid Mech.* **147**, 133.

Bergmann, R., I. Roth, and M. K. Hudson: 1988, 'Linear stability of the H^+ - O^+ two-stream interaction in a magnetized plasma'. *J. Geophys. Res.* **93**, 4005.

Berkey, F. T., M. B. Silevitch, and N. R. Parsons: 1980, 'Time sequence analysis of flickering auroras 1. application of fourier analysis'. *J. Geophys. Res.* **85**, 6827.

Berman, R. H., D. J. Tetreault, and T. H. Dupree: 1985, 'Simulation of phase space density hole growth and the development of intermittent plasma turbulence'. *Phys. Fluids* **28**, 1055.

Bernstein, I. B., J. M. Greene, and M. D. Kruskal: 1957, 'Exact nonlinear plasma oscillations'. *Phys. Rev.* **108**, 546.

Berthomier, M., R. Pottelette, M. Malingre, and Y. Khotyaintsev: 2000, 'Electron-acoustic solitons in an electron-beam plasma system'. *Physics of Plasmas* **7**, 2987.

Beuermann, K., L. Stella, and J. Patterson: 1987, 'Einstein observations of EF Eridani (2A 0311-227) - The textbook example of AM Herculis-type systems'. *Astrophys. J.* **316**, 360.

Bingham, R., D. A. Bryant, and D. S. Hall: 1984, 'A wave model for the aurora'. *Geophys. Res. Lett.* **11**, 327.

Birn, J. and M. Hesse: 1991, 'The substorm current wedge and field-aligned currents in MHD simulations of magnetotail reconnection'. *J. Geophys. Res.* **96**, 1611.

Birn, J. and M. Hesse: 1996, 'Details of current disruption and diversion in simulations of magnetotail dynamics'. *J. Geophys. Res.* **101**, 15345.

Birn, J., M. Hesse, G. Haerendel, W. Baumjohann, and K. Shiokawa: 1999, 'Flow braking and the substorm current wedge'. *J. Geophys. Res.* **104**, 19895.

Birn, J., M. Hesse, and K. Schindler: 1993, 'What determines the latitudinal extent of the auroral acceleration region?'. *J. Atmos. Terr. Phys.* **55**, 1171.

Blanc, M., J. L. Horwitz, J. Blake, I. A. Daglis, J. Lemaire, M. B. Moldwin, S. Orsini, R. Thorne, and R. A. Wolf: 1999, 'Source and loss processes in the inner magnetosphere'. In: B. Hultqvist, M. Øieroset, G. Paschmann, and R. Treumann (eds.): *Magnetospheric Plasma Sources and Losses*, Vol. 6 of *Space Science Series of ISSI*. Dordrecht, Boston and London: Kluwer Academic Publishers, Chapt. 4, p. 137.

Block, L. P.: 1972, 'Potential double layers in the ionosphere'. *Cosmic Electrodyn.* **3**, 349.

Block, L. P. and C.-G. Fälthammar: 1990, 'The role of magnetic-field-aligned electric fields in auroral acceleration'. *J. Geophys. Res..* **95**, 5877.

Boehm, M. H., C. W. Carlson, J. P. McFadden, J. H. Clemmons, and F. S. Mozer: 1990, 'High-resolution sounding rocket observations of large-amplitude Alfvén waves'. *J. Geophys. Res.* **95**, 12157.

Boehm, M. H., J. Clemmons, J.-E. Wahlund, A. Eriksson, L. Eliasson, L. Blomberg, P. Kintner, and H. Höfner: 1995, 'Observations of an upward-directed electron beam with the perpendicular temperature of the cold ionosphere'. *Geophys. Res. Lett.* **22**, 2103.

Boehm, M. H., D. M. Klumpar, E. Möbius, L. M. Kistler, J. P.McFadden, C. W.Carlson, and R. E. Ergun: 1999, 'Fast auroral snapshot observations observations of bouncing ion distributions: Field-line length measurements'. *J. Geophys. Res.* **104**, 2343.

Bonnell, J., R. C. Elphic, S. Palfrey, R. J. Strangeway, W. K. Peterson, D. Klumpar, C. W. Carlson, R. E. Ergun, and J. P. McFadden: 1999, 'Observations of polar cap arcs on FAST'. *J. Geophys. Res.* **104**, 12669.

Bonnell, J., P. M. Kintner, J. E. Wahlund, K. A. Lynch, and R. L. Arnoldy: 1996, 'Interferometric Determination of Broadband ELF Wave Phase Velocity within a region of transverse ion acceleration'. *Geophys. Res. Lett.* **23**, 3297.

Borovsky, J. and J. W. Bonnell: 2001, 'The DC electrical coupling of flow vortices and flow channels in the magnetosphere to the resistive ionosphere'. *J. Geophys. Res.* **106**, 28967.

Borovsky, J., M. F. Thomsen, R. C. Elphis, T. E. Cayton, and D. J. McComas: 1998, 'The transport of plasma sheet material from the distant tail to geosynchronous orbit'. *J. Geophys. Res.* **103**, 20297.

Borovsky, J. E.: 1993a, 'Auroral Arc Thicknesses as Predicted by Various Theories'. *J. Geophys. Res.* **98**, 6101.

Borovsky, J. E.: 1993b, 'The strong-double-layer model of auroral arcs: An assessment'. In: *Auroral Plasma Dynamics*. Washington, D.C.: American Geophysical Union, p. 113.

Borovsky, J. E., D. Suszcynsky, H. DeHaven, and M. Buchwald: 1991, 'Measuring the thickness of auroral curtains'. *Arctic* **44**, 231.

Bosqued, J.-M.: 1987, 'AUREOL-3 results on ion precipitation'. *Phys. Scr.* **T18**, 158.

Bosqued, J.-M., J. Maurel, J. Sauvaud, R. Kovrazhkin, and Y. Galperin: 1986, 'Observations of auroral electron inverted-V structures by the AUREOL-3 satellite'. *Planet. Space Sci.* **34**, 255.

Boström, R.: 1964, 'A model of the auroral electrojets'. *J. Geophys. Res.* **69**, 4983–4999.

Boström, R.: 1973, *Electrodynamics of the ionosphere*, p. 181. Copenhagen: Scandanavian University Books.

Boström, R., G. Gustafsson, B. Holback, G. Holmgren, H. Koskinen, and P. Kintner: 1988, 'Characteristics of solitary waves and weak double layers in the magnetospheric plasma'. *Phys. Rev. Lett.* **61**, 82–85.

Boström, R., B. Holback, G. Holmgren, and H. Koskinen: 1989, 'Solitary structures in the magnetospheric plasma observed by Viking'. *Physica Scripta* **39**, 782–786.

Broadfoot, A. L., S. K. Atreya, J. L. Bertaux, J. E. Blamont, A. Dessler, T. Donahue, W. Forrester, D. Hall, F. Herbert, J. Holberg, D. Hunten, V. Krasnopolsky, S. Linick, J. Lunine, J.-C. McConnell, H. Moos, B. Sandel. N. Schneider, D. Shemansky, G. Smith, D. Strobel, and R. V. Yelle: 1989, 'Ultraviolet spectrometer observations of Neptune and Triton'. *Science* **246**, 1459.

Broadfoot, A. L., F. Herbert, J. B. Holberg, D. M. Hunten, S. Kumar, B. R. Sandel, D. E. Shemansky, A. J. Dessler, S. Linick, and R. Springer: 1986, 'Ultraviolet spectrometer observations of Uranus'. *Science* **23**, 74.

Broadfoot, A. L., B. R. Sandel, D. E. Shemansky, J. B. Holberg, and G. R. Smith: 1981, 'Extreme ultraviolet observations from Voyager 1 encounter with Saturn'. *Science* **233**, 74.

Brown, R. R.: 1966, 'Electron precipitation in the auroral zone'. *Space Sci. Rev.* **5**, 311.

Bryant, D. A.: 1983, 'Magnetospheric processes reflected in the pulsating aurora'. In: *Chapman Conference on Waves in Magnetospheric Plasma*. Kona Coast, Hawaii.

Bryant, D. A.: 2002, 'The roles of static and dynamic electric fields in the auroral acceleration region'. *J. Geophys. Res.* **107**, 10.1029/2001JA900162.

Buchert, S., G. Haerendel, and W. Baumjohann: 1990, 'A model for the electric fields and currents during a strong Ps 6 pulsation event'. *J. Geophys. Res.* **95**, 3733.

Büchner, J. and L. Zelenyi: 1989, 'Regular and chaotic charged particle motion in magnetotail like field reversals, 1. Basic theory of trapped motion'. *J. Geophys. Res.* **94**, 11821.

Burch, J., P. Reiff, and M. Sugiura: 1983, 'Upward electron beams measured by DE-1: a primary source of dayside region-1 Birkeland currents'. *Geophys. Res. Lett.* **10**, 753.

Burch, J. L., W. Lennartsson, W. B. Hanson, R. A. Heelis, J. H. Hoffman, and R. A. Hoffman: 1976, 'Properties of spikelike shear flow reversals observed in the auroral plasma by Atmosphere Explorer C'. *J. Geophys. Res.* **81**, 3886.

Burch, J. L., P. H. Reiff, J. D. Menietti, J. D. Winningham, R. A. Heelis, W. B. Hanson, S. D. Shawhan, E. G. Shelley, M. Sugiura, and D. R. Weimer: 1985, 'IMF By-dependent plasma flow and Birkeland currents in the dayside magnetosphere. I - Dynamics Explorer observations'. *J. Geophys. Res.* **90**, 1577.

Burke, W. J., M. C. Kelley, R. C. Sagalyn, M. Smiddy, and S. T. Lai: 1979, 'Polar cap electric field structures with a northward interplanetary magnetic field'. *Geophys. Res. Lett.* **6**, 21.

Burton, R., R. McPherron, and C. Russell: 1975, 'An empirical relationship between interplanetary conditions and Dst'. *J. Geophys. Res.* **80**, 4204.

Bythrow, P., W. Burke, T. Potemra, L. Zanetti, and A. Lui: 1985, 'Ionospheric evidence for irregular reconnection and turbulent plasma flows in the magnetotail during periods of northward IMF'. *J. Geophys. Res.* **90**, 5319.

Bythrow, P. F., T. A. Potemra, R. E. Erlandson, L. J. Zanetti, and D. M. Klumpar: 1988, 'Birkeland currents and charged particles in the high-latitude prenoon region: A new interpretation'. *J. Geophys. Res.* **90**.

Cahill, L., R. Greenwald, and E. Nielsen: 1978, 'Auroral radar and rocket double-probe observations of the electric field across the Harang discontinuity'. *Geophys. Res. Lett.* **5**, 687.

Calvert, W.: 1981, 'The auroral plasma cavity'. *Geophys. Res. Lett.* **8**, 919.

Carlson, C. W. and M. C. Kelley: 1977, 'Observation and interpretation of particle and electric field measurements inside and adjacent to an active auroral arc'. *J. Geophys. Res.* **82**, 2349.

Carlson, C. W., J. P. McFadden, R. E. Ergun, M. Temerin, W. Peria, F. S. Mozer, D. M. Klumpar, E. G. Shelley, W. K. Peterson, E. Möbius, R. Elphic, R. Strangeway, C. A. Cattell, and R. Pfaff: 1998a, 'FAST observations in the downward auroral current region: Energetic upgoing electron beams, parallel potential drops, and ion heating'. *Geophys. Res. Lett.* **25**, 2017.

Carlson, C. W., R. F. Pfaff, and J. G. Watzin: 1998b, 'Fast Auroral Snapshot (FAST) mission'. *Geophys. Res. Lett.* **25**, 2013.

Carlson, H. C.: 1990, 'Dynamics of the quiet polar cap'. *J. Geomagn. Geoelectricity* **42**, 697.

Carr, T. D., M. D. Desch, and J. K. Alexander: 1983. In: A. J. Dessler (ed.): *Physics of the Jovian Magnetosphere*. Cambridge, UK: Cambridge Univ. Press, p. 226.

Cattell, C.: 1981, 'The relationship of field-aligned currents to electrostatic ion cyclotron waves'. *J. Geophys. Res.* **86**, 3641.

Cattell, C., R. Bergmann, K. Sigsbee, C. Carlson, C. Chaston, R. Ergun, J. McFadden, F. S. Mozer, M. Temerin, R. Strangeway, R. Elphic, L. Kistler, E. Möbius, L. Tang, D. Klumpar, and R. Pfaff: 1998, 'The association of electrostatic ion cyclotron waves, ion and electron beams and field-aligned currents: FAST observations of an auroral zone crossing near midnight'. *Geophys. Res. Lett.* **25**, 2053.

Cattell, C., L. Johnson, R. Bergmann, D. Klumpar, C. Carlsson, J. P. McFadden, R. Strangeway, R. Ergun, K. Sigsbee, and R. Pfaff: 2002, 'FAST observations of discrete electrostatic waves in association with downgoing ion beams in the auroral zone'. *J. Geophys. Res.* **107**, in press.

Cattell, C., M. Somoza, G. Siscoe, and J. Roeder: 1996, 'GEM tests of a boundary layer model of field-aligned currents using S3-3 data'. *J. Geophys. Res.* **101**, 4953.

Cattell, C. A., S. Chari, and M. A. Temerin: 1991, 'An S3-3 satellite study of the effects of the solar cycle on the auroral acceleration process'. *J. Geophys. Res.* **96**, 17903.

Cattell, C. A., R. L. Lysak, R. B. Torbert, and F. S. Mozer: 1979, 'Observations of differences between regions of current flowing into and out of the ionosphere'. *Geophys. Res. Lett.* **6**, 621.

Cattell, C. A., T. Nguyen, M. Temerin, W. Lennartsson, and W. Peterson: 1993, 'Effects of solar cycle on auroral particle acceleration'. In: R. L. Lysak (ed.): *Auroral Plasma Dynamics*, Vol. 80 of *Geophysical Monograph Series*. Washington DC: AGU, p. 219.

Chang, S.-W., J. D. Scudder, J. B. Sigwarth, L. A. Frank, N. C. Maynard, W. J. Burke, W. K. Peterson, E. G. Shelley, R. Friedel, J. B. Blake, R. A. Greenwald, R. P. Lepping, G. J. Sofko, J.-P. Villain, and M. Lester: 1998, 'A comparison of a model for the theta aurora with observations from Polar, Wind, and SuperDARN'. *J. Geophys. Res.* **103**, 17367.

Chang, T., G. B. Crew, N. Herkowitz, J. R. Jasperse, J. M. Retterer, and J. D. Winningham: 1986, 'Transverse acceleration of oxygen ions by electromagnetic ion cyclotron resonance with broad band left-hand polarized waves'. *Geophys. Res. Lett.* **13**, 636.

Chaston, C. C., J. W. Bonnell, J. P. McFadden, R. E. Ergun, and C. W. Carlson: 2002a, 'Electromagnetic ion-cyclotron waves at proton cyclotron harmonics'. *J. Geophys. Res.* **107**, in press.

Chaston, C. C., J. W. Bonnell, L. M. Peticolas, C. W. Carlson, and J. P. McFadden: 2002b, 'Driven Alfvén waves and electron acceleration: a FAST case study'. *Geophys. Res. Lett.* **29**, 10.1029/2001GL013842.

Chaston, C. C., C. W. Carlson, R. E. Ergun, and J. P. McFadden: 2000, 'Alfvén waves, density cavities and electron acceleration observed from the FAST spacecraft'. *Physica Scripta* **T84**, 64.

Chaston, C. C., C. W. Carlson, W. J. Peria, R. E. Ergun, and J. P.McFadden: 1999, 'FAST observations of inertial Alfvén waves in the dayside aurora'. *Geophys. Res. Lett.* **26**, 647.

Chaston, C. C., R. E. Ergun, G. T. Delory, W. Peria, M. Temerin, C. Cattell, R. Strangeway, J. P. McFadden, C. W. Carlson, R. C. Elphic, D. M. Klumpar, W. K. Peterson, E. Möbius, and R. Pfaff: 1998, 'Characteristics of electromagnetic proton cyclotron waves along auroral field lines observed by FAST in regions of upward current'. *Geophys. Res. Lett.* **25**, 2057.

Chaston, C. C., L. M. Peticolas, J. W. Bonnell, C. W. Carlson, W. J. Peria, R. E. Ergun, and J. P. McFadden: 2002c, 'The width and brightness of auroral arcs driven by inertial Alfvén waves'. *J. Geophys. Res.* **107**, in press.

Chen, C. and R. Wolf: 1993, 'Interpretation of high-speed flows in the plasma sheet'. *J. Geophys. Res.* **98**, 21409.

Chen, C. and R. Wolf: 1999, 'Theory of thin-filament motion in Earth's magnetotail and its application to bursty bulk flows'. *J. Geophys. Res.* **104**, 14613.

Chen, F. F.: 1984, *Plasma Physics and Controlled Fusion*. New York and London: Plenum Press, second edition.

Chen, L. and A. Hasegawa: 1974, 'Plasma heating by spatial resonance of Alfvén wave'. *Physics of Fluids* **17**, 1399.

Chen, L. and A. Hasegawa: 1974, 'A theory of long-period magnetic pulsations, 1. Steady state excitation of field line resonance'. *J. Geophys. Res.* **79**, 1024.

Chen, M. W. and M. Schultz: 2001, 'Simulations of diffuse aurora with plasma sheet electrons in pitch angle diffusion less than everywhere strong'. *J. Geophys. Res.* **106**, 28949.

Cheng, A.: 1987, 'Proton and oxygen plasmas at Uranus'. *J. Geophys. Res.* **92**, 15309.

Cheng, C. Z. and G. Knorr: 1976, 'The integration of the Vlasov equation in configuration space'. *J. Comput. Phys.* **22**, 330.

Chiu, Y. T. and J. M. Cornwall: 1980, 'Electrostatic Model of a Quiet Auroral Arc'. *J. Geophys. Res.* **85**, 543–556.

Chiu, Y. T. and M. Schulz: 1978, 'Self-consistent particle and parallel electrostatic field distributions in the magnetospheric-ionospheric auroral region'. *J. Geophys. Res.* **83**, 629–642.

Christon, S. P., D. J. Williams, D. G. Mitchell, C. Y. Huang, and L. A. Frank: 1991, 'Spectral characteristics of plasma sheetion and electron populations during disturbed geomagnetic condition'. *J. Geophys. Res.* **96**, 1.

Clarke, J., H. Moos, S. Atreya, and A. Lane: 1981, 'UE detection of bursts of HLya emission from Saturn'. *Nature* **29**, 226.

Clarke, J. T.: 1982, 'Detection of auroral hydrogen Lyman-alpha emission from Uranus'. *Astrophys. J.* **263**, L105.

Clarke, J. T.: 1996, 'Far-ultraviolet imaging of Jupiter's aurora and the Io footprint'. *Science* **274**, 404.

Clarke, J. T., M. K. Hudson, and Y. L. Yung: 1987, 'The excitation of the far ultraviolet electroglow emissions on Uranus, Saturn and Jupiter'. *J. Geophys. Res.* **92**, 15139.

Cogger, L., J. Murphree, S. Ismail, and C. Anger: 1977, 'Characteristics of dayside 5577 Å and 3914 Å aurora'. *J. Geophys. Res.* **82**, 413.

Collin, H. L., W. K. Peterson, O. W. Lennartsson, and J. F. Drake: 1998, 'The seasonal variation of auroral ion beams'. *Geophys. Res. Lett.* **25**, 4071.

Collin, H. L., W. K. Peterson, and E. G. Shelley: 1987, 'Solar cycle variation of some mass dependent characteristics of upflowing beams of terrestrial ions'. *J. Geophys. Res.* **92**, 4757.

Collin, H. L., R. D. Sharp, E. Shelley, and R. G. Johnson: 1981, 'Some general characteristics of upflowing ion beams over the auroral zone and their relationship to auroral electrons'. *J. Geophys. Res.* **86**, 6820.

Connerney, J. E. P., M. H. Acuna, and N. F. Ness: 1987, 'The magnetic field of Uranus'. *J. Geophys. Res.* **92**, 15329.

Coroniti, F., W. Kurth, F. Scarf, S. Krimigis, C. Kennel, and D. Gurnett: 1987, 'Whistler mode emissions in the Uranian radiation belts'. *J. Geophys. Res.* **92**, 15234.

Cowley, S. W. H. and E. J. Burke: 2001, 'Origin of the main oval on Jupiter's coupled magnetosphere-ionosphere system'. *Planet. Space Sci.* **19**, 1067–1088.

Cowley, S. W. H., J. P. Morelli, and M. Lockwood: 1991, 'Dependence of convection flows and particle precipitation in the high-latitude dayside ionosphere on the X and Y components of the interplanetary magnetic field'. *J. Geophys. Res.* **96**.

Craven, J. and L. Frank: 1991, 'Diagnosis of auroral dynamics using global auroral imaging with emphasis on large-scale evolutions'. In: C.-I. Meng, M. J. Rycroft, and L. A. Frank (eds.): *Auroral Physics*. Cambridge University Press.

Craven, J., J. Murphree, L. Frank, and L. Cogger: 1991, 'Simultaneous optical observations of transpolar arcs in the two polar caps with DE 1 and Viking'. *J. Geophys. Res.* **96**, 2297.

Craven, J. D., L. Frank, C. Russell, E. Smith, and R. Lepping: 1986, 'Global auroral responses to magnetospheric compressions by shocks in the solar wind: Two case studies'. In: Y. Kamide and J. A. Slavin (eds.): *Solar Wind-Magnetosphere Coupling*. p. 377.

Cresswell, G. and T. N. Davis: 1966, 'Observations on pulsating auroras'. *J. Geophys. Res.* **71**, 3155.

Crooker, N. U.: 1988, 'Mapping the merging potential from the magnetopause to the ionosphere through the dayside cusp'. *J. Geophys. Res.* **93**.

Cumnock, J. A., J. R. Sharber, R. A. Hellis, M. R. Hairston, and J. D. Craven: 1997, 'Evolution of the global aurora during positive IMF B_z and varying IMF B_y conditions'. *J. Geophys. Res.* **102**, 17489.

Curtis, S.: 1985, 'Possible nightside source dominance in nonthermal radio emissions from Uranus'. *Nature* **31**, 47.

Daglis, I., R. Thorne, W. Baumjohann, and S. Orsini: 1999, 'The terrestrial ring current: Origin, formation, and decay'. *Rev. Geophys.* **37**, 407.

Daglis, I. A., S. Livi, E. T. Sarris, and B. Wilken: 1994, 'Energy density of ionospheric and solar wind origin ions in the near-Earth magnetotail during substorms'. *J. Geophys. Res.* **99**, 5691.

D'Angelo, N.: 1980, 'Field-aligned currents and large-scale magnetospheric electric fields'. *Ann. Geophysicae* **36**.

Darius, J. and K. H. Fricke: 1981, *Observations of outer planets at Lyman alpha*, Vol. 2171 of *NASA Conf. Publ.*, pp. 85–90. NASA, Washington, D.C.

Davidson, G. T.: 1990, 'Pitch-angle diffusion and the origin of temporal and spatial structures in morningside aurorae'. *Space Sci. Rev.* **53**, 45.

Davis, T.: 1960, 'The morphology of the polar aurora'. *J. Geophys. Res.* **65**, 3497.

Davis, T. N.: 1962, 'The morphology of the auroral displays of 1957-1958 2. Detail analysis of Alaska data and analysis of high-latitude data'. *J. Geophys. Res.* **67**, 75.

Davis, T. N.: 1963, 'Negative correlation between polar-cap visual activity and magnetic activity'. *J. Geophys. Res.* **68**, 4447.

Davis, T. N.: 1978, 'Observed characteristics of auroral forms'. *Space Sci. Rev.* **22**, 77.

Davis, T. N.: 1992, *The Aurora Watcher's Handbook*. Fairbanks, Alaska: University of Alaske Press.

Davis, T. N. and G. T. Hicks: 1964, 'Telvision cinematography of auroras and preliminary measurements of auroral velocities'. *J. Geophys. Res.* **69**, 1931.

de la Beaujardière, O., C. Leger, D. Alcayde, and J. Fontanari: 1991, 'Seasonal dependence of high-latitude electric fields'. *J. Geophys. Res.* **96**, 5723.

de la Beaujardière, O., L. R. Lyons, J. M. Ruohoniemi, E. Friis-Christensen, C. Danielson, F. R. Rich, and P. T. Newell: 1994, 'Quiet-time intensifications along the poleward auroral boundary near midnight'. *J. Geophys. Res.* **99**, 287.

de la Beaujardière, O., R. Vondrak, and M. Baron: 1977, 'Radar observations of electric fields and currents associated with auroral arcs'. *J. Geophys. Res.* **82**, 5051.

de la Beaujardière, O., J. Watermann, P. T. Newell, and F. J. Rich: 1993, 'Relationship between Birkeland current regions, particle precipitation, and electric field'. *J. Geophys. Res.* **96**.

Deehr, C. S., G. G. Sivjee, A. Egeland, K. Henriksen, P. Sandholt, R. Smith, P. Sweeney, C. Duncan, and J. Gilmer: 1980, 'Ground-based observations of F-Region aurora associated with the magnetospheric cusp'. *J. Geophys. Res.* **85**, 2185.

Delory, G. T., R. E. Ergun, E. M. Klementis, C. W. Carlson, and J. P. McFadden: 1997, 'Measurements of short wavelength VLF bursts in the auroral ionosphere: A case for electromagnetic mode conversion?'. *Geophys. Res. Lett.* **24**, 1131.

Demekhov, A. G. and V. Y. Trakhtengerts: 1994, 'A mechanism for formation of pulsating aurorae'. *J. Geophys. Res.* **99**, 5831.

Deng, W., T.L.Killeen, A.G.Burns, R. Roble, J. Slavin, and L. Wharton: 1993, 'The effects of neutral inertia on ionospheric currents in the high-latitude thermosphere following a geomagnetic storm'. *J. Geophys. Res.* **98**, 7775.

Doe, R. A., J. F. Vickrey, and M. Mendillo: 1995, 'Electrodynamic model for the formation of auroral ionospheric cavities'. *J. Geophys. Res.* **100**, 9683.

Dombeck, J., C. Cattell, J. Crumley, W. K. Peterson, H. L. Collin, and C. Kletzing: 2001, 'Observed trends in auroral zone ion mode solitary wave structure characteristics using data from Polar '. *J. Geophys. Res.* **106**, 19013.

Donovan, E. F.: 1993, 'Modeling the magnetic effects of field-aligned currents'. *J. Geophys. Res.* **98**, 13529.

Dubouloz, N., M. Bouhram, C. Senior, D. C. Delcourt, M. Malingre, and J. A. Sauvaud: 2001, 'Spatial structure of the cusp/cleft ion fountain: A case study using a magnetic conjugacy between Interball Auroral Probe and a pair of SuperDARN radars'. *J. Geophys. Res.* **106**, 261.

Dubouloz, N., R. A. Treumann, and R. Pottelette: 1993, 'Turbulence generated by a gas of electron-acoustic solitons'. *J. Geophys. Res.* **98**, 17415.

Dum, C. T. and T. H. Dupree: 1970, 'Non-linear stabilisation of high frequency instabilities in a magnetic field'. *Phys. Fluids* **13**, 2064.

Dungey, J. W.: 1955, 'Electrodynamics of the outer atmosphere'. In: *Proceedings of the Ionosphere Conference, Physical Society of London.* p. 225.

Dungey, J. W.: 1961, 'Interplanetary magnetic field and the auroral zones'. *Phys. Rev. Lett* **6**, 47.

Dupree, T. H.: 1982, 'Theory of phase space holes'. *Phys. Fluids* **25**, 277.

Durrance, S. T. and J. T. Clarke: 1984, *Lyman-alpha aurora*, Vol. 2330 of *NASA Conf. Publ.*, p. 559. NASA, Washington, D.C.

Eather, R. H.: 1980, *Majestic lights: The aurora in science, history, and the arts*. Washington, D. C., USA: American Geophysical Union.

Ellis, P. and D. J. Southwood: 1983, 'Reflection of Alfvén waves by non-uniform ionospheres'. *Planet. Space Sci.* **31**, 107.

Elphic, R., J. Bonnell, R. Strangeway, C. Carlson, M. Temerin, J. McFadden, R. Ergun, and W. Peria: 2000, 'FAST observations of upward accelerated electron beams and the downward field-aligned current region'. In: S.-I. Ohtani (ed.): *Magnetospheric Current Systems*, Vol. 118 of *Geophysical Monograph Series*. Washington, D. C., USA: American Geophysical Union, p. 173.

Elphic, R. C., P. A. Mutch, and C. T. Russell: 1985, 'Observations of field-aligned currents at the plasma sheet boundary : An ISEE-1 and 2 survey'. *Geophys. Res. Lett.* **12**, 631.

Elphinstone, R., D. Hearn, L. Cogger, J. Murphree, A. Wright, I. Sandahl, S. Ohtani, P. Newell, D. Klumpar, M. Shapshak, T. Potemra, K. Mursula, and J. Sauvaud: 1995, 'The double oval UV auroral distribution 2. The most poleward arc system and the dynamics of the magnetotail'. *J. Geophys. Res.* **100**, 12093.

Elphinstone, R., D. Hearn, J. Murphree, L. Cogger, M. Johnson, and H. Vo: 1993, 'Some UV dayside aurora morphologies'. In: R. L. Lysak (ed.): *Auroral Plasma Dynamics*, Vol. 80 of *Geophysical Monograph Series*. Washington, D. C., USA: American Geophysical Union, p. 31.

Elphinstone, R. D., K. Jankowska, J. S. Murphree, and L. L. Cogger: 1990, 'The configuration of the auroral distribution of imterplanetary magnetic field B_z northward, 1. IMF B_x and B_y dependence as observed by the Viking satellite'. *J. Geophys. Res.* **95**, 5791.

Elphinstone, R. D., J. S. Murphree, and L. L. Cogger: 1996, 'What is a global auroral substorm'. *Rev. Geophys. Space Phys.* **34**, 169.

Elvey, C.: 1957, 'Problems of auroral morphology'. *Proc. Nat. Acad. Sci. of the U.S.* **43**, 63.

Ergun, R. E., C. W. Carlson, J. P. McFadden, J. H. Clemmons, and M. H. Boehm: 1991, 'Evidence of a transverse Langmuir modulational instability in a space plasma'. *Geophys. Res. Lett.* **18**, 1177.

Ergun, R. E., C. W. Carlson, J. P. McFadden, G. T. Delory, R. J. Strangeway, and P. L. Pritchett: 2000, 'Electron-cyclotron maser driven by charged-particle acceleration from magnetic-field-aligned electric fields'. *Astrophys. J.* **538**, 456.

Ergun, R. E., C. W. Carlson, J. P. McFadden, F. S. Mozer, G. T. Delory, W. Peria, C. C. Chaston, M. Temerin, R. Elphic, R. J. Strangeway, R. Pfaff, C. A. Cattell, D. Klumpar, E. Shelly, W. Peterson, E. Möbius, and L. Kistler: 1998a, 'FAST satellite wave observations in the AKR source region'. *Geophys. Res. Lett.* **25**, 2061.

Ergun, R. E., C. W. Carlson, J. P. McFadden, F. S. Mozer, G. T. Delory, W. Peria, C. C. Chaston, M. Temerin, R. Elphic, R. J. Strangeway, R. Pfaff, C. A. Cattell, D. Klumpar, E. Shelly, W. Peterson, E. Möbius, and L. Kistler: 1998b, 'FAST satellite observations of electric field structures in the auroral zone'. *Geophys. Res. Lett.* **25**, 2025.

Ergun, R. E., C. W. Carlson, J. P. McFadden, F. S. Mozer, G. T. Delory, W. Peria, C. C. Chaston, M. Temerin, R. Elphic, R. J. Strangeway, R. Pfaff, C. A. Cattell, D. Klumpar, E. Shelly, W. Peterson, E. Möbius, and L. Kistler: 1998c, 'FAST satellite observations of large-amplitude solitary structures'. *Geophys. Res. Lett.* **25**, 2041.

Ergun, R. E., C. W. Carlson, J. P. McFadden, F. S. Mozer, L. Muschietti, and I. Roth: 1998d, 'Debye-scale plasma structures associated with magnetic-field-aligned electric fields'. *Phys. Rev. Lett.* **81**, 826.

Ergun, R. E., C. W. Carlson, J. P. McFadden, F. S. Mozer, and R. J. Strangeway: 2002, 'Parallel Electric Fields in Discrete Arcs'. *Geophys. Res. Lett.* **28**, in press.

Ergun, R. E., C. W. Carlson, J. P. McFadden, R. J. Strangeway, M. V. Goldman, and D. L. Newman: 2001, 'Electron Phase-Space Holes and the VLF Saucer Source Region'. *Geophys. Res. Lett.* **28**, 3805.

Erlandson, R., L. Zanetti, T. Potemra, P. Bythrow, and R. Lundin: 1988, 'IMF By dependence of region 1 Birkeland currents near noon'. *J. Geophys. Res.* **93**, 9804.

Escoubet, C. P., M. F. Smith, S. F. Fung, P. C. Anderson, R. A. Hoffman, E. M. Basinska, and J. M. Bosqued: 1992, 'Staircase ion signature in the polar cusp - A case study'. *Geophys. Res. Lett.* **19**, 1735.

Evans, D.: 1985, 'The characteristics of a persistent auroral arc at high laittudes in the 1400 MLT sector'. In: J. Holtet and A. Egeland (eds.): *The Polar Cusp*. D. Reidel, p. 99.

Evans, D. S.: 1974, 'Precipitation electron fluxes formed by a magnetic-field-aligned potential difference'. *J. Geophys. Res.* **79**, 2853.

Evans, D. S.: 1994, 'A study of intense auroral electron precipitation events'. In: *Proceedings of the Second Space Environment Center Users Conference, Boulder, CO.*

Fairfield, D. and L. Cahill, Jr.: 1966, 'Transition region magnetic field and polar magnetic disturbances'. *J. Geophys. Res.* **71**, 155.

Fairfield, D. H., N. A. Tsyganenko, A. V. Usmanov, and M. Malkov: 1994, 'A large magnetosphere magnetic field database'. *J. Geophys. Res.* **99**, 11319.

Fedder, J. A. and J. G. Lyon: 1995, 'The Earth's magnetosphere is 165 R_E, long: or self-consistent currents, convection, magnetospheric structure and processes for northward interplanetary magnetic field'. *J. Geophys. Res.* **100**, 3623.

Fejer, J. A.: 1953, 'Semidiurnal currents and electron drift in the ionosphere'. *J. Atmos. Terr. Phys.* **4**, 184.

Feldstein, Y. and Y. Galperin: 1985, 'The auroral luminosity structure in the high-latitude upper atmosphere: its dynamics and relationship to the large-scale structure of the Earth's magnetosphere'. *Rev. Geophys. Space Phys.* **23**, 217.

Feldstein, Y. I.: 1963, 'Some problems concerning the morphology of auroras and magnetic disturbances at high latitudes'. *Geomag. Aero., Eng. Transl.* **3**, 183.

Feldstein, Y. I.: 1974, 'Night-time aurora and its relation to the magnetosphere'. *Ann. Geophysicae* **30**, 259.

Feldstein, Y. I., L. I. Gromova1, J. Woch, I. Sandahl, L. Blomberg, G. Marklund, and C.-I. Meng: 2001, 'Structure of the auroral precipitation region in the dawn sector: Relationship to convection reversal boundaries and field-aligned currents'. *Ann. Geophysicae* **19**, 495.

Feldstein, Y. I. and G. V. Starkov: 1967, 'Dynamics of auroral belt and polar geomagnetic disturbances'. *Planet. Space Sci.* **15**, 209.

Feldstein, Y. I. and G. V. Starkov: 1970, 'The auroral oval and the boundary of closed field lines of geomagnetic field'. *Planet. Space Sci.* **18**, 501.

Forget, B., J.-C. Cerisier, A. Berthelier, and J.-J. Berthelier: 1991, 'Ionospheric closure of small-scale Birkeland currents'. *J. Geophys. Res.* **96**, 1843.

Foster, J. C., J. M. Holt, R. G. Musgrove, and D. S. Evans: 1986, 'Ionospheric convection associated with discrete levels of particle precipitation'. *Geophys. Res. Lett.* **13**, 656.

Frank, L. and J. Craven: 1988, 'Imaging results from Dynamics Explorer 1'. *Rev. Geophys* **2**, 249.

Frank, L., J. Craven, J. Burch, and J. Winningham: 1982, 'Polar views of the Earth's aurora with Dynamics Explorer'. *J. Geophys. Res.* **87**, 1001.

Frank, L. A., J. D. Craven, D. A. Gurnett, S. D. Shawhan, J. L. Burch, J. D. Winningham, C. R. Chappell, J. H. Waite, N. C. Maynard, and M. Sugiura: 1986, 'The theta aurora'. *J. Geophys. Res.* **91**, 3177.

Franz, J. R., P. M. Kintner, and J. S. Pickett: 1998, 'Polar observations of coherent electric field structures'. *Geophys. Res. Lett.* **25**, 1277.

Franz, J. R., P. M. Kintner, C. Seyler, J. S. Pickett, and J. D. Scudder: 2000, 'On the perpendicular scale of electron phase-space holes'. *Geophys. Res. Lett.* **27**, 169.

Frey, H. U., G. Haerendel, D. Knudsen, S. Buchert, and O. H. Bauer: 1996, 'Optical and radar observations of the motion of auroral arcs'. *J. Atmos. Terr. Phys.* **58**, 57.

Fried, B. D. and S. D. Conte: 1961, *The Plasma Dispersion Function*. New York: Academic Press.

Friis-Christensen, E., Y. Kamide, A. D. Richmond, and S. Matsushita: 1985, 'Interplanetary magnetic field control of high-latitude electric fields and currents from Greenland magnetometer data'. *J. Geophys. Res.* **90**.

Fujii, R., R. Hoffman, P. Anderson, J. Craven, M. Sugiura, L. Frank, and N. Maynard: 1994, 'Electrodynamic parameters in the nighttime sector during auroral substorms'. *J. Geophys. Res.* **99**, 6093.

Fujii, R., T. Iijima, T. Potemra, and M. Sugiura: 1981, 'Seasonal dependence of large-scale field-aligned currents'. *Geophys. Res. Lett.* **8**, 1103.

Fukushima, N.: 1976, 'Generalized theorem of no ground magnetic effect of vertical currents connected with Pedersen currents in the uniform conducting ionosphere'. *Rep. Ionos. Space Res. Jpn.* **30**, 35.

Fuller-Rowell, T. J. and D. S. Evans: 1987, 'Height-integrated Pedersen and Hall conductivity patterns inferred from the TIROS-NOAA satellite data'. *J. Geophys. Res.* **92**, 7606.

Galperin, Y. and Y. Feldstein: 1991, *Auroral luminosity and its relationship to magnetospheric plasma domains*, p. 207. Cambridge University Press.

Galperin, Y. and Y. Feldstein: 1996, 'Mapping of the precipitation regions to the plasma sheet'. *J. Geophys. Geomagn.* **48**, 857.

Galperin, Y., N. Jorjio, I. Ivanov, I. Karpinsky, E. Lein, T. Mularchik, B. Polenov, V. Temny, N. Fedorova, B. Khazanov, A. Shifrin, and F. Shuiskaya: 1970, 'Studies of geoactive corpuscles and photoelectrons on the Cosmos-261 satellite. 2. Measurements of low-energy electrons'. *Cosmic Research* **8**, 108.

Galperin, Y., V. Ponomarev, and A. Zosimova: 1974, 'Plasma convection in the polar ionosphere'. *Ann. Geophysicae* **30**, 1.

Galperin, Y. I.: 2002, 'Polarization Jet: Characteristics and a model'. *Ann. Geophysicae* **20**, 391.

Ganguli, G., M. J. Keskinen, H. Romero, R. Heelis, T. Moore, and C. Pollock: 1994, 'Coupling of microprocesses and macroprocesses due to velocity shear: An application to the low-altitude ionosphere'. *J. Geophys. Res.* **99**, 8873.

Gary, S. P.: 1993, *Theory of Space Plasma Microinstabilities*. New York: Cambridge.

Gérard, J., D. Grodent, R. Prange, J. H. White, G. R. Gladstone, V. Dols, F. Paresce, A. Storrs, L. Ben Jaffel, and K. A. Franke: 1994, 'A remarkable auroral event on Jupiter observed in the ultraviolet with the Hubble Space Telescope'. *Science* **26**, 1675.

Ghielmetti, A., R. Johnson, R. Sharp, and E. Shelley: 1978, 'The latitudinal, diurnal and altitudinal distributions of energetic upflowing ions of ionospheric origin'. *Geophys. Res. Lett.* **5**, 59.

Gjerloev, J. and R. Hoffman: 2000, 'Height-integrated conductivity in auroral substorms, 1. Data'. *J. Geophys. Res.* **105**, 215.

Gjerloev, J. W. and R. A. Hoffman: 2002, 'Currents in auroral substorms'. *J. Geophys. Res.* **107**, 10.1029/2001JA000194.

Glassmeier, K. H.: 1995, 'ULF Pulsations'. In: H. Volland (ed.): *Handbook of Atmospheric Electrodynamics Vol II*. Boca Raton, FL: CRC Press, Inc., p. 463.

Glassmeier, K. H., H. Volpers, and W. Baumjohann: 1984, 'Ionospheric Joule dissipation as a damping mechanism for high-latitude ULF pulsations: observational evidence'. *Planet. Space Sci.* **32**, 1463.

Goertz, C. K. and R. W. Boswell: 1979, 'Magnetosphere-ionosphere coupling'. *J. Geophys. Res.* **84**, 7239.

Goldman, M. V., D. L. Newman, F. Crary, and M. M. Oppenheim: 2000, 'Turbulence driven by two-stream instability in a magnetized plasma'. *Phys. Plasmas* **7**, 1732.

Goldman, M. V., M. M. Oppenheim, and D. L. Newman: 1999, 'Nonlinear two-stream instabilities as an explanation for auroral bipolar wave structures'. *Geophys. Res. Lett.* **26**, 1821.

Gorney, D., A. Clarke, D. Croley, J. Fennell, J. Luhmann, and P. Mizera: 1981, 'The distribution of Ion Beams and Conics below 8000 km'. *J. Geophys. Res.* **83**, 86.

Gorney, D. J., Y. T. Chiu, and D. R. Croley: 1985, 'Trapping of ion conics by downward parallel electric fields'. *J. Geophys. Res.* **90**, 4205.

Gray, P. C., M. K. Hudson, W. Lotko, and R. Bergmann: 1991, 'Decay of ion beam driven acoustic waves into ion holes'. *Geophys. Res. Lett.* **18**, 1675.

Greenwald, R. and A. Walker: 1980, 'Energetics of long-period resonant hydromangetic waves'. *Geophys. Res. Lett.* **7**, 745.

Greifinger, C. and P. Greifinger: 1968, 'Theory of hydromagnetic propagation in the ionospheric waveguide'. *J. Geophys. Res.* **73**, 7473.

Gulkis, S. and T. Carr: 1987, 'The main source of radio emission from the magnetosphere of Uranus'. *J. Geophys. Res.* **92**, 15159.

Gurnett, D. A.: 1966, 'Satellite observations of VLF hiss'. *J. Geophys. Res.* **71**, 5599.

Gurnett, D. A.: 1972, 'INJUN 5 Observations of Magnetospheric Electric Fields and Plasma Convection'. In: B. M. McCormac (ed.): *Earth's Magnetospheric Processes*, Vol. 32 of *ASSL*. p. 233.

Gurnett, D. A.: 1974, 'The earth as a radio source'. *J. Geophys. Res.* **79**, 4227.

Gurnett, D. A. and R. R. Anderson: 1981, 'The kilometric radio emission spectrum: Relationship to auroral acceleration processes'. In: S.-I. Akasofu and J. R. Kan (eds.): *Physics of Auroral Arc Formation*, Vol. 25 of *Geophysical Monograph Series*. Washington D. C., USA: AGU, p. 341.

Gurnett, D. A., W. S. Kurth, G. B. Hospodarsky, A. M. Persoon, P. Zarka, A. Lecacheux, S. J. Bolton, M. D. Desch, W. M. Farrell, M. L. Kaiser, H.-P. Ladreiter, H. O. Rucker, P. Galopeau, P. Louarn, D. T. Young, W. R. Pryor, and M. K. Dougherty: 2002, 'Control of Jupiter's radio emission and aurorae by the solar wind'. *Nature* **415**, 985.

Gusev, M. G. and O. A. Troshichev: 1986, 'Hook-shaped arcs in dayside polar cap and their relation to the IMF'. *Planet. Space Sci.* **34**, 489.

Gussenhoven, M.: 1982, 'Extremely high latitude auroras'. *J. Geophys. Res.* **87**, 2401.

Gussenhoven, M., D. Hardy, and N. Heinemann: 1983, 'Systematics of the equatorward diffuse auroral boundary'. *J. Geophys. Res.* **88**, 5692–5708.

Gussenhoven, M. S. and E. G. Mullen: 1989, 'Simultaneous relativistic electron and auroral particle access to the polar caps during interplanetary magnetic field B_z northward: A scenario for an open field line source of auroral particles'. *J. Geophys. Res.* **94**, 17121.

Gustafsson, G., W. Baumjohann, and I. Iversen: 1981, 'Multi-method observations and modeling of the three-dimensional currents associates with a very strong Ps 6 event'. *J. Geophys* **49**, 138.

Gustafsson, G., N. Papitashvili, and V. Papitashvili: 1992, 'A revised corrected geomagnetic coordinate system for Epochs 1985 and 1990'. *J. Atmos. Terr. Phys.* **54**, 1609.

Haerendel, G.: 1983, 'An Alfvén wave model of auroral arcs'. In: B. Hultqvist and T. Hagfors (eds.): *High-Latitude Space Plasma Physics*. New York, USA, p. 515.

Haerendel, G.: 1989, 'Cosmic linear accelerators'. In: *Proceedings of an International School and Workshop on Plasma Astrophysics*. p. 37.

Haerendel, G.: 1990, 'Field-aligned currents in the Earth's magnetosphere'. In: C. T. Russell, E. R. Priest, and L. C. Lee (eds.): *Physics of Magnetic Flux Ropes*, Vol. 58 of *Geophysical Monograph Series*. Washington, D. C., USA: American Geophysical Union, p. 539.

Haerendel, G.: 1992, 'Disruption, ballooning or auroral avalanche - on the cause of substorms'. In: *Proceedings of the First International Conference on Substorms, Kiruna, Sweden, 23-27 March 1992*. p. 417.

Haerendel, G.: 1994, 'Acceleration from Field-Aligned Potential Drops'. *Ap. J. Supp. Series* **90**, 765.

Haerendel, G.: 1999, 'Origin and dynamics of thin auroral arcs'. *Adv. Space. Res.* **23**, 1637.

Haerendel, G.: 2000, 'Outstanding Issues in Understanding the Dynamics of the Inner Plasma Sheet and Ring Current During Storms and Substorms'. *Adv. Space. Res.* **25**, 2379.

Haerendel, G.: 2001a, 'Auroral acceleration in astrophysical plasmas'. *Physics of Plasmas* **8**, 2365.

Haerendel, G.: 2001b, 'Mechanisms of auroral particle acceleration'. *Phys. Chem. Earth C* **26**, 25.

Haerendel, G., S. Buchert, C. L. Hoz, B. Raaf, and E. Rieger: 1993, 'On the Proper Motion of Auroral Arcs'. *J. Geophys. Res.* **98**, 6087.

Haerendel, G. and R. Lüst: 1970, 'Electric Fields in the Ionosphere and Magnetosphere'. In: B. M. McCormac (ed.): *Particles and Field in the Magnetosphere*, Vol. 17 of *ASSL*. p. 213.

Haerendel, G., B. U. Olipitz, S. Buchert, O. H. Bauer, E. Rieger, and C. La Hoz: 1996, 'Optical and radar observations of auroral arcs with emphasis on small-scale structures'. *J. Atmos. Terr. Phys.* **58**, 71.

Haerendel, G., E. Rieger, A. Valenzuela, H. Foeppl, H. C. Stenbaek-Nielsen, and E. M. Wescott: 1976, 'First observation of electrostatic acceleration of barium ions into the magnetosphere'. In: *ESA European Programmes on Sounding-Rocket and Balloon Research in the Auroral Zone*. p. 203.

Hallinan, T.: 1976, 'Auroral Spirals, II: Theory'. *J. Geophys. Res.* **81**, 3959.

Hallinan, T. and T. Davis: 1970, 'Small-scale auroral arc distortions'. *Planet. Space Sci.* **18**, 1735.

Hallinan, T., J. Kimball, D. Osborne, and C. Deehr: 1998, 'Spectra of Type-B red lower borders'. *J. Geophys. Res.* **103**, 11635.

Hallinan, T., J. Kimball, H. Stenbaek-Nielsen, and C. Deehr: 1997, 'Spectroscopic evidence for suprathermal electrons in Enhanced Auroras'. *J. Geophys. Res.* **102**, 7501.

Hallinan, T., H. Stenbaek-Nielsen, and C. Deehr: 1985, 'Enhanced aurora'. *J. Geophys. Res.* **90**, 8461.

Hallinan, T. and H. C. Stenbaek-Nielsen: 2001, 'The connection between auroral acceleration and auroral morphology'. *Phys. Chem. Earth* **26**, 169.

Hallinan, T. J.: 1981, 'The distribution of vorticity in auroral arcs'. In: *Physics of auroral arc formation; Proceedings of the Chapman Conference on Formation of Auroral Arcs*, Vol. 25 of *Geophysical Monograph Series*. American Geophysical Union, p. 42.

Hallinan, T. J., J. Kimball, H. C. Stenbaek-Nielsen, K. Lynch, R. Arnoldy, J. Bonnell, and P. Kintner: 2001, 'Relation between optical emissions, particles, electric fields, and Alfvén waves in a multiple rayed arc'. *J. Geophys. Res.* **106**, 15445.

Hamrin, M., M. André, P. Norqvist, and K. Rönnmark: 2000, 'The importance of a dark ionosphere for ion heating and auroral arc formation'. *Geophys. Res. Lett.* **27**, 1635.

Harang, L.: 1946, 'The mean field of disturbance of polar geomagnetic storms'. *Terr. Magn. Atmos. Electr.* **51**, 353.

Hardy, D., M. Gussenhoven, and E. Holeman: 1985, 'A statistical model of auroral electron precipitation'. *J. Geophys. Res.* **90**, 4229.

Hardy, D. A., M. S. Gussenhoven, R. Raistrick, and W. J. McNeil: 1987, 'Statistical and functional representations of the pattern of auroral energy flux, number flux, and conductivity'. *J. Geophys. Res.* **92**, 12275.

Harel, M., R. Wolf, R. Spiro, P. Reiff, C.-K. Chen, W. Burke, F. Rich, and M. Smiddy: 1981, 'Quantitative simulation of a magnetospheric substorm. 2. Comparison with observations'. *J. Geophys. Res.* **86**, 2242.

Hasegawa, A.: 1976, 'Particle acceleration by MHD surface wave and formation of aurora'. *J. Geophys. Res.* **81**, 5083.

Hasegawa, A. and T. Sato.: 1979, 'Generation of field-aligned current during substorm'. In: S.-I. Akasofu (ed.): *Dynamics of the Magnetosphere.* Dordrecht, p. 529.

Heacock, R. and R. Hunsucker: 1981, 'Type Pi1-2 magnetic field pulsations'. *Space Sci. Rev.* **28**, 191.

Heikkila, W. J.: 1984, 'Magnetospheric topology fields and currents'. In: *Magnetospheric Currents*, Vol. 28 of *Geophysical Monograph Series.* Washington, D. C., USA: American Geophysical Union, p. 208.

Henderson, M. G., J. S. Murphree, and G. D. Reeves: 1998, 'Are north-south aligned auroral structures the ionospheric manifestations of bursty bulk flows?'. *Geophys. Res. Lett.* **25**, 3737.

Heppner, J. P.: 1972, 'Harang discontinuity in auroral belt ionospheric currents'. *Geophys. Publ.* **29**.

Heppner, J. P. and N. C. Maynard: 1987, 'Empirical high-latitude electric field models'. *J. Geophys. Res.* **92**, 4467.

Herbert, F. and B. Sandel: 1994, 'The Uranian aurora and its relationship to the magnetosphere'. *J. Geophys. Res.* **99**, 4143.

Herbert, F., B. Sandel, R. Yelle, J. Holberg, A. Broadfoot, D. Shemansky, S. Altreya, and P. Romani: 1987, 'EUV occultations observed by Voyager 2'. *J. Geophys. Res.* **92**, 15093.

Hesse, M., J. Birn, and K. Schindler: 1990, 'A self-consistent two-dimensional resistive fluid theory of field-aligned potential structures including charge separation and magnetic and velocity shear'. *J. Geophys. Res.* **95**, 18929.

Hesse, M. and K. Schindler: 1988, 'A theoretical foundation of general magnetic reconnection'. *J. Geophys. Res.* **93**, 5559.

Hesse, M. and D. Winske: 1998, 'Electron dissipation in collisionless magnetic reconnection'. *J. Geophys. Res.* **103**, 26479.

Hesse, M., D. Winske, and M. M. Kuznetsova: 1995, 'Hybrid modeling of collisionless reconnection in two-dimensional current sheets: Simulations'. *J. Geophys. Res.* **100**, 21815.

Hilgers, A., B. Holback, G. Holmgren, and R. Boström: 1992, 'Probe measurements of low plasma densities with applications to the auroral acceleration region and auroral kilometric radiation sources'. *J. Geophys. Res.* **97**, 8631.

Hilmer, R. V. and G.-H. Voigt: 1995, 'A magnetospheric magnetic field model with flexible internal current systems driven by independent physical parameters'. *J. Geophys. Res.* **100**, 5613.

Hirahara, M., T. Mukai, T. Nagai, N. Kaya, H. Hayakawa, and H. Fukunishi: 1996, 'Two types of ion energy dispersions observed in the nightside auroral regions during geomagnetically disturbed periods'. *J. Geophys. Res.* **101**, 7749.

Hirahara, M., T. Mukai, E. Sagawa, N. Kaya, and H. Hayakawa: 1997a, 'Multiple energy-dispersed ion precipitations in the low-latitude auroral oval: Evidence of E x B drift effect and upward flowing ion contribution'. *J. Geophys. Res.* **102**, 2513.

Hirahara, M., A. Yamazaki, K. Seki, T. Mukai, E. Sagawa, N. Kaya, and H. Hayakawa: 1997b, 'Characteristics of downward flowing ion energy dispersions observed in the low-altitude central plasma sheet by Akebono and DMSP'. *J. Geophys. Res.* **102**, 4821.

Hoffman, R. A., R. Fujii, and M. Sugiura: 1994, 'Characteristics of the fieldaligned current system in the nighttime sector during auroral substorms'. *J. Geophys. Res.* **99**, 21303.

Hoffman, R. A., M. Sugiura, N. C. Maynard, R. M. Candey, J. D. Craven, and L. A. Frank: 1988, 'Electrodynamic patterns in the polar region during periods of extreme magnetic quiescence'. *J. Geophys. Res.* **93**.

Hones, E., T. Pytte, and H. West: 1984, 'Association of geomegnetic activity with plasma sheet thinning and expansion: a statistical study'. *J. Geophys. Res.* **89**, 5471.

Hones, E. J.: 1979, 'Plasma flow in the magnetotail and its implications for substorm theories'. In: S.-I. Akasofu (ed.): *Dynamics of the Magnetosphere*. D. Reidel, Dordrecht, p. 545.

Hones, Jr., E. W., J. D. Craven, L. A. Frank, D. S. Evans, and P. T. Newell: 1989, 'The horse-collar aurora: a frequent pattern of the aurora in quiet times'. *Geophys. Res. Lett.* **16**, 37.

Huang, C. Y., J. D. Craven, and L. A. Frank: 1989, 'Simultaneous observations of a theta aurora and associated magnetotail plasmas'. *J. Geophys. Res.* **94**, 10137.

Huang, C. Y., L. A. Frank, W. K. Peterson, D. J. William, W. Lennartsson, D. G. Mitchell, R. C. Elphic, and C. T. Russell: 1987, 'Filamentary structures in the magnetotail lobes'. *J. Geophys. Res.* **92**, 2349.

Hudson, M. K., J. T. Clarke, and J. A. Warren: 1989, 'Ionospheric dynamo theory for production of far ultraviolet emissions from Uranus'. *J. Geophys. Res.* **94**, 6517.

Hudson, M. K., W. Lotko, I. Roth, and E. Witt: 1983, 'Solitary waves and double layers on auroral field lines'. *J. Geophys. Res.* **88**, 916.

Hughes, W. J.: 1994, 'Magnetospheric ULF waves: a tutorial with a historical perspective'. In: M. J. Engebretson, K. Takahashi, and M. Scholer (eds.): *Solar Wind Sources of Magnetospheric Ultra-Low-Frequency Waves*, Vol. 81 of *AGU Monograph Series*. Washington, D.C., USA: American Geophysical Union, p. 1.

Hui, C. H. and C. E. Seyler: 1992, 'Electron acceleration by Alfvén waves in the magnetosphere'. *J. Geophys. Res.* **97**, 3953.

Hultqvist, B.: 1971, 'On the production of a magnetic-field-aligned electric field by the interaction between the hot magnetospheric plasma and the cold ionosphere'. *Planet. Space Sci.* **19**, 749.

Hultqvist, B.: 1988, 'On the acceleration of electrons and positive ions in the same direction along magnetic field lines by parallel electric fields'. *J. Geophys. Res.* **93**, 9777.

Hultqvist, B.: 1996, 'On the acceleration of positive ions by high-altitude, large-amplitude electric field fluctuations'. *J. Geophys. Res.* **101**, 27111.

Hultqvist, B.: 2002, 'Downward ion acceleration at auroral latitudes: Cause of parallel electric field'. *Ann. Geophysicae* **20**, 1117 – 1136.

Hultqvist, B., R. Lundin, K. Stasiewicz, L. Block, and P.-A. Lindqvist: 1988, 'Simultaneous observation of upward moving field-aligned energetic electrons and ions on auroral zone field lines'. *J. Geophys. Res.* **93**, 9765.

Hultqvist, B., M. Øieroset, G. Paschmann, and R. Treumann (eds.): 1999, *Magnetospheric Plasma Sources and Losses*, Vol. 6 of *Space Science Series of ISSI*. Dordrecht, Holland: Kluwer Academis Publishers.

Ichimaru, S.: 1986, *Plasma Physics: An Introduction to Statistical Physics of Charged Particles*. Menlo Park, CA: Benjamin Cummings.

Ieda, A., D. H. Fairfield, T. Mukai, Y. Saito, S. Kokubun, K. Liou, C.-I. Meng, G. K. Parks, and M. J. Brittnacher: 2001, 'Plasmoid ejection and auroral brightenings'. *J. Geophys. Res.* **106**, 3845.

Ieda, A., S. Machida, T. Mukai, Y. Saito, T. Yamamoto, A. Nishida, T. Terasawa, and S. Kokubun: 1998, 'Statistical analysis of the plasmoid evolution with Geotail observations'. *J. Geophys. Res.* **103**, 4453.

Iijima, T. and T. A. Potemra: 1976, 'The amplitude distribution of field-alignedcurrents at northern high latitudes observed by Triad'. *J. Geophys. Res.* **81**, 2165.

Iijima, T. and T. A. Potemra: 1978, 'Large-scale characteristics of field-aligned currents associated with substorms'. *J. Geophys. Res.* **83**, 599.

Iijima, T., T. A. Potemra, L. J. Zanetti, and P. F. Bythrow: 1984, 'Large-scale Birkeland currents in the dayside polar region during strongly northward IMF: A new Birkeland current system'. *J. Geophys. Res.* **89**, 7441.

Inhester, B., W. Baumjohann, R. A. Greenwald, and E. Nielsen: 1981, 'Joint twodimensional observations of ground magnetic and ionospheric electric fields associated with auroral zone currents:

3. Auroral zone currents during the passage of a westward traveling surge'. *Journal of Geophysics* **49**, 155.

Ismail, S., D. D. Wallis, and L. L. Cogger: 1977, 'Characteristics of polar cap sun-aligned arcs'. *J. Geophys. Res.* **82**, 4741.

Ivchenko, N., G. Marklund, K. Lynch, D. Pietrowski, R. Torbert, F. Primdahl, and A. Ranta: 1999, 'Quasiperiodic oscillations observed at the edge of an auroral arc by Auroral Turbulence 2'. *Geophys. Res. Lett.* **26**, 3365.

James, H. G.: 1976, 'VLF saucers'. *J. Geophys. Res.* **81**, 501.

Janhunen, P. and H. E. J. Koskinen: 1997, 'The closure of Region-1 field-aligned current in MHD simulation'. *Geophys. Res. Lett.* **24**, 1419.

Janhunen, P. and A. Olsson: 2000, 'New model for auroral acceleration: O-shaped potential structure cooperating with waves'. *Ann. Geophysicae* **18**, 596.

Janhunen, P., A. Olsson, F. S. Mozer, and H. Laakso: 1999, 'How does the U-shaped potential close above the acceleration region? A study using Polar data'. *Ann. Geophysicae* **17**, 1276.

Janhunen, P., A. Olsson, W. K. Peterson, H. Laakso, J. S. Pickett, T. I. Pulkkinen, and C. T. Russell: 2001, 'A study of inverted-V auroral acceleration mechanisms using Polar/Fast auroral snapshot conjunctions'. *J. Geophys. Res.* **106**, 18995.

Jankowska, K., R. Elphinstone, J. S. Murphree, L. L. Cogger, and D. Hearn: 1990, 'The configuration of the auroral distribution for interplanetary magnetic field Bz northward: ionospheric convection consistent with Viking observations'. *J. Geophys. Res.* **95**, 5805.

Jasperse, J. R.: 1998, 'Ion heating, electron acceleration, and the self-consistent parallel E-field in downward auroral current regions'. *Geophys. Res. Lett.* **25**, 3485.

Jasperse, J. R. and N. Grossbard: 2000, 'The Alfvén-Fälthammar formula for the parallel E field and its analogue in downward auroral-current regions'. *IEEE Trans. on Plasma Science*.

Johnson, M. T., J. R. Wygant, C. Cattell, F. S. Mozer, M. Temerin, and J. Scudder: 2001, 'Observations of the seasonal dependence of the thermal plasma density in the southern hemisphere auroral zone and polar cap at 1 Re'. *J. Geophys. Res.* **106**, 19023.

Johnstone, A. D.: 1978, 'Pulsating Aurora'. *Nature* **274**, 119–126.

Johnstone, A. D.: 1983, 'The mechanisms of pulsating aurora'. *Ann. Geophysicae* **1**, 397.

Johnstone, A. D. and J. D. Winningham: 1982, 'Satellite observations of suprathermal electron bursts'. *J. Geophys. Res.* **87**, 2321.

Jørgensen, A., H. Spence, T. Hughes, and D. McDiarmid: 1999, 'A study of omega bands and Ps6 pulsations on the ground, at low altitude and at geostationary orbit'. *J. Geophys. Res.* **104**, 14705.

Judge, D. L., F.-M. Wu, and R. W. Carlson: 1980, 'Ultraviolet photometer observations of the Saturnian system'. *Science* **207**, 431.

Kallio, E., T. Pulkkinen, H. Koskinen, A. Viljanen, J. Slavin, and K. Ogilvie: 2000, 'Loading-unloading processes in the nightside ionosphere'. *Geophys. Res. Lett.* **27**, 1627.

Kamide, Y. and S.-I. Akasofu: 1975, 'The auroral electrojet and global auroral features'. *J. Geophys. Res.* **80**, 3585.

Kan, J.: 1993, 'A global Magnetosphere-Ionosphere Coupling model of substorms'. *J. Geophys. Res.* **98**, 17263.

Kan, J., L. Zhu, and S.-I. Akasofu: 1988, 'A theory of substorms: Onset and subsidence'. *J. Geophys. Res.* **93**, 5624.

Kan, J. R. and W. J. Burke: 1985, 'A theoretical model of polar cap arcs'. *J. Geophys. Res.* **90**, 4171.

Kan, J. R. and L. C. Lee: 1980, 'Theory of imperfect magnetosphere-ionosphere coupling'. *Geophys. Res. Lett.* **7**, 633.

Karlsson, T. and G. Marklund: 1996, 'A statistical study of intense low altitude electric fields observed by Freja'. *Geophys. Res. Lett.* **23**, 1005.

Karlsson, T., G. Marklund, L. G. Blomberg, and A. Malkki: 1998, 'Subauroral electric fields observed by the Freja satellite: A statistical study'. *J. Geophys. Res.* **103**, 4327.

Karlsson, T. and G. T. Marklund: 1998, 'Simulations of effects of small-scale auroral currents closure in the return current region'. *Physics of Space Plasmas* **15**, 401.

Kasahara, Y., T. Hosada, T. Mukai, S. Watanabe, I. Kimura, H. Kojima, and R. Nitsu: 2001, 'ELF/VLF waves correlated with the transversely accelerated ions in the auroral region observed by Akebono'. *J. Geophys. Res.* **106**, 21123.

Kaufmann, R. L., P. B. Dusenbery, B. J. Thomas, and R. L. Arnoldy: 1978, 'Auroral electron distribution function'. *J. Geophys. Res.* **83**, 586.

Kaufmann, R. L. and P. M. Kintner: 1982, 'Upgoing ion beams. I - Microscopic analysis'. *J. Geophys. Res.* **87**, 10487.

Kaufmann, R. L. and D. Larson: 1989, 'Electric field mapping and auroral Birkeland currents'. *J. Geophys. Res.* **94**, 15307.

Kauristie, K., V. A. Sergeev, M. V. Kubyshkina, T. I. Pulkkinen, V. Angelopoulos, T. Phan, R. P. Lin, and J. A. Slavin: 2000, 'A conjugate study of Wind and ground-based observations during transient plasma sheet flows'. *J. Geophys. Res.* **105**, 10677.

Kawasaki, K. and G. Rostoker: 1979, 'Perturbation magnetic fields and current systems associated with eastward drifting auroral structures'. *J. Geophys. Res.* **84**, 1464.

Keiling, A., J. Wygant, C. Cattell, M. Johnson, M. Temerin, F. Mozer, C. Kletzing, J. Scudder, C. Russell, and W. Peterson: 2001, 'Properties of large electric fields in the plasma sheet at 4-7 RE measured with Polar'. *J. Geophys. Res.* **106**, 5779.

Kelley, M. C.: 1989, *The Earth's Ionosphere*, Vol. 43 of *International Geophysics Series*. San Diego: Academic Press.

Kellogg, P. J. and S. J. Monson: 1979, 'Radio emissions from the aurora'. *Geophys. Res. Lett.* **6**, 297.

Kennel, C., F. Scarf, R. Fredricks, J. McGehee, and F. Coroniti: 1970, 'VLF electric field observations in the magnetosphere'. *J. Geophys. Res.* **75**, 6136.

Kepko, L. and M. Kivelson: 2000, 'Generation of Pi2 pulsations by bursty bulk flows'. *J. Geophys. Res.* **104**, 25021.

Khalipov, V., Y. I. Galperin, Y. Lissakov, J. Crasnier, L. Nikolaenko, V. Sinitsin, and J.-A. Sauvaud: 1977, 'Diffuse auroral zone: 2. Formation and dynamics of the polar edge of the subauroral ionospheric trough in the evening sector'. *Cosmic Research* **15**, 708.

Khalipov, V. L., Y. I. Galperin, A. E. Stepanov, and L. V. Shestakova: 2001, 'Formation of the Polarisation Jet during substorm expansion phase: Results of ground-based measurements'. *Cosmic Research* **39**, 500.

Kimball, J. and T. Hallinan: 1998a, 'A morphological study of black vortex streets'. *J. Geophys. Res.* **103**, 14683.

Kimball, J. and T. Hallinan: 1998b, 'Observations of the relationship of black auroral patches to other types of aurora'. *J. Geophys. Res.* **103**, 14671–14682.

Kindel, J. M. and C. F. Kennel: 1971, 'Topside current instabilities'. *J. Geophys. Res.* **76**, 3055.

Kintner, P. M., J. Bonnell, R. Arnoldy, K. Lynch, C. Pollock, and T. Moore: 1996, 'SCIFER — Transverse Ion Acceleration and Plasma Waves'. *Geophys. Res. Lett.* **23**, 1873.

Kintner, P. M., J. Franz, P. Schuck, and E. Klatt: 2000, 'Interferometric coherency determination of wavelength or what are broadband ELF waves?'. *J. Geophys. Res.* **105**, 21237.

Kintner, P. M., M. C. Kelley, and F. S. Mozer: 1978, 'Electrostatic hydrogen cyclotron waves near one Earth radius altitude in the polar magnetosphere'. *Geophys. Res. Lett.* **5**, 139.

Kintner, P. M., M. C. Kelley, R. D. Sharp, A. G. Ghielmetti, M. Temerin, C. A. Cattell, and P. Mizera: 1979, 'Simultaneous observations of energetic (keV) upstreaming ions and EHC waves'. *J. Geophys. Res.* **84**, 7201.

Kintner, P. M., J. Vago, S. Chesney, R. L. Arnoldy, K. A. Lynch, T. E. Moore, and C. J. Pollock: 1992, 'Localized Lower Hybrid Acceleration of Ionospheric Plasma'. *Phys. Rev. Lett.* **68**, 2448.

Kirkwood, S., H. J. Opgenoorth, and J. Murphree: 1988, 'Ionospheric conductivities, electric fields and currents associated with auroral substorms measured by the EISCAT radar'. *Planet. Space Sci.* **36**, 1359.

Kivelson, M. G. and C. T. Russell: 1995, *Introduction to Space Physics*. Cambridge: Cambridge University Press.

Kivelson, M. G. and D. J. Southwood: 1985, 'Resonant ULF waves: A new interpretation'. *Geophys. Res. Lett.* **12**, 49.

Kivelson, M. G. and D. J. Southwood: 1986, 'Coupling of global magnetospheric MHD eignemodes to field line resonances'. *J. Geophys. Res.* **91**, 4345.

Kletzing, C., C. A. Cattell, F. S. Mozer, S.-I. Akasofu, and K. Makita: 1983, 'Evidence for electrostatic shocks as the source of discrete auroral arcs'. *J. Geophys. Res.* **88**, 4105.

Kletzing, C. A.: 1994, 'Electron acceleration by kinetic Alfvén waves'. *J. Geophys. Res.* **99**, 11095.

Kletzing, C. A. and S. Hu: 2001, 'Alfvén wave generated electron time dispersion'. *Geophys. Res. Lett.* **28**, 693.

Klumpar, D. and W. Heikkila: 1982, 'Electrons in the ionospheric source cone: evidence for runaway electrons as carriers of downward Birkeland currents'. *Geophys. Res. Lett.* **9**, 873.

Klumpar, D. M.: 1993, 'Statistical distributions of the auroral electron albedo in the magnetosphere'. In: R. L. Lysak (ed.): *Auroral Plasma Dynamics*, Vol. 80 of *Geophysical Monograph Series*. Washington D.C., USA: American Geophysical Union, p. 163.

Klumpar, D. M., R. J. Strangeway, C. Carlson, J. P. McFadden, and M. A. Temerin: 1999, 'Latitude and local time distribution of downward directed ion beams in the auroral ionosphere,'. *EOS Trans. AGU, Spring Meet. Suppl.* **80**.

Knight, S.: 1973, 'Parallel electric fields'. *Planet. Space Sci.* **21**, 741.

Knipp, D. J., B. A. Emery, A. D. Richmond, N. U. Crooker, M. R. Hairston, J. A. Cumnock, W. F. Denig, F. J. Rich, O. de la Beaujardiere, J. M. Ruohoniemi, A. S. Rodger, G. Crowley, B. H. Ahn, D. S. Evans, T. J. Fuller-Rowell, E. Friis-Christensen, M. Lockwood, H. W. Kroehl, C. G. Maclennan, A. McEwin, R. J. Pellinen, R. J. Morris, G. B. Burns, V. Papitashvili, A. Zaitzev, O. Troshichev, N. Sato, P. Sutcliffe, and L. Tomlinson: 1993, 'Ionospheric convection response to slow, strong variations in a northward interplanetary magentic field: A case study for January 14, 1988'. *J. Geophys. Res.* **98**, 19273.

Knudsen, D. J.: 1996, 'Spatial modulation of electron energy and density by nonlinear stationary inertial Alfvén waves'. *J. Geophys. Res.* **101**, 10761.

Knudsen, D. J., E. Donovan, L. Cogger, B. Jackel, and W. Shaw: 2001, 'Width and structure of mesoscale optical arcs'. *Geophys. Res. Lett.* **28**, 705.

Knudsen, D. J., M. C. Kelley, and J. F. Vickrey: 1992, 'Alfvén Waves in the Auroral Ionosphere: A Numerical Model Compared With Measurements'. *J. Geophys. Res.* **97**, 77.

Knudsen, D. J. and J.-E. Wahlund: 1998, 'Core ion flux bursts within solitary kinetic Alfvén waves'. *J. Geophys. Res.* **203**, 4157.

Knudsen, D. J., B. A. Whalen, T. Abe, and A. Yau: 1994a, 'Temporal Evolution and Spatial Dispersion of Ion Conics: Evidence for a Polar Cusp Heating Wall'. In: J. L. Burch and J. J. Hunter Waite (eds.): *Solar System Plasmas in Space and Time*, Vol. 84 of *Geophysical Monograph Series*. Washington, D. C., USA: American Geophysical Union, p. 163.

Knudsen, D. J., B. A. Whalen, A. W. Yau, M. J. Greffen, A. I. Eriksson, N. Lloyd, M. Boehm, J. Clemmons, and L. G. Blomberg: 1994b, 'Sub-kilometer thermal plasma structure near 1750 km altitude in the polar cusp/cleft'. *Geophys. Res. Lett.* **21**, 1907.

Koechlin, F. and A. Samain: 1971, 'New method of magnetic pumping for plasma heatin'. *Phys. Rev. Lett.* **26**, 490.

Koepke, M. E., J. J. Carroll III, and M. W. Zintl: 1999, 'Laboratory simulation of broadband ELF waves in the auroral ionosphere'. *J. Geophys. Res.* **104**, 14397.

Kornilova, T. A., M. I. Pudovkin, and G. V. Starkov: 1990, 'Fine auroral structure near the poleward boundary of the auroral bulge during breakup phase'. *Geomagn. Aeron.* **30**, 250.

Koskik, J.-C.: 1998, 'Quantitative magnetic field model including magnetospheric ring current'. *J. Geophys. Res.* **94**, 12021.

Koskinen, H., R. Lopez, R. Pellinen, T. Pulkkinen, D. Baker, and T. Bösinger: 1993, 'Pseudobreakup and substorm growth phase in the ionosphere and magentosphere'. *J. Geophys. Res.* **98**, 5801.

Koskinen, H. E. J. and T. I. Pulkkinen: 1995, 'Midnight velocity shear zone and the concept of Harang discontinuity'. *J. Geophys. Res.* **100**, 9539.

Kovrazhkin, R., J.-M. Bosqued, L. Zelenyi, and N. Jorjio: 1987, 'Reconnection and plasma acceleration evidences manifestation at 0.5 million km distance in the Earth's magnetic tail'. *Letters to ZETP (in Russian)* **45**, 377.

Kozyra, J., A. Nagy, and D. Slater: 1997, 'High-altitude energy source(s) for stable auroral red arcs'. *Rev. Geophys.* **35**, 155.

Krall, A. and B. Trivelpiece: 1973, *Principles of Plasma Physics*. McGraw-Hill.

Krimigis, S., T. Armstrong, W. Axford, C. Bostrom, A. Cheng, G. Gloeckler, D. Hamilton, E. Keath, L. Lanzerotti, B. Mauk, and J. Van Allen: 1989, 'Hot plasma and energetic particles in Neptune's magnetosphere'. *Science* **24**, 1483.

Kullen, A.: 1999, 'The connection between transpolar arcs and magentotail rotation'. *Geophys. Res. Lett.* **27**, 73.

Kumamoto, A. and H. Oya: 1998, 'Asymmetry of occurrence-frequency and intensity of AKR between summer polar region and winter polar region sources'. *Geophys. Res. Lett.* **25**, 2369.

Kunitake, M. and T. Oguti: 1984, 'Spatial-temporal characteristics of flickering spots in flickering auroras'. *J. Geophys. Geomagn.* **36**, 121.

Kurth, W. S., M. M. Baumback, and D. A. Gurnett: 1975, 'Direction-finding measurements of auroral kilometric radiation'. *J. Geophys. Res.* **80**, 2764.

LaBelle, J. and P. M. Kintner: 1989, 'The measurement of wavelength in space plasmas'. *Review of Geophysics* **27**, 495.

LaBelle, J., P. M. Kintner, A. W. Yau, and B. A. Whalen: 1986, 'Large Amplitude Wave Packets Observed in the Ionosphere in Association With Transverse Ion Acceleration'. *J. Geophys. Res.* **91**, 7113.

LaBelle, J., K. L. McAdams, and M. L. Trimpi: 1999, 'High-frequency and time resolution rocket observations of structured low- and medium-frequency whistler mode emissions in the auroral ionosphere'. *J. Geophys. Res.* **104**, 28101.

LaBelle, J. and R. A. Treumann: 2002, 'Auroral radio emissions, 1. Hisses, Roars and Bursts'. *Space Sci. Rev.* p. in press.

Langmuir, I.: 1929, 'The interaction of electron and positive ion space charges in cathode sheaths'. *Phys. Rev.* **33**, 954.

Lassen, K. and C. Danielsen: 1978, 'Quiet time pattern of auroral arcs for different directions of the interplanetary magnetic field in the Y-Z plane'. *J. Geophys. Res.* **83**, 5277.

Lassen, K. and C. Danielsen: 1989, 'Distribution of auroral arcs during quiet geomagnetic conditions'. *J. Geophys. Res.* **94**, 2587.

Le Quéau, D. R., R. Pellat, and A. Roux: 1984, 'Direct generation of the auroral kilometric radiation by the maser synchroton instability: Physical discussion of the mechanism and parametric study'. *J. Geophys. Res.* **89**, 2841.

Lee, D.-H. and R. L. Lysak: 1991, 'Impulsive excitation of ULF waves in the three-dimensional dipole model: the initial results'. *J. Geophys. Res.* **96**, 3479.

Lee, D.-H. and R. L. Lysak: 1994, 'Numerical studies on ULF wave structures in the dipole model'. In: M. Engebretson, K. Takahashi, and M. Scholer (eds.): *Solar Wind Sources of Magnetospheric Ultra-Low-Frequency Waves*, Vol. 81 of *AGU Monograph Series*. Washington, D. C.: American Geophysical Union, p. 293.

Lee, K. F.: 1972, 'Ion cyclotron instability in current-carrying plasmas with anisotropic temperatures'. *J. Plasma. Phys.* **8**, 379.

Lessard, M. R. and D. J. Knudsen: 2001, 'Ionospheric Reflection of Small-Scale Alfvén Waves'. *Geophys. Res. Lett.* **28**, 3573.

Li, X., I. Roth, M. Temerin, J. Wygant, M. K. Hudson, and J. B. Blake: 1993, 'Simulation of the prompt energization and transport of radiation belt particles during the March 23,1991 SSC'. *Geophys. Res. Lett.* **20**, 2423.

Li, X. and M. Temerin: 1992, 'Ponderomotive effects on ion acceleration in the auroral zone'. *Geophys. Res. Lett.* **25**, 4449.

Liou, K., P. Newell, C.-I. Meng, M. Brittnacher, and G. Parks: 1997, 'Synoptic auroral distribution: A survey using Polar ultraviolet imagery'. *J. Geophys. Res.* **102**, 27197.

Liou, K., P. Newell, C.-I. Meng, T. Sotirelis, M. Brittnacher, and P. G: 1999, 'Source region of 1500 MLT auroral bright spots: Simultaneous Polar UVI images and DMSP particle data'. *J. Geophys. Res.* **104**.

Lockwood, M. and S. Cowley: 1999, 'Comment on "A statistical study of the ionospheric convection response to changing interplanetary magnetic field conditions using the assimilative mappic of ionospheric electrodynamics technique" by A.J. Ridley et al.'. *J. Geophys. Res.* **104**, 4387.

Lockwood, M. and M. F. Smith: 1992, 'The variation of reconnection rate at the dayside magnetopause and cusp ion precipitation'. *J. Geophys. Res.* **97**, 14841.

Lockwood, M., J. Waite, J. H., T. E. Moore, C. R. Chappell, and M. O. Chandler: 1985, 'The cleft ion fountain'. *J. Geophys. Res.* **90**, 9736.

Lonnqvist, H., M. Andre, L. Matson, A. Bahnsen, L. G. Blomberg, and R. E. Erlandson: 1993, 'Generation of VLF saucer emissions observed by the Viking satellite'. *J. Geophys. Res.* **98**, 13565.

Lorentzen, D., C. S. Deehr, J. I. Minow, R. W. Smith, H.-. Nielsen, F.Sigernes, R. Arnoldy, and K. Lynch: 1996, 'SCIFER-Dayside auroral signature of magnetosphericenergetic electrons'. *Geophys. Res. Lett.* **23**, 1885.

Lotko, W.: 1986, 'Diffusive acceleration of auroral primaries'. *J. Geophys. Res.* **91**, 191.

Lotko, W., B. Sonnerup, and R. L. Lysak: 1987, 'Nonsteady boundary layer flow including ionospheric drag and parallel electric fields'. *J. Geophys. Res.* **92**, 8635.

Lotko, W., A. Streltsov, and C. Carlson: 1998, 'Discrete auroral arc, electrostatic shock and suprathermal electrons powered by dispersive, anomalously resistive field line resonance'. *Geophys. Res. Lett.* **25**, 4449.

Louarn, P., A. Roux, H. de Feraudy, D. L. Quau, M. Andr, and L. Matson: 1990, 'Trapped electrons as a free energy source for the auroral kilometric radiation'. *J. Geophys. Res.* **95**, 5983.

Louarn, P., J.-E. Wahlund, T. Chust, H. D. Feraudy, A. Roux, B. Holback, P. O. Dovner, A. I. Erikson, and G. Holmgren: 1994, 'Observations of kinetic Alfven waves by the Freja spacecraft'. *Geophys. Res. Lett.* **21**, 1847.

Lu, G., D. Baker, R. McPherron, C. Farrugia, D. Lummerzheim, J. Ruohoniemi, F. Rich, D. Evans, R. Lepping, M. Brittnacher, X. Li, R. Greenwald, G. Sofko, J. Villain, M. Lester, J. Thayer, T. Moretto, D. Milling, O. Troshichev, A. Zaitzev, V. Odintzov, G. Makarov, and K. Hayashi: 1998, 'Global energy deposition during the January 1997 magnetic cloud event'. *J. Geophys. Res.* **103**, 11685.

Lu, G., M. Brittnacher, G. Parks, and D. Lummerzheim: 2000, 'On the magnetospheric source regions of substorm-related field-aligned currents and auroral precipitation'. *J. Geophys. Res.* **105**, 18483.

Lu, G., A. Richmond, B. Emery, and R. Roble: 1995, 'Magnetosphere-ionosphere-thermosphere coupling: Effect of neutral winds on energy transfer and field-aligned current'. *J. Geophys. Res.* **100**, 19643.

Lu, G., A. D. Richmond, B. A. Emery, P. H. Reiff, O. de la Beaujardiere, F. J. Rich, W. F. Denig, H. W. Kroehl, L. R. Lyons, J. M. Ruohoniemi, E. Friis-Christensen, H. Opgenoorth, M. A. L. Persson, R. P. Lepping, A. S. Rodger, T. Hughes, A. McEwin, S. Dennis, R. Morris, G. Burns, and L. Tomlison: 1994, 'Interhemispheric asymmetry of the high-latitude ionospheric convection pattern'. *J. Geophys. Res.* **99**.

Lu, G., G. Siscoe, A. Richmond, T. Pulkkinen, N. Tsyganenko, H. Singer, and B. Emery: 1997, 'Mapping of the ionospheric field-aligned currents to the equatorial magnetosphere'. *J. Geophys. Res.* **102**, 14467.

Lu, G., N. A. Tsyganenko, A. T. Y. Lui, H. J. Singer, T. Nagai, and S. Kokubun: 1999, 'Modeling of time-evolving magnetic fields during substorms'. *J. Geophys. Res.* **104**, 12327.

Lühr, H., A. Aylward, S. Buchert, A. Pajunpää, K. Pajunpää, T. Holmboe, and S. M. Zalewski: 1998, 'Westward moving dynamic substorm features observed with the IMAGE magnetometer network and other ground-based instruments'. *Ann. Geophysicae* **16**, 425.

Lühr, H. and K. Schlegel: 1994, 'Combined measurements of EISCAT and the EISCAT magnetometer cross to study omega bands'. *J. Geophys. Res.* **99**, 8951.

Lui, A.: 1996, 'Current disruption in the Earth's magnetosphere: Observations and models'. *J. Geophys. Res.* **101**, 13067.

Lui, A. and C. Anger: 1973, 'A uniform belt of diffuse auroral emission seen by the ISIS-2 scanning photometer'. *Planet. Space Sci.* **21**, 799.

Lui, A., D. Williams, R. M. Entire, S. Ohtani, L. Zanetti, W. Bristow, R. Greenwald, P. Newell, S. Christon, T. Mukai, K. Tsuruda, Y. Yamamoto, S. Kokobun, H. Matsumoto, H. Kojima, T. Murata, D. Fairfield, R. Lepping, J. Samson, G. Rostoker, G. Reeves, A. Rodger, and H. Singer: 1998, 'Multipoint study of a substorm on February 9, 1995'. *J. Geophys. Res.* **103**, 17333.

Lund, E. J., J. LaBelle, R. B. Torbert, K. Liou, W. Peria, C. A. Kletzing, M. C. Kelley, S. D. Baker, F. Primdahl, H. C. Stenbaek-Nielsen, A. Ranta, G. Haerendel, and H. U. Frey: 1995, 'Observation of electromagnetic oxygen cyclotron waves in a flickering aurora'. *Geophys. Res. Lett.* **22**, 2465.

Lund, E. J., E. Möbius, L. Tang, L. M. Kistler, M. A. Popecki, D. M. Klumpar, W. K. Peterson, E. G. Shelley, B. Klecker, D. Hovestadt, M. Temerin, R. E. Ergun, J. P. McFadden, C. W. Carlson, F. S. Mozer, R. C. Elphic, R. J. Strangeway, C. A. Cattell, and R. F. Pfaff: 1998, 'FAST observations of preferentially accelerated He^+ in association with auroral electromagnetic ion cyclotron waves'. *Geophys. Res. Lett.* **25**, 2049.

Lund, E. J., E. Möbius, C. W. Carlson, R. E. Ergun, L. M. Kistler, B. Klecker, D. M. Klumpar, J. P. McFadden, M. A. Popecki, R. J. Strangeway, and Y. K. Tung: 2000, 'Transverse ion accereration mechanisms in the aurora at solar minimum: occurrence distributions'. *J. Atmos. Terr. Phys.* **62**, 467.

Lund, E. J., E. Möbius, R. E. Ergun, and C. W. Carlson: 1999, 'Mass-dependent effects in ion conic production: the role of parallel electric fields'. *Geophys. Res. Lett.* **26**, 3593.

Lundin, R., A. Zakharov, R. Pellinen, H. Borg, B. Hultqvist, N. Pissarenko, and E. Dubinin: 1989, 'First composition measurements of the ionospsheric plasma escape from Mars'. *Nature* **34**, 609.

Lynch, K. A., R. L. Arnoldy, J. Bonnell, and P. M. Kintner: 1996, 'The AMICIST Auroral sounding rocket—A comparison of transverse ion acceleration methods'. *Geophys. Res. Lett.* **23**, 3293.

Lynch, K. A., R. L. Arnoldy, P. M. Kintner, P. Schuck, J. Bonnell, and V. Coffey: 1999a, 'Auroral Ion Acceleration from Lower Hybrid Solitary Structures: A Summary of Sounding Rocket Observations'. *J. Geophys. Res.* **104**, 28515.

Lynch, K. A., R. L. Arnoldy, P. M. Kintner, and J. L. Vago: 1994, 'Electron Distribution Function Behavior During localized transverse ion acceleration events in the topside auroral zone'. *J. Geophys. Res.* **99**, 2227.

Lynch, K. A., J. W. Bonnell, C. W. Carlson, and W. J. Peria: 2002, 'Return current region aurora: $E_{\|}$, j_z, particle energization and BBELF wave activity'. *J. Geophys. Res.* **107**, submitted.

Lynch, K. A., D. Pietrowski, R. B. Torbert, N. Ivchenko, G. Marklund, and F. Primdahl: 1999b, 'Multiple-point electron measurements in a nightside auroral arc: Auroral Turbulence II particle observations'. *Geophys. Res. Lett.* **26**, 3361.

Lyons, L.: 1974, 'Electron diffusion driven by magnetospheric electrostatic waves'. *J. Geophys. Res.* **79**, 575.

Lyons, L. and J. Fennell: 1986, 'Characteristics of auroral electron precipitation on the morningside'. *J. Geophys. Res.* **91**, 11225.

Lyons, L. and D. Williams: 1984, *Quantitative Aspects of Magnetospheric Physics*. Dordrecht, Holland: D. Reidel Pub. Co.

Lyons, L. R.: 1980, 'Generation of large-scale regions of auroral currents, electric potentials, and precipitation by the divergence of the convection electric field'. *J. Geophys. Res.* **85**, 17.

Lyons, L. R.: 1981, 'The field-aligned current versus electric potential relation and auroral electrodynamics'. In: *Physics of Auroral Arc Formation*, Vol. 25. Washington D.C.: American Geophysical Union, p. 252.

Lyons, L. R.: 1985, 'A simple model for polar cap convection patterns and generation of θ-aurora'. *J. Geophys. Res.* **90**, 1561.

Lyons, L. R., H. E. J. Koskinen, J. B. Blake, A. Egeland, M. Hirahara, M. Øieroset, P. E. Sandholt, and K. Shiokawa: 1999a, 'Processes leading to plasma losses into the high-latitude atmosphere'. In: B. Hultqvist, M. Øieroset, G. Paschmann, and R. Treumann (eds.): *Magnetospheric Plasma Sources and Losses*, Vol. 6 of *Space Science Series of ISSI*. Dordrecht, Boston and London: Kluwer Academic Publishers, Chapt. 3, p. 85.

Lyons, L. R., G. Lu, O. de la Beaujardière, and F. J. Rich: 1996, 'Synoptic maps of polar caps for stable interplanetary magnetic field intervals during January 1992 geospace environment modeling campaign'. *J. Geophys. Res.* **101**, 27283.

Lyons, L. R., T. Nagai, G. T. Blanchard, J. C. Samson, T. Yamamoto, T. Mukai, A. Nishida, and S. Kokubun: 1999b, 'Association between Geotail plasma flows and auroral poleward boundary intensifications observed by CANOPUS photometers'. *J. Geophys. Res.* **104**, 4485.

Lyons, L. R., T. Nagai, G. T. Blanchard, J. C. Samson, T. Yamamoto, T. Mukai, A. Nishida, and S. Kokubun: 1999c, 'Association between Geotail plasma flows and auroral poleward boundary intensifications observed by CANOPUS photometers'. *J. Geophys. Res.* **104**, 4485.

Lysak, R.: 1990, 'Electrodynamic coupling of the magnetosphere and ionosphere'. *Space Sci. Rev.* **52**, 33.

Lysak, R.: 1991, 'Feedback Instability of the Ionospheric Resonant Cavity'. *J. Geophys. Res.* **96**, 1553.

Lysak, R.: 1993, 'Generalized model of the ionospheric Alfvén resonator'. In: R. Lysak (ed.): *Auroral Plasma Dynamics*, Vol. 80 of *Geophysical Monograph Series*. Washington, DC: American Geophysical Union, p. 121.

Lysak, R.: 1998, 'The relationship between electrostatic shocks and kinetic Alfvń waves'. *Geophys. Res. Lett.* **25**, 2089.

Lysak, R. and C. W. Carlson: 1981, 'Effect of microscopic turbulence on magnetosphere-ionosphere coupling'. *Geophys. Res. Lett.* **8**, 269.

Lysak, R. and M. Hudson: 1979, 'Coherent anomalous resistivity in the region of electrostatic shocks'. *Geophys. Res. Lett.* **6**, 661.

Lysak, R. and M. K. Hudson: 1987, 'Effect of double layers on magnetosphere-ionosphere coupling'. *Lasers and Particle Beams* **5**, 351.

Lysak, R., M. K. Hudson, and M. Temerin: 1980, 'Ion heating by strong electrostatic ion cyclotron turbulence'. *J. Geophys. Res.* **85**, 678.

Lysak, R. and W. Lotko: 1996, 'On the kinetic dispersion relation for shear Alfvn waves'. *J. Geophys. Res.*. **101**, 5085.

Lysak, R. and Y. Song: 1996, 'Coupling of Kelvin-Helmholtz and current-sheet instabilities to the ionosphere: A dynamic theory of auroral spirals'. *J. Geophys. Res.* **101**, 15411.

Lysak, R. L.: 1985, 'Auroral electrodynamics with current and voltage generators'. *J. Geophys. Res.*. **90**, 4178.

Lysak, R. L. and C. T. Dum: 1983, 'Dynamics of magnetosphere-ionosphere coupling including turbulent transport'. *J. Geophys. Res.* **88**, 365.

Lysak, R. L. and Y. Song: 2000, 'A Three-Dimensional Model of the Propagation of Alfvn Waves Through the Auroral Ionosphere: First Results'. Technical Report UMSI 2000/206, University of Minnesota Supercomputing Institute.

Möbius, E., L. Tang, L. M. Kistler, M. Popecki, E. J. Lund, D. Klumpar, W. Peterson, E. G. Shelley, B. Klecker, D. Hovestadt, C. W. Carlson, R. Ergun, J. P. McFadden, F. Mozer, M. Temerin, C. Cattell, R. Elphic, R. Strangeway, and R. Pfaff: 1998, 'Species-dependent energies in upward-directed ion beams over auroral arcs as observed with FAST TEAMS'. *Geophys. Res. Lett.* **25**, 2029.

Maggs, J. and T. Davis: 1968, 'Measurements of the thickness of auroral structures'. *Planet. Space Sci.* **1**, 205.

Maggs, J. E.: 1976, 'Coherent Generation of VLF Hiss'. *J. Geophys. Res.* **81**, 1707.

Mäkelä, J., A. Mälkki, H. Koskinen, M. Boehm, B. Holback, and L. Eliasson: 1998, 'Observations of meso-scale auroral plasma cavity crossings with the Freja satellite'. *J. Geophys. Res.* **103**, 9391.

Makita, K., C.-I. Meng, and S.-I. Akasofu: 1991, 'Transpolar aurora their particle precipitation and IMF B_y component'. *J. Geophys. Res.* **96**, 14085.

Mälkki, A., A. I. Eriksson, P. Dovner, R. Bostrom, B. Holback, G. Holmgren, and H. E. J. Koskinen: 1993, 'A statistical survey of auroral solitary waves and weak double layers. I - Occurrence and net voltage'. *J. Geophys. Res.* **98**, 15521.

Mallinckrodt, A. J. and C. W. Carlson: 1978, 'Relations between transverse electric fields and field-aligned currents'. *J. Geophys. Res.*. **83**, 1426.

Mann, I. R., A. N. Wright, and P. S. Cally: 1995, 'Coupling of magnetospheric cavity modes to field line resonances: a study of resonant widths'. *J. Geophys. Res.* **100**, 19441.

Marklund, G.: 1984, 'Auroral arc classification scheme based on the observed arc-associated electric field patterns'. *Planet. Space Sci.* **32**, 193.

Marklund, G., L. Blomberg, C.-C. Fälthammar, and P.-A. Lindqvist: 1994, 'On intense diverging electic fields associated with black aurora'. *Geophys. Res. Lett.* **21**, 1859.

Marklund, G., T. Karlsson, L. Blomberg, P.-A. Lindqvist, C.-G. Fälthammar, M. Johnson, J. Murphree, L. Andersson, L. Eliasson, H. Opgenoorth, and L. Zanetti: 1998, 'Observations of the electric field fine structure associated with the westward traveling surge and large-scale auroral spirals'. *J. Geophys. Res.* **103**, 4125.

Marklund, G., T. Karlsson, and J. Clemmons: 1997, 'On low altitude particle acceleration and intense electric fields and their relationship to black aurora'. *J. Geophys. Res.* **102**, 17509.

Marklund, G. T.: 1993, 'Viking investigations of auroral electrodynamic parameters'. *J. Geophys. Res.* **98**, 1691.

Marklund, G. T. and L. G. Blomberg: 1991, 'On the influence of localized electric fields and field-aligned currents associated with polar arcs on the global potential distribution'. *J. Geophys. Res.* **96**, 13977.

Marklund, G. T., L. G. Blomberg, J. S. Murphree, R. D. Elphinstone, L. J. Zanetti, R. E. Erlandson, I. Sandahl, O. de la Beaujardiére, H. Opgenoorth, and F. J. Rich: 1991, 'On the electrodynamical state of the auroral ionosphere during northward interplanetary magnetic field: A transpolar arc case study'. *J. Geophys. Res.* **96**, 9567.

Marklund, G. T., N. Ivchenko, T. Karlsson, A. Fazakerley, M. Dunlop, P.-A. Lindqvist, S. Buchert, C. Owen, M. Taylor, A. Vaivads, P. Carter, M. André, and A. Balogh: 2001, 'Temporal evolution of acceleration structures in the auroral return current region'. *Nature* **414**, 724.

Marklund, G. T., I. Sandahl, and H. Opgenoorth: 1982, 'A study of the dynamics of a discrete auroral arc'. *Planet. Space Sci.* **30**, 179.

Mathie, R. A. and I. R. Mann: 2000, 'Observations of anomalously low frequency Alfvén continuum in an abnormally expanded magnetosphere'. *Geophys. Res. Lett.* **27**, 4017.

Matsumoto, H., H. Kojima, T. Miyatake, Y. Omura, M. Okada, I. Nagano, and M. Tsutsui: 1994, 'Electrostatic solitary waves (ESW) in the magnetotail: BEN wave forms observed by GEOTAIL'. *Geophys. Res. Lett.* **21**, 2915.

Mauk, B., S. Krimigis, E. Keath, A. Cheng, T. Armstrong, L. Lanzerotti, G. Gloeckler, and D. Hamilton: 1987, 'The hot plasma and radiation environment of the Uranian magnetosphere'. *J. Geophys. Res.* **92**, 15283.

Mawson, D. and C. Chree: 1925, 'Records of the aurora polaris, Vol 2: Magnetism and Aurora'. In: *Australasian Antarctic Expedition, 1911-1914*, Scientific Reports Series B. A. J. Kent, Sydney, Australia, Govt. Printer, p. 11.

McAdams, K. L., J. LaBelle, M. L. Trimpi, P. M. Kintner, and R. A. Arnoldy: 1999, 'Rocket observations of banded structure in waves near the Langmuir frequency in the auroral ionosphere'. *J. Geophys. Res.* **104**, 28109.

McCrea, I. W. and M. Lockwood: 1997, 'Incoherent Scatter Radars'. In: *Satellite-Ground Based Coordination Sourcebook*. p. 239.

McDiarmid, I. B., J. R. Burrows, and M. D. Wilson: 1979, 'Large-scale magnetic field perturbations and particle measurements at 1400 km on the dayside'. *J. Geophys. Res.* **84**.

McFadden, J. P., C. W. Carlson, and M. H. Boehm: 1986, 'Field-aligned electron precipitation at the edge of an arc'. *J. Geophys. Res.* **91**, 1723.

McFadden, J. P., C. W. Carlson, and M. H. Boehm: 1990, 'Structure of an energetic narrow discrete arc'. *J. Geophys. Res.* **95**, 6533.

McFadden, J. P., C. W. Carlson, M. H. Boehm, and T. J. Hallinan: 1987, 'Field-aligned electron flux oscillations that produce flickering aurora'. *J. Geophys. Res.* **92**, 11133.

McFadden, J. P., C. W. Carlson, and R. E. Ergun: 1999a, 'Microstructure of the auroral acceleration region as observed by FAST'. *J. Geophys. Res.* **104**, 14453.

McFadden, J. P., C. W. Carlson, R. E. Ergun, D. M. Klumpar, and E. Möbius: 1999b, 'Ion and electron characteristics in auroral density cavities associated with ion beams: No evidence for cold ionospheric plasma'. *J. Geophys. Res.* **104**, 14671.

McFadden, J. P., C. W. Carlson, R. E. Ergun, F. S. Mozer, L. Muschiette, I. Roth, and E. Möbius: 2002, 'FAST Observations of Ion Solitary Waves'. *J. Geophys. Res.* **107**, submitted.

McIlwain, C.: 1960, 'Direct measurement of particles producing visible aurora'. *J. Geophys. Res.* **65**, 2727.

McIlwain, C.: 1961, 'Coordinates for Mapping the Distribution of Magnetically Trapped Particles'. *J. Geophys. Res.* **66**, 3681.

McIlwain, C. E.: 1975, 'Auroral electron beams near the magnetic equator'. In: *Physics of Hot Plasma in the Magnetosphere*. New York: Plenum, p. 91.

McPherron, R.: 1997, 'The role of substorms in the generation of magnetic storms'. In: B. Tsurutani, W. Gonzalez, Y. Kamide, and J. Arballo (eds.): *Magnetic storms, Geophysical Monograph 98*. AGU, Washington, D.C., p. 131.

McPherron, R. L., A. Nishida, and C. T. Russell: 1987, 'Is near-Earth current sheet thinning the cause of auroral substorm expansion onset? (extended abstract)'. In: Y. Kamide and R. A. Wolf (eds.): *Quantitative Modelling of Magnetosphere Ionosphere Coupling Processes*. Kyoto Sangyo Univ., Kyoto, Japan, p. 252.

McPherron, R. L., C. T. Russell, and M. A. Aubry: 1973, 'Satellite studies of magnetospheric substorms on August 15, 1968, 9, Phenomenological model for substorms'. *J. Geophys. Res.* **78**, 3131.

Melrose, D. B., K. G. Rönmark, and R. G. Hewitt: 1982, 'Terrestrial kilometric radiation: The cyclotron theory'. *J. Geophys. Res.* **87**, 5140.

Meng, C.-I.: 1981, 'Electron precipitation in the midday auroral oval'. *J. Geophys. Res.* **86**, 2149.

Meng, C.-I. and R. Lundin: 1986, 'Auroral morphology of the midday oval'. *J. Geophys. Res.* **91**, 1572.

Meng, C.-I., B. Mauk, and C. McIlwain: 1979, 'Electron precipitation of evening diffuse aurora and its conjugate electron fluxes near the magnetic equator'. *Geophys. Research* **84**, 2545.

Meng, C.-I., A. Snyder Jr., and H. Kroehl: 1978, 'Observations of auroral westward traveling surges and electron precipitations'. *J. Geophys. Res.* **83**, 575.

Menietti, J. and J. Burch: 1987, 'DE 1 observations of theta aurora plasma source regions and Birkeland current charge carriers'. *J. Geophys. Res.* **92**, 7503.

Mitchell, C., J. Maggs, S. Vincena, and W. Gekelman: 2002, 'A laboratory investigation of Alfvén wave refraction in parallel β gradient'. *J. Geophys. Res.* p. in press.

Miyake, T., Y. Omura, H. Matsumoto, and H. Kojima: 1998, 'Two-dimensional computer simulations of electrostatic solitary waves observed by Geotail spacecraft'. *J. Geophys. Res.* **103**, 11841.

Miyake, W., T. Mukai, and N. Kaya: 2000, 'Interplanetary magnetic field control of dayside ion conics'. *J. Geophys. Res.* **105**, 23339.

Moffatt, H. K.: 1969, 'The degree of knottedness of tangled vortex lines'. *J. Fluid Mech.* **35**, 117.

Montgomery, D. C.: 1971, *Theory of Unmagnetized Plasma*. New York: Gordon and Breach.

Montgomery, D. C. and G. Joyce: 1969, 'Shock-like solutions of the electrostatic Vlasov equation'. *Journal of Plasma Physics* **3**, 1.

Moore, T. E., R. Lundin, D. Alcayde, S. B. André, Ganguli, M. Temerein, and A. Yau: 1999, 'Source and loss processes in the inner magnetosphere'. In: B. Hultqvist, M. Øieroset, G. Paschmann, and R. Treumann (eds.): *Magnetospheric Plasma Sources and Losses*, Vol. 6 of *Space Science Series of ISSI*. Dordrecht, Boston and London: Kluwer Academic Publishers, Chapt. 2, p. 137.

Mozer, F. S., C. W. Carlson, M. K. Hudson, R. B. Torbert, B. Parady, J. Yatteau, and M. C. Kelley: 1977, 'Observations of paired electrostatic shocks in the polar magnetosphere'. *Phys. Rev. Lett.* **38**, 292.

Mozer, F. S., C. A. Catell, M. Temerin, R. B. Torbert, S. Von Glinski, M. Wolddorf, and J. Wygant: 1979, 'The DC and AC electric field, plasma density, plasma temperature, and field aligned current experiments on the S3-3 satellite'. *J. Geophys. Res.* **84**, 5875.

Mozer, F. S., C. A. Cattell, M. K. Hudson, R. L. Lysak, M. Temerin, and R. B. Torbert: 1980, 'Satellite measurements and theories of low altitude auroral particle acceleration'. *Space Sci. Rev.* **27**, 155.

Mozer, F. S., R. E. Ergun, M. Temerin, C. Cattell, J. Dombeck, and J. Wygant: 1997, 'New features of time domain electric-field structures in the auroral acceleration region'. *Phys. Rev. Lett.* **79**, 1281.

Mozer, F. S. and A. Hull: 2001, 'Origin and geometry of upward parallel electric fields in the auroral acceleration region'. *J. Geophys. Res.* **106**, 5763.

Mozer, F. S. and C. A. Kletzing: 1998, 'Direct observation of large, quasi-static, parallel electric fields in the auroral acceleration region'. *Geophys. Res. Lett.* **25**, 1629.

Mozer, F. S. and R. B. Torbert: 1980, 'An average parallel electric field deduced from the latitude and altitude variations of the perpendicular electric field below 8000 kilometers'. *Geophys. Res. Lett.* **7**, 219.

Mravlag, E., M. Scourfield, A. Walker, P. Stucliffe, and E. Nielsen: 1991, 'Simultaneous observations of omega band related phenomena in both hemispheres'. *J. Atmos. Terr. Phys.* **53**, 309.

Murphree, J., R. Elphinstone, L. Cogger, and D. Wallis: 1989, 'Short-term dynamics of the high-latitude auroral distribution'. *J. Geophys. Res.* **94**, 6969.

Murphree, J. S., L. L. Cogger, C. D. Anger, D. D. Wallis, and G. G. Shepherd: 1987, 'Oval intensifications associated with polar arcs'. *Geophys. Res. Lett.* **14**, 403.

Muschietti, L., R. E. Ergun, I. Roth, and C. W. Carlson: 1999a, 'Correction to 'Phase-space electron holes along magnetic field lines''. *Geophys. Res. Lett.* **26**, 1093.

Muschietti, L., R. E. Ergun, I. Roth, and C. W. Carlson: 1999b, 'Phase-space electron holes along magnetic field lines'. *Geophys. Res. Lett.* **26**, 1093.

Muschietti, L., I. Roth, C. W. Carlson, and R. E. Ergun: 2000, 'Transverse Instability of Magnetized Electron Holes'. *Phys. Rev. Lett.* **85**, 94.

Muschietti, L., I. Roth, R. E. Ergun, and C. W. Carlson: 1999c, 'Analysis and simulation of BGK electron holes'. *Nonlinear Processes in Geophysics* **6**, 211.

Nagai, T, M. F., Y. Saito, S. Machida, T. Terasawa, R. Nakamura, T. Yamamoto, T. Mukai, A. Nishida, and S. Kokubun: 1998, 'Structure and dynamics of magnetic reconnection for substorm onsets with Geotail observations'. *J. Geophys. Res.* **103**, 4419.

Nakamura, R., D. Baker, T. Yamamoto, R. Belian, E. Bering, J. Benbrook, and J. Theal: 1994, 'Particle and field signatures during pseudobreakup and major expansion onset'. *J. Geophys. Res.* **99**, 207.

Nakamura, R., W. Baumjohann, M. Brittnacher, V. A. Sergeev, M. Kubyshkina, T.Mukai, and K. Liou: 2001a, 'Flow Bursts and Auroral Activations: Onset Timing and Foot Point Location'. *J. Geophys. Res.* **106**, 10777.

Nakamura, R., W. Baumjohann, R. Schödel, M. Brittnacher, V. A. Sergeev, M. Kubyshkina, T. Mukai, and K. Liou: 2001b, 'Earthward Flow Bursts, Auroral Streamers, and Small Expansions'. *J. Geophys. Res.* **106**, 10791.

Ness, N., M. Acuna, K. Behannon, L. Burlaga, J. Connerney, R. Lepping, and F. Neubauer: 1986, 'Magnetic field of Uranus'. *Science* **23**, 85.

Ness, N., M. Acuna, L. Burlaga, J. Connerney, R. Lepping, and F. Neubauer: 1989, 'Magnetic fields at Neptune'. *Science* **24**, 1473.

Ness, N., M. Acuna, R. Lepping, J. Connerney, K. Behannon, L. Burlaga, and F. Neubauer: 1981, 'Magnetic field studies by Voyager 1: Preliminary results at Saturn'. *Science* **21**, 211.

Nevanlinna, H. and T. I. Pulkkinen: 1998, 'Solar cycle correlations of substorm and auroral occurrence frequencies'. *Geophys. Res. Lett.* **25**, 3087.

Newell, P., Y. Feldstein, Y. Galperin, and C.-I. Meng: 1996a, 'The morphology of nightside precipitation'. *J. Geophys. Res.* **101**, 10737.

Newell, P., K. Liou, C.-I. Meng, M. Brittnacher, and G. Parks: 1999, 'Dynamics of double-theta aurora: Polar UVI study of January 10-11'. *J. Geophys. Res.* **104**, 95.

Newell, P., K. Lyons, and C. Meng: 1996b, 'A large survey of electron acceleration events'. *J. Geophys. Res.* **101**, 2599.

Newell, P., C. Meng, and S. Wing: 1998, 'Relation to solar activity of intense aurorae in sunlight and in darkness'. *Nature* **393**, 342.

Newell, P. and C.-I. Meng: 1994, 'Ionospheric projections of magnetospheric regions under low and high solar wind pressure conditions'. *J. Geophys. Res.* **99**, 273.

Newell, P. and C.-I. Meng: 1995, 'Cusp low-energy ion cutoffs: A survey and implications for merging'. *J. Geophys. Res.* **100**, 21943.

Newell, P., C.-I. Meng, and K. Lyons: 1996c, 'Suppression of discrete aurora by sunlight'. *Nature* **38**, 766.

Newell, P. T.: 2000, 'Reconsidering the inverted-V particle signature: Relative frequency of large-scale electron acceleration events'. *J. Geophys. Res.* **105**, 15779.

Newman, D. L., M. V. Goldman, M. Spector, and F. Perez: 2001, 'Dynamics and instability of electron phase-space tubes'. *Phys. Rev. Lett.* **86**, 1239.

Nicholson, D. R.: 1983, *Introduction to Plasma Theory*. New York: J. Wiley and Sons.

Nishida, A. and T. Ogino: 1998, 'Convection and reconnection in the magnetotail'. In: *New perspectives on magnetotail dynamics*, Vol. 105 of *Geophysical Monograph Series*. Washington, D.C., USA: American Geophysical Union, p. 61.

Norqvist, P., M. André, L. Eliasson, A. I. Eriksson, L. Blomberg, H. Lühr, and J. H. Clemmons: 1996, 'Ion cyclotron heating in the dayside magnetosphere'. *J. Geophys. Res.* **101**, 13179.

Norqvist, P., M. André, and M. Tyrland: 1998, 'A statistical study of ion energization mechanisms in the auroral region'. *J. Geophys. Res.* **103**, 23459.

Northrop, T. G.: 1963, *The Adiabatic Motion of Charged Particles*. New York: Wiley Interscience.

Obara, T., M. Kitayama, T. Mukai, N. Kaya, J. S. Murphree, and L. L. Cogger: 1988, 'Simultaneous observations of sun-aligned polar cap arcs in both hemispheres by EXOS-C and Viking'. *Geophys. Res. Lett.* **15**, 713.

Obara, T., T. Mukai, H. Hayakawa, A. Matsuoka, K. Tsuruda, K. Fukui, J. V. Rodriguez, and C. E. Valladares: 1996a, 'Simultaneous satellite and ground based observations of polar cap phenomena'. *J. Geomagn. Geoelectricity* **48**, 935.

Obara, T., T. Mukai, H. Hayakawa, A. Nishida, K. Tsuruda, S. Machida, and H. Fukunishi: 1993, 'Akebono (EXOS-D) observations of small-scale electromagnetic signatures relating to polar cap precipitation'. *J. Geophys. Res.* **98**, 11153.

Obara, T., T. Mukai, H. Hayakawa, K. Tsuruda, A. Matsuoka, and A. Nishida: 1996b, 'Signature of electric field associated with localized electron precipitation in the polar cap region – Akebono (EXOS-D) results'. *J. Geomagn. Geoelectricity* **48**, 327.

Obara, T., T. Mukai, H. Hayakawa, K. Tsuruda, A. Matsuoka, A. Nishida, H. Fukunishi, L. Zhu, J. Sojka, and D. Crain: 1996c, 'Multiple polar cap arcs: Akebono (EXOS D) observations'. *Radio Science* **31**, 645.

Ogawa, T. and T. Sato: 1971, 'New mechanism of auroral arcs'. *Planet. Space Sci.* **19**, 1393.

Oguti, T.: 1974, 'Rotational deformations and related drift motions of auroral arcs'. *J. Geophys. Res.* **79**, 3861.

Oguti, T.: 1975, 'Two-tiered auroral band'. *J. Atmos. Terr. Phys.* **37**, 1501.

Ohtani, S., T. A. Potemra, P. T. Newell, L. J. Zanetti, T. Iijima, M. Watanabe, M. Yamauchi, R. D. Elphinstone, O. de la Beaujardiere, and L. G. Blomberg: 1995, 'Simultaneous prenoon and postnoon observations of three field-aligned current systems from Viking and DMSP-F7'. *J. Geophys. Res.* **100**, 119.

Øieroset, M., T. D. Phan, R. P. Lin, and B. U. O. Sonnerup: 2000, 'Walen and variance analyses of high-speed flows observed by Wind in the midtail plasma sheet: Evidence for reconnection'. *J. Geophys. Res.* **105**, 25247.

Øieroset, M., M. Yamauchi, L. Liszka, and B. Hultqvist: 1998, 'Increase in energetic ion outflow from the ionosphere during substorms: A statistical study'. In: S. Kokubun and Y. Kamide (eds.): *Substorms-4*.

Øieroset, M., M. Yamauchi, L. Liszka, and B. Hultqvist: 1999, 'Energetic ion outflow from the dayside ionosphere: Categorization, classification, and statistical study'. *J. Geophys. Res.* **104**, 24915.

Olsson, A., L. Andersson, A. I. Eriksson, J. Clemmons, R. E. Erlandsson, G. Reeves, T. Huges, and J. S. Murphee: 1998, 'Freja studies of the current-voltage relation in substorm-related events'. *J. Geophys. Res.* **103**, 4285.

Olsson, A., A. Eriksson, and P. Janhunen: 1996a, 'On the current-volatge relationship in auroral breakups and westwards-travelling surges'. *Ann. Geophysicae* **14**, 1265.

Olsson, A., M. Persson, H. Opgenoorth, and S. Kirkwood: 1996b, 'Particle precipitation in auroral breakups and westward traveling surges'. *J. Geophys. Res.* **101**, 24661.

Omura, Y., H. Kojima, and H. Matsumoto: 1994, 'Computer simulation of electrostatic solitary waves: A nonlinear model of broadband electrostatic noise'. *Geophys. Res. Lett.* **21**, 2923.

Onsager, T. and T. Mukai: 1995, 'Low altitude signature of the plasma sheet boundary layer: Observations and model'. *Geophys. Res. Lett.* **22**, 855.

Onsager, T. G., M. F. Thomsen, R. C. Elphic, and J. T. Gosling: 1991, 'Model of electron and ion distributions in the plasma sheet boundary layer'. *J. Geophys. Res.* **96**, 20999.

Opgenoorth, H., R. Pellinen, W. Baumjohann, E. Nielsen, G. Marklund, and L. Eliasson: 1983a, 'Three-dimensional current flow and particle precipitation in a westward traveling surge (observed during the Barium-GEOS rocket experiment)'. *J. Geophys. Res.* **88**, 3183.

Opgenoorth, H. J., I. Haggstrom, P. J. S. Williams, and G. O. L. Jones: 1990, 'Regions of strongly enhanced perpendicular electric fields adjacent to auroral arcs'. *J. Atmos. Terr. Phys.* **52**, 449.

Opgenoorth, H. J. O., K. Kaila, E. Nielsen, and W. Baumjohann: 1983b, 'haracteristics of eastward drifting omega bands in the morning sector of the auroral oval'. *J. Geophys. Res.* **88**, 9171.

Oppenheim, M. M., D. L. Newman, and M. V. Goldman: 1999, 'Evolution of electron phase-space holes in a 2D magnetized plasma'. *Phys. Rev. Lett.* **83**, 2344.

Otto, A. and G. Birk: 1992, 'The dynamical evolution of small-scale auroral arc phenomena due to a resistive instability'. *J. Geophys. Res.* **97**, 8391.

Papadopoulos, K.: 1977, 'A review of anomalous resistivity in the ionosphere'. *Rev. Geophys. Space Phys.* **15**, 113.

Papitashvili, V. O., B. A. Belov, D. S. Faermark, Y. I. Feldstein, S. A. Golyshev, L. I. Gromova, and A. E. Levitin: 1994, 'Electric potential patterns in the northern and southern polar regions parameterized by the interplanetary magnetic field'. *J. Geophys. Res.* **99**, 13251.

Pellinen, R. and W. Heikkila: 1984, 'Inductive electric fields in the magnetotail and their relation to auroral and substorm phenomena'. *Space Sci. Rev.* **37**, 1–61.

Peria, W., C. Carlson, R. Ergun, J. McFadden, J. Bonnell, R. Elphic, and R. Strangeway: 2000, 'Characteristics of field-aligned currents near the auroral acceleration region: FAST observations'. In: S.-I. Ohtani (ed.): *Magnetospheric Current Systems*, Vol. 118 of *Geophysical Monograph Series*. Washington DC: American Geophysical Union, p. 181.

Peroomian, V., M. Ashour-Abdalla, and L. M. Zelenyi: 2000, 'Dynamical properties of self-consistent magnetotail configurations'. *J. Geophys. Res.* **105**, 18807.

Perreault, P. and S.-I. Akasofu: 1978, 'A study of geomagnetic storms'. *Geophys. J. R. Astr. Soc.* **54**, 547.

Persoon, A. D. G., W. K. Peterson, J. J. H. Waite, J. Burch, and J. L. Green: 1988, 'Electron density depletions in the nightside auroral zone'. *J. Geophys. Res.* **93**, 1871.

Peterson, W. K. and E. G. Shelley: 1984, 'Origin of the plasma in a cross-polar cap auroral feature (theta aurora)'. *J. Geophys. Res.* **89**, 6729.

Peticolas, L. M., T. J. Hallinan, H. C. Stenbaek-Nielsen, J. W. Bonnell, and C. W. Carlson: 2001, 'A study of black aurora from aircraft-based optical observations and plasma measurements on FAST'. *J. Geophys. Res.* **106**, submitted.

Petrukovich, A. A., V. A. Sergeev, L. M. Zelenyi, T. Mukai, T. Yamamoto, S. Kokubun, K. Shiokawa, C. S. Deehr, E. Y. Budnick, J. Buchner, A. O. Fedorov, V. P. Grigorieva, T. J. Hughes, N. F. Pissarenko, S. A. Romanov, and I. Sandahl: 1998, 'Two spacecraft observations of a reconnection pulse during an auroral breakup'. *J. Geophys. Res.* **103**, 47.

Peymirat, C., A.D.Richmond, B. Emery, and R. roble: 1998, 'A magnetosphere-thermosphere-ionosphere electrodynamics general circulation model'. *J. Geophys. Res.* **103**, 17467.

Pinçon, J.-L., P. M. Kintner, P. W. Schuck, and C. E. Seyler: 1997, 'Observation and analysis of lower hybrid solitary structures as rotating eigenmodes'. *J. Geophys. Res.* **102**, 17283.

Pokhotelov, D., W. Lotko, and A. V. Streltsov: 2002, 'Effects of the seasonal asymmetry in ionospheric Pedersen conductance on the appearance of discrete aurora'. *Geophys. Res. Lett.* **29**, 10.1029/2001GL01401.

Potemra, T. A., T. Iijima, and N. A. Saflekos: 1980, *Large-scale characteristics of Birkeland currents*, Dynamics of the Magnetosphere. Norwell, Massachusetts: D. Reidel.

Potemra, T. A., L. J. Zanetti, P. F. Bythrow, A. T. Y. Lui, and T. Iijima: 1984, 'B(y)-dependent convection patterns during northward interplanetary magnetic field'. *J. Geophys. Res.* **89**, 9753.

Pottelette, R., R. E. Ergun, R. A. Treumann, M. Berthomier, C. W. Carlson, J. P. McFadden, and I. Roth: 1999, 'Modulated electron-acoustic waves in auroral density cavities : FAST observations'. *Geophys. Res. Lett.* **26**, 2629.

Pottelette, R. and R. A. Treumann: 1998, 'Impulsive broadband electrostatic noise in the cleft: A signature of dayside reconnection'. *J. Geophys. Res.* **103**, 9299.

Pottelette, R., R. A. Treumann, M. Berthomier, and J. Jasperse: 2002, 'Characteristics of downward electron acceleration processes as deduced from AKR fine structures'. *Nonlinear Processes in Geophysics* **9**, in press.

Pottelette, R., R. A. Treumann, and N. Dubouloz: 1992, 'Generation of auroral kilometric radiation in upper hybrid wave-lower hybrid soliton interaction'. *J. Geophys. Res.* **97**, 12029.

Prangé, R., D. Rego, L. Pallier, J. Connerney, and P. Zarka: 1998, 'Detailed study of FUV Jovian auroral Features with the post-COSTAR HST faint object camera'. *J. Geophys. Res.* **103**, 20195.

Pritchett, P. L.: 1984a, 'Relativistic dispersion and the generation of auroral kilometric radiation'. *Geophys. Res. Lett.* **11**, 143.

Pritchett, P. L.: 1984b, 'Relativistic dispersion, the cyclotron maser instability, and auroral kilometric radiation'. *J. Geophys. Res.* **89**, 8957.

Pritchett, P. L. and F. V. Coroniti: 1994, 'Convection and the formation of thin current sheets in the near-Earth plasma sheet'. *Geophys. Res. Lett.* **21**, 1587.

Pritchett, P. L. and F. V. Coroniti: 1995, 'Formation of thin current sheets during plasma sheet convection'. *J. Geophys. Res.* **100**, 23551.

Pritchett, P. L. and R. J. Strangeway: 1985, 'A simulation study of kilometric radiation generation along an auroral field line'. *J. Geophys. Res.* **90**, 9650.

Pritchett, P. L., R. J. Strangeway, C. W. Carlson, R. E. Ergun, J. P. McFadden, and G. T. Delory: 1999, 'Free energy source and frequency bandwith for the auroral kilometric radiation'. *J. Geophys. Res.* **104**, 10326.

Proehl, J. A., W. Lotko, I. Kouznetsov, and S. D. Geimer: 2002, 'Ultra-low-frequency magnetohydrodynamics in boundary-constrained geomagnetic flux coordinates'. *J. Geophys. Res.* **107**, 10.1029/2001JA000135.

Pryor, W. R., J. M. Ajello, W. K. Tobiska, D. E. Shemansky, G. K. James, C. W. Hord, S. K. Stephens, R. A. West, A. I. F. Stewart, W. E. McClintock, K. E. Simmons, A. R. Hendrix, and D. A. Miller: 1998, 'Galileo ultraviolett spectrometer observations of Jupiter's auroral spectrum from 1600-3200 C5'. *J. Geophys. Res.* **103**, 20149.

Pulkkinen, T.: 1996, 'Psedobreakup or substorm?'. In: *Third International Conference on Substorms (ICS-3), ESA-SP-389*. p. 285.

Pulkkinen, T., R. Pellinen, H. Opgenoorth, J. Murphree, V. Petrov, A. Zaitsev, and E. Friis-Christensen: 1991, 'Auroral signatures of substorm recovery phase: A case study'. In: J. Kan, T. A. Potemra, S. Kokubun, and T. Iijima (eds.): *Magnetospheric Substorms*, Vol. 64 of *Geophysical Monograph Series*. AGU, Washington D.C., p. 333.

Pulkkinen, T. I., D. Baker, P. K. Toivanen, R. Pellinen, R. H. W. Friedel, and A. Korth: 1994, 'Magnetospheric field and current distributions during the substorm recovery phase'. *J. Geophys. Res.* **99**, 10955.

Pulkkinen, T. I., D. N. Baker, L. A. Frank, J. B. Sigwarth, H. J. Opgenoorth, R. Greenwald, E. Friis-Christensen, T. Mukai, R. Nakamura, H. Singer, G. D. Reeves, and M. Lester: 1998, 'Two Substorm Intensifications Compared: Onset, Expansion and Global Consequences'. *J. Geophys. Res.* **103**, 15.

Pulkkinen, T. I., D. N. Baker, J. S. Murphree, and L. A. Frank: 1995, 'Mapping of the auroral oval and individual arcs during substorms'. *J. Geophys. Res.* **100**, 21987.

Pulkkinen, T. I., D. N. Baker, R. J. Pellinen, J. Büchner, H. E. J. Koskinen, R. E. Lopez, R. L. Dyson, and L. A. Frank: 1992, 'Particle scattering and current sheet stability in the geomagnetic tail during the substorm growth phase'. *J. Geophys. Res.* **97**, 19283.

Pulkkinen, T. I., M. V. Kubyshkina, D. N. Baker, L. L. Cogger, S. Kokubun, T. Mukai, H. J. Singer, J. A. Slavin, and L. Zelenyi: 2000, 'Magnetotail currents during the growth phase and local auroral breakup'. In: S.-I. Ohtani, R. Fujii, M. Hesse, and R. L. Lysak (eds.): *Magnetospheric Current Systems*, Vol. 118 of *Geophysical Monograph Series*. Washington DC: AGU, p. 81.

Pulkkinen, T. I. and M. Wiltberger: 1999, 'Global magnetospheric response to IMF driving: ISTP observations, empirical modeling, and MHD simulation'. *Phys. Chem. Earth (C)* **24**, 153.

Raeder, J., R. J. Walker, and M. Ashour-Abdalla: 1995, 'The structure of the distant geomagnetic tail during long periods of northward IMF'. *Geophys. Res. Lett.* **22**, 349.

Rairden, R. L. and S. B. Mende: 1989, 'Properties of 6300-Å auroral emission at south pole'. *J. Geophys. Res.* **94**, 1402.

Rankin, R., F. Fenrich, and V. Tikhonchuk: 2000, 'Shear Alfvén waves on stretched magnetic field lines near midnight in Earth's magnetosphere'. *Geophys. Res. Lett.* **27**, 3265.

Rankin, R., P. Frycz, V. Tikhonchuk, and J. Samson: 1995, 'Ponderomotive saturation of magnetospheric field line resonances'. *Geophys. Res. Lett.* **2**, 1741.

Rankin, R., B. G. Harrold, J. C. Samson, and P. Frycz: 1993, 'The nonlinear evolution of field line resonances in the Earth's magnetosphere'. *J. Geophys. Res.* **98**, 5839.

Rankin, R., J. Samson, and V. Tikhonchuk: 1999a, 'Discrete auroral arcs and nonlinear dispersive field line resonances'. *Geophys. Res. Lett.* **26**, 663.

Rankin, R., J. C. Samson, and V. T. Tikhonchuk: 1999b, 'Parallel electric fields in dispersive shear Alfvén waves in the dipolar magnetosphere'. *Geophys. Res. Lett.* **26**, 3601.

Rasmussen, J. J.: 1982, 'Effects of trapped particles on strongly nonlinear electron plasma waves'. *Physica Scripta* **T2/1**, 29.

Rawer, K., D. Bilitza, and S. Ramakrishnan: 1978, 'Goals and status of the International Reference Ionosphere'. *Rev. Geophys. Space Phys.* **16**, 177.

Reasoner, D. L. and W. J. Burke: 1973, 'Measurement of the Lunar Photoelectron Layer in the Geomagnetic Tail'. In: B. M. McCormac (ed.): *Photon and Particle Interactions with Surfaces in Space*, Vol. 37 of *ASSL*. p. 369.

Redsun, M., M. Temerin, and F. S. Mozer: 1985, 'Classification of auroral electrostatic shocks by their ion and electron associations'. *J. Geophys. Res.* **90**, 9615.

Rees, M. H.: 1963, 'Auroral ionization and excitation by incident energetic electrons'. *Planet. Space Sci.* **11**, 1209.

Rees, M. H.: 1989, *Physics and Chemistry of the Upper Atmosphere*. New York: Cambridge.

Rees, M. H. and D. Luckey: 1974, 'Auroral electron energy derived from ratio of spectroscoptic emissions, 1. Model computations'. *J. Geophys. Res.* **79**, 51810.

Reeves, G.: 1998, 'New perspectives on substorm injections'. In: S. Kokubun and Y. Kamide (eds.): *Substorms 4*. Tokyo: Terra Scientific Publishing Company, p. 785.

Reiff, P. H.: 1984, 'Models of auroral-zone conductances'. In: T. A. Potemra (ed.): *Magnetospheric Currents*, Vol. 28 of *Geophysical Monograph Series*. Washington, DC: American Geophysical Union, p. 180.

Reiff, P. H. and J. L. Burch: 1985, 'IMF B(y)-dependent plasma flow and Birkeland currents in the dayside magnetosphere. II - A global model for northward and southward IMF'. *J. Geophys. Res.* **90**, 1595.

Rich, F. J. and M. S. Gussenhoven: 1987, 'The absence of Region 1/Region 2 field-aligned currents during prolonged quiet times'. *Geophys. Res. Lett.* **14**, 689.

Rich, F. J. and M. Hairston: 1994, 'Large-scale convection patterns observed by DMSP'. *J. Geophys. Res.* **99**, 3827.

Rich, F. J. and N. C. Maynard: 1989, 'Consequences of using simple analytical functions for the high-latitude convection electric field'. *J. Geophys. Res.* **94**.

Richmond, A.: 1987. In: S.-I. Akasofu and Y. Kamide (eds.): *The Solar Wind and the Earth*. Terra Sci. Publ. Comp., Japan, p. 123.

Richmond, A. D. and Y. Kamide: 1988, 'Mapping electrodynamic features of the high-latitude ionosphere from localized observations: Technique'. *J. Geophys. Res.* **93**, 5741.

Ridley, A., C. Clauer, G. Lu, and V. Papitashvili: 1998, 'A statistical study of the ionospheric convection response to changing interplanetary magnetic field conditions using the assimilative mappic of ionospheric electrodynamics technique'. *J. Geophys. Res.* **103**, 4023.

Ridley, A., G. Lu, C. Clauer, and V. Papitashvili: 1999, 'Reply'. *J. Geophys. Res.* **104**, 4393.

Rieger, E. and H. Marschhäuser: 1990. In: *Max91 Workshop No. 3*. Boulder, Co., p. 68.

Roach, F. E. and L. L. Smith: 1967, 'The worldwide morphology of the atomic oxygen nightglows'. In: B. M. McCormack (ed.): *Aurora and Airglow*. Dordrecht-Holland.

Roble, R., E. Ridley, A. Richmond, and R. Dickinson: 1988, 'A coupled thermosphere/ionosphere general circulation model'. *Geophys. Res. Lett.* **15**, 1325.

Rodriguez, J. V., C. E. Valladares, K. Fukui, and H. A. Gallagher Jr.: 1997, 'Antisunward decay of polar cap arcs'. *J. Geophys. Res.* **102**, 27227.

Roeder, J. and H. Koons: 1989, 'A survey of electron cyclotron waves in the magnetosphere and the diffuse auroral electron precipitation'. *J. Geophys. Res.* **94**, 2529.

Rönnmark, K.: 1999, 'Electron acceleration in the auroral current circuit'. *Geophys. Res. Lett.* **26**, 983.

Rönnmark, K. and M. Hamrin: 2000, 'Auroral electron acceleration by Alfvén waves and electrostatic fields'. *J. Geophys. Res.* **105**, 25333.

Rostoker, G., A. T. Y. Lui, C. D. Anger, and J. S. Murphree: 1987a, 'North-south structures in the midnight sector auroras as viewed by the Viking imager'. *Geophys. Res. Lett.* **14**, 407.

Rostoker, G., A. Vallance-Jones, R. L. Gattinger, C. D. Anger, and J. S. Murphree: 1987b, 'The development of the substorm expansive phase: the eye of the substorm'. *Geophys. Res. Lett.* **14**, 399.

Rothwell, P. L., M. B. Silevitch, L. P. Block, and C. G. Fälthammar: 1994, 'O(+) phase bunching and auroral arc structure'. *J. Geophys. Res.* **99**, 2461.

Roux, A., A. Hilgers, H. de Feraudy, D. Le Queau, P. Louarn, S. Perraut, A. Bahnsen, M. Jespersen, E. Ungstrup, and M. Andre: 1993, 'Auroral kilometric radiation sources - In situ and remote observations from Viking'. *J. Geophys. Res.* **98**, 11657.

Royrvik, O. and T. Davis: 1977, 'Pulsting aurora: Local and global morphology'. *J. Geophys. Res.* **82**, 4720.

Ruohoniemi, J. and R. Greenwald: 1996, 'Statistical patterns of high-latitude convection obtained from Goose Bay HF radar observations'. *J. Geophys. Res.* **101**, 21743.

Ruohoniemi, J. R., R. Greenwald, K. B. Baker, and J. Samson: 1991, 'HF radar observations of Pc5 field line resonances in the midnight/early morning MLT sector'. *J. Geophys. Res.* **96**, 15697.

Saeki, K., P. Michelsen, H. L. Pecseli, and J. J. Rasmussen: 1994, 'Formation and Coalescence of Electron Solitary Holes'. *Phys. Rev. Lett.* **42**, 501.

Sagdeev, R. Z.: 1966, 'Cooperative phenomena and shock waves in collisionless plasmas'. *Rev. Plasma Phys.* **4**.

Saito, T.: 1978, 'Long period irregular micropulsations, Pi 3'. *Space Sci. Rev.* **21**, 427.

Samson, J., L. L. Cogger, and Q. Pao: 1996, 'Observations of field-line resonances, auroral arcs, and auroral vortex structures'. *J. Geophys. Res.* **101**, 17373.

Samson, J. C., T. J. Hughes, F. Creutzberg, D. D. Wallis, R. A. Greenwald, and J. M. Ruohoneimi: 1991, 'Observations of a detached, discrete auraral arc in association with field line resonances'. *J. Geophys. Res.* **96**, 15683.

Samson, J. C. and G. Rostoker: 1972, 'Latitude-dependent characteristics of high-latitude Pc4 and Pc5 micropulsations'. *J. Geophys. Res.* **77**, 6133.

Sandel, B. and A. Broadfoot: 1981, 'Morphology of Saturn's aurora'. *Nature* **292**, 779.

Sandel, B. R., D. E. Shemansky, A. L. Broadfoot, J. B. Holberg, G. R. Smith, J. C. McConnell, D. F. Strobel, S. K. Atreya, T. M. Donahue, H. W. Moos, D. M. Hunten, R. B. Pomphrey, and S. Linick: 1982, 'Extreme ultraviolet observations from the Voyager 2 encounter with Saturn'. *Science* **215**, 548.

Sandholt, P., M. Lockwood, T. Oguti, S. Cowley, K. Freeman, B. Lybekk, A. Egeland, and D. Willis: 1990, 'Midday auroral breakup events and related energy and momentum transfer from the magnetosheath'. *J. Geophys. Res.* **95**, 1039–1060.

Sandholt, P. E., C. J. Farrugia, J. Moen, O. Noraberg, B. Lybekk, and T. S. T. Hansen: 1998, 'A classification of dayside auroral forms and activities as a function of interplantary magnetic field orientation'. *J. Geophys. Res.* **103**, 23325.

Sanny, J., R. L. McPherron, C. T. Russell, D. N. Baker, T. I. Pulkkinen, and A. Nishida: 1994, 'Growth-Phase thinning of the near-Earth current sheet during the CDAW-6 substorm'. *J. Geophys. Res.* **99**, 5805.

Sato, T.: 1978, 'A Theory of Quiet Auroral Arcs'. *J. Geophys. Res.* **83**, 1042.

Sato, T. and T. E. Holzer: 1973, 'Quiet Auroral Arcs and Electrodynamic Coupling between the Ionosphere and Magnetosphere, 1'. *J. Geophys. Res.* **78**, 7314.

Satoh, R., J. Connerney, and R. Baron: 1996, 'Emission source model of Jupiter's H_3^+ aurorare: A generalized inverse analysis of images'. *Icarus* **12**, 1.

Saunders, M. A.: 1992, 'The morphology of dayside Birkeland currents'. *J. Atmos. Terr. Phys.* **54**.

Sauvaud, J. A., D. Popescu, D. C. Delcourt, G. K. Parks, M. Brittnacher, V. A. Sergeev, R. A. Kovrazhkin, T. Mukai, and S. Kokubun: 1999, 'Sporadic plasma sheet injections into the high altitude auroral bulge - satellite observations'. *J. Geophys. Res.* **104**, 28565.

Sazhin, S. S., K. Bullough, and M. Hayakawa: 1993, 'Auroral hiss - A review'. *Planet. Space Sci.* **41**, 153.

Schamel, H.: 1982a, 'Kinetic theory of phase space vortices and double layers'. *Physica Scripta* **T2/1**, 228.

Schamel, H.: 1982b, 'Stability of electron vortex structures in phase space'. *Phys. Rev. Lett.* **48**, 481.

Schardt, A. and F. McDonald: 1983, 'The flux and source of energetic protons in Saturn's inner magnetosphere'. *J. Geophys. Res.* **88**, 8923.

Schindler, K. and J. Birn: 1978, 'Magnetospheric physics'. *Phys. Reports* **47**, 109.

Schindler, K. and J. Birn: 2002, 'Models of two-dimensional embedded thin current sheets from Vlasov theory'. *J. Geophys. Res.* **107**, 10.1029/2001JA000304.

Schindler, K., M. Hesse, and J. Birn: 1988, 'General magnetic reconnection, parallel electric fields and helicity'. *J. Geophys. Res.* **93**, 5547.

Schödel, R., W. Baumjohann, R. Nakamura, V.A.Sergeev, and T. Mukai: 2001, 'Rapid flux transport in the central plasma sheet'. *J. Geophys. Res.* **106**, 301.

Scholer, M.: 1970, 'On the motion of artificial ion clouds in the magnetosphere'. *Planet. Space Sci.* **18**, 977.

Schriver, D.: 1999, 'Particle simulation of the auroral zone showing parallel electric field, waves and plasma acceleration'. *J. Geophys. Res.* **104**, 14655.

Schuck, P. W., C. E. Seyler, J. L. Pincon, J. W. Bonnell, and P. M. Kintner: 1998, 'Theory, simulation, and observation of discrete eigenmodes associated with lower hybrid solitary structures'. *J. Geophys. Res.* **103**, 6935.

Schulz, M. and L. J. Lanzerotti: 1974, *Particle diffusion in the radiation belts*. New York and Berlin: Springer Verlag.

Schumaker, T., M. Gussenhoven, D. Hardy, and R. Carovillano: 1989, 'The relationship between diffuse auroral and plasma sheet electron distributions near local midnight'. *J. Geophys. Res.* **94**, 10061.

Sckopke, N.: 1979, 'External plasma flow'. In: B. Battrick (ed.): *Magnetospheric Boundary Layers*. Paris: ESA SP-148, p. 37.

Semenov, V. S., Y. V. Bogdanova, R. P. Rijnbeek, and M. J. Buchan: 1999, 'A new mechanism for interpreting the motion of auroral arcs in the nightside ionosphere'. *Geophys. Res. Lett.* **26**, 2367.

Sergeev, V., C.-I. M. K. Liou, P. Newell, M. Brittnacher, G. Parks, and G. Reeves: 1999, 'Development of auroral streamers in association with localized impulsive injections to the inner magnetotail'. *Geophys. Res. Lett.* **26**, 417.

Sergeev, V., M. V. Kubyshkina, K. Liou, P. T. Newell, G. Parks, R. Nakamura, and T. Mukai: 2001, 'Substorm and convection bay compared: Auroral and magnetotail dynamics during convection bay'. *J. Geophys. Res.* **106**, 18843.

Sergeev, V., M. Malkov, and K. Mursula: 1993, 'Testing the isotropic boundary algorithm method to evaluate the magnetic field configuration in the tail'. *J. Geophys. Res.* **98**, 7609.

Sergeev, V., R. Pellinen, and T. Pulkkinen: 1996a, 'Steady magnetospheric convection: A review of recent results'. *Space Sci. Rev.* p. 551.

Sergeev, V., J.-A. Sauvaud, D. Popescu, R. Kovrazhkin, K. Liou, P. Newell, M. Brittnacher, G. Parks, R. Nakamura, T. Mukai, and G. Reeves: 2000a, 'Multiple-spacecraft observations of a narrow transient plasma jet in the Earth's plasma sheet'. *Geophys. Res. Lett.* **27**, 851.

Sergeev, V., J.-A. Sauvaud, D. Popescu, R. A. Kovrazhkin, V. N. Lutsenko, L. M. Zelenyi, M. Syrjäsuo, A. Viljanen, T. Pulkkinen, K. Kudela, S. Kokubun, and T. Mukai: 2000b, 'Plasma sheet ion injections into the auroral bulge: correlative study of spacecraft and ground observations'. *J. Geophys. Res.* **105**, 18465.

Sergeev, V., E. M. Sazhina, N. A. Tsyganenko, J. A. Lundblad, and F. Soraas: 1983, 'Pitch-angle scattering of energetic protons in the magnetotail current sheet as the dominant source of their isotropic precipitation into the nightside ionosphere'. *Planet. Space Sci.* **31**, 1147.

Sergeev, V. A., T. I. Pulkkinen, and R. J. Pellinen: 1996b, 'Coupled-mode scenario for the magnetospheric dynamics'. *J. Geophys. Res.* **101**, 13047.

Sergeev, V. A. and A. G. Yahnin: 1979, 'The features of auroral bulge expansion'. *Planet. Space Sci.* **27**, 1429.

Seyler, C. E.: 1990, 'A Mathematical Model of the Structure and Evolution of Small-Scale Discrete Auroral Arcs'. *J. Geophys. Res.* **95**, 17199.

Seyler, C. E., A. E. Clark, J. Bonnell, and J.-E. Wahlund: 1998, 'Electrostatic broadband ELF wave emission by Alfveń wave breaking'. *J. Geophys. Res.* **103**, 7027.

Seyler, C. E. and J.-E. Wahlund: 1996, 'Theory of nearly perpendicular electrostatic waves and comparison to Freja satellite observations'. *J. Geophys. Res.* **101**, 21795.

Shawhan, S. D., C.-G. Falthämmar, and L. P. Block: 1978, 'On the nature of large auroral zone electric fields at 1-R_E altitude'. *J. Geophys. Res.* **83**, 1049.

Shelley, E. G., R. D. Sharp, and R. G. Johnson: 1976, 'Satellite observations of an ionospheric acceleration mechanism'. *Geophys. Res. Lett.* **3**, 654.

Shepherd, S. G., J. LaBelle, R. A. Doe, M. McCready, and A. T. Weatherwax: 1998, 'Ionospheric structure and the generation of auroral roar'. *J. Geophys. Res.* **103**, 29253.

Shiokawa, K., W. Baumjohann, G. Haerendel, G. Paschmann, J. F. Fennell, E. Friis-Christensen, H. Lühr, G. D. Reeves, C. T. Russell, P. R. Sutcliffe, and K. Takahashi: 1998, 'High-speed ion flow, substorm current wedge, and multiple Pi 2 pulsations'. *J. Geophys. Res.* **103**, 4491.

Shue, J. H., P. T. Newell, K. Liou, and C. I. Meng: 2000, 'Influence of interplanetary magnetic field on global auroral patterns'. *J. Geophys. Res.* **106**.

Sigernes, F., J. Moen, M. Øieroset, B. Lybekk, J. Holtet, D. Lorentzen, C. Deehr, and R. W. Smith: 1996, 'Auroral prenoon height measurements in relation to SCIFER'. *Geophys. Res. Lett.* **23**, 1889.

Singer, H. J., D. J. Southwood, R. J. Walker, and M. G. Kivelson: 1981, 'Alfvén wave resonances in a realistic magnetospheric magnetic field geometry'. *J. Geophys. Res.* **86**, 4589.

Sittler, Jr., E., K. Ogilivie, and R. Sellsnick: 1987, 'Survey of electrons in the Uranian magnetosphere'. *J. Geophys. Res.* **92**, 15263.

Smith, R. L., L. Kimura, J. Vigneron, and J. Katsufrakis: 1966, 'Lower hybrid resonance noise and a new ionospheric duct'. *J. Geophys. Res.* **71**, 1925.

Snyder, A. L., S.-I. Akasofu, and T. N. Davis: 1974, 'Auroral substorms observed from above the north polar region by a satellite'. *J. Geophys. Res.* **79**, 1393.

Solovyev, S., D. Baishev, E. Barkova, M. Engebretson, J. Posch, W. Hughes, K. Yumoto, and V. Pilipenko: 1999, 'Structure of disturbences in the dayside and nightside ionosphere during periods of negative interplanetary magnetic field Bz'. *J. Geophys. Res.* **104**, 28019.

Song, Y.: 1998, 'Theoretical constraints on mechanisms for the substorm current wedge'. In: S. Kokubun and Y. Kamide (eds.): *Substorms-4*. Tokyo: Terra Scientific Publ. Comp., p. 543.

Song, Y. and R. L. Lysak: 1989a, 'Dynamo effect of 3-d time-dependent reconnection in the dayside magnetopause'. *Geophys. Res. Lett.* **16**, 911.

Song, Y. and R. L. Lysak: 1989b, 'Evaluation of twist helicity in FTE flux tubes'. *J. Geophys. Res.* **94**, 5273.

Song, Y. and R. L. Lysak: 1994, 'Alfvénon, driven reconnection and the direct generation of field-aligned current'. *Geophys. Res. Lett.* **21**, 1755.

Song, Y. and R. L. Lysak: 2000, 'Paradigm transition in cosmic plasma physics, magnetic reconnection and the generation of field-aligned current'. In: S.-I. Ohtani (ed.): *Magnetospheric Current Systems*, Vol. 118 of *Geophysical Monograph Series*. Washington, D.C.: American Geophysical Union, p. 11.

Song, Y. and R. L. Lysak: 2001, 'Towards a new paradigm: from a quasi-steady description to a dynamical description of the magnetosphere'. *Space Sci. Rev.* **95**, 273.

Sonnerup, B. U. O.: 1980, 'Theory of the low latitude boundary layer'. *J. Geophys. Res.* **85**, 2017.

Southwood, D. and R. Wolf: 1978, 'An assessment of the role of precipitation in magnetospheric convection'. *J. Geophys. Res.* **83**, 5227.

Southwood, D. J.: 1974, 'Some features of field line resonances in the magnetosphere'. *Planet. Space Sci.* **22**, 483.

Spann, J. F., M. Brittnacher, R. Elsen, G. A. Germany, and G. K. Parks: 1998, 'Initial response and complex polar cap structures of the aurora in response to the January 10, 1997 magnetic cloud'. *Geophys. Res. Lett.* **25**, 2577.

Spiro, R., R. Heelis, and W. Hanson: 1979, 'Rapid subauroral ion drifts observed by Atmospheric Explorer C'. *Geophys. Res. Lett.* **6**, 660.

Spiro, R. W., P. H. Reiff, and L. J. Maher: 1982, 'Precipitating electron energy flux and auroral zone conductances: an empirical model'. *J. Geophys. Res..* **87**, 8215.

Stasiewicz, K., P. Bellan, C. Chaston, C. Kletzing, R. Lysak, J. Maggs, O. Pokhotelov, C. Seyler, P. Shukla, L. Stenflo, A. Streltsov, and J.-E. Wahlund: 2000, 'Small Scale Alfvénic Structure in the Aurora'. *Space Science Reviews* **92**, 423.

Stasiewicz, K., Y. Khotyaintsev, M. Berthomier, and J.-E. Wahlund: 2000, 'Identification of widespread turbulence of dispersive Alfvén waves'. *Geophys. Res. Lett.* **27**, 173.

Stenbaek-Nielsen, H. and T. Hallinan: 1979, 'Pulsating auroras: Evidence for non-collisional thermalization of precipitating electrons'. *J. Geophys. Res.* **84**, 3257.

Stenbaek-Nielsen, H., T. Hallinan, D. Osborne, J. Kimball, C. Chaston, J. McFadden, G. Delory, M. Temerin, and C. Carlson: 1998, 'Aircraft observations conjugate to FAST'. *Geophys. Res. Lett.* **25**, 2073.

Stern, D. P.: 1981, 'One-dimensional models of quasi-neutral parallel electric fields'. *J. Geophys. Res.* **86**, 5839.

Stiles, G. S., J. C. Foster, and J. R. Doupnik: 1980, 'Prolonged radar observations of an auroral arc'. *J. Geophys. Res.* **85**, 1223.

Stix, T.: 1992, *Waves in plasmas*. New York: American Institute of Physics.

Størmer, C.: 1946, 'Frequency of 12,330 measured heights of aurora'. *Terr. Magn. Atmos. Electr.* **51**, 501.

Størmer, C.: 1955, *The polar aurora*. Oxford: Clarendon Press.

Strangeway, R. J., L. Kepko, R. C. Elphic, C. W. Carlson, R. E. Ergun, J. P. McFadden, W. J. Peria, G. T. Delory, C. C. Chaston, M. Temerin, C. A. Cattell, E. Möbius, L. Kistler, D. M. Klumpar, W. K. Petersen, E. Shelley, and R. Pfaff: 1998, 'FAST observations of VLF waves in the auroral zone: Evidence of very low plasma densities'. *Geophys. Res. Lett.* **25**, 2065.

Streltsov, A. and W. Lotko: 1995, 'Dispersive field line resonances on auroral field lines'. *J. Geophys. Res.* **100**, 19457.

Streltsov, A. and W. Lotko: 1996, 'The fine structure of dispersive, nonradiative field line resonance layers'. *J. Geophys. Res.* **101**, 5343.

Streltsov, A. and W. Lotko: 1999, 'Small-scale, electrostatic auroral structures and Alfvén waves'. *J. Geophys. Res.* **104**, 4411.

Streltsov, A., W. Lotko, J. R. Johnson, and C. Z. Cheng: 1998, 'Small-scale, dispersive field line resonances in the hot magnetospheric plasma'. *J. Geophys. Res.* **103**, 26559.

Streltsov, A., W. Lotko, A. Keiling, and J. Wygant: 2002, 'Numerical modeling of Alfvén waves observed by Polar spacecraft in the nightside plasmasheet boundary layer'. *J. Geophys. Res.* **107**, –in press.

Swift, D. W.: 1972, 'Effective height-integrated conductivity of the ionosphere'. *J. Geophys. Res.* **77**, 1279.

Swift, D. W.: 1975, 'On the formation of auroral arcs and acceleration of auroral electrons'. *J. Geophys. Res.* **80**, 2096.

Swift, D. W.: 1979, 'An equipotential model for auroral arcs: The theory of two-dimensional laminar electrostatic shocks'. *J. Geophys. Res.* **84**, 6427.

Swift, D. W.: 1981, 'Mechanisms for auroral precipitation: A review,'. *Rev. Geophys. Space Pysics* **19**, 185.

Syrjäsuo, M., T. I. Pulkkinen, R. P. Pellinen, P. Janhunen, K. Kauristie, A. Viljanen, H. J. Opgenoorth, P. Karlsson, S. Wallman, P. Eglitis, E. Nielsen, and C. Thomas: 1998, 'Observations of substorm electrodynamics using the MIRACLE network'. In: *Proc. of the 4th International Conference on Substorms, ICS4*. Tokyo, Japan, p. 111.

Taguchi, S., M. Sugiura, J. D. Winningham, and J. A. Slavin: 1993, 'Characterization of the IMF B_Y-dependent field-aligned currents in the cleft region based on DE 2 observations'. *J. Geophys. Res.* **98**.

Takahashi, K. and E. W. Hones: 1988, 'ISEE 1 and 2 observations of ion distributions at the plasma sheet-tail lobe boundary'. *J. Geophys. Res.* **93**, 8558.

Takahashi, K. and R. L. McPherron: 1984, 'Standing hydromagnetic oscillations in the magnetosphere'. *Planet. Space Sci.* **32**, 1343.

Tanskanen, E. I., H. E. J. Koskinen, T. I. Pulkkinen, J. A. Slavin, and K. Ogilvie: 2002, 'Dissipation to the Joule heating: Isolated and storm-time substorms'. *Adv. Space. Res.* **30**, in press.

Temerin, M.: 1986, 'Evidence for a large bulk ion conic heating region'. *Geophys. Res. Lett.* **13**, 1059.

Temerin, M., C. Carlson, and J. McFadden: 2001, 'The low-altitude extent of the auroral acceleration region in the upward current region as determined by the upward accelerated ion beams'. *Geophys. Res. Lett.* **28**, in press.

Temerin, M. and C. W. Carlson: 1998a, 'Asymmetry of occurrence-frequency and intensity of AKR between summer polar region and winter polar region sources'. *Geophys. Res. Lett.* **25**, 2369.

Temerin, M. and C. W. Carlson: 1998b, 'Current-voltage relationship in the downward auroral current region'. *Geophys. Res. Lett.* **25**, 2365.

Temerin, M., C. A. Cattell, R.Lysak, M. Hudson., R. B. Torbert, F. S. Mozer, R. Sharp, and P. Kintner: 1984, 'The small scale structure of electrostatic shocks'. *J. Geophys. Res.* **86**, 11278.

Temerin, M., K. Cerny, W. Lotko, and F. S. Mozer: 1982, 'Observations of double layers and solitary waves in the auroral plasma'. *Phys. Rev. Lett.* **48**, 1175.

Temerin, M., J. McFadden, M. Boehm, C. W. Carlson, and W. Lotko: 1986, 'Production of flickering aurora and field-aligned electron flux by electromagnetic ion cyclotron waves'. *J. Geophys. Res.* **91**, 5769.

Temerin, M., M. Woldroff, and F. S. Mozer: 1979, 'Nonlinear steepening of the electrostatic ion cyclotron wave'. *Phys. Rev. Lett.* **43**, 1941.

Tetreault, D. J.: 1988, 'Growing ion holes as the cause of auroral double layers'. *Geophys. Res. Lett.* **15**, 164.

Thompson, B. J. and R. L. Lysak: 1996, 'Electron acceleration by inertial Alfvén waves'. *J. Geophys. Res.* **101**, 5359.

Thompson, B. J. and R. L. Lysak: 1996, 'Electron acceleration by inertial Alfven waves'. *J. Geophys. Res.* **101**, 5359.

Timofeev, E. and Y. Galperin: 1991, 'Convection and currents in stable auroral arcs and inverted-V's'. *J. Geophys. Geomagn.* **43**, 259.

Toivanen, P., H. Koskinen, and T. Pulkkinen: 1998, 'Mapping between the ionospheric andd the tail electric fields in a time-dependent Earth's magnetosphere'. *J. Geophys. Res.* **103**, 9153.

Toivanen, P. K.: 1997, 'Effects of the large-scale electric field on particle drifts in the near-Earth tail'. *J. Geophys. Res.* **102**, 2405.

Trakhtengertz, V. Y. and A. Y. Feldstein: 1984, 'Quiet auroral arcs: ionospheric effect of magnetospheric convection stratification'. *Planet. Space Sci.* **32**, 127.

Trakhtengertz, V. Y. and A. Y. Feldstein: 1991, 'Turbulent Alfvén boundary layer in the polar ionosphere, I, Excitation conditions and energetics'. *J. Geophys. Res..* **96**, 19363.

Trauger, J., J. Clarke, G. Ballester, R. Evans, J. Burrows, D. Crisp, J. Gallagher III, R. Griffiths, J. Hester, J. Hoessel, J. Holtzman, J.-E. Krist, J. Mould, R. Sahai, P. Scowen, K. Stapelfeldt, and A. Watson: 1998, 'Saturn's hydrogen aurora: Wide field and planetary camera 2 imaging from the Hubble Space Telescope'. *J. Geophys. Res.* **103**, 20237.

Treumann, R. A.: 1999, 'Generalized-Lorentzian thermodynamics'. *Physica Scripta* **59**, 204.

Treumann, R. A. and W. Baumjohann: 1997, *Advanced Space Plasma Physics*. London: Imperial College Press, second edition.

Trondsen, T. and L. Cogger: 1997, 'High-resolution television observations of black aurora'. *J. Geophys. Res.* **102**, 363.

Trondsen, T. S. and L. L. Cogger: 1998, 'A survey of small-scale spatially periodic distortions of auroral forms'. *J. Geophys. Res.* **103**, 9405.

Troshichev, O. A., M. G. Gusev, S. V. Nickolashkin, and V. P. Samsonov: 1988, 'Features of the polar cap aurorae in the southern polar region'. *Planet. Space Sci.* **36**, 429.

Tsyganenko, N.: 1987, 'Global quantitative models of the geomagnetic field in the cislunar magnetosphere for different disturbance levels'. *Planet. Space Sci.* **35**, 1347.

Tsyganenko, N.: 1989, 'A magnetospheric magnetic field model with a warped tail current sheet'. *Planet. Space Sci.* **37**, 5.

Tsyganenko, N., D. Stern, and Z. Kaymaz: 1993, 'Birkeland currents in the plasma sheet'. *J. Geophys. Res.* **98**, 19455.

Tsyganenko, N. A.: 1990, 'Quantitative models of the magnetospheric magnetic field: Methods and results'. *Space Sci. Rev.* **54**.

Tsyganenko, N. A.: 1995, 'Modeling the Earth's magnetospheric magnetic field confined within a realistic magnetopause'. *J. Geophys. Res.* **100**, 5599.

Tsyganenko, N. A.: 2000, 'Modeling the inner magnetosphere: The asymmetric ring current and Region 2 Birkeland currents revisited'. *J. Geophys. Res.* **105**, 27739.

Tsyganenko, N. A. and A. V. Usmanov: 1982, 'Determination of the magnetospheric current system parameters and development of experimental geomagnetic field models based on data from IMP and HEOS satellites'. *Planet. Space Sci.*. **30**, 985.

Tung, Y.-K., C. W. Carlson, J. P. McFadden, D. M. Klumpar, G. K. Parks, W. J. Peria, and K. Liou: 2001, 'Auroral polar cap boundary ion conic outflow observed on FAST'. *J. Geophys. Res.* **106**, 3603.

Turikov, V. A.: 1984, 'Electron phase space holes as localized BGK solutions'. *Physica Scripta* **30**, 73.

Ungstrup, E., A. Bahnsen, H. K. Wong, M. André, and L. Matson: 1990, 'Energy source and generation mechanism for auroral kilometric radiation'. *J. Geophys. Res.* **95**, 5957.

Uspensky, M. V., R. J. Pellinen, W. Baumjohann, G. V. Starkov, E. Nielsen, G. J. Sofko, and K. U. Kaila: 1983, 'Spatial variations of ionospheric conductivity and radar auroral amplitude in the eastward electrojet region during pre-substorm conditions'. *Journal of Geophysics* **52**.

Vago, J. L., P. M. Kintner, S. W. Chesney, R. L. Arnoldy, K. A. Lynch, T. E. Moore, and C. J. Pollock: 1992, 'Transverse ion acceleration by localized lower hybrid waves in the topside auroral ionosphere'. *J. Geophys. Res.* **97**, 16935.

Valladares, C. E., H. C. Carlson, and K. Fukui: 1994, 'Interplanetary magnetic field dependency of stable Sun-aligned polar cap arcs'. *J. Geophys. Res.* **99**, 6247.

Vallance Jones, A.: 1974, *Aurora.* Dordrecht-Holland/Boston-U.S.A.: D. Reidel Publ.Comp.

Van Allen, J. A., G. H. Ludwig, E. C. Ray, and C. E. McIlwain: 1958, 'Observations of high intensity radiation by satellites 1958 Alpha and gamma'. *Jet Propul.* **28**, 588.

van den Heuvel, E. P. J. and S. A. Rappaport (eds.): 1992, 'X-Ray binaries and recycled pulsars'.

Vasyliunas, V. M.: 1970, 'Mathematical models of magnetospheric convection and its coupling to the ionosphere'. In: B. M. McCormac (ed.): *Particles and Fields in the Magnetosphere*, Vol. 17 of *ASSL*. p. 60.

Vasyliunas, V. M.: 1984, 'Fundamentals of current description'. In: T. A. Potemra (ed.): *Magnetospheric Currents*, Vol. 28 of *Geophysical Monograph Series*. Washington, D. C., USA: American Geophysical Union, p. 63.

Vickrey, J. F., R. R. Vondrak, and S. J. Matthews: 1981, 'The diurnal and latitudinal variation of auroral zone ionospheric conductivity'. *J. Geophys. Res.*. **86**, 65.

Villalon, E. and W. J. Burke: 1995, 'Pitch-angle scattering of diffuse auroral electrons by whistler-mode waves'. *J. Geophys. Res.* **100**, 19361.

Vogt, J.: 1997, 'Elektrodynamischer Kopplung von polarer Ionosphere und äqutorialer Magnetosphäre,'. Ph.D. thesis, Technische Universität Braunschweig, Braunschweig, Germany.

Vogt, J.: 2002, 'Alfvén wave coupling in the auroral current circuit'. *Surveys in Geophysics* **23**, 335.

Vogt, J. and K.-H. Glassmeier: 2000, 'On the location of trapped particle populations in quadrupole magnetospheres'. *J. Geophys. Res.* **105**, 13063.

Vogt, J. and G. Haerendel: 1998, 'Reflection and Transmission of Alfvén Waves at the Auroral Acceleration Region'. *Geophys. Res. Lett.* **25**, 277.

Vogt, J., G. Haerendel, and K.-H. Glassmeier: 1999, 'A model for the reflection of Alfvén waves at the source region of the Birkeland current system: The tau generator'. *J. Geophys. Res.* **104**, 269.

Volwerk, M., P. Louarn, T. Chust, A. Roux, H. de Feraudy, and B. Holback: 1996, 'Solitary kinetic Alfvén waves: A study of their Poynting flux'. *J. Geophys. Res.* **101**, 13335.

Voronkov, I., J. C. Samson, E. Friedrich, R. Rankin, V. T. Tikhonchuk, and E. F. Donovan: 2000, 'On the distinction between, and the relevance of, auroral breakup andsubstorm expansive phase onset'. In: *Fifth International Conference on Substorms, ESA-SP-443*. p. 249.

Vovchenko, V. V., Y. I. Galperin, D. V. Chugunin, and N. Dubouloz: 2000, 'A new population of the cusp suprathermal protons formed at mid-altitudes: INTERBALL-2 measurements'. *Cosmic Research (English version)* **38**, 585.

Wagner, J., R. Sydora, T. Tajima, T. Hallinan, L. Lee, and S.-I. Akasofu: 1983, 'Small-scale auroral arc deformations'. *J. Geophys. Res.* **88**, 8013.

Wahlund, J.-E., A. I. E. Eriksson, B. Holback, M. H. Boehm, J. Bonnell, P. M. Kintner, C. E. Seyler, J. H. Clemmons, L. Eliasson, D. J. Knudsen, P. Norqvist, and L. J. Zanetti: 1998, 'Broadband ELF plasma emission during auroral energization: 1. Slow ion acoustic waves'. *J. Geophys. Res.* **103**, 4343.

Wahlund, J.-E., P. Louarn, T. Chust, H. de Feraudy, A. Roux, B. Holback, P.-O. Dovner, and G. Holmgren: 1994, 'On Ion Acoustic Turbulence and the Nonlinear Evolution of Kinetic Alfvén Waves in Aurora'. *Geophys. Res. Lett.* **21**, 1831.

Waite, J. H., M. O. Chandler, R. V. Yelle, B. R. Sandel, and T. E. Cravens: 1988, 'Suprathermal electron processes in the upper atmosophere of Uranus: Aurora and electroglow'. *J. Geophys. Res.* **93**, 14295.

Walker, A. D. M., R. A. Greenwald, W. F. Stuart, and C. A. Green: 1979, 'STARE auroral radar observations of Pc5 geomagnetic pulsations'. *J. Geophys. Res.* **84**, 3373.

Walker, A. D. M., J. M. Ruohoniemi, K. B. Baker, R. A. Greenwald, and J. C. Samson: 1992, 'Spatial and temporal behavior of ULF pulsations observed by the Goose Bay HF radar'. *J. Geophys. Res.* **97**, 15411.

Walker, R. J., T. Ogino, J. Raeder, and M. Ashour-Abdalla: 1993, 'A global magnetohydrodynamic simulation of the magnetosphere when the interplanetary magnetic field is southward: The onset of magnetotail reconnection'. *J. Geophys. Res.* **98**, 17235.

Wang, P. and G. Siscoe: 1995, 'Some early descriptions of aurorae in China'. *Ann. Geophysicae* **13**, 517.

Warner, B.: 1995, *Cataclysmic variable stars*. Cambridge, New York: Cambridge University Press.

Warwick, J.: 1961, 'Dynamic spectra of Jupiter's decametric emission'. *Astrophys. J.* **13**, 41.

Warwick, J., D. Evans, G. Peltzer, R. Peltzer, J. Romig, C. Sawyer, A. Riddle, A. Schweitzer, M. Desch, M. Kaiser, W. Farrell, T. Carr, I. de Pater, D. Saelin, S. Gulkis, R. Poynter, A. Boischot, F. Genova, Y. Leblanc, A. Lecacheux, B. M. Pedersen, and P. Zarka: 1989, 'Voyager planetary radio astronomy at Neptune'. *Science* **24**, 1498.

Warwick, J., D. Evans, J. Romig, C. Sawyer, M. Desch, M. Kaiser, J. Alexander, T. Carr, D. Saelin, S. Gulkis, R. Paynter, M. Augier, A. Boischot, Y. Leblanc, A. Lecachaux, B. Pedersen, and P. Zarka: 1986, 'Voyager 2 radio observations of Uranus'. *Science* **23**, 102.

Warwick, J. W.: 1967, 'Radiophysics of Jupiter'. *Space Sci. Rev.* **6**, 841.

Watanabe, M., T. Ijima, M. Nakagawa, T. Potemra, L. Zanetti, S. Ohtani, and P. Newell: 1998, 'Field-aligned current systems in the magnetospheric ground state'. *J. Geophys. Res.* **103**, 6853.

Weatherwax, A. T., J. LaBelle, M. L. Trimpi, and R. Brittain: 1993, 'Ground-based observations of radio emissions near $2f_{ce}$ and $3f_{ce}$ in the auroral zone'. *Geophys. Res. Lett.* **20**, 1447.

Weatherwax, A. T., J. LaBelle, M. L. Trimpi, R. A. Treumann, J. Minow, and C. Deehr: 1995, 'Statistical and case studies of radio emissions observed near $2f_{ce}$ and $3f_{ce}$ in the auroral zone'. *J. Geophys. Res.* **100**, 774.

Weber, E. J. and J. Buchau: 1981, 'Polar cap F-layer auroras'. *Geophys. Res. Lett.* **8**, 125.

Wei, C. Q., J. C. Samson, R. Rankin, and P. Frycz: 1994, 'Electron inertial effects on geomagnetic field line resonances'. *J. Geophys. Res.* **99**, 11265.

Weimer, D.: 1995, 'Models of high-latitude electric potentials derived with a least error fit of spherical harmonic coefficients'. *J. Geophys. Res.* **100**, 19595.

Weimer, D.: 1999, 'Substorm influence on the ionospheric electric potentials and currents'. *J. Geophys. Res.* **104**, 185.

Weimer, D.: 2001, 'An improved model of ionospheric electric potentials including substorm perturbations and application to the Geospace Environment Modeling November 24, 1996, event'. *J. Geophys. Res.* **106**, 407.

Weimer, D., J. Craven, L. Frank, W. Hanson, N. Maynard, R. Hoffman, and J. Slavin: 1994, 'Satellite measurements through the center of a substorm surge'. *J. Geophys. Res.* **99**, 23639.

Weimer, D. and D. Gurnett: 1993, 'Large-amplitude auroral electric fields measured with DE 1'. *J. Geophys. Res.* **98**, 13557.

Weimer, D. R., C. K. Goertz, D. A. Gurnett, N. C. Maynard, and J. L. Burch: 1985, 'Auroral zone electric fields from DE 1 and DE 2 at magnetic conjunctions'. *J. Geophys. Res.* **90**, 7479.

Wescott, E. M., J. D. Stolarik, and J. P. Heppner: 1970, 'Auroral and Polar Cap Electric Fields from Barium Releases'. In: B. M. McCormac (ed.): *Particles and Field in the Magnetosphere*, Vol. 17 of *ASSL*. p. 229.

Whalen, B. A. and I. B. McDiarmid: 1972, 'Observations of magnetic-field-aligned auroral-electron precipitation'. *J. Geophys. Res.* **77**, 191.

Whalen, B. A., S. Watanabe, and A. W. Yau: 1991, 'Observations in the transverse ion energization region'. *Geophys. Res. Lett.* **18**, 725.

Whalen, J. and I. B. M. Diarmid: 1973, 'Pitch angle diffusion of low-energy auroral electrons'. *J. Geophys. Res.* **78**, 1608.

Wild, J., T. Yeoman, P. Eglitis, and H. Opgenoorth: 2000, 'Multi-instrument observations of the electric and magnetic field structure of omega bands'. *Ann. Geophysicae* **18**, 99.

Williams, P. J. S., G. O. L. Jones, B. Jones, H. J. Opgenoorth, and I. Häggström: 1990, 'High resolution measurements of magnetospheric electric fields'. *J. Atmos. Terr. Phys.* **52**, 439.

Winckler, J. R. and R. J. Nemzek: 1993, 'Observations of Pulsation Phase of Auroras Observed at Minneapolis During the Peak of Solar Cycle 22'. In: *Auroral Plasma Dynamics*, Vol. 89 of *Geophysical Monograph Series*. Washington, DC: American Geophysical Union, p. 1.

Winglee, R. M. and P. L. Pritchett: 1986, 'The generation of low-frequency electrostatic waves in association with auroral kilometric radiation'. *J. Geophys. Res.* **91**, 13531.

Winningham, J., J. Burch, and R. Frahm: 1984, 'Bands of ions and angular V's: a conjugate manifestation of ionospheric ion acceleration'. *J. Geophys. Res.* **89**, 1749.

Wu, C. S. and L. C. Lee: 1979, 'A theory of terrestrial kilometric radiation'. *Astrophys. J.* **230**, 621.

Wülser, J.-P., R. C. Cannfield, and E. Rieger: 1990. In: *Max91 Workshop No. 3*. Boulder, Co., p. 149.

Wygant, J. R., A. Keiling, C. A. Cattell, M. Johnson, R. L. Lysak, M. Temerin, F. S. Mozer, C. A. Kletzing, J. D. Scudder, W. Peterson, C. T. Russell, G. Parks, M. Brittnacher, G. Germany, and J. Spann: 2000, 'Polar spacecraft based comparisons of intense electric fields and Poynting flux near and within the plasma sheet-tail lobe boundary to UVI images: An energy source for the aurora'. *J. Geophys. Res.* **105**, 18675–18692.

Wygant, J. R., A. Keiling, C. A. Cattell, M. Temerin, F. S. Mozer, C. A. Kletzing, J. D. Scudder, A. V. Streltsov, W. Lotko, C. T. Russell, and R. L.Lysak: 2002, 'Evidence for kinetic Alfvén waves and parallel electron energization at 4-6 R_E altitudes in the plasma sheer boundary layer'. *J. Geophys. Res.* **107**, 10.1029/2001JA900113.

Yahnin, A., M. Malkov, V. Sergeev, R. Pellinen, O. Aulamo, S. Vennerstrom, E. Friis-Christensen, K. Lassen, C. Danielsen, J. Craven, C. Deehr, and L. Frank: 1994, 'Features of steady magnetospheric convection'. *J. Geophys. Res.* **99**, 4039.

Yahnin, A. G., T. Bosinger, J. Kangas, , and R. D. Belian: 1990, 'Some implications on substorm dynamics inferred from correlations between multiple flux peaks of drifting proton clouds and ground observations'. *Ann. Geophysicae* **8**, 327.

Yamagishi, H., Y. Fujita, N. Sato, M. Nishino, P. Stauning, R. Liu, and T. Saemundsson: 2000, 'Inter-hemispheric conjugacy of auroral poleward expansion observed by conjugate imaging riometers at 67 and 75-77deg invariant latitude'. *Adv. Polar Atmos. Res.* **14**, 12.

Yamamoto, T., S. Inoue, and C.-I. Meng: 1997, 'Formation of auroral omega bands in the paired region 1 and region 2 field-aligned current system'. *J. Geophys. Res.* **102**, 2531.

Yau, A. W. and M. André: 1997, 'Source of ion outflow in the high latitude ionosphere'. *Space Sci. Rev.* **37**, 1.

Yau, A. W., B. A. Whalen, W. K. Peterson, and E. G. Shelley: 1984, 'Distribution of upflowing ionospheric ions in the high-altitude polar cap and auroral ionosphere'. *J. Geophys. Res.* **89**, 5507.

Zarka, P.: 1992, 'The auroral radio emissions from planetary magnetospheres: What do we know, what don't we know, what do we learn from them?'. *Adv. Space Res.* **12**, 115.

Zarka, P., B. P. Ryabov, V. B. Ryabov, R. Prangé, M. Abada-Simon, T. Farges, and L. Denis: 1997, 'Ground-based high sensitivity radio astronomy at decameter wavelengths'. In: H. O. Rucker, S. J. Bauer, and A. Lecacheux (eds.): *Planetary Radio Emissions IV*. Vienna: Austrian Acad. Sci. Pres, p. 101.

Zarka, P., R. A. Treumann, B. P. Ryabov, and V. B. Ryabov: 2001, 'Magnetically-Driven Planetary Radio Emissions and Application to Extrasolar Planets'. *Astrophysics and Space Science* **277**, 293.

Zelenyi, L., R. Kovrazhkin, and J.-M. Bosqued: 1990, 'Velocity-dispersed ion beams in the nightside auroral zone: AUREOL-3 observations'. *J. Geophys. Res.* **95**, 12119.

Zhou, X.-Y. and B. T. Tsurutani: 1999, 'Rapid intensification and propagation of the dayside aurora: Large scale interplanetary pressure pulses (fast shocks)'. *Geophys. Res. Lett.* **26**, 1097.

Zhu, L., R. W. Schunk, and J. J. Sojka: 1997, 'Polar cap arcs: A review'. *J. Atmos. Terr. Phys.* **59**, 1087.

Zhu, L., C. E. Valladares, J. J. Sojka, R. W. Schunk, and D. J. Crain: 1996, 'Model-observation comparison study of multiple polar cap arcs'. *J. Geophys. Res.* **101**, 323.

Zhu, X. and M. G. Kivelson: 1988, 'Analytic formulation and quantitative solutions of the coupled ULF wave problem'. *J. Geophys. Res.* **93**, 8602.

Ziesolleck, C. and D. McDiarmid: 1995, 'Statistical survey of auroral latitude Pc5 spectral and polarization characteristics'. *J. Geophys. Res.* **100**, 19299.

Author Affiliations and Addresses

Olaf Amm
Finnish Meteorological Institute
Dept. of Geophysics
P.O. Box 503
00101 Helsinki, Finland
olaf.amm@fmi.fi

Joachim Birn
MS D466
Los Alamos National Laboratory
Los Alamos, NM 87545, USA
jbirn@lanl.gov

John Bonnell
Space Sciences Laboratory
University of California
Berkeley, CA 94720, USA
jbonnell@ssl.berkeley.edu
Acknowledges NASA support under grant NAG5-3596.

Joseph E. Borovsky
MS D466
Los Alamos National Laboratory
Los Alamos, NM 87545, USA
jborovsky@lanl.gov

Jim F. Carbary
The Johns Hopkins University
Applied Physics Laboratory
Johns Hopkins Rd
Laurel, MD 20723-6099, USA
jim.carbary@jhuapl.edu

Charles W. Carlson
Space Sciences Laboratory
University of California
Berkeley, CA 94720, USA
cwc@ssl.berkeley.edu
Acknowledges NASA support under grant NAG5-3596.

Cynthia Cattell
University of Minnesota
Tate Lab of Physics
116 Church St SE
Minneapolis, MN 55455, USA
cattell@belka.space.umn.edu
Acknowledges support by Research Corporation and NASA grant NAG5-3596.

Robert Ergun
University of Colorado
LASP
1234 Innovation Drive
Campus Box 530
Boulder, CO 80303, USA
robert.ergun@lasp.colorado.edu
Acknowledges NASA support under grant NAG5-3596.

Anders Eriksson
Swedish Institute of Space Physics
Uppsala Division
75591 Uppsala, Sweden
aie@irfu.se

Yuri I. Galperin[1]
Space Research Institute
Russian Academy of Sciences
Profsoyuznaya St. 84/32
117810 Moscow, Russia

Stein Haaland
ISSI
Hallerstr. 6
3012 Bern, Switzerland
stein.haaland@issi.unibe.ch
Acknowledges support from the Norwegian Academy of Science and Letters, and the Fridtjof Nansen Fund for Scientific Research

Thomas Hallinan
Geophysical Institute
University of Alaska
College, AK 99701, USA
hallinan@gi.alaska.edu

Gerhard Haerendel
International University Bremen
Campus Ring 1
28725 Bremen, Germany
hae@iu-bremen.de

Bengt Hultqvist
Swedish Institute of Space Physics
P.O.Box 812
98128 Kiruna, Sweden
hultqv@irf.se

Tomas Karlsson
Royal Institute of Technology
Dept. of Plasma Physics
10044 Stockholm, Sweden
tkarlsson@plama.kth.se

David J. Knudsen
University of Calgary
2500 University Dr NW
Calgary AB T2N IN4, Canada
knudsen@phys.ucalgary.ca
Acknowledges support from the Natural Sciences and Engineering Research Council of Canada.

Hannu E. J. Koskinen
University of Helsinki
Department of Physical Sciences
P.O. Box 64
00014 University of Helsinki, Finland
also at
Finnish Meteorological Institute
Geophysical Research
P.O. Box 503
00101 Helsinki, Finland
hannu.koskinen@fmi.fi

William Lotko
Dartmouth College
Thayer School of Engineering
Hanover, NH 03755, USA
lotko@dartmouth.edu

Gang Lu
National Center for Atmospheric Research
Boulder, CO 80307, USA
ganglu@ncar.ucar.edu

Kristina A. Lynch
Dartmouth College
Department of Physics and Astronomy, Hanover, NH 03755, USA
kristina.lynch@dartmouth.edu

Robert L. Lysak
University of Minnesota
School of Physics and Astronomy
116 Church St SE
Minneapolis, MN 55455, USA
bob@aurora.space.umn.edu
Acknowledges NASA support under grant NAG5-9254.

Goran Marklund
Royal Institute of Technology
Dept. of Plasma Physics
10044 Stockholm, Sweden
marklund@plasma.kth.se
Acknowledges support by the Swedish National Space Board.

[1]Deceased 28. December 2001

James McFadden
Space Sciences Lab
University of California
Berkeley, CA 94720, USA
mcfadden@ssl.berkeley.edu
Acknowledges NASA support under grant NAG5-3596.

Arnaud Masson
Research and Scientific Support Department
ESTEC
P.O.Box 299
2200 Noordwijk ZH, The Netherlands
arnaud.masson@esa.int

Ching Meng
The Johns Hopkins University
Applied Physics Laboratory
Johns Hopkins Rd
Laurel, MD 20723-6099, USA
ching.meng@jhuapl.edu

Hermann Opgenoorth
Swedish Institute of Space Physics
Uppsala Division
75591 Uppsala, Sweden
opg@irfu.se

Götz Paschmann
ISSI
Hallerstr. 6
3012 Bern, Switzerland
also at
Max-Planck-Institut für extraterrestrische Physik
Giessenbachstrasse
85748 Garching, Germany
gep@mpe.mpg.de

Raymond Pottelette
CETP
4, Av. de Neptune
94107 St. Maur-des-Fosses, France
pottelet@cetp.ipsl.fr

Tuija I. Pulkkinen
Finnish Meteorological Institute
Geophysical Research
P.O. Box 503
00101 Helsinki, Finland
tuija.pulkkinen@fmi.fi

Victor Sergeev
University of St. Petersburg
Inst. of Physics
198504 St. Petersburg, Russia
victor@geo.phys.spbu.ru

Yan Song
University of Minnesota
School of Physics and Astronomy
116 Church St SE
Minneapolis, MN 55455, USA
yan@fields.space.umn.edu
Acknowledges NASA support under grant NAG5-9254.

Rudolf Treumann
Max-Planck-Institut für extraterrestrische Physik
Giessenbachstrasse
85748 Garching, Germany
tre@mpe.mpg.de

Index

Space Science Series of ISSI

1. R. von Steiger, R. Lallement and M.A. Lee (eds.): *The Heliosphere in the Local Interstellar Medium*. 1996 ISBN 0-7923-4320-4
2. B. Hultqvist and M. Øieroset (eds.): *Transport Across the Boundaries of the Magnetosphere*. 1997 ISBN 0-7923-4788-9
3. L.A. Fisk, J.R. Jokipii, G.M. Simnett, R. von Steiger and K.-P. Wenzel (eds.): *Cosmic Rays in the Heliosphere*. 1998 ISBN 0-7923-5069-3
4. N. Prantzos, M. Tosi and R. von Steiger (eds.): *Primordial Nuclei and Their Galactic Evolution*. 1998 ISBN 0-7923-5114-2
5. C. Fröhlich, M.C.E. Huber, S.K. Solanki and R. von Steiger (eds.): *Solar Composition and its Evolution – From Core to Corona*. 1998 ISBN 0-7923-5496-6
6. B. Hultqvist, M. Øieroset, Goetz Paschmann and R. Treumann (eds.): *Magnetospheric Plasma Sources and Losses*. 1999 ISBN 0-7923-5846-5
7. A. Balogh, J.T. Gosling, J.R. Jokipii, R. Kallenbach and H. Kunow (eds.): *Co-rotating Interaction Regions*. 1999 ISBN 0-7923-6080-X
8. K. Altwegg, P. Ehrenfreund, J. Geiss and W. Huebner (eds.): *Composition and Origin of Cometary Materials*. 1999 ISBN 0-7923-6154-7
9. W. Benz, R. Kallenbach and G.W. Lugmair (eds.): *From Dust to Terrestrial Planets*. 2000 ISBN 0-7923-6467-8
10. J.W. Bieber, E. Eroshenko, P. Evenson, E.O. Flückiger and R. Kallenbach (eds.): *Cosmic Rays and Earth*. 2000 ISBN 0-7923-6712-X
11. E. Friis-Christensen, C. Fröhlich, J.D. Haigh, M. Schüssler and R. von Steiger (eds.): *Solar Variability and Climate*. 2000 ISBN 0-7923-6741-3
12. R. Kallenbach, J. Geiss and W.K. Hartmann (eds.): *Chronology and Evolution of Mars*. 2001 ISBN -7923-7051-1
13. R. Diehl, E. Parizot, R. Kallenbach and R. von Steiger (eds.): *The Astrophysics of Galactic Cosmic Rays*. 2001 ISBN 0-7923-7051-1
14. Ph. Jetzer, K. Pretzl and R. von Steiger (eds.): *Matter in the Universe*. 2001 ISBN -1-4020-0666-7

Kluwer Academic Publishers – Dordrecht / Boston / London

www.ingramcontent.com/pod-product-compliance
Lightning Source LLC
Chambersburg PA
CBHW070803200526
45355CB00003B/140
* 9 7 8 1 4 0 2 0 0 9 6 3 1 *